PRESTRESSED CONCRETE

Dr. Shrikant B. Vanakudre
Principal, SDMCET, Dharwad

Mr. Ashish A. Yaligar
Department of Civil Engineering
SDMCET, Dharwad

KHANNA BOOK PUBLISHING CO. (P) LTD.
Publisher of Engineering and Computer Books
4C/4344, Ansari Road, Darya Ganj, New Delhi-110002
Phone : 011-23244447-48 **Mobile:** +91-9910909320
E-mail : contact@khannabooks.com
Website : www.khannabooks.com

Dear Readers,

To prevent the piracy, this book is secured with HIGH SECURITY HOLOGRAM on the front title cover. In case you don't find the hologram on the front cover title, please write us to at contact@khannabooks.com or whatsapp us at +91-99109 09320 and avail special gift voucher for yourself.

Specimen of Hologram on front Cover title:

Moreover, there is a SPECIAL DISCOUNT COUPON for you with EVERY HOLOGRAM.

How to avail this SPECIAL DISCOUNT:

Step 1: Scratch the hologram
Step 2: Under the scratch area, your "coupon code" is available
Step 3: Logon to www.khannabooks.com
Step 4: Use your "coupon code" in the shopping cart and get your copy at a special discount
Step 5: Enjoy your reading!

Price: ₹ 399.00

ISBN: 978-93-86173-31-7

Prestressed Concrete
(Materials, Analysis and Design)
by Shrikant B. Vanakudre, Ashish A. Yaligar

First Edition: 2018
Reprint: 2019, 2021, **2023**

To view complete list of books, Please scan the QR Code:

Published by:
Khanna Book Publishing Co. (P) Ltd.
CIN: U22110DL1998PTC095547

Visit us at: **www.khannabooks.com**
Write us at: **contact@khannabooks.com**

Printed in India by: City Press, Delhi.

Preface to the First Edition

I am pleased to present the First Edition of my book titled 'Prestressed Concrete', conforming to the latest IS Code 1343. This book comprises of 15 Chapters, with 397 pages and deals with valuable examples.

This book is quite useful for B. E. / B.Tech. / AMIE students and also students appearing for Competitive Examination in 'Prestressed Concrete'. Attempts have been made in this book, to convert the regular class-notes which are being taught in the class rooms since many years into a better understandable form.

I feel it is an appropriate opportunity to thank my Teacher Sri Srichand of S.J.C.E., Mysore, who taught me this subject during my undergraduate program. Some of the examples taught by my Teacher are also incorporated in this book, to recall his memories and also for the benefit of the Readers.

With the kind blessings of Poojya Dr. D. Veerendra Heggadeji, I could bring out this Book. Meantime, I will remember the support of Late Sri Jinendra Prasad, Former Secretary, Sri Jeevandhar Kumar, Secretary of SDME Society, Dharwad. I will also not forget to remember Dr. S. S. Bhavikatti, former Head of the Department of Civil Engineering, S. D. M. College of Engineering and Technology, Dharwad, who inspired me to write this book.

Most importantly, I would like to thank my parents, wife Premalatha and children for their constant support. I am also thankful to all my friends in the profession for their suggestions in the preparation of this book.

My thanks to Co-author Mr. Ashish A. Yaligar, Research Scholar, Department of Civil Engineering, for his contribution in writing this book.

Though utmost care has been taken while preparing the book and there is always a scope for improvement. Hence, any suggestions are most welcome and I would be very happy to correct it whereever necessary.

Wishing all the readers a very Happy Reading.

Dr. SHRIKANT B. VANAKUDRE

Preface to the First Edition

I am pleased to present the First Edition of my book titled "Prestressed Concrete" conforming to the latest IS Code 1343. This book comprises of 13 Chapters with 397 pages and [illegible] solved examples.

This book is quite useful for B.E. / B.Tech / AMIE students and also students appearing for Competitive Examination in Prestressed Concrete. Attempts have been made in this book to convert the regular class notes which are being taught in the class rooms since [illegible] in a better readable form.

I feel it is an appropriate opportunity to thank my Teacher Sri Srikanth of S.J.C.E. Mysore, who taught me this subject during my undergraduate program. Some of the examples taught by my Teacher are also incorporated in this book, to recall his memories and also for the benefit of the Readers.

With the kind blessings of Pooja Dr. D. Veerendra Heggade, I could bring out this Book. Meanwhile I will remember the support of Late Sri Jinendra Prasad, Former Secretary, Sri Jeevandhar Kumar, Secretary of SDME Society, Dharwad. I will also not forget to remember Dr. S. Shivakumar, former Head of the Department of Civil Engineering, S. D. M. College of Engineering and Technology, Dharwad, who motivated me to write this book.

Most importantly, I would like to thank my parents, wife Premalatha and children for their constant support. I am also thankful to all my friends in the profession for their suggestions in the preparation of this book.

My thanks to Geetha M. & Ashok V. Malgat, Research Scholar, Department of Civil Engineering, for his contribution in writing this book.

Though utmost care has been taken while preparing the book and there is always a scope for improvement. Hence, any suggestions are most welcome and I would be very happy to incorporate them if found necessary.

Dr. SHRIKANT B. VANAKUDRE

Contents

List of Abbreviations

BBRV = Post–tensioning system found by Birkenmaier, Brandestini and Ros
BM = Bending Moment
BMD = Bending Moment Diagram
CCL = Post–tensioning system
CGC = Centre of Gravity of Concrete or centroid of concrete
CGS = Centre of Gravity of Steel or centroid of pre–stressing steel
DL = Dead Load
FF = Floor Finish
FIP = International Federation for Pre–stressing
FRP = Fibre Reinforced Polymeric Reinforcements
GGBFS = Ground Granulated Blast Furnace Slag
HT = High Tension
HYSD = High Yielding Strength Deformation
IRC = Indian Road Congress
LL = Live Load
MPa = Mega Pascal
NA = Neutral Axis
PCI = Precast/Pre–stressed Concrete Institute
PCS–2A = Precast Sleeper made in India
PM = Primary Moment
PSC = Pre–stressed Concrete
RC = Reinforced Concrete
RCC = Reinforced Cement Concrete
RM = Resultant Moment
SF = Shear Force
SM = Secondary Moment (Parasitic Bending Moment)
UDL = Uniformly Distributed Load
UK = United Kingdom
USA = United States of America
USSR = Union of Soviet Socialist Republics
WKT = We Know That
w.r.t = with respect to

List of Symbols

$2y_o$ = Depth of the equivalent prism or side of end block or side of the surrounding prism
$2y_{po}$ = Depth of the anchorage plate or side of loaded area or side of loaded or punching area
A = Cross sectional area of the concrete member
a = Vertical deflection at any point of the grid floor
A_1 = Area of half of the flange
A_{br} = Bearing area
A_c = Cross sectional area of concrete per unit length along the circumference
A_e = Equivalent concrete area
A_p = Area of pre–stressing tendons
A_{pf} = Area of pre–stressing steel for flange
A_{pun} = Punching area
A_{pw} = Area of pre–stressing steel for web
A_{st} = Cross sectional area of wire winding in mm^2
A_{sv} = Total area of the legs of stirrups
$A_{sv,min}$ = Minimum amount of stirrups
A_{svf} = Amount of horizontal reinforcement in the flange
A_t = Area of section above CGC
b = Breadth of the section
b_1 = Centre to centre distance between corner bars in the direction of width
b_f = Breadth of the flange
b_w = Breadth of the web for flanged sections
C = Creep strain/Resultant thrust
C_c = Creep coefficient which varies from 2 to 3
d = Effective depth
D = Total depth of the section
d_1 = Centre to centre distance between corner bars in the direction of depth
D_f = Depth of the flange
d_p = Depth of CGS from the extreme compression fiber
d_s = Depth of centroid of non–pre–stressed steel
e = Eccentricity of pre–stressing force or eccentricity of the cable
E or E_c = Modulus of elasticity of concrete
e_c = Strain in concrete
E_{ci} = Initial modulus of elasticity of concrete
E_s = Modulus of elasticity of steel
e_{sh} = Shrinkage strain in concrete
f_1' = Stress at top fibre

f'_2 = Stress at bottom fibre
F = Effective pre–stressing force
F_a = Allowable direct stress
F_b = Allowable compressive stress in bending
F_{bst} = Bursting tensile force
f_c = Actual compressive strength in concrete
f_{ck} = Characteristic compressive strength of concrete cubes in MPa
f_{cL} = Minimum longitudinal pre–stress
f_{cp} = Average intensity of pre–stress at the centroid
f_{cr} = Modulus of rupture & flexural tensile strength of concrete in MPa
f_{cT} = Minimum transverse pre–stress
f_{ct} = Permissible compressive stress in concrete at transfer
f_{cu} = Cube strength of concrete at transfer in MPa
f_{cy} = Cylinder compressive strength
f_{inf} = Pre–stress in concrete developed at bottom fibres
fl_1 or f_{cw} = Stress in the top most concrete fibres at working load
fl_2 or f_{tw} = Stress in the bottom most concrete fibres at working load
$f_{min, w}$ = Permissible stress in concrete under working pressure
f_{n1} = Principal compressive stress
f_{n2} = Principal tension or diagonal tension due to the combined action of shear and flexure
F_o = Initial pre–stressing force due to tensioned steel and dead load moment
f_p = Characteristic strength of pre–stressing steel
f_{pe} = Effective pre–stress in tendons after losses
f_{pi} = Stress in tendon at transfer
f_{pk} = Characteristic strength of pre–stressing steel
f_{pt} = Magnitude of the compressive stress in concrete at the level of CGS due to pre–stress only
f_{pu} = Tensile strength of wire winding in N/mm^2
f_{se} = Effective stress in steel after winding for compatibility of strains
f_{si} = Initial pre–stress in the tensioned steel
f_{st} = Stress in wires at transfer in N/mm^2
f_{sup} = Pre–stress in concrete developed at top fibres
f_t = Direct tensile strength of concrete or maximum principal tensile stress
f_{u1} or f_{tt} = Stresses in the top most concrete fibres at transfer of pre–stress
f_{u2} or f_{ct} = Stresses in the bottom most concrete fibres at transfer of pre–stress
f_v = Transverse tensile stress
$f_{v(max)}$ = Maximum transverse tensile stress
f_y = Yield stress of steel cylinder in N/mm^2 or yield stress of the stirrups
g = Dead load
I = Second moment of area / moment of inertia of section about its centroid
I_e = Moment of inertia of equivalent concrete
k = Coefficient for wave effect varying from 15×10^{-4} to 50×10^{-4} per metre

K = Radius of gyration
K_p = Coefficient of passive earth pressure
l = Length of tendon
L = Length of the pavement slab or Length of the section
L_t = Transmission length
M = External moment at a section due to the weight of the beam
m = Modular ratio
M' = Moment acting on the composite section which is generally due to imposed loads
M_e = Equivalent ultimate bending moment
M_G = Dead load moment
M^I = Moment due to pre–stressing force
M_L = Bending moment due to superimposed load or due to live load
M_{Ln} = Maximum negative live load moment
M_{Lp} = Maximum positive live load moment
M_o = Moment initiating a flexural crack
M_r = Range of Moments
M_T = Total bending moment
M_u = Moment due to ultimate loads at the design section
n = Number of turns of circumferential wire windings per m length of pipe/ Number of bars
N_d = Hoop tension developed under working pressure
N_t = Hoop compression due to radial pressure of wires
P = Equivalent to an axial force
p = External pressure due to wire wrapping
P_e = Effective pre–stressing force
P_i = Initial pre–stressing force
P_k = Applied compressive force on the end block (tendon jacking force)
P_o = Pre–stressing force in steel at the tensioning end acting in the direction of the tangent to the curve of the cable
P_t = Pre–stress at time t
P_u = Bursting pressure in N/mm^2
q = Live load
R = Internal radius of pipe
r = Radius of the curve
R_o = External radius
s = Spacing of cables or spacing of the wires at the given section in mm
S_v = Spacing of stirrup reinforcement
t = Thickness of the pavement slab or pipe shell or wall in mm
T = Torsional moment applied to a cross–section under ultimate load conditions
t_0 = Time at commencement of drying
T_c = Torsional moment carried by concrete
t_s = Thickness of the steel cylinder in mm

V = Shearing force at a cross–section calculated for the specified ultimate load
V_a = Shear capacity of aggregate interlock
V_c = Shear capacity of a concrete contribution
V_{co} = Shear causing web shear cracking at CGC
V_{cr} = Shear corresponding to flexure shear cracking
V_{cw} = Ultimate shear resistance of the support sections uncracked in flexure
V_{cz} = Shear carried by un–cracked concrete
V_d = Shear capacity of dowel action
V_{DL} = Shear due to dead load
V_f = Shear forces in flanges
V_{LL} = Shear due to live load
V_p = Vertical component of pre–stressing force in inclined tendons
V_p = Vertical component of the pre–stressing force
V_s = Shear carried by stirrups
V_u = Factored shear at ultimate loads
V_u = Shear force at a section due to ultimate loads
V_{uR} = Ultimate resistance for shear
W_{cr} = Cracking load
w_d = Total UDL per unit area
w_t = Average radial pressure of wires at transfer at a given section in N/mm^2
W_w = Hydrostatic pressure
x = Pitch of the wire winding
x_u = Neutral axis depth
y = Depth of the CGS from CGC
y = Distance of centre of gravity of the section from outmost concrete fibre
$\overline{y}$ = Distance of centroid of half of the flange from the neutral axis at CGC
Y = Position of centroid of the section from the extreme fibre
y_b or y_2 = Distance of bottom fibre from the centroid of the section
y_o = Side of end block
y_{po} = Side of loaded area
y_t or y_1 = Distance of top fibre from the centroid of the section
Z_b = Section modulus of the bottom fibres
Z_b' = Section modulus of the bottom fibre of the composite section
Z_t = Section modulus of the top fibres
α = Curvature angle in radian through which the tangent to the cable profile at the section or obliquity of the cables
α_e = Modular ratio
β = Constant depending upon the details of strand and wire
γ_m = Material safety factor
δ = Deflection at the centre
δ_a = Amount by which the anchorage slips
δ_{ip} = Initial deflection due to pre–stress
δ_{pt} = Initial deflection due to transverse loads

ε_0 = Strain corresponding to f_{ck} = 0.002

ε_c = Compressive strain

$\varepsilon_{cr,ult}$ = Ultimate creep strain

ε_{cu} = Ultimate compressive strain = 0.0035

ε_{el} = Elastic creep strain

ε_{sh} = Ultimate shrinkage strain

η = Loss ratio

θ = Change of slope from one end to the other end of the tendon in radian or creep coefficient

μ = Coefficient of friction in curve or coefficient of friction between the tendons and the duct or coefficient of friction between steel and concrete

ν_c = Poisson's ratio of concrete

ν_s = Poisson's ratio of steel

σ_1 = Principal tensile stress

σ_2 = Principal compressive stress

τ_{av} = Average shear stress

τ_c = Ultimate shear stress capacity of concrete

$\tau_{c,max}$ = Maximum ultimate shear stress capacity of concrete

τ_f = Variation of shear stress in a flange

$\tau_{f,max}$ = Maximum shear stress in the flange

τ = Shear stress

ϕ = Diameter of wire/strand/cable

ϕ_{mt} = Curvature produced by transverse loads

ϕ_{pt} = Curvature produced by pre–stress

ϕ_a = Nominal outside dia. of connection sleeve (mm) for male cone

ϕ_c = Female cone diameter

ϕ_e = Nominal outside dia. of connection sleeve (mm) for female cone

CHAPTER

1

Introduction, Materials and Systems of Pre–stressing

1.1 INTRODUCTION

1.1.1 Basic Concept

A pre–stressed concrete structure is different from a traditional reinforced concrete structure due to the application of an initial load on the structure earlier to its use. The initial load or 'pre–stress' is applied to enable the structure to counteract the stresses arising during its service life.

The pre–stressing of a structure is not the only instance of pre–stressing. The concept of pre–stressing existed before the applications in concrete. Two examples of pre–stressing before the development of pre–stressed concrete are as follows.

• Force Fitting of Metal Bands on Wooden Barrels

The metal bands induce a state of initial hoop compression, to counteract the hoop tension caused by filling of liquid in the barrels as shown in Fig. 1.1.

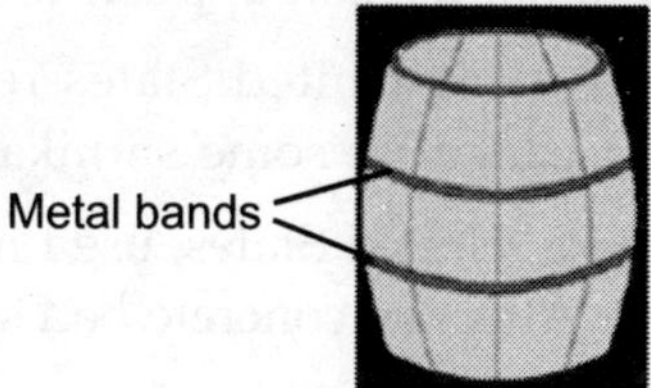

Fig. 1.1 Force Fitting of Metal Bands on Wooden Barrels

• Pre–tensioning the Spokes in a Bicycle Wheel

The pre–tension of a spoke in a bicycle wheel is applied to such an extent that there will be a residual tension in the spoke as shown in Fig. 1.2.

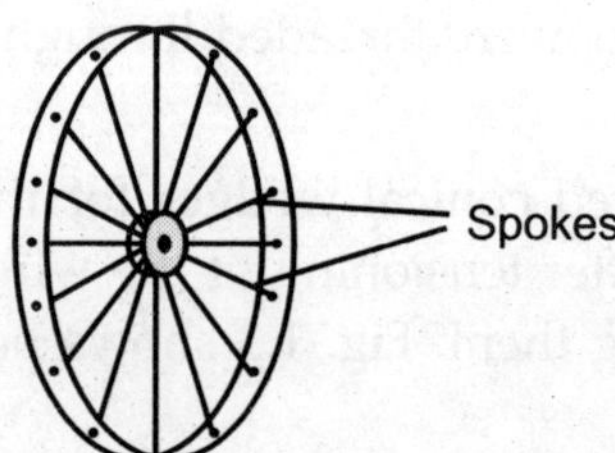

Fig. 1.2 Pre–tensioning the Spokes in a Bicycle Wheel

For concrete, internal stresses are induced (by means of tensioned steel) for the following reasons.

- The tensile strength of concrete is just about 8% to 14% of its compressive strength.
- Cracks tend to develop at early stages of loading in flexural members such as beams and slabs.
- To prevent such cracks, compressive force can be suitably applied in the perpendicular direction.
- Pre–stressing enhances the bending, shear and torsional capacities of the flexural members.
- Pipes and liquid storage tanks having hoop tensile stresses can be effectively fought by circular pre–stressing.

1.1.2 Historical Background

The present state of development in the field of pre–stressed concrete is due to the continuous research done by engineers and scientists in this field last century.

- In 1872, P. H. Jackson, an engineer of San Francisco, California, is reported to have filed a patent for a method of pre–stressing arches by tightening steel rods against artificial stress for construction of vaulted roofs and the slab and again in 1886, applied for a patent for "construction of artificial stone and concrete pavements", in which pre–stress is introduced by tensioning reinforcing rods set in sleeves.
- Around the year 1888, C. F. W. Doehring of Germany secured patent of concrete reinforced with steel with initial tension applied to it before the slab was loaded.
- In the year 1908, C. R. Steiner of United States recovered some of the losses by retightening the reinforcing bars after some shrinkage and creep takes place.
- In 1925, R. E. Dill of Nebraska, United States, used high strength steel bars coated to prevent bond with concrete. After the concrete had set, the steel rode were tensioned and anchored to the concrete by nuts.
- In 1928, the French engineer Eugene Freyssinet put his theories on pre–stressing after carried out tests on a larger tie member, pre–stressed with steel wires tensioned after the concrete had set and anchored by wedges in steel plates.
- In 1934, F. O. Anderegg of United States applied pre–stressing to burnt clay building block. High tensile steel ties were threaded through holes in the blocks, stressed and grouted in position.
- In 1939, Freyssinet prepared conical wedges for the end anchorages and designed double acting jacks, and after tensioning of the wires, thrust the male cones into the female cones for anchoring them. Fig. 1.3 shows portrait of Eugene Freyssinet.

Fig. 1.3 Portrait of Eugene Freyssinet

- Prof. G. Magnel of Belgium in 1940 developed the Magnel system in which two wires were stretched at a time and anchored with simple metal wedge at the ends.

During the Second World War, applications of pre–stressed and precast concrete increased fluently. Few persons involved in developing pre–stressed concrete are Y. Guyon, (France) built numerous pre–stressed concrete bridges in western and central Europe. P. W. Abeles, (England) introduced the concept of partial pre–stressing. F. Leonhardt, (Germany), V. Mikhailor, (Russia) and T. Y. Lin, (USA) are famous in the field of pre–stressed concrete.

France and Belgium led the development of pre–stressed concrete, followed quickly by England, Germany, Switzerland and Holland. During the period of 1949 to 1953, more than 350 bridges out of a total of 500 bridges in Germany were pre–stressed.

In United States, besides linear pre–stressing for pre–stressed concrete beams and slabs, circular pre–stressing applied to storage tanks was developed. This was introduced by preload company and during the period of 1935 to 1953, more than 700 pre–stressed concrete tanks were constructed.

In 1949, the construction of famous Philadelphia Walnut Lane Bridge was started in United States. The first pre–stressed concrete bridge in United States was completed in 1950, in Madison County, Tennessee.

By the middle of 1951, about 175 bridges and 50 buildings had been constructed in Europe with pre–stressed concrete.

An organization formed, International Federation for Pre–stressing (FIP), in Europe in 1952. The Precast/Pre–stressed Concrete Institute (PCI) was established in USA in 1954.

In India, pre–stressed concrete first applied on the army tank garages in Meerut in the year 1940. The field of road bridges has accounted for 80% of the activity in pre–stressed concrete, the first pre–stressed concrete bridge constructed in India being the ones of Assam rail links in the year 1947. The Pamban Road Bridge at Rameshwaram, Tamilnadu, remains a best example for pre–stressed concrete girders as shown in Fig. 1.4.

Fig. 1.4 Pamban Road Bridge at Rameshwaram, Tamil Nadu

1.1.3 Principles of Pre–stressing Concrete

Pre–stressed concrete may be defined as concrete in which permanent internal compressive stresses are deliberately introduced to counteract to the desired degree the undesirable tensile stresses (and consequent cracking) which would result from given external loadings. In concrete members the pre–stress is commonly introduced by tensioning the reinforcing steel.

In the most common form of pre–stressed concrete structures, pre–stressing reinforcement provides initial stresses, controlled in magnitude and direction for counterbalancing to a desired degree, stresses resulting from external loading. Three different concepts may be applied to explain and analyze the behavior of pre–stressed concrete.

These concepts are

1. Stress concept, 2. Strength concept, 3. Load balancing concept

- **Stress Concept:** This Concept is due to Eugene Freyssinet. According to this, pre–stressed concrete is considered essentially as a concrete structure with tendons, supplying the pre–compression, which transforms a brittle material like concrete into an elastic one.

  In pre–stressed concrete, main material concrete is visualized as being subjected to two systems of forces namely, internal pre–stress and external load, with the tensile stress due to external load counteracted by the compressive stress due to pre–stress. The cracking of concrete due to load is prevented or delayed by the pre–compression provided by the tendons. So long as there are no cracks, the stress, strains and deflection due to the two systems of forces can be considered separately and superimposed as necessary.

- **Strength Concept:** The second point of view is to consider steel and concrete together, with steel taking tension and concrete taking compression.

  In pre–stressed concrete, in which high tensile steel used has to be elongated a great deal before its strength is fully utilized. If the high strength tensile steel is simply embedded in concrete, as in the case of reinforced concrete, the surrounding concrete has to crack before the full strength of steel is developed. It is therefore necessary to pre–tension the steel with respect to concrete. By pre–tensioning and anchoring the steel against concrete, desirable stresses and strain in both materials, namely, compressive stresses and strains in concrete and tensile stresses and strain in steel are produced. This combined action allows the safe and economical utilization of the two materials which cannot achieve by simply embedding steel in concrete as in case of ordinary reinforced concrete.

- **Load Balancing Concept:** This method, which was developed by T. Y. Lin, transformed a pre–stressed concrete structure into a non–pre–stressed one, subjected only to the unbalanced portion of loading.

  While this concept often represents the simplest approach to pre–stressed analysis and design, its advantage over the other two concepts is not significant for statically determinate structures. However, its application to statically indeterminate beams and frames not only saves the time, but presents a realistic approach which helps the engineer to visualize the effect of pre–stressing. The horizon of pre–stressing will be greatly extended when this method is applied to slabs, grids, thin shells and folded plates. This method is also adaptable to a combination of pre–stressed and reinforced concrete where varying amounts of live load is to be carried by non–pre–stressed reinforcement.

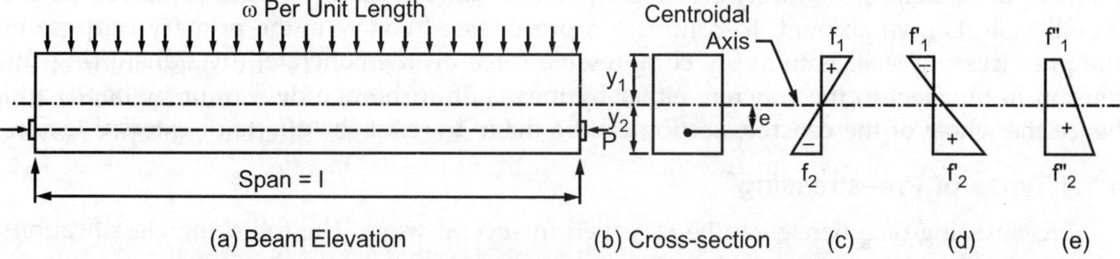

Fig. 1.5 Stress Distributions across Pre–stressed Concrete Section

Consider a simple rectangular beam as shown in Fig. 1.5 (a) and cross-section as shown in Fig. 1.5 (b). If it is a plain concrete beam; the maximum load the beam could carry will be small, being the load that causes a tensile stress f_2 at the bottom equal to the tensile strength of concrete in flexure (modulus of rupture value, f_{cr}). The ultimate load will thus be the cracking load W_{cr} which could be obtained from,

$$\frac{W_{cr} \times l^2}{8} \times \frac{Y}{I} = f_2 = f_{cr} \qquad ...(1.1)$$

As the tensile strength of concrete is small being only about on tenth of its compressive strength, the cracking load will be small and the compressive strength will remain largely unutilized. In RC beams this lacuna is overcome by introducing longitudinal steel reinforcement in the tension zone and the load carrying capacity of the beam is made greater than W_{cr} as the steel takes up the tension. (However, under working load, all normally designed RC beams will crack at those sections where the tensile stress due to dead and live load bending moments reaches or exceeds the modulus of rupture value of concrete). If the same plain concrete beam is pre–stressed by introducing an eccentric force, P, a stress distribution is created as shown in Fig. 1.5 (d).

It must be noted that since the pre–stressing force is eccentric, it is equivalent to an axial force P and a hogging bending moment = Pe. Hence the distribution of normal stress in any section in concrete is given by

$$\text{Stress at top fibre, } f_1' = \frac{P}{A} - \frac{Pey_1}{I} \text{ (taking compressive stress as positive)} \quad ...(1.2)$$

$$\text{Stress at bottom fibre, } f_2' = \frac{P}{A} - \frac{Pey_2}{I} \quad ...(1.3)$$

The important observation to be made here is that the pre–stressing force P causes a compressive stress (Fig. 1.5 (d)) on the zone in which applied load causes tension (Fig. 1.5 (c)). It is possible to increase the applied load W until the stress f_2 is counteracted without causing net tensile stress at the bottom. Thus pre–stressing improves the load carrying capacity of the beam and by suitably designing P and e, crack free structure under working load is obtained.

There are important differences in the design principles between reinforced concrete and pre–stressed concrete members. In the design of RC members, the tensile strength of concrete (in the tension zone) is neglected and the tensile forces caused by the bending moment are taken to b entirely resisted by reinforcement usually exerts no forces on the member on its own account. In contrast, in pre–stressed concrete, the primary purpose of the pre–stressing steel is to apply compressive force on the concrete, (by transferring the tension in the steel to the concrete either by means of anchorage devices or by bond) and hence the whole of the concrete section can be made to resist the effects of external loads.

1.1.4 Types of Pre–stressing

Pre–stressing of concrete can be classified in several ways. The following classifications are discussed.

❖ Source of Pre–stressing Force

This classification is based on the method by which the pre–stressing force is generated. There are three sources of pre–stressing force: Mechanical, hydraulic and electrical.

• Hydraulic Pre–stressing

This is the simplest type of pre–stressing, producing large pre–stressing forces. The hydraulic jack used for the tensioning of tendons, having calibrated a pressure gauge which indicates the magnitude of force developed during the tensioning.

• Mechanical Pre–stressing

In this type of pre–stressing, the devices includes weights with or without lever transmission, geared transmission in conjunction with pulley blocks, screw jacks with or without gear drives and wire–winding machines. This type of pre–stressing is adopted for mass scale production.

• Electrical Pre–stressing

In this type of pre–stressing, the steel wires are electrically heated and anchored before placing concrete in the moulds. This type of pre–stressing is also known as thermo–electric pre–stressing.

❖ External or Internal Pre–stressing

This type is based on the location of the pre–stressing tendon with respect to the concrete section.

• *External Pre–stressing*

When the pre–stressing is achieved by elements located outside the concrete, it is called external pre–stressing. The tendons can placed outside the member (for example in I–girders or walls) or inside the hollow space of a box girder. This technique is used in bridges and strengthening of buildings. Fig. 1.6 shows, the box girder of a bridge is pre–stressed with tendons that lie outside the concrete.

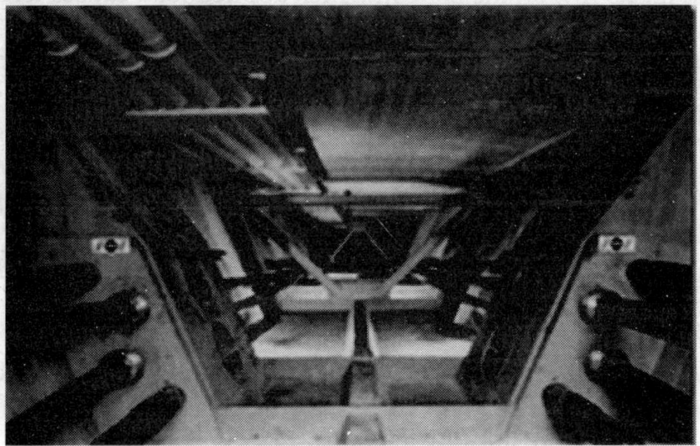

Fig. 1.6 External Pre–stressing of a Box Girder

• *Internal Pre–stressing*

When the pre–stressing is achieved by elements placed inside the concrete member (commonly, by embedded tendons), it is called internal pre–stressing. Most of the applications of pre–stressing are internal pre–stressing. Fig. 1.7 shows, concrete will be cast around the ducts for placing the tendons.

Fig. 1.7 Internal Pre–stressing of a Box Girder

❖ Pre–tensioning or Post–tensioning

This is the most precious classification and is based on the sequence of casting the concrete and applying tension to the tendons.

• *Pre–tensioning*

The tendons are under tension before casting of the concrete. The pre–compression is transmitted from steel to concrete through bond over the transmission length near the ends. Fig. 1.8 shows, manufactured pre–tensioned electric poles.

Fig. 1.8 Pre–tensioned Electric Poles

• *Post–tensioning*

The tension is applied to the tendons (located in a duct) after the setting of concrete. The pre–compression is transmitted from steel to concrete by means of anchorage device (at the end blocks). Fig. 1.9 shows a post–tensioned box girder of a bridge.

Fig. 1.9 Post–tensioning of a Box Girder

❖ Linear or Circular Pre–stressing

This type is based on the shape of the member pre–stressed.

• *Linear Pre–stressing*

When the pre–stressed members are straight or flat, in the direction of pre–stressing, the pre–stressing is called linear pre–stressing. For example, pre–stressing of beams, piles, poles and slabs. The profile of the pre–stressing tendon may be curved. Fig. 1.10 shows linearly pre–stressed railway sleepers.

Fig. 1.10 Linearly Pre–stressed Railway Sleepers

• *Circular Pre–stressing*

When the pre–stressed members are curved, in the direction of pre–stressing, the pre–stressing is called circular pre–stressing. Example: circumferential pre–stressing of tanks, silos, pipes and similar structures. Fig. 1.11 shows the nuclear reactor containment structure which is circularly pre–stressed.

Fig. 1.11 Circularly Pre–stressed Containment Structure, Kaiga Atomic Power Station, Karnataka

❖ Full, Limited or Partial Pre–stressing

Based on the amount of pre–stressing force, three types of pre–stressing are defined.

• *Full Pre–stressing*

When the level of pre–stressing is such that no tensile stress is allowed in concrete under service loads, it is called full pre–stressing (Type 1, as per IS: 1343–2012).

• *Limited Pre–stressing*

When the level of pre–stressing is such that the tensile stress under service loads is within the cracking stress of concrete, it is called limited pre–stressing (Type 2).

• *Partial Pre–stressing*

When the level of pre–stressing is such that under tensile stresses due to service loads, the crack width is within the allowable limit, it is called partial pre–stressing (Type 3).

❖ Uni–axial, Biaxial or Multi–axial Pre–stressing

Thisbifurcation based on the directions of pre–stressing a member.

• *Uni–axial Pre–stressing*

When the pre–stressing tendons are parallel to one axis, it is called uni–axial pre–stressing. Example: Longitudinal pre–stressing of beams.

• *Biaxial Pre–stressing*

When there are pre–stressing tendons parallel to two axes, it is called biaxial pre–stressing. Example : Pre–stressing of slabs.

• *Multi–axial Pre–stressing*

When the pre–stressing tendons are parallel to more than two axes, it is called multi-axial pre–stressing. Example : Pre–stressing of domes.

1.1.5 Advantages of Pre–stressed Concrete

Pre–stressed concrete structures are free from cracks, even hair cracks, under full working load. This improves their durability which is of great value particularly in the case of marine and other structures exposed to atmospheres or to aggressive ground conditions. Because of compression due to pre–stress, principal tensile stresses are reduced and shear resistance is achieved without the use of large webs or heavy reinforcements. Since less main and shear steels is used, it is possible to use and place much better concrete than can normally be used in RC works. The cross section of the member can be smaller since the whole of the concrete is put to structural use. It is therefore possible to design beams of longer spans using comparatively shallow and slender sections. Pre–stressed construction not only saves the material, it reduces dead loads, with consequent benefits. By controlling the amount of pre–stress, a structure could be made either flexible or rigid without affecting the ultimate resistance. Flexible structures such as pre–stressed fender–piles of wharfs or jetties are obviously more resilient and can absorb considerable energy of impact before failure and thus show superior performance under seismic and dynamic conditions. Because of the above said advantages, pre–stressed concrete can be widely used in long span girders, pipes, railway sleepers, shell structures, folded plates, nuclear reactor vessels, dams, multi–storeyed frames, piles, etc.

1.1.6 Applications of Pre–stressed Concrete

It is a well–established fact that the basic economy of pre–stressed concrete lies in its high strength to weight and strength to cost ratios, its resistance to fire and cracking, and its versatility and adaptability.

The use of pre–stressed concrete has revolutionized the entire building industry in the USSR, USA, UK, Japan and the continent. Pre–stressed concrete building components comprising hollow cored and ribbed slabs were widely used in Russia. Single and double T–units and channel sections are popular in the USA for the construction of floors in the buildings. Fig. 1.12 shows the use of precast pre–stressed double T–beams for floor construction[8]. Typical pre–stressed concrete flat slab floor construction using the lift slab technique is shown in Fig. 1.13

Fig. 1.12 Precast Pre–stressed Double T–floor Beams

Fig. 1.13 Thirteen Storey Apartment Building with Precast Post–tensioned in Slab Construction, 203mm Thick Lightweight Concrete Slabs of 8.6m Span, San Francisco

Fig. 1.14 Post–tensioned Pre–stress Concrete Twin Box Girder Bridge Construction Using Segmentally Cast Cantilever Method

Pre–stressed concrete is ideally suited for long–span bridge construction. A typical twin box girder bridge construction, using the segmentally cast cantilever method, is shown in Fig. 1.14. The present trend is to adopt pre–stressed concrete for long span cable–stayed bridges which are aesthetically superior and economical in comparison with steel bridges. A typical cable–stayed bridge of 365m main span, constructed at Tampa Bay, Florida, USA, is shown in Fig. 1.15. The longest precast pre–stressed concrete cable–stayed box girder, Chaco–Corrientes Bridge constructed in Argentina, South America, is shown in Fig. 1.16.

Fig. 1.15 Sunshine Sky Bridge, Tampa Bay, Florida, 365m Main Span Cable Stayed Bridge

Fig. 1.16 Chaco–Corrientes Bridge, Argentina, the Longest Precast Pre–stressed Concrete Cable Stayed Box Girder Bridge in South America

Typical use of pre–stressed concrete simple span box girders for the bay area rapid–transit system in San Francisco, California, is shown in Fig. 1.17.

Fig. 1.17 Precast Pre–stressed Box Girders of 21m Span and 3.35m Wide, California Guide Ways

Typical use of pre–stressed concrete in Stratfjord "B" Condeep offshore oil drilling platform Norway is shown in Fig. 1.18.

Fig. 1.18 Stratfjord "B" Condeep Offshore Oil Drilling Platform Norway

1.2 MATERIALS FOR PRE–STRESSING

1.2.1 Introduction

Most effective use of pre–stressed concrete is achieved by using high strength concrete and high tensile steel. It is important to know the properties of these materials.

1.2.2 Concrete for Pre–stressing

Pre–stressed concrete requires concrete of higher compressive strength for several reasons. The total pre–stressing force in high tensile steel tendons is usually very large. This large force is transferred to the ends of a member by bond in the case of pre–tensioning and by bearing in the case of post–tensioning. In practice neither the bond length nor the bearing area can be very large. Therefore the permissible bond and bearing stresses of the concrete at the time pre–stress is transferred must be high. Since pre–stress is normally transferred to concrete before it attains its 28 day strength, it is mandatory that the concrete used in pre–stressed work must be of high quality to offer the required bond and bearing resistance. Further the higher the permissible compressive stress of the concrete, the smaller is the cross sectional area of the member required to a given externally applied load. In turn, the smaller the cross section of a member, the smaller is its dead weight and this results in saving in pre–stressing steel and supporting structures. High strength concrete is less liable to shrinkage cracks which sometimes occur in low strength concrete before the application of pre–stress. High strength concrete has higher modulus of elasticity and smaller creep strain, resulting in smaller loss of pre–stress. Due to all these reasons the codes of various countries prescribed certain minimum grade of concrete. The minimum grades of concrete to be adopted for post–tensioning and pre–tensioning works as per IS: 1343–2012 Code is M30 and M40 respectively.

• Objects of using High Strength Concrete in PSC Construction

1. The main object of PSC is to ensure that members of relatively smaller sizes can be used for very large spans. This can be achieved provided concrete can withstand very high pre–stresses. Thus without the use of high strength concrete, the very purpose of pre–stressing of concrete will not be served.
2. The anchorage fixtures used in post–tensioned construction are designed on the basis of high strength concrete.
3. The effect of creep and shrinkage are relatively less in high strength concrete.

1.2.2.1 Cement

Cement is a combined mixture of lime stone and clay which is heated in kiln up–to 1400–1600°C. The types of cement permitted by IS: 1343–2012 (Clause 4.1) for pre–stress applications are the following. The information is revised as per IS: 456–2000.

- Ordinary Portland cement (IS: 8112–2013)
- Portland slag cement (IS: 455–1989, but with no more than 50% slag content).
- Rapid–hardening Portland cement (IS: 8041–1990).

1.2.2.2 Aggregates

The coarse aggregate are granular materials obtained from rocks and crushed stones. They may be also obtained from synthetic material like slag, shale, fly ash and clay for use in light–weight concrete.

The sand obtained from river beds or quarries is used as fine aggregate. The fine aggregate along with the hydrated cement paste fill the space between the coarse aggregate.

The important properties of aggregates are, shape and texture, size gradation, moisture content, specific gravity, unit weight, durability and absence of deleterious materials.

The requirements of aggregate are from Section 4.2 of IS: 1343–2012.

1.2.2.3 Water

The water should satisfy the requirements of Section 5.4 of IS: 456–2000.

Water fit for drinking is generally considered fit for making concrete. Water is free from acids, oils, alkalies, vegetables or other organic impurities. Soft waters also produce weaker concrete. Water has two functions in a concrete mix. Firstly, it reacts chemically with the cement to form a cement paste in which the inert aggregates are held in suspension until the cement paste has hardened. Secondly, it serves as a vehicle or lubricant in the mixture of fine aggregates and cement.

1.2.2.4 Admixtures

IS: 1343–2012 allows using admixtures that conform to IS: 9103–1999. The admixtures are divided as chemical and mineral admixtures.

The common chemical admixtures are, air–entraining admixtures, water reducing admixtures, set retarding admixtures, set accelerating admixtures, water reducing and set retarding admixtures, water reducing and set accelerating admixtures.

The common mineral admixtures are as, fly ash, GGBFS, silica fumes, rice husk ash, metakoline.

1.2.2.5 Grades of Concrete

Grades of concrete are explained in Table 1 of IS: 1343–2012 and shown in Table 1.1.

Table 1.1 Grades of Concrete

Grade	Specified characteristic compressive strength at 28 days (N/mm²)
M30	30
M35	35
M40	40
M45	45
M50	50
M55	55
M60	60

Note: In the designation of concrete mix, letter M refers to the mix and the number to the specified compressive strength of 15cm cube at 28 days expressed in N/mm^2.

1.2.2.6 Characteristic Strength of Concrete (f_{ck})

The compressive strength of concrete is termed as characteristic compressive strength of 15 cm size cubes tested at 28 days of curing (f_{ck}). This concept assumes a normal distribution of the strengths of the samples of concrete. The sampling and strength test of concrete are as per Section 15 of IS: 1343–2012.

The minimum grades of concrete for pre–stressed applications are as follows.

- 30MPa for post–tensioned members
- 40MPa for pre–tensioned members.

The maximum grade of concrete is 60MPa.

The increase in strength with age as given in IS: 1343–2012, is not observed in present day concrete that gains substantial strength in 28 days. Hence, the age factor given in Clause 5.2.1 should not be used. It has been removed from IS: 456–2000.

1.2.2.7 Tensile Strength of Concrete (f_{cr})

The tensile strength of concrete can be expressed as follows.

- Flexural tensile strength is measured by the testing beams under 2 point loading (also called 4 point loading including the reactions).
- Splitting tensile strength is measured by testing cylinders under diametral compression.
- Direct tensile strength is measured by testing rectangular specimens under direct tension.

Code recommends using an estimate of the flexural tensile strength from the compressive strength as,

$$f_{cr} = 0.7\sqrt{f_{ck}} \qquad \text{...(1.4)}$$

Where,

f_{cr} = Flexural tensile strength in MPa

f_{ck} = Characteristic compressive strength of cubes in MPa.

1.2.2.8 Stress v/s Strain Curves for Concrete

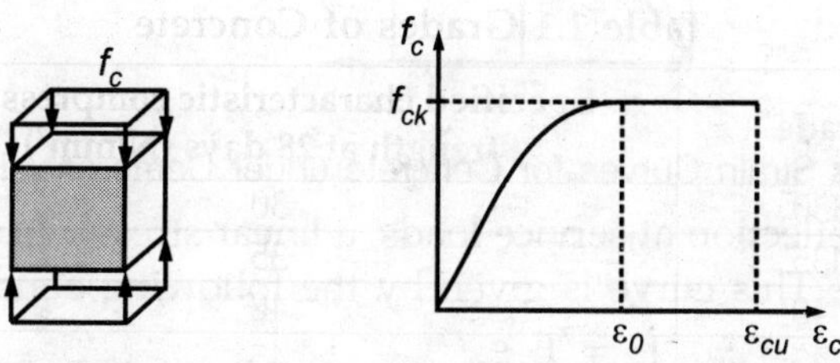

Fig. 1.19 (a) Concrete Cube under Compression, b) Design Stress v/s Strain Curve for Concrete under Compression due to Flexure

• Curve under Uniaxial Compression

The stress v/s strain behaviour of concrete under uniaxial compression is initially linear (stress α strain) and elastic (strain is recovered at unloading). With the generation of micro cracks, the behaviour becomes nonlinear and inelastic. After the specimen reach the peak stress, the resisting stress decreases with increase in the strain.

IS: 1343–2012, recommends a parabolic characteristic stress v/s strain curve (proposed by Hognestad), for concrete under uniaxial compression as shown in Fig. 1.19 (Fig. 3 in the Code).

The equation for the design curve under compression due to flexure is,

For $\varepsilon_c \le \varepsilon_0$

$$f_c = f_{ck}\left[2\left(\frac{\varepsilon_c}{\varepsilon_0}\right) - \left(\frac{\varepsilon_c}{\varepsilon_0}\right)^2\right] \qquad \text{...(1.5)}$$

$$f_c = f_{ck}\, d$$

For $\varepsilon_c < \varepsilon_c \le \varepsilon_{cu}$

$$f_c = f_{ck} \qquad \text{...(1.6)}$$

Where,

f_c = Compressive stress

f_{ck} = Characteristic compressive strength of cubes

ε_c = Compressive strain

ε_0 = Strain corresponding to f_{ck} = 0.002

ε_{cu} = Ultimate compressive strain = 0.0035

For concrete under compression due to axial load, the ultimate strain is restricted to 0.002. From the characteristic curve, the design curve is defined by multiplying the stress with a size factor of 0.67 and dividing the stress by a material safety factor of γ_m = 1.5. The design curve is used in the calculation of ultimate strength. Fig. 1.20 shows the two curves.

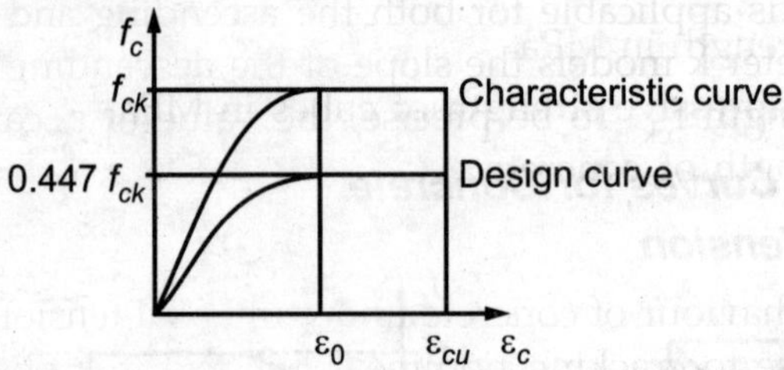

Fig. 1.20 Stress v/s Strain Curves for Concrete under Compression due to Flexure

In the calculation of deflection at service loads, a linear stress v/s strain curve is assumed up to the allowable stress. This curve is given by the following equation.

$$f_c = E_c \varepsilon_c \quad \text{...(1.7)}$$

Note that, the size factor and the material safety factor are not used in the elastic modulus E_c. For high strength concrete (say M100 grade of concrete and above) under uniaxial compression, the ascending and descending branches are steep as shown in Fig. 1.21.

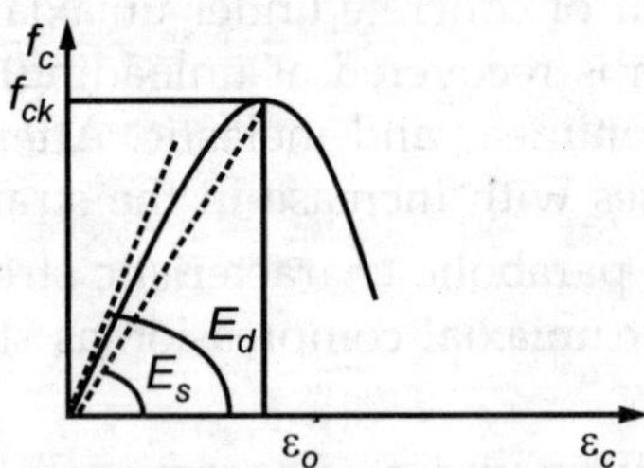

Fig. 1.21 Stress v/s Strain Curves for High Strength Concrete under Compression

The equation proposed by Thorenfeldt, Tomaxzewicz and Jensen is appropriate for high strength concrete.

$$f_c = f_{ck} \frac{n\left(\dfrac{\varepsilon_c}{\varepsilon_0}\right)}{n-1+\left(\dfrac{\varepsilon_c}{\varepsilon_0}\right)^{nk}} \quad \text{...(1.8)}$$

The variables in the previous equation are as follows.

f_c = Compressive stress

f_{ck} = Characteristic compressive strength of cubes in N/mm²

ε_c = Compressive strain ε_0 = strain corresponding to f_{ck}

k = 1 for $\varepsilon_c \le \varepsilon_0$ = 0.67 + (f_{ck}/77.5) for $\varepsilon_c > \varepsilon_0$. The value of k should be greater than 1.

n = $E_{ci}/(E_{ci} - E_s)$

E_{ci} = Initial modulus

E_s = Secant modulus at $f_{ck} = f_{ck} / \varepsilon_c$.

The previous equation is applicable for both the ascending and descending branches of the curve. Also, the parameter k models the slope of the descending branch, which increases with the characteristic strength f_{ck}. To be precise, the value of ε_0 can be considered to vary with the compressive strength of concrete.

• *Curve under Uniaxial Tension*

The stress v/s strain behaviour of concrete under uniaxial tension is linear elastic initially as shown in Fig. 1.22. Close to cracking nonlinear behaviour is observed. In calculation of deflections of flexural members at service loads, the nonlinearity is neglected and a linear elastic behaviour $f_c = E_c\varepsilon_c$ is assumed. In the analysis of ultimate strength, the tensile strength of concrete is usually neglected.

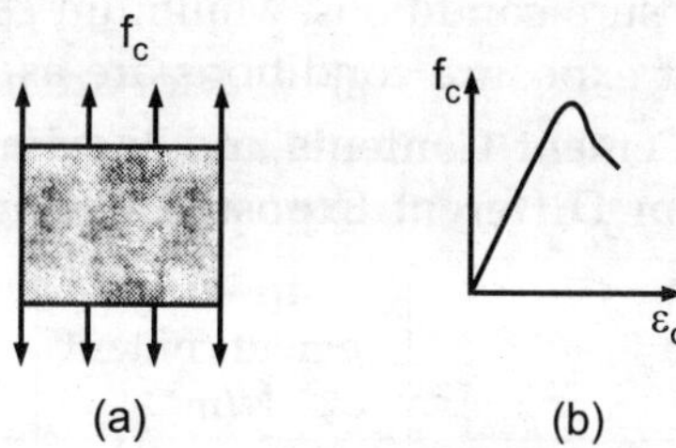

Fig. 1.22 (a) Concrete Panel under Tension, (b) Stress v/s Strain Curve for Concrete under Tension

1.2.2.9 Stiffness of Concrete

The stiffness of concrete is required to estimate the deflection of members. The stiffness is given by the modulus of elasticity. For a non–linear stress (f_c) versus strain (ε_c) behaviour of concrete the modulus can be initial, tangential or secant modulus.

IS: 1343–2012 recommends a secant modulus at a stress level of about $0.3f_{ck}$. The modulus is expressed in terms of the characteristic compressive strength and not the design compressive strength.

In the absence of test data, the modulus of elasticity for short term loading (neglecting the effect of creep) is given by the following equation.

$$E_c = 5700\sqrt{f_{ck}} \qquad ...(1.9)$$

Where,

E_c = Short term static modulus of elasticity in MPa

f_{ck} = Characteristic compressive strength of concrete.

The above expression is updated as per IS: 456–2000.

1.2.2.10 Durability

The durability of concrete is important regarding the life cycle cost of a structure. The life cycle cost includes not only the initial cost of the materials and labour, but also the cost of maintenance and repair.

In recent years emphasis has been laid on the durability issues of concrete. This is reflected in the enhanced section on durability (Section 8) in IS: 456–2000. It is expected that the revised version of IS: 1343 will also have similar importance on durability.

The durability of concrete is defined as its ability to resist weathering action, chemical attack, abrasion, or any other process of deterioration. The common durability problems in concrete are sulphate and other chemical attacks of concrete, alkali–aggregate reaction, freezing and thawing damage in cold regions, corrosion of steel bars or tendons.

The durability of concrete is related to its water tightness or permeability. Hence, the concrete should have low permeability and there should be sufficient cover to reinforcing bars. The selection of proper materials and good quality control are necessary for durability of concrete.

The durability is addressed in IS: 1343–2012 in Section 7. In Appendix A, there are guidelines on durability. Table 1.2 specifies the maximum water–cement ratio and the minimum cement content for different exposure conditions. Minimum cement contents and maximum water–cement ratios for different exposure conditions are as follows.

Table 1.2 Minimum Cement Contents and Maximum Water–Cement Ratios for Different Exposure Conditions

Exposure	Minimum cement content N/m^3	Maximum water–cement ratio
Mild – For example, completely protected against weather or aggressive conditions, except for a brief period exposure to normal weather conditions during constructions.	3000	0.65
Moderate – For example, sheltered from heavy and wind driven rain and against freezing, whilst saturated with water, buried concrete in soil and concrete continuously in water.	3000	0.55
Severe – For example, exposed to sea water, alternate wetting and drying to freezing whilst wet subjected to heavy consideration or corrosive fumes.	3600	0.45

Note: The minimum cement content prescribed above is based on 20mm nominal maximum size of aggregate. For 40mm aggregates, minimum cement content could be reduced by about 10% only for the severe exposure condition. For 12.5mm aggregates, the minimum cement content should be increased by about 10% for moderate and severe exposure conditions.

Similarly the code also specifies the requirements for concrete exposed to sulphate attack.

1.2.2.11 Creep of Concrete

Creep of concrete is defined as the increase in deformation with time under constant load. Due to the creep of concrete, the pre–stress in the tendon is reduced with time. Hence, the study of creep is important in pre–stressed concrete to calculate the loss in pre–stress.

The creep occurs due to two causes. 1. Rearrangement of hydrated cement paste (especially the layered products) 2. Expulsion of water from voids under load.

If a concrete specimen is subjected to slow compressive loading, the stress v/s strain curve is elongated along the strain axis as compared to the curve for fast loading. This can be explained in terms of creep. If the load is sustained at a level, the increase in strain due to creep will lead to a shift from the fast loading curve to the slow loading curve (Fig. 1.23).

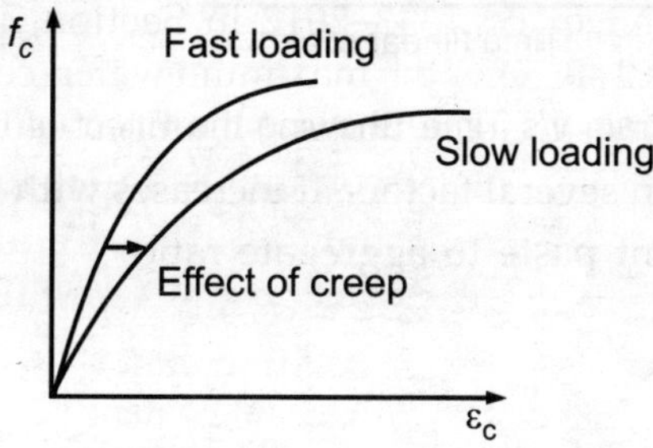

Fig. 1.23 Stress v/s Strain Curves for Concrete under Compression

Creep is explained in terms of the strain that occurs in addition to the elastic strain due to the applied loads. If the applied loads are close to the service loads, the creep strain increases at a decreasing rate with time. The ultimate creep strain is found to be proportional to the elastic strain. The ratio of the ultimate creep strain to the elastic strain is called the creep coefficient θ.

For stress in concrete less than about one–third of the characteristic strength, the ultimate creep strain is given as follows.

$$\varepsilon_{cr,ult} = \theta\varepsilon_{el} \quad ...(1.10)$$

The variation of strain with time, under constant axial compressive stress, is represented in the following Fig. 1.24.

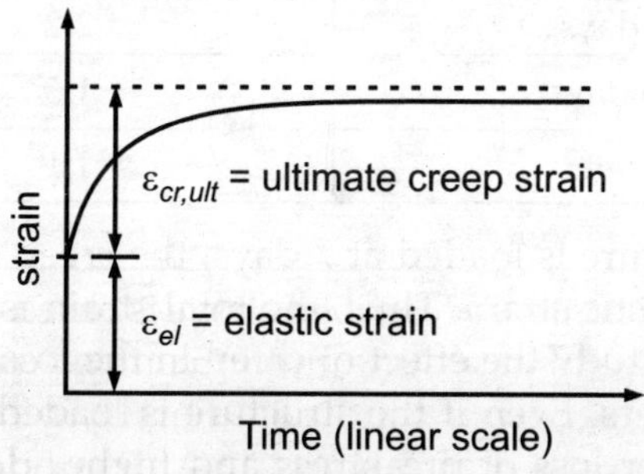

Fig. 1.24 Strain v/s Time for Concrete under Compression

If the load is removed, the elastic strain is suddenly recovered. However the recovered elastic strain is less than the initial elastic strain, as the elastic modulus increases with age.

There is reduction of strain due to creep recovery which is less than the creep strain. There is some residual strain which cannot be recovered (Fig. 1.25).

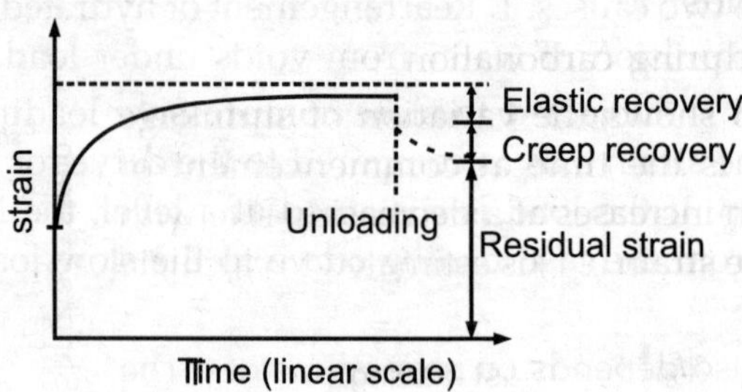

Fig. 1.25 Strain v/s Time Showing the Effect of Unloading

The creep strain depends on several factors. It increases with the increase in the following.

- Cement content (cement paste to aggregate ratio)
- Water–cement ratio
- Air entrainment
- Ambient temperature.

The creep strain decreases with the increase in the following.

- Age of concrete at the time of loading.
- Relative humidity
- Volume to surface area ratio.

The creep strain is also depends on the type of aggregate.

IS: 1343–2012 (Section 5.2.5), gives guidelines to estimate the ultimate creep strain. It is a simplified estimate where only one factor has been considered. The factor is age of loading of the pre–stressed concrete structure. The creep coefficient θ is provided for different age of loading in the following Table 1.3.

Table 1.3 Creep Coefficient θ for Different Age of Loading

Age at loading	Creep coefficient
7 days	2.2
28 days	1.6
1 year	1.1

It showed that, if the structure is loaded at 7 days, the creep coefficient is 2.2. Means, the creep strain is 2.2 times the elastic strain. Thus, the total strain is more than thrice the elastic strain. Hence, it is essential to study the effect of creep in the loss of pre–stress and deflection of pre–stressed flexural members. Even if the structure is loaded at 28 days, the creep strain is substantial. This gives higher loss of pre–stress and higher deflection.

Cure the concrete sufficiently and delays in the application of load provide long term benefits with regards to durability, loss of pre–stress and deflection.

1.2.2.12 Shrinkage

Shrinkage of concrete is defined as the contraction due to moisture loss. The study of shrinkage is also necessary in pre–stressed concrete to evaluate the loss in pre–stress. Shrinkage occurs due to two causes.

1. Water loss from voids
2. Volume reduction during carbonation

The following Fig. 1.26 shows the variation of shrinkage strain with time. Here, t_0 is the time at commencement of drying. The shrinkage strain increases at a decreasing rate with time. The ultimate shrinkage strain (ε_{sh}) is estimated to calculate the loss in pre–stress.

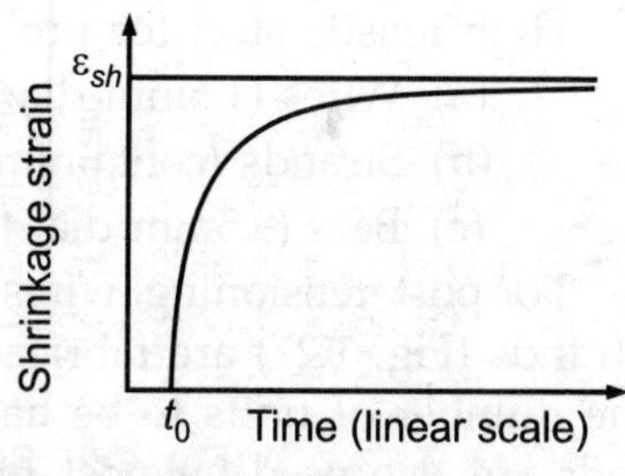

Fig. 1.26 Shrinkage Strain v/s Time

Like creep, shrinkage also depends on several factors. The shrinkage strain increases with the increase in the following variables.

- Ambient temperature
- Temperature gradient in the members
- Water–to–cement ratio
- Cement content.

The shrinkage strain decreases with the increase in the following variables.

- Age of concrete at commencement of drying
- Relative humidity
- Volume to surface area ratio.

The shrinkage strain also depends on the type of aggregate.

IS: 1343–2012 (Section 5.2.4), gives guidelines to estimate the shrinkage strain. The simplified estimate of the ultimate shrinkage strain (ε_{sh}).

$$\text{For pre–tension } \varepsilon_{sh} = 0.0003 \quad \text{...(1.11)}$$

$$\text{For post–tension } \varepsilon_{sh} = \frac{0.0002}{\log_{10}(t+2)} \quad \text{...(1.12)}$$

Where, t is the age at transfer in days. Note that for post–tension, t is the age at transfer in days which approximates the curing time.

With increasing age at transfer, the shrinkage strain reduces. As mentioned before, cure the concrete sufficiently and delaying the application of load provides long term benefits with regards to durability and loss of pre–stress.

1.2.3 Steel for Pre–stressing

For pre–stressing concrete members, the desirable properties of steel are:

(a) Small quantities should be capable of producing large working forces.

(b) Small changes in length etc. should produce little effect on these working forces.

Since high tensile steel possess these characteristics, it is almost the universal material for pre–stressing concrete members. The ultimate tensile strength of steel used in pre–stressed concrete work varies between 1000MPa (large diameter bars) and 2200MPa (small diameter wires). These high values of tensile strength can be achieved through one or more of the following three procedures.

(a) Adjusting the composition of steel
(b) Cold working: either by rolling, stretching or drawing
(c) Special heat treatment

High tensile steel for pre–stressing usually takes one of the following forms.

(a) Wires (1.5mm dia. to 8mm dia.)
(b) Strands (containing 2 to 7 wires)
(c) Bars (9.5mm dia. to 32mm dia.)

For post–tensioning, wires are widely employed: they are grouped in parallel, into cables. Strands (Fig. 1.27) are fabricated in the factory by twisting wires together, thus decreasing the number of units to be handled in the tensioning operations. Strands and high tensile rods are also used for post–tensioning. For pre–tensioning work, small diameter wires (not greater than 3mm and when smooth and not greater than 5mm when corrugated) are desirable because the pre–stress has to be transfer to concrete by means of bond only. In USA, 7 wire strands have been extensively used for pre–tensioning. Although strands cost more than wires of the same tensile strength, their better bonding characteristics makes them especially suitable for pre–tensioning. Mild steel is not suitable for pre–stressing because the loss of strain due to shrinkage and creep of concrete and other causes will nullify the available small strain in steel.

(a) Two wire strand (b) Seven wire strand

Fig. 1.27 Section of Strands

• Objects of using High Strength Steel in PSC Construction

1. The young's modulus of elasticity is almost the same for all kinds of steel, is such the amount of elastic elongation is directly proportional to the strength of steel. It is therefore in order to have adequate amount of elongation. It is necessary to use only high strength steel in PSC construction.
2. If ordinary mild steel is employed in PSC construction. It can be tensioned to not more than 200N/mm^2. This initial stress would be almost lost due to various losses of pre–stress. As such the amount of effective pre–stress available in tensioned steel would be practically zero.

1.2.3.1 Specifications of IS: 1343–2012

The pre–stressing steel may be any one of the following.

(i) Plain hard drawn steel wire [IS: 1785 (Part 1)–1983[6] and IS: 1785 (Part 2)–1983[7]]
(ii) Cold–drawn indented wire [IS: 6003–1983[11]]
(iii) High tensile steel bar [IS: 2090–1983[8]]
(iv) Uncoated stress relieved strand [IS: 6006–1983[12]]

❖ Requirements for Cold–drawn Stress Relieved Plain Wire (Table 1.4)

Table 1.4 Requirements for Cold–drawn Stress Relieved Plain Wire

Nominal dia. (mm)	Tolerance (mm)	Tensile strength (N/mm^2)	0.2% proof strength (N/mm^2)	Minimum elongation on 200mm gauge length (mm)
2.5	±0.025	2010	1706	2.5
3.0	±0.05	1863	1581	2.5
4.0	±0.05	1716	1458	3.0
5.0	±0.05	1569	1334	4.0
7.0	±0.05	1520	1250	4.0
8.0	±0.05	1373	1167	4.0

Note: Plain hard drawn steel wire is the most common type of steel used in our country for pre–stressed concrete. The stress v/s strain curve of this type of wire is shown in Fig. 1.28.

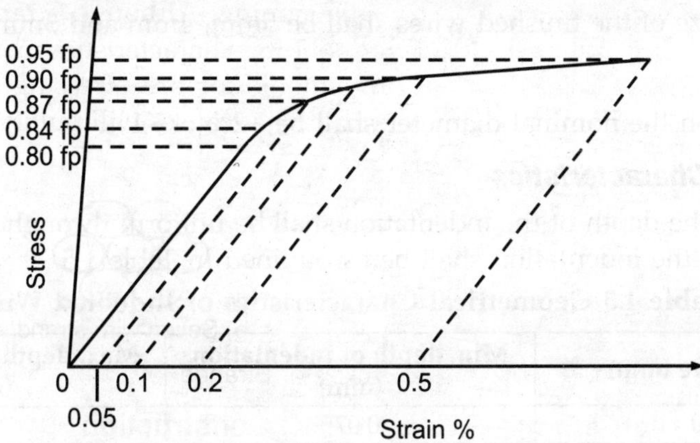

Fig. 1.28 Stress v/s Strain Curve for Pre–stressing Steel

❖ Specifications for Indented Wire for Pre–stressed Concrete

(a) Indentation

Indentation is the creation of a depression or a series of depressions of particular shape and dimension on the surface of the wire by some mechanical process or otherwise, without altering the mechanical and chemical properties (Either elliptical or circular indentations are made. Refer Fig. 1.29).

(b) Nominal Diameter

The nominal diameter of the indented wire is equivalent to the diameter of a plain wire having the same weight per unit length as the indented wire. (The nominal diameter shall be determined by weighing twelve test pieces of same length equal to at–least five pitch lengths containing all indentations)

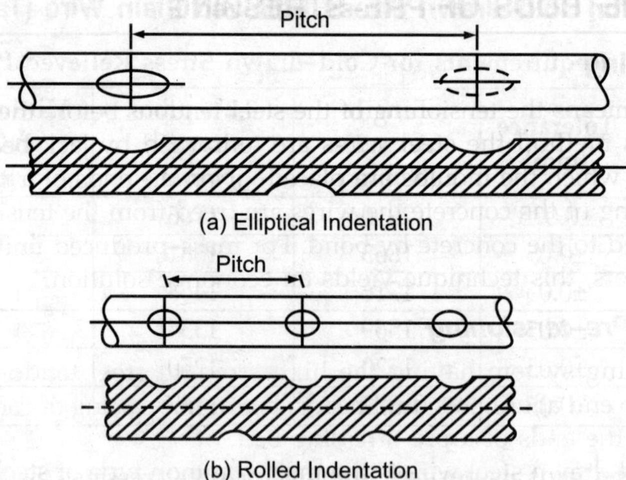

Fig. 1.29 Illustrative Sketch of Typical Indentations of Indented Wire

(c) Nominal size

The nominal size of the finished wires shall be 5mm, 4mm and 3mm.

(d) Tolerance

The tolerance on the nominal diameter shall be +2% or –1%.

(e) Geometrical Characteristics

The pitch and the depth of the indentation shall be uniform throughout the length of the wire. The depth of the indentation shall be as outlined in Table 1.5.

Table 1.5 Geometrical Characteristics of Indented Wires

Dia. of wire (mm)	Min. depth of indentation (mm)	Max. depth of indentation (mm)
5.00	0.075	0.15
4.00, 3.00	0.05	0.10

(f) Mechanical Properties (Table 1.6)

Table 1.6 Mechanical Properties of Indented Wires

Nominal dia. (mm)	Min. tensile strength (MPa)	0.2% Proof stress (MPa)	Min. % elongation (Gauge length 200mm)
5.00	1569.1	1333.6	3.5
4.00	1716.2	1458.6	3.0
3.00	1863.3	1583.8	2.5

(g) Ductility

The wires shall withstand the reverse bend test (Described in IS: 6003–1983).

(h) Relaxation

The relaxation stress in the wire when tested as per IS: 6003–1983 code shall not exceed 68.6MPa at the end of 1000 hours.

1.3 GENERAL METHODS OF PRE–STRESSING

1.3.1 Pre–tensioning

Pre–tensioning means the tensioning of the steel tendons before the concrete is placed in the moulds. In this method the steel wires are tensioned by jack bearing either on rigid abutments between which the moulds are placed, or on the ends of a rigid mould. After the setting and hardening of the concrete the wires are freed from the tensioning device and the forces are transferred to the concrete by bond. For mass–produced units such as joists, floor slabs, railway sleepers, this technique yields an economic solution.

1.3.1.1 Stages of Pre–tensioning

- Pre–tensioning system having the high strength steel tendons, which are pulled between two end abutments (bulkheads) before the casting of concrete. The abutments are fixed at the ends of a pre–stressing bed.
- After concrete gains its desired strength for pre–stressing, the tendons are cut loose from the abutments.
- The pre–stress is transferred to the concrete from the tendons, due to the bond between them. During the transfer of pre–stress, the member undergoes elastic shortening. If the tendons are located eccentrically, the member is likely to bend and deflect (camber).

The various steps of the pre–tensioning operation are as follows.

- Anchoring of tendons against the end abutments
- Placing of jacks
- Applying tension to the tendons
- Casting of concrete
- Cutting of the tendons.

When cutting the tendons, the pre–stress is transferred to the concrete with elastic shortening and camber of the member.

The steps are shown schematically in the below Fig. 1.30.

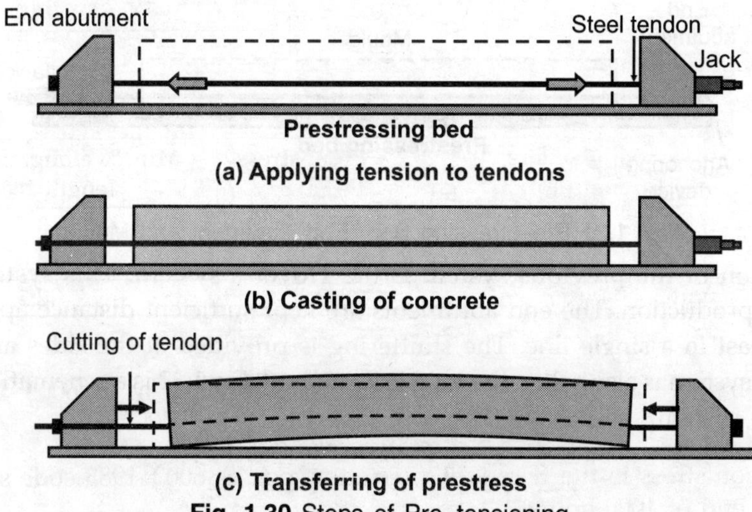

Fig. 1.30 Steps of Pre–tensioning

1.3.1.2 Advantages of Pre–tensioning

The advantages of pre–tensioning as compared to post–tensioning are,

- Excellent and reliable bond can be obtained between the tensioned steel and the surrounded concrete along the length.
- As pre–tensioning is carried out in factory, proper curing and satisfactory supervision are achieved.
- Pre–tensioning is suitable for precast members produced in bulk.
- In pre–tensioning large anchorage device is not present.

1.3.1.3 Disadvantages of Pre–tensioning

The relative disadvantages are,

- A pre–stressing bed is required for the pre–tensioning operation.
- Waiting period in the pre–stressing bed, before the concrete gains adequate strength.
- There should be good bond between concrete and steel over the transmission length.

1.3.1.4 Devices

The essential devices for pre–tensioning are as follows.

- Pre–stressing bed
- End abutments
- Shuttering / mould
- Jack
- Anchoring device
- Harping device (optional)

Pre–stressing Bed, End Abutments and Mould

Fig. 1.31 shows the devices.

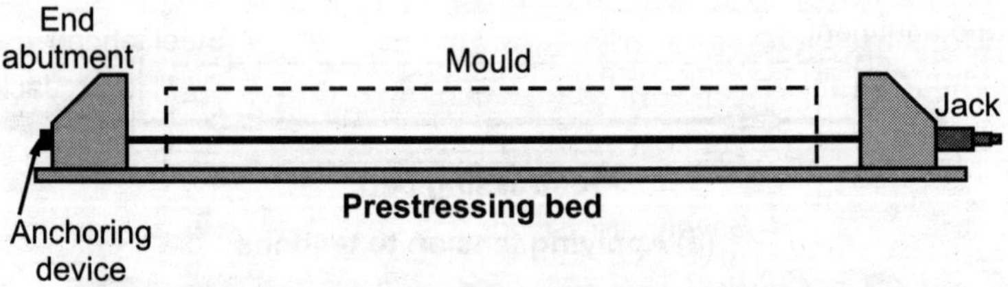

Fig. 1.31 Pre–stressing Bed, End Abutment and Mould

An extension of the previous system is the Hoyer's system. This system is generally used for mass production. The end abutments are kept sufficient distance apart, and several members are cast in a single line. The shuttering is provided at the sides and between the members. This system is also called the long line method. Fig.1.32 is a schematic representation of the Hoyer's system.

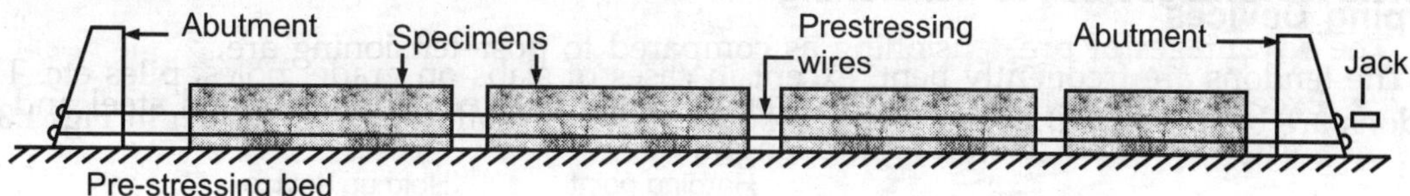

Fig. 1.32 Schematic Representation of Hoyer's System

Jacks

The jacks are used to apply tension to the tendons. Hydraulic jacks are commonly used. These jacks work on oil pressure generated by a pump. The principle behind the design of jacks is Pascal's law. The load applied by a jack is measured by the pressure reading from a gauge attached to the oil inflow or by a separate load cell. Fig. 1.33 shows a double acting hydraulic jack with a load cell.

Fig. 1.33 A Double Acting Hydraulic Jack with a Load Cell

Anchoring Devices

Anchoring devices are done on the wedge and friction principle. In pre–tensioned members, the tendons are to be held in tension during the casting and hardening of concrete. Here simple and cheap quick–release grips are generally adopted. Fig. 1.34 provides examples of anchoring devices.

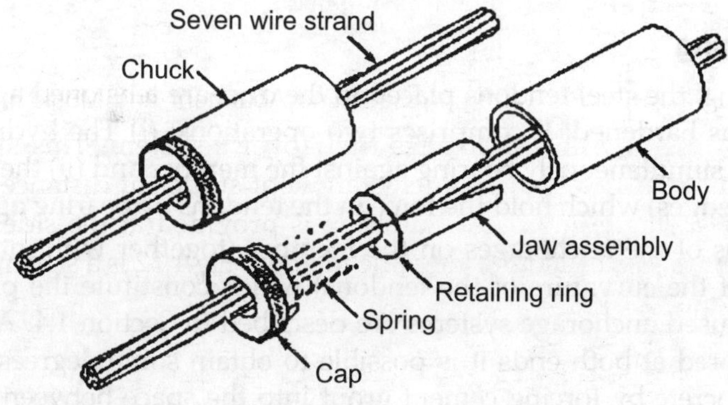

Fig. 1.34 Chuck Assembly for Anchoring Tendons

Harping Devices

The tendons are frequently bent, except in cases of slabs on grade, poles, piles etc. The tendons are bent (harped) in between the supports with shallow sag as shown in Fig. 1.35.

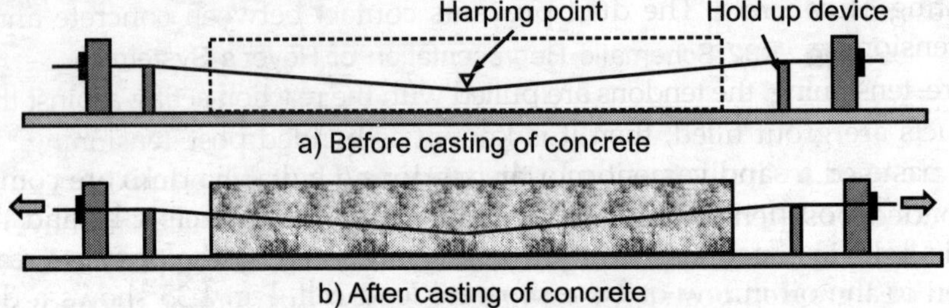

Fig. 1.35 Harping of Tendons

The tendons are harped using special hold-down devices as shown in Fig. 1.36.

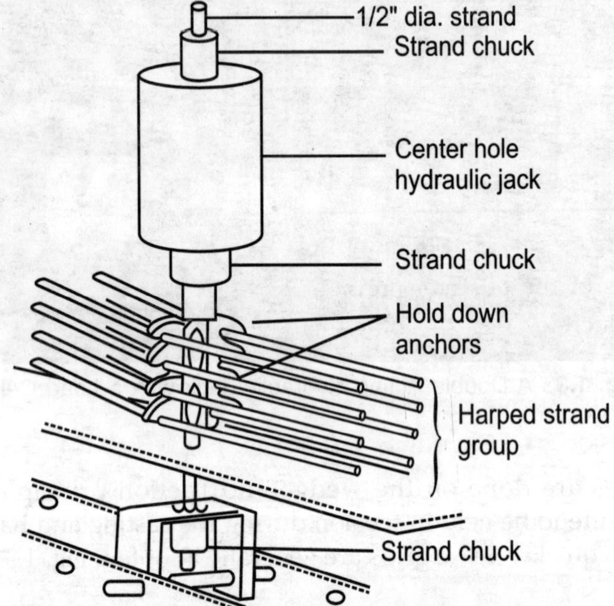

Fig. 1.36 Hold–down Anchor for Harping of Tendons

1.3.2 Post–tensioning

In post–tensioning, the steel tendons placed in the duct are tensioned against the concrete after the concrete has hardened. It comprises two operations: (i) The hydraulic jack tension the steel wires while simultaneously bearing against the member and (ii) the jacks are replaced by anchorages (or wedges) which hold the force in the tendons by bearing against the member.

It is the reactions of the anchorages on the member (together with internal uplift forces produced because of the curvature of the tendons) which constitute the pre–stressing force. Some of the widely used anchorage systems are described in Section 1.4. After the tensioned steel has been anchored at both ends it is possible to obtain some degrees of bond between the steel and the concrete by forcing cement grout into the space between them in the duct.

1.3.2.1 Stages of Post–tensioning

Post–tensioning systems have ducts for tendons (or strands). Those are placed along with the reinforcement before the casting of concrete. The tendons are placed in the ducts after the casting of concrete. The duct prevents contact between concrete and the tendons during the tensioning.

Unlike pre–tensioning, the tendons are pulled with the reaction acting against the set concrete.

When ducts are grout filled, then it is known as bonded post–tensioning. The grout is a neat cement paste or a sand cement mortar containing suitable admixture.

In un–bonded post–tensioning, as the name suggests, the ducts are never grouted and the tendon is held in tension solely by the end anchorages. Fig. 1.37 shows a schematic representation of a grouted post–tensioned member. The duct profile depends on the support conditions. For a simply supported member, the duct has a sagging profile between the ends. For a continuous member, the duct sags in the span and hogs over the support.

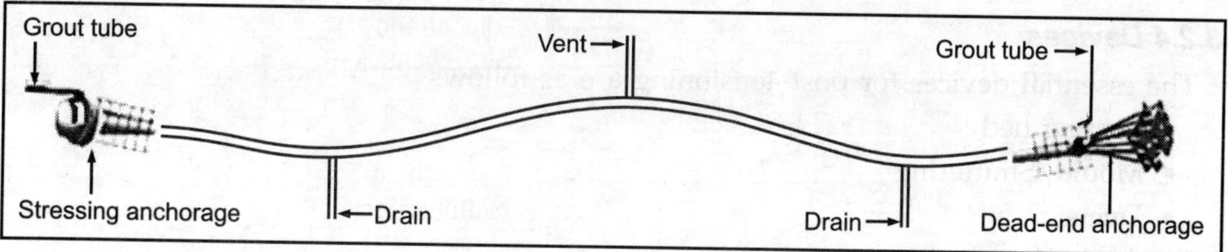

Fig. 1.37 Post–tensioning

The various steps of the post–tensioning operation are as follows.

- Casting of concrete
- Placement of the tendons
- Placement of the anchorage block and jack
- Applying tension to the tendons
- Seating of the wedges
- Cutting of the tendons

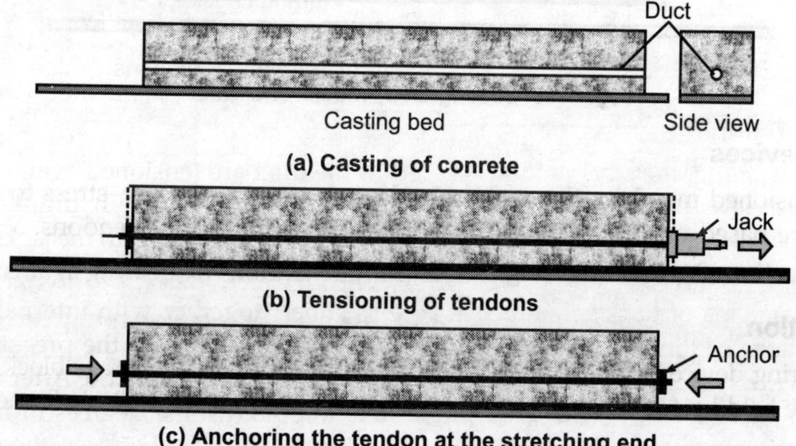

(a) Casting of conrete

(b) Tensioning of tendons

(c) Anchoring the tendon at the stretching end

Fig. 1.38 Steps of Post–tensioning

The steps are shown schematically in the Fig. 1.38. After anchoring a tendon at one end, the tension is applied at the other end by a jack. The tensioning of tendons and pre–compression of concrete occur simultaneously. A system of self–equilibrating forces develops after the stretching of the tendons.

1.3.2.2 Advantages of Post–tensioning

The relative advantages of post–tensioning as compared to pre–tensioning are as follows.

- Post–tensioning is suitable for heavy cast–in–place members.
- The waiting period in the casting bed is less.
- The transfer of pre–stress is independent of transmission length.

1.3.2.3 Disadvantage of Post–tensioning

The relative disadvantage of post–tensioning as compared to pre–tensioning is the requirement of anchorage device and grouting equipment.

1.3.2.4 Devices

The essential devices for post–tensioning are as follows.

- Casting bed
- Mould/Shuttering
- Ducts
- Anchoring devices
- Jacks
- Couplers (optional)
- Grouting equipment (optional).

Casting Bed, Mould and Ducts

Fig. 1.39 shows the devices.

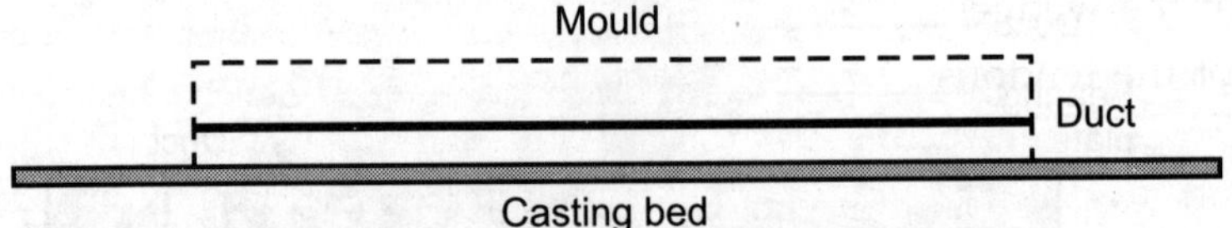

Fig. 1.39 Casting Bed, Mould and Duct

Anchoring Devices

In post–tensioned members the anchoring devices transfer the pre–stress to the concrete. The devices are based on the following principles of anchoring the tendons.

1. Wedge action, 2. Direct bearing, 3. Looping the wires

❖ Wedge Action

The anchoring device based on wedge action comprises of an anchorage block and wedges. The strands are held by frictional grip of the wedges in the anchorage block. Some examples of systems based on the wedge–action are Freyssinet, Gifford–Udall, Anderson and Magnel–Blaton anchorages. Fig. 1.40, 1.41 and 1.42 show some patented anchoring devices.

Fig. 1.40 Freyssinet "T" System Anchorage Cones

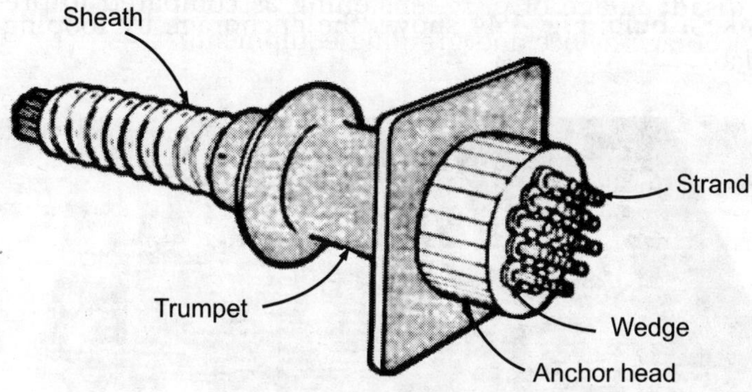

Fig. 1.41 Anchoring Devices

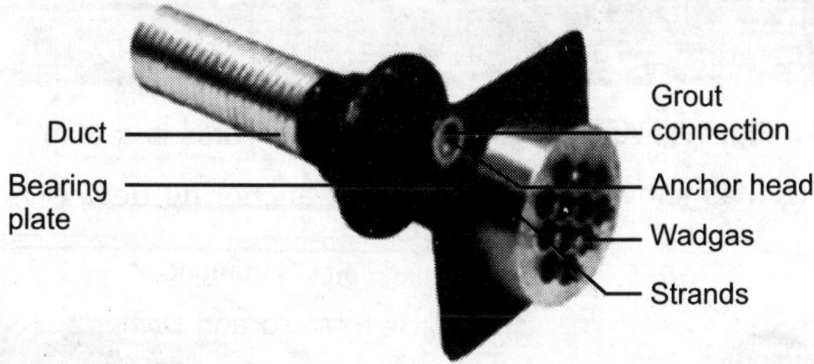

Fig. 1.42 Anchoring Devices

❖ Direct Bearing

The rivet or bolt heads or button heads formed at the end of the wires directly bear against a block. The BBRV post–tensioning system and the Prescon system are based on this principle. Fig. 1.43 shows the anchoring by direct bearing.

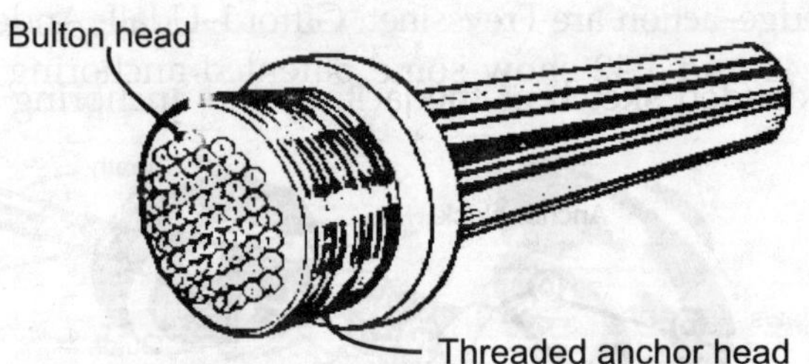

Fig. 1.43 Anchoring with Button Heads

❖ Looping the Wires

The Baur–Leonhardt system, Leoba system and also the Dwidag single–bar anchorage system, work on this principle where the wires are looped around the concrete. The wires are looped to make a bulb. Fig. 1.44 shows the anchorage by looping of the wires in a post–tensioned slab.

Fig. 1.44 Anchorage by Looping the Wires in a Slab

The following Fig. 1.45 shows the final form of anchoring device

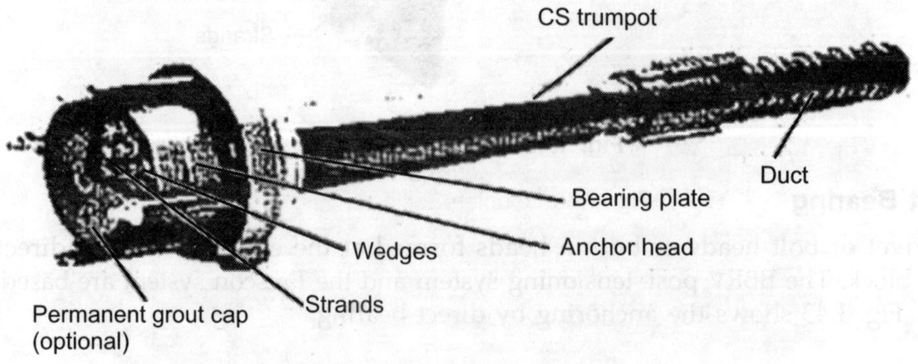

Fig. 1.45 Final Form of an Anchoring Device

Jacks

Fig. 1.46 shows an extruded sketch of the jacking and anchoring with wedges.

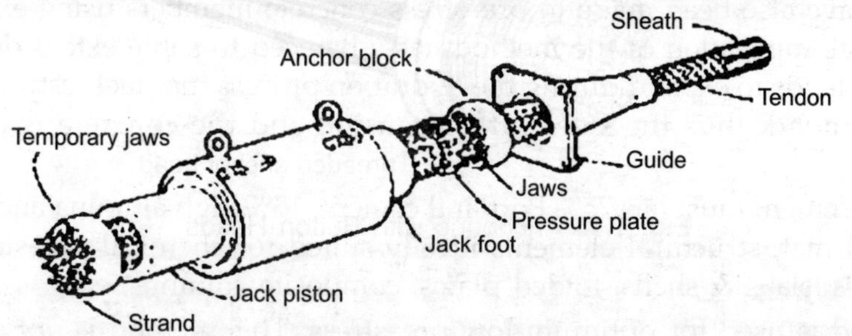

Fig. 1.46 Jacking and Anchoring with Wedges

Couplers

The couplers are used to connect strands or bars. They are located at the junction of the members, for example at or near columns in post–tensioned slabs, on piers in post–tensioned bridge decks.

The couplers are tested to transmit the full capacity of the strands or bars. A few types of couplers are shown in Fig. 1.47, 1.48 and 1.49.

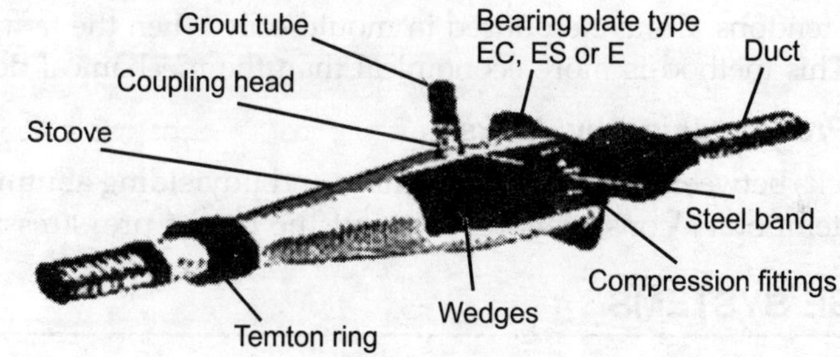

Fig. 1.47 Coupler for Strands

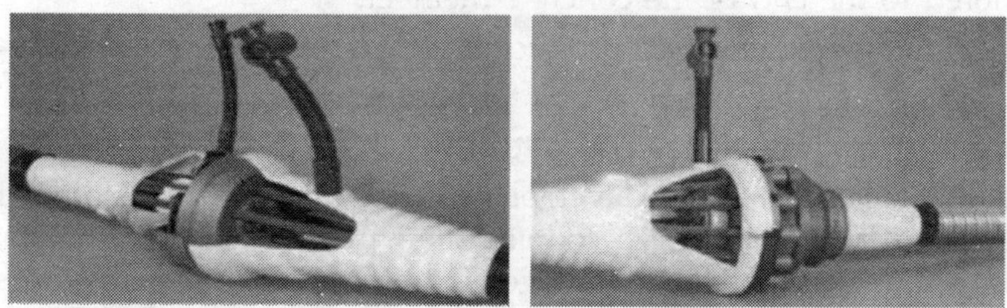

Fig. 1.48 Couplers for Strands

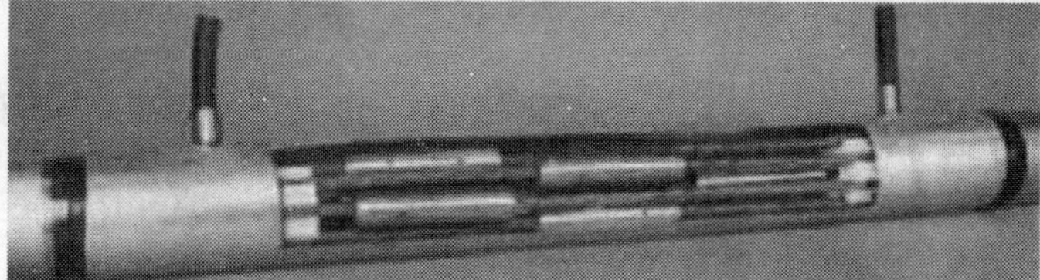

Fig. 1.49 Couplers for Strands

1.3.3 Other Methods of Pre–stressing

1.3.3.1 Chemical Pre–stressing

Attempts have also been made to pre–stress concrete members using expanding cement and the practical application of the method has advanced to some extent only in the USSR. As the concrete tends to expand during the hydration process the steel restrains the expansion because of the bond, thus the steel suffers tension and the concrete is correspondingly pre–stressed.

Expanding cement must has 75% Portland cement, 15% high alumina and 10% gypsum. It has been found that, structural elements ideally suited for chemical pre–stressing includes, pipes, thin walls, slabs & shells, folded plates, composite columns, etc.

This method is used for optimum low pre–stress. This method is not suitable for high degree of pre–stress and high percentage of steel where mechanical pre–stressing can be used.

1.3.3.2 Electro–thermal Pre–stressing

This is by heated tendons achieved by passing electric current in high tensile wires and hence the name. The process consists of electrically heated bars up to temperature 300–400ºC within 3–5 min. the bars undergo an elongation of 0.4–0.5%. The bars, after cooling try to shorten but are checked by fixed anchors at two ends.

The cooling period is reckoned to be 12–15 minutes. By this method we get stress up to 500–600N/mm^2 in tendons. Concrete poured in mould only when the temperature of tendon is less than 90ºC. This method is more economical than the mechanical devices.

1.3.3.3 External Pre–stressing by Jacks

By using flat jacks between the ends of a member and unyielding abutments, pre–stressing has been done in temporary constructions (without the use of pre–stressing steel).

1.4 ANCHORAGE SYSTEMS

There are essentially three principles underlying the ways in which post–tensioned steel wires are anchored to the end of the concrete member.

(i) The principle of wedge action producing a frictional grip on the wires.

(ii) Direct bearing of rivet or bolt heads formed at the end of the wires.

(iii) Looping of wires around the concrete.

Many reliable systems have been developed based on the principles of wedge action and of direct bearing. The last principle, i.e., looping the wires around the concrete, has not been widely applied, although it has its advantages. Three popular pre–stressing systems using anchorages based on wedge action are Freyssinet system, the Magnel–Blaton system and Gifford–Udall system. The Lee–McCall system makes use of the direct bearing anchorage.

1.4.1 The Freyssinet System (Developed by French Engineer Freyssinet)

In this system, each anchorage unit consists of a cylinder with a conical interior (called female cone) through which wire pass, and against the walls of which the wires are wedged by a conical plug lined longitudinally with grooves (called male cone) to receive them.

At present Freyssinet anchorage cones are manufactured in India in the following sizes.

(a) 12 ϕ 5mm cones suitable for cables with 12 wires of 5mm dia. HT steel (High strength steel).

(b) 12 ϕ 7mm cones suitable for cables with 12 wires of 7mm dia. HT wires.

(c) 12 ϕ 8mm cones suitable for cables with 12 wires of 8mm dia. HT wires.

(d) 6T13 cones suitable for cables of 6 strands of 13mm HT steel.

(e) 12T13 anchorage suitable for cables of 12 Nos. HT strands of 13mm dia.

Each anchorage unit consists of a female cone and a male cone. The female cone is a heavily reinforced cement mortar cylinder with a conical opening at the center. At one end, a mild steel tube is embedded to provide a water tight connection between the sheathing and the cone (Fig. 1.50). The male cone is essentially truncated cone with grooves along the entire length; each groove accommodates one tendon of the Freyssinet cable. A central hallow tube provides access for grouting of the cable after completing the pre–stressing operations.

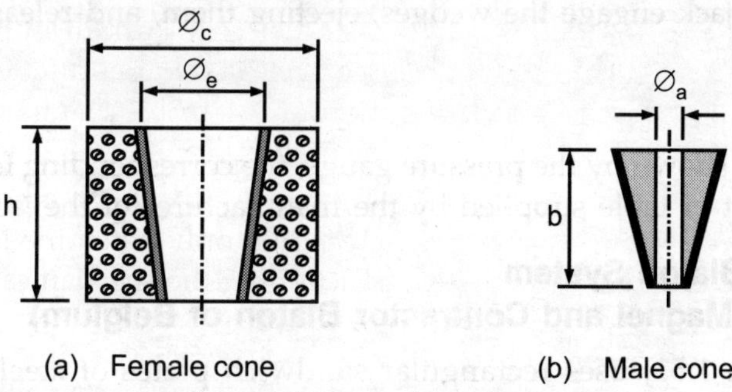

Fig. 1.50 Male and Female Cones

1.4.1.1 Female Cone

Table 1.7 Female Cone

Type of cone (outside dia.) (mm)	ϕ_c dia. (mm)	h (mm)	ϕ_e = Nominal outside dia. of connection sleeve (mm)
12ϕ5	96	100	32
12 ϕ 5 Ext	120	100	32
12 ϕ 7	120	120	38
12 ϕ 7 Ext	150	125	38
12 ϕ 8	150	125	51
6 T 13	150	125	51

Normally female cones are embedded in concrete with proper reinforcement to cater for bursting forces. However special female cones for external use are available and are indicated as external cones in the Table 1.7. (Refer note for external cone, at 1.3.1.2)

1.4.1.2 Male Cone

Table 1.8 Male Cone

Type of cone	ϕ_a (mm)	b (mm)
12 ϕ 5	8	74
12 ϕ 7	11	95
12 ϕ 8	11	108
6 T 13	10	10

Note: Male cones for 7mm and 8mm wires and 13mm strands are coated with carborundum to give additional grip. Table 1.8 shows different types of male cones.

- ***Dewedging***

Emptying the tension chamber and resuming pumping into the blocking chamber (which communicates with the release chamber) the jack is retracted and, at the end of its stroke, the shoulders of the jack engage the wedges, ejecting them, and release the wires from the jack.

- ***Pressure Gauge***

From the reading shown by the pressure gauge, the corresponding force applied is known from calibration chart or table supplied by the manufacturer of the jack.

1.4.2 The Magnel–Blaton System (Due to Prof. Magnel and Contractor Blaton of Belgium)

This system (Fig. 1.51) uses rectangular sandwich plates of steel which have tapered notches to receive the wedges. Wires of 5mm or 7mm are gripped between the grooves of the wedges and the sandwich plates. The most commonly used plates are those for 8 wires. In this system each cable consists of even number of wires up to 64 Nos. Depending upon the number of wires in a cable, 1 to 8 of the sandwich plates are laid one over the other, which react against a cast–steel distribution plate placed between them and the concrete. The drawn into performed ducts made by means of the member, the spacing of wires is maintained by spacers kept at regular intervals of 1.25mm to 2.5mm.

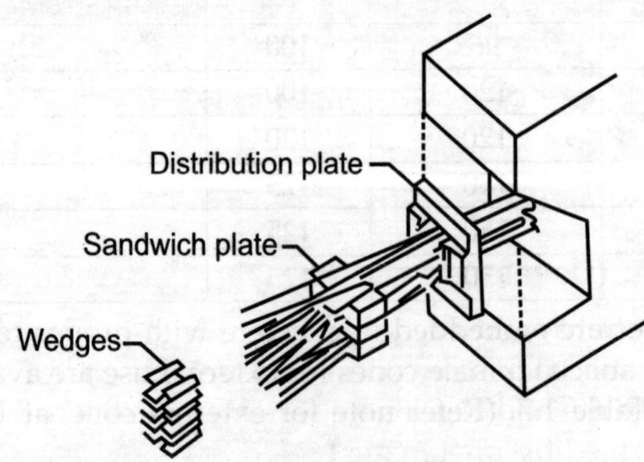

Fig. 1.51 View of Anchorage of the Magnel–Blaton Post–tensioning System

The wires are stretched two at a time by a hydraulic jack designed for the purpose. After tensioning is over, the wedge for the two wires is driven in. After the stressing of all two wires of a cable is completed in this manner the duct is grouted. After all the cables have been stressed, the anchorage units are covered by concrete for protection.

The sizes of the anchorage units depend upon the number of wires used.

1.4.3 The Gifford–Udall System

This is one of the systems widely used in India as shown in Fig. 1.52. With the jack of this system is incorporated a pump, a pressure gauge, an extension rule and supporting legs, so that stressing can be carried out by one man.

Each wire is anchored by standard CCL anchor grips, which consists of two semicircular hard steel wedges within a barrel, which is slipped over the wire. The inner face of the wedges has serrations which indent the wire, combining a wedge action with a mechanical grip to give a positive anchorage. The CCL grips bear on a drilled steel plate (anchor plate) through which the wire passes; this in turn bears on a square thrust ring cast into the concrete so as to distribute the force uniformly. A somewhat similar grip, heavy duty automatic grip, is fixed temporarily to the wire and pulled back by the jack. The heavy duty grips are recovered and reused.

The anchorage may be either plate anchorage or tube anchorage and the dimensions of the units depend on the number of wires used in the cable. More details of this system could be obtained from Killick Nixon & Co. Ltd, Bombay.

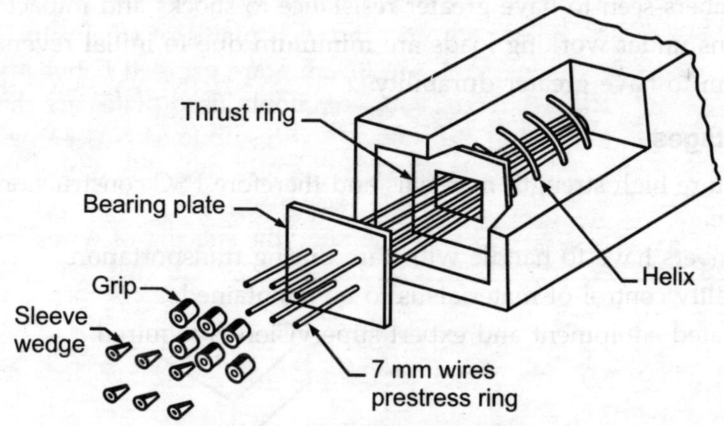

Fig. 1.52 View of Anchorage of The Gifford–Udall Pre–stressing System

1.4.4 Lee–Mccall System: (Developed By Donovan Lee of England)

This system is used to stretch and anchor high tensile steel bars of dia. 12mm to 32mm. Each rod is pulled by hydraulic jack screwed to the threaded end of the bar. The anchorage unit consists of high strength nut and bearing on steel plate on the end of the beam (Fig. 1.53). The duct is circular and formed by pneumatic tube or by flexible corrugated metal tube.

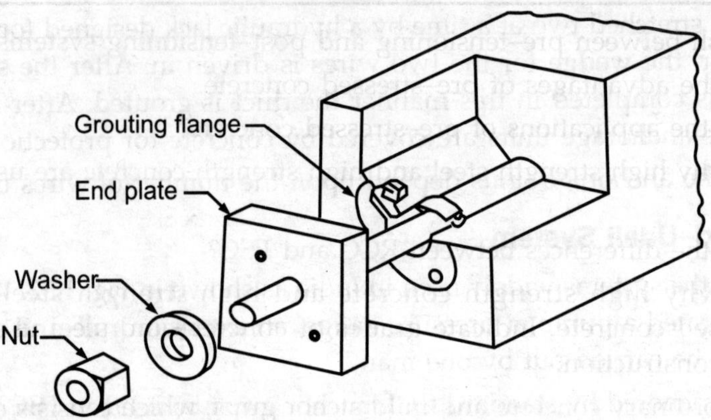

Fig. 1.53 Screwed Anchorage for High Tensile Steel Bars

1.5 ADVANTAGES AND DISADVANTAGES OF PRE–STRESSED CONCRETE CONSTRUCTION

1.5.1 Advantages

1. It is ideally suited for very large span for which RCC members would be of abnormal sizes.
2. Due to reduction in size, the dead load is reduced and therefore the supporting structure can also be of relatively of smaller in size.
3. PSC members seen to have greater resistance to shocks and impact loads.
4. Deflections under working loads are minimum due to initial reversed curvature.
5. They seem to have greater durability.

1.5.2 Disadvantages

1. They require high strength materials and therefore PSC construction relatively more expensive.
2. PSC members have to handle with care during transportation.
3. Strict quality control of materials is to be maintained.
4. Sophisticated equipment and expert supervision is required.

REVIEW QUESTIONS

1.1 Explain the basic concept of pre–stressing.

1.2 Mention the historical background of pre–stressing.

1.3 Explain principles of pre–stressing with the help of stress, strength and load balancing concepts.

1.4 Explain the different types of pre–stressing.

1.5 Distinguish between pre–tensioning and post–tensioning systems.

1.6 Mention the advantages of pre–stressed concrete.

1.7 What are the applications of pre–stressed concrete?

1.8 Explain why high strength steel and high strength concrete are used in pre–stressed concrete.

1.9 What are the differences between RCC and PSC?

1.10 Explain why high strength concrete and high strength steel are essential for pre–stressed concrete. Indicate grades of concrete and steel used in pre–stressed concrete construction.

1.11 Why the ordinary concrete and mild steel cannot be used in PSC?

1.12 Explain the necessity of high strength concrete in pre–stressed concrete works.

1.13 Explain the general requirements of cement, aggregates, water and admixtures for pre–stressing as per IS: 1343–2012.

1.14 Sketch the stress v/s strain curves for high strength concrete and high strength steel and explain their salient features.

1.15 Explain the stiffness requirements of pre–stressed concrete.

1.16 Discuss the durability properties of pre–stressed concrete.

1.17 Explain creep and shrinkage requirements of pre–stressed concrete as per IS: 1343–2012.

1.18 Mention the general specifications for pre–stressing steel as per IS: 1343–2012.

1.19 Discuss about the necessity of adoption of high tensile steel and high strength concrete in pre–stressed concrete structures.

1.20 Explain with neat sketches the different types of pre–stressing adopted.

1.21 Explain the different stages of pre–tensioning method of pre–stressing with the relative advantages and disadvantages.

1.22 List the various pre–tensioning devices used in pre–stressed concrete.

1.23 Explain the different stages of post–tensioning method of pre–stressing with the relative advantages and disadvantages.

1.24 List the various post–tensioning devices used in pre–stressed concrete.

1.25 Explain with a neat sketch "Hoyer's long line" system of pre–stressing.

1.26 Explain Magnel–Blaton system of pre–stressing.

1.27 Explain with sketches Freyssinet's system of post–tensioning.

1.28 Explain with neat sketches Gifford–Udall system of pre–stressing.

1.29 List the various anchorage systems of pre–stressing. Explain any one of the system with neat sketches.

1.30 Distinguish between pre–tensioned and post–tensioned members and discuss their relative advantages and disadvantages.

OBJECTIVE QUESTIONS

1.1 Minimum grade of concrete for pre–tensioning method of pre–stressing is

a. M20 b. M30
c. M40 d. M50

1.2 Minimum grade of concrete for post–tensioning method of pre–stressing is

a. M20 b. M30
c. M40 d. M501.

1.3 Stress concept of pre–stressing was introduced by

a. T. Y. Lin b. Eugene Freyssinet
c. Nilson Arthur H. d. Magnel–Blaton

1.4 Conical wedges for the end anchorages were developed by

a. T. Y. Lin b. Eugene Freyssinet
c. Nilson Arthur H. d. Magnel–Blaton

1.5 Load balancing concept of pre–stressing was introduced by

a. T. Y. Lin b. Eugene Freyssinet
c. Magnel–Blaton d. Nilson Arthur H.

1.6 Maximum grade of concrete for pre–stressing is

a. M30 b. M40
c. M50 d. M60

1.7 Tensile strength of pre–stressed concrete can be determined by using

a. $f_{cr} = 0.6\sqrt{f_{ck}}$ b. $f_{cr} = 0.7\sqrt{f_{ck}}$
c. $f_{cr} = 0.8\sqrt{f_{ck}}$ d. $f_{cr} = 0.9\sqrt{f_{ck}}$

1.8 Modulus of elasticity of the pre–stressed concrete is given by

a. $E_c = 5700\sqrt{f_{ck}}$ b. $E_c = 5000\sqrt{f_{ck}}$
c. $E_c = 5300\sqrt{f_{ck}}$ d. $E_c = 4900\sqrt{f_{ck}}$

1.9 Minimum cement content for mild exposure condition of pre–stressed concrete is

a. $3000N/m^3$ b. $3200N/m^3$
c. $3400N/m^3$ d. $3600N/m^3$

1.10 Maximum water–cement ratio for moderate exposure condition of the pre–stressed concrete is

a. 0.45 b. 0.50
c. 0.55 d. 0.65

1.11 Creep coefficient for 28 days age of loading is given by

a. 2.2 b. 1.6
c. 1.4 d. 1.1

1.12 Shrinkage strain for the pre–tensioned concrete is given by

a. $\varepsilon_{sh} = 0.0003$ b. $\varepsilon_{sh} = 0.0004$
c. $\varepsilon_{sh} = 0.0005$ d. $\varepsilon_{sh} = 0.0006$

1.13 The nominal maximum coarse aggregate size for the pre–stressed concrete is limited by the lowest of the following quantities
a. 1/4 times the minimum thickness of the member
b. Spacing between the tendons/strands minus 5mm
c. 40mm
d. All of above

1.14 For PSC permissible tension in large diameter bars is
a. 800MPa b. 1000MPa
c. 1200MPa d. 1400MPa

1.15 For PSC permissible tension in small diameter wires is
a. 1800MPa b. 2000MPa
c. 2200MPa d. 2400MPa

1.16 The nominal diameter for the finished wires shall be
a. 5mm b. 4mm
c. 3mm d. All of above

1.17 Minimum depth of indentation for 5mm dia. wire is
a. 0.075mm b. 0.080mm
c. 0.085mm d. 0.090mm

1.18 Freyssinet system of post–tensioning developed in the year
a. 1928 b. 1934
c. 1939 d. 1942

1.19 Fluted male cone and female anchorage system exists in
a. the Freyssinet system b. the Magnel–Blaton system
c. the Gifford–Udall system d. Lee–McCall System

1.20 The concept used in the principles of pre–stressed concrete is
a. stress concept b. strength concept
c. load balancing concept d. All of above

1.21 Circular pre–stressing is preferable for
a. silos b. tanks
c. pipes d. All of above

1.22 Linear pre–stressing preferable for
a. beams b. piles
c. slabs d. All of above

1.23 The admixture used in pre–stressed concrete should follow the specifications of
a. IS: 9103–1999 b. IS: 456–2000
c. IS: 1343–2012 d. All of above

1.24 IS: 1343–2012 recommends a secant modulus at a stress level of
a. $0.3f_{ck}$ b. $0.35f_{ck}$
c. $0.45f_{ck}$ d. $0.5f_{ck}$

1.25 Maximum water–cement ratio for mild exposure of pre–stressed concrete is

a. 0.45 b. 0.50 c. 0.55 d. 0.65

1.26 Maximum water–cement ratio for severe exposure condition of the pre–stressed concrete is

a. 0.45 b. 0.50 c. 0.55 d. 0.65

1.27 Minimum cement content for moderate exposure condition of the pre–stressed concrete is

a. 3000N/m^3 b. 3200N/m^3
c. 3400N/m^3 d. 3600N/m^3

1.28 Minimum cement content for severe exposure condition of the pre–stressed concrete is

a. 3000N/m^3 b. 3200N/m^3
c. 3400N/m^3 d. 3600N/m^3

1.29 Creep coefficient for 7 days age of loading is given by

a. 2.2 b. 1.6 c. 1.4 d. 1.1

1.30 Creep coefficient for 1 year age of loading is given by

a. 2.2 b. 1.6 c. 1.4 d. 1.1

1.31 High tensile steel for pre–stressing usually form of

a. wires (1.5mm dia. to 8mm dia.) b. strands (containing 2 to 7 wires)
c. bars (9.5mm dia. to 32mm dia.) d. All of above

1.32 Plain hard drawn steel wire conforming to

a. IS: 1785 (Part 1)–1983 and IS: 1785 (Part 2)–1983
b. IS: 6003–1983
c. IS: 2090–1983
d. IS: 6006–1983

1.33 Cold–drawn indented wire conforming to

a. IS: 1785 (Part 1)–1983 and IS: 1785 (Part 2)–1983
b. IS: 6003–1983 c. IS: 2090–1983
d. IS: 6006–1983

1.34 High tensile steel bar conforming to

a. IS: 1785 (Part 1)–1983 and IS: 1785 (Part 2)–1983
b. IS: 6003–1983 c. IS: 2090–1983
d. IS: 6006–1983

1.35 Uncoated stress relieved strand conforming to

a. IS: 1785 (Part 1)–1983 and IS: 1785 (Part 2)–1983
b. IS: 6003–1983 c. IS: 2090–1983
d. IS: 6006–1983

1.36 Minimum depth of indentation for 4mm dia. wire is

a. 0.065mm b. 0.060mm
c. 0.055mm d. 0.050mm

1.37 Minimum depth of indentation for 3mm dia. wire is
a. 0.065mm b. 0.060mm
c. 0.050mm d. 0.055mm

1.38 Maximum depth of indentation for 5mm dia. wire is
a. 0.15mm b. 0.10mm
c. 0.20mm d. 0.25mm

1.39 Minimum depth of indentation for 4mm dia. wire is
a. 0.15mm b. 0.10mm
c. 0.20mm d. 0.25mm

1.40 Minimum depth of indentation for 3mm dia. wire is
a. 0.15mm b. 0.10mm
c. 0.20mm d. 0.25mm

1.41 For the ductility of pre–stressing steel, it should withstand .... as per IS: 6003–1983
a. reverse bent test b. elongation test
c. Both a & b d. Neither a nor b

1.42 The relaxation stress in the wire when tested as per IS: 6003–1983 code shall not exceed .... at the end of 1000 hours.
a. 72.8MPa b. 68.6MPa
c. 64.4MPa d. 62.2MPa

1.43 Various steps of the pre–tensioning operation are as follows
1. Anchoring of tendons against the end abutments
2. Placing of jacks 3. Applying tension to the tendons
4. Casting of concrete
5. Cutting of the tendons. Arrange them, from start to end
a. 1,2,3,4,5 b. 1,3,4,2,5
c. 2,4,5,1,3 d. 2,5,4,3,1

1.44 Long line method of pre–stressing is also known as
a. the Freyssinet system b. the Magnel–Blaton system
c. Hoyer's system d. Lee–McCall System

1.45 The various stages of the post–tensioning operation are summarized as follows
1. Casting of concrete 2. Placement of the tendons
3. Placement of the anchorage block and jack
4. Applying tension to the tendons
5. Seating of the wedges
6. Cutting of the tendons. Arrange them, from start to end
a. 1,2,3,4,5,6 b. 1,3,4,2,5,6
c. 2,4,5,1,6,3 d. 2,5,4,3,6,1

1.46. The principle of anchoring the tendons is depends upon

a. wedge action b. direct bearing

c. looping the wires d. All of above

1.47 The BBRV post–tensioning system and the Prescon system of pre–stressing are based on the principle is ....

a. wedge action b. direct bearing

c. looping the wires d. All of above

1.48 The Baur–Leonhardt system, Leoba system and also the Dwidag single–bar anchorage system, works on the principle of ....

a. wedge action b. direct bearing

c. looping the wires d. All of above

1.49 Freyssinet, Gifford–Udall, Anderson and Magnel–Blaton anchorages, works on principle of

a. wedge action b. direct bearing

c. looping the wires d. All of above

1.50 Freyssinet anchorage cones are manufactured in India are

a. 12 ϕ 8mm cones suitable for cables with 12 wires of 8mm dia. HT wires

b. 6T13 cones suitable for cables of 6 strands of 13mm HT steel

c. 12T13 anchorage suitable for cables of 12 Nos. HT strands of 13mm dia.

d. All of above

Answers

1.1 c	1.2 b	1.3 b	1.4 b	1.5 a
1.6 d	1.7 b	1.8 a	1.9 a	1.10 c
1.11 b	1.12 a	1.13 d	1.14 b	1.15 c
1.16 d	1.17 a	1.18 c	1.19 a	1.20 d
1.21 d	1.22 d	1.23 a	1.24 a	1.25 d
1.26 a	1.27 a	1.28 d	1.29 a	1.30 d
1.31 d	1.32 a	1.33 b	1.34 c	1.35 d
1.36 d	1.37 c	1.38 a	1.39 b	1.40 b
1.41 a	1.42 b	1.43 a	1.44 c	1.45 a
1.46 d	1.47 b	1.48 c	1.49 a	1.50 d

❑

CHAPTER 2

Analysis of Sections for Flexure

2.1 BASIC ASSUMPTIONS

The analysis of stresses developed in a pre–stressed concrete structural element is based on the following assumptions:

(i) Concrete is homogeneous elastic material,

(ii) Within the range of working stresses, both the concrete and steel behave elastically, notwithstanding the small amount of creep which occurs in both the materials under sustained loading, and

(iii) The plane section before bending is assumed to remain plane even after bending, which implies a linear strain across the depth of the member.

As long as tensile stresses do not exceed the limit of modulus of rupture of concrete (corresponding to the stage of visible cracking of concrete), any change in the loading of the member results in a change of stress in the concrete only, the sole function of the pre–stressing tendon being to impart and maintain the pre–stress in the concrete. Up to the stage of visible cracking on concrete, the changes in the stress of steel, the loading being negligibly small, are generally not considered in the computations.

2.2 INTRODUCTION

Pre–stressed concrete is the only structural material in which, it is possible to eliminate tension in concrete under any stage of loading. Very high strength concrete and steel are employed in PSC construction. The minimum cube strength required for concrete to be used in PSC construction is $30N/mm^2$. The permissible tension in steel is of order $1000N/mm^2$.

· Sign Convention for Stresses in Concrete

So far as analysis of PSC structure is concerned, we shall treat compressive stress in concrete as +ve and tensile stress as –ve.

Following are the notations for the analysis of pre–stress

F_o = Pre–stressing force, e = Eccentricity of pre–stressing force,

M^I = Moment due to pre–stressing force = $F_o e$,

A = Cross sectional area of the concrete member,

I = Second moment of area / moment of inertia of section about its centroid,

Z_t and Z_b = Section modulus of the top and bottom fibres,

f_{sup} and f_{inf} = Pre–stress in concrete developed at top and bottom fibres,

y_t = y_1= Distance of top fibre from the centroid of the section,

y_b = y_2= Distance of bottom fibre from the centroid of the section,

K = Radius of gyration

2.2.1 Resultant Direct Stress in any Fibre in a Given PSC Section

Case (A): When the Tendon is Placed Concentrically

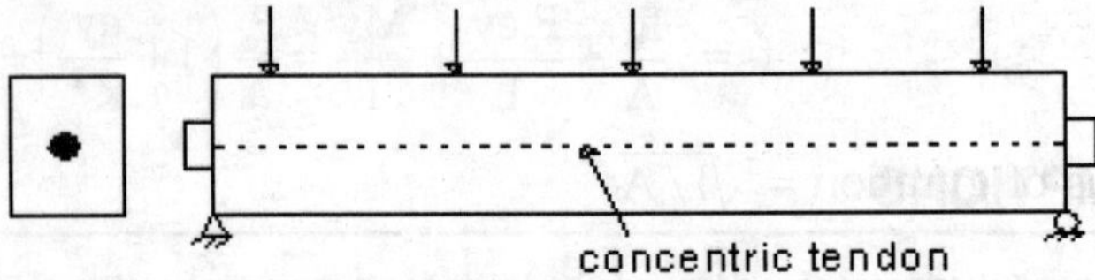

Fig. 2.1 Rectangular Beam Pre–stressed by a Tendon through its Centroidal Axis

Let us consider a simply supported rectangular beam pre–stressed by a tendon through its centroidal axis as shown in Fig. 2.1. Due to the pre–stressing force, a uniform compressive stress of $f = \frac{F_o}{A}$ will be produced across the section that has an area of cross-section 'A'. If 'M' is the external moment at a section due to the weight of the beam, then the stress at any point across the section is given by,

$$\frac{M}{I} = \frac{f}{y} = \frac{My}{I} \qquad ...(2.1)$$

Where y is the distance from the centroidal axis and I is the moment of inertia of the section. Thus the resultant stress distribution is given by,

$$f = \frac{F_o}{A} + \frac{My}{I} \qquad ...(2.2)$$

Eq. 2.2 can also be written as, $f = \frac{F_o}{A} + \frac{M}{Z}$ ...(2.3)

Where, Z = Section modulus and is given by Z = I/y

Case (b): When the Tendon is Placed Eccentrically w.r.t. the Centroid of the Concrete Section

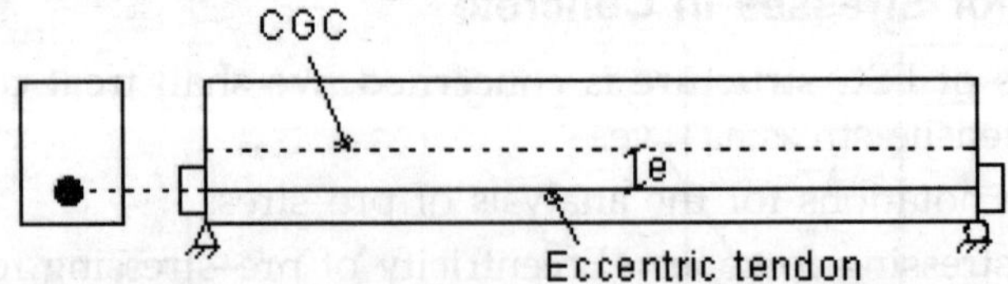

Fig. 2.2 Rectangular Beam Pre–stressed by an Eccentric Tendon

Here resultant compressive force 'F_o' in the concrete acts at the centroid of the tendon, which is at a distance 'e' from CGC as shown in Fig. 2.2. Due to an eccentric pre–stressing force, the concrete is subjected to a moment as well as a direct load. The moment M^I, produced by the pre–stressing force is '$F_o e$'

The stress due to this moment is

$$f = \frac{M^{I}}{I} = \frac{F_o ey}{I} = \frac{F_o e}{Z} \quad ...(2.4)$$

Where, Z = Section modulus and is given by Z = I/y

Along with moment due to pre–stressing force, if external moment M due to the weight of the beam, is also applied then, the resultant stress distribution is given by,

$$f = \frac{F_o}{A} + \frac{F_o ey}{1} + \frac{My}{I} = \frac{F_o}{A}\left(1 + \frac{ey}{K^2}\right) + \frac{My}{I} \quad ...(2.5)$$

Where K = Radius of gyration = $\sqrt{I/A}$ ...(2.6)

Eq. 2.5 can also be written as $f = \frac{F_o}{A} + \frac{F_o e}{Z} + \frac{M}{Z}$ ...(2.7)

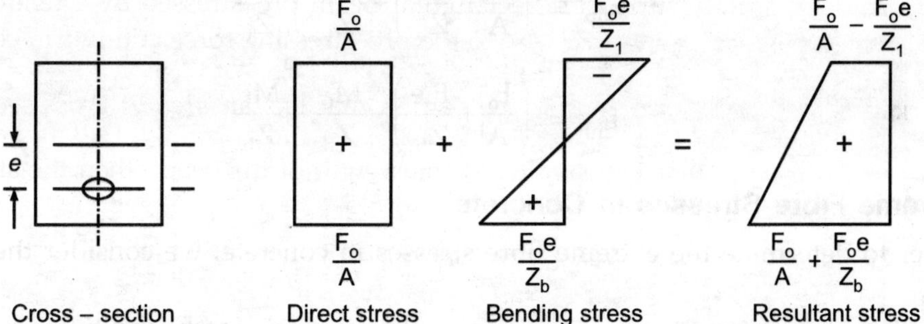

Fig. 2.3 Stresses in Beam Pre–stressed by an Eccentric Tendon

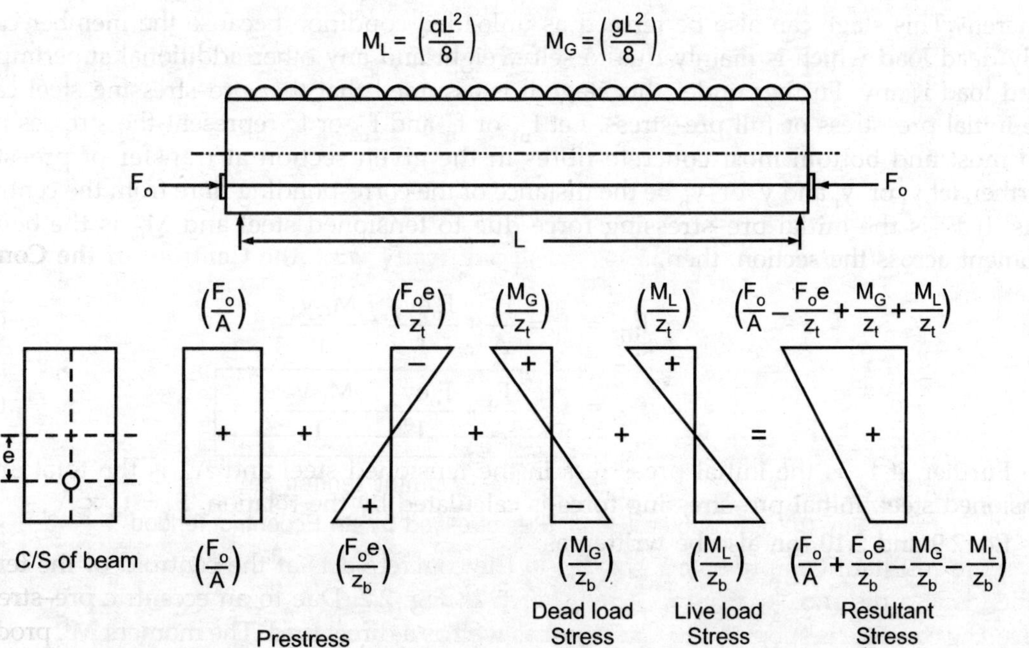

Fig. 2.4 Stresses due to Eccentric Pre–stressing, Dead and Live loads

The resultant stresses (Fig. 2.3) due to only pre–stressing moment at the top and bottom fibres are given by,

$$f_{sup} = \left(\frac{F_o}{A} - \frac{F_o e}{Z_t}\right) \quad ...(2.8)$$

$$f_{inf} = \left(\frac{F_o}{A} + \frac{F_o e}{Z_b}\right) \quad ...(2.9)$$

Simply supported beam which supports uniformly distributed live load and dead loads of intensity q and g as shown in Fig. 2.4. These loads produces live load moment M_L and dead load moment M_G. The beam is pre–stressed by a tendon carrying pre–stressing force F_o at an eccentricity of e. The resultant stresses due to pre–stressing moment and external moment (due to self–weight of beam and due to imposed loads) at the top and bottom fibres are given by,

$$f_{sup} = \left(\frac{F_o}{A} - \frac{F_o e}{Z_b}\right) + \frac{M_G}{Z_t} + \frac{M_L}{Z_t} \quad ...(2.10)$$

$$f_{inf} = \left(\frac{F_o}{A} + \frac{F_o e}{Z_b}\right) - \frac{M_G}{Z_b} - \frac{M_L}{Z_b} \quad ...(2.11)$$

2.2.2 Extreme Fibre Stresses in Concrete

In order to determine the extreme fibre stresses in concrete, we consider the following two cases.

(a) Transfer of pre–stress (b) Working load condition

(a) Transfer of Pre–stress: This is the stage when the pre–stressing force is transferred to concrete. This stage can also be termed as unloaded condition because the member carries only dead load which is mainly due to self–weight and any other additional superimposed dead load if any. Further under this stage it is assumed that the pre–stressing steel carries the initial pre–stress or full pre–stress. Let f_{u1} or f_{tt} and f_{u2} or f_{ct} represent the stresses in the top most and bottom most concrete fibres in the given section at transfer of pre–stress. Further, let y_1 or y_t and y_2 or y_b be the distance of the corresponding fibre from the centroidal axis. If F_o is the initial pre–stressing force due to tensioned steel and M_G is the bending moment across the section, then,

$$f_{u1} = f_{tt} = \frac{F_o}{A} + \frac{F_o e y_1}{I} + \frac{M_G y_1}{I} \quad ...(2.12)$$

$$f_{u2} = f_{ct} = \frac{F_o}{A} + \frac{F_o e y_2}{I} + \frac{M_G y_2}{I} \quad ...(2.13)$$

Further, if f_{si} is the initial pre–stress in the tensioned steel and A_{st} is the total area of tensioned steel. Initial pre–stressing force is calculated by the relation, $F_o = f_{si} \times A_{st}$.

Eq. 2.9 and 2.10 can also be written as,

$$\left(f_{sup} + \frac{M_G}{Z_t}\right) \geq f_{tt} \quad \text{...(2.14)}$$

$$\left(f_{inf} - \frac{M_G}{Z_b}\right) \leq f_{ct} \quad \text{...(2.15)}$$

Where, $f_{sup} = \left(\frac{F_o}{A} - \frac{F_o e}{Z_t}\right)$, $f_{inf} = \left(\frac{F_o}{A} + \frac{F_o e}{Z_b}\right)$,

$$Z_t = \frac{I}{y_t} = \frac{I}{y_1} \text{ and } Z_b = \frac{I}{y_b} = \frac{I}{y_2}$$

(b) Working Load Condition: Under this condition the member is assumed to carry the total designed load. Further the effective pre–stressing force F available in tensioned steel is about 80% of F_o due to various losses of pre–stress. Let M_T is the total BM across the section. So $M_T = M_G + M_L$, where M_L is the bending moment due to superimposed load which may be due to live load or only dead load or a combination of live and dead load. If fl_1or f_{cw} and fl_2 or f_{tw} represents the stress in the topmost and bottom fibres at working load, then,

$$fl_1 = f_{cw} = \frac{F}{A} + \frac{Fey_1}{I} + \frac{M_T y_1}{I} \quad \text{...(2.16)}$$

$$fl_2 = f_{tw} = \frac{F}{A} + \frac{Fey_2}{I} + \frac{M_T y_2}{I} \quad \text{...(2.17)}$$

Loss Ratio or Loss Factor (η): It is defined as the ratio of the effective pre–stressing force (F) to the initial pre–stressing force (F_o).

$$\eta = \frac{F}{F_o} \text{ or } F = \eta F_o \quad \text{...(2.18)}$$

For Ex: If the loss of pre–stress is 16% then η = 0.84. Further, if f_{se} is the effective pre–stress in the tensioned steel, then, $F = f_{se} A_{st}$

Hence we can also write, $$\eta = \frac{f_{se}}{f_{si}} \quad \text{...(2.19)}$$

Note: All the distances measured above the centroidal axis shall be treated as positive. Eq. 2.13 and 2.14 can also be written as,

$$fl_1 = f_{cw} = \frac{\eta F_o}{A} + \frac{\eta F_o ey_1}{I} + \frac{M_G y_1}{I} + \frac{M_L y_1}{I} \quad \text{...(2.20)}$$

$$fl_2 = f_{tw} = \frac{\eta F_o}{A} + \frac{\eta F_o ey_2}{I} + \frac{M_G y_2}{I} + \frac{M_L y_2}{I} \quad \text{...(2.21)}$$

$$\left(\eta f_{sup} + \frac{M_G}{Z_t} + \frac{M_L}{Z_t}\right) \leq f_{cw} \quad \text{...(2.22)}$$

$$\left(\eta f_{inf} - \frac{M_G}{Z_b} - \frac{M_L}{Z_b}\right) \geq f_{tw} \quad ...(2.23)$$

Where, $f_{sup} = \left(\frac{F_o}{A} - \frac{F_o e}{Z_t}\right)$, $f_{inf} = \left(\frac{F_o}{A} + \frac{F_o e}{Z_b}\right)$, $Z_t = \frac{I}{y_t} = \frac{I}{y_1}$ and $Z_b = \frac{I}{y_b} = \frac{I}{y_2}$

2.3 SOLVED EXAMPLES

Problem 2.1

A simply supported beam of span 10m is rectangular in section 300mm × 600mm. It is pre–stressed with an initial pre–stressing force of 700kN which is located at 100mm from the soffit. The beam carries a load 12kN/m in addition to its own weight. Assume the loss of pre–stress as 15%. Determine the stress distribution across the mid–span section. Also draw the stress variation diagram under the following stages of loading (a) at the transfer of pre–stress, (b) at working load.

Solution:

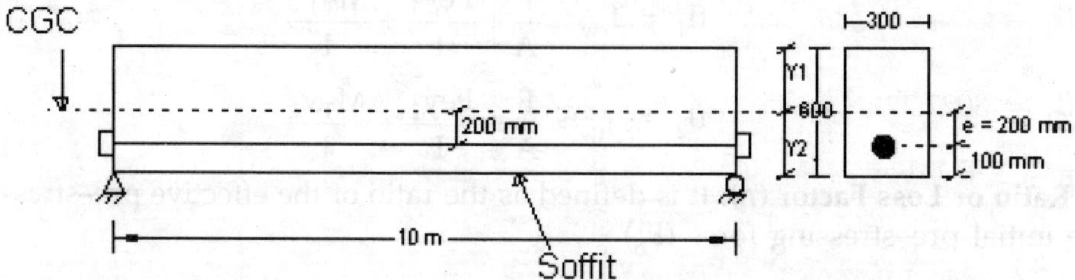

Fig. 2.5 Longitudinal and Cross–section of Simply Supported Beam

Position of extreme fibres,

y_1 = 300mm = 0.3m (above NA), y_2 = 300mm = –0.3m (below NA),

e = 200mm = –0.2m (below NA), A = 300×600 = $18 \times 10^4 mm^2 = 0.18m^2$, Span = L = 10m,

Self–weight = g = 0.18 × 24 = 4.32kN/m, Unit weight of concrete = $24kN/m^3$,

Super imposed load = q = 12kN/m, Initial pre–stressing force (F_o) = 700kN.

Bending moment due to self–weight, $M_G = \frac{gL^2}{8} = \frac{4.32 \times 10^2}{8} = 54kNm$

Bending moment due to super imposed load, $M_L = \frac{gL^2}{8} = \frac{12 \times 10^2}{8} = 150kNm$

Total BM, $(M_T) = M_G + M_L = 54 + 150 = 204kNm$

Moment of inertia, $I = \frac{b \times d^3}{12} = \frac{300 \times 600^3}{12} = 5.4 \times 10^9 mm^4 = 5.4 \times 10^{-3} m^4$

Case (1): At the Transfer of Pre–stress

Let f_{u1} and f_{u2} be the extreme fibre stresses in concrete at transfer of pre–stress.

$$f_{u1} = \frac{F_o}{A}+\frac{F_o ey_1}{I}+\frac{M_G y_1}{I}=\left(\frac{700}{0.18}\right)+\left(\frac{700\times(-0.2)\times 0.3}{5.4\times 10^{-3}}\right)+\left(\frac{54\times 0.3}{5.4\times 10^{-3}}\right)$$

$$f_{u1} = -888.89\text{kN/m}^2 = 0.889\text{N/mm}^2 \text{ (Tension)}$$

$$f_{u2} = \frac{F_o}{A}+\frac{F_o ey_2}{I}+\frac{M_G y_2}{I}=\left(\frac{700}{0.18}\right)+\left(\frac{700\times(-0.2)\times(-0.3)}{5.4\times 10^{-3}}\right)+\left(\frac{54\times(-0.3)}{5.4\times 10^{-3}}\right)$$

$$f_{u2} = 8666.65\text{kN/m}^2 = 8.666\text{N/mm}^2 \text{ (Compression)}$$

Case (2): The Working Load Condition

Let fl_1 and fl_2 be the extreme fibre stresses in concrete at working load.

$$\text{Loss of pre–stress} = 15\% \text{ (given)}$$

$$\therefore \quad \text{Loss ratio} = \eta = 100 - 15 = 85\% = 0.85$$

$$F = \eta F_o = 0.85 \times 700 = 595\text{kN}$$

$$fl_1 = \frac{F}{A}+\frac{Fey_1}{1}+\frac{M_T y_1}{I}=\left(\frac{595}{0.18}\right)+\left(\frac{595\times(-0.2)\times 0.3}{5.4\times 10^{-3}}\right)+\left(\frac{204\times 0.3}{5.4\times 10^{-3}}\right)$$

$$fl_1 = 8027.72\text{kN/m}^2 = 8.027\text{N/mm}^2 \text{ (Compression)}$$

$$fl_2 = \frac{F}{A}+\frac{Fey_2}{1}+\frac{M_T y_2}{I}=\left(\frac{595}{0.18}\right)+\left(\frac{595\times(-0.2)\times 0.3}{5.4\times 10^{-3}}\right)+\left(\frac{204\times(-0.3)}{5.4\times 10^{-3}}\right)$$

$$fl_2 = 1416.72\text{kN/m}^2 = 1.416\text{N/mm}^2 \text{ (Tension)}$$

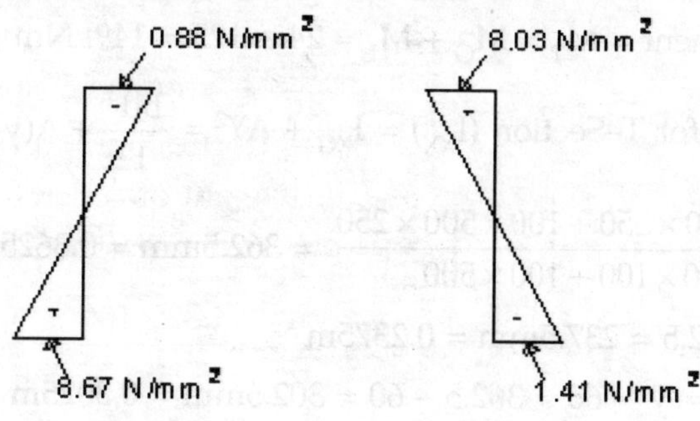

Fig. 2.6 Stress Variation Diagram

Problem 2.2

Fig. 2.7 shows the mid–span of simply supported girder of span 10m. The beam is required to carry a load of 10kN/m in addition to its own weight. Determine initial pre–stressing force required to nullify the bottom fibre stresses, when the beam is fully loaded. Hence, determine the extreme fibre stress at transfer of pre–stress and maximum compressive stress in concrete at working condition. Assume loss factor as 0.85.

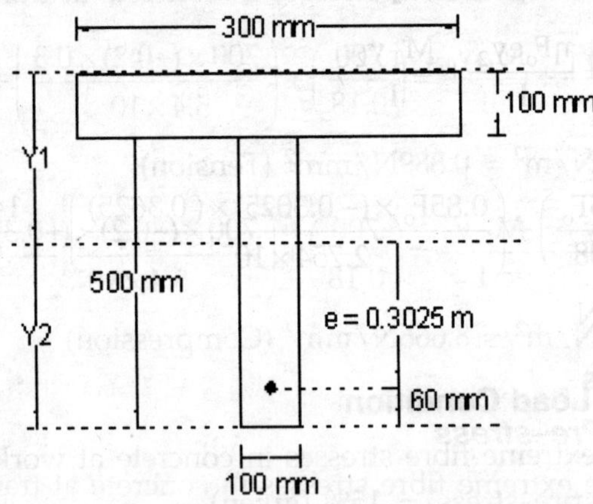

Fig. 2.7 Cross–section of Simply Supported Girder

Solution:

Span = L = 10m, Super imposed load = q = 10kN/m

Area of cross section = A = 300 × 100 + 500 × 100 = 80,000mm^2 = 0.08m^2

Self–weight = g = 0.08 × 24 = 1.92kN/m

Bending moment due to self–weight, $M_G = \frac{gL^2}{8} = \frac{1.92 \times 10^2}{8} = 24\text{kNm}$

Bending moment due to super imposed load, $M_L = \frac{qL^2}{8} = \frac{10 \times 10^2}{8} = 125\text{kNm}$

Total Bending moment = $M_T = M_G + M_L$ = 24 + 125 = 149kNm

Moment of inertia for T–Section $(I_{XX}) = I_{XG} + AY^2 = \frac{bd^3}{12} + A(\bar{y} - y)^2$

$$\bar{y} = \frac{\Sigma ay}{A} = \frac{300 \times 100 \times 550 + 100 \times 500 \times 250}{300 \times 100 + 100 \times 500} = 362.5\text{mm} = 0.3625\text{m} = y_2$$

y_1 = 500 + 100 – 362.5 = 237.5mm = 0.2375m

∴ Eccentricity, e = $\bar{y}$ – 60 = 362.5 – 60 = 302.5mm = 0.3025m

$$I_{xx} = \frac{300 \times 100^3}{12} + 300 \times 100 \times (362.5 - 550)^2 + \frac{300 \times 500^3}{12} + 100 \times 500 \times (362.5 - 250)^2$$

$I_{xx} = 2.754 \times 10^{-3}\text{m}^4$

Under the fully loaded condition to nullify the bottom fibre stress the required initial pre–stressing force is calculated as follows. Make fl_2 = 0.

$$fl_2 = \frac{\eta F_o}{A} + \frac{\eta F_o e y_2}{1} + \frac{M_T y_2}{I}$$

$$0 = \frac{\eta F_o}{A} + \frac{\eta F_o e y_2}{I} + \frac{M_T y_2}{I},$$

$$\eta = 0.85$$

$$0 = \left(\frac{0.85F_o}{0.08}\right) + \left(\frac{0.85F_o \times (-0.3025) \times (0.3625)}{2.754 \times 10^{-3}}\right) + \left(\frac{149 \times (-0.3625)}{2.754 \times 10^{-3}}\right)$$

$$F_o = 441\text{kN}$$

- **Extreme Fibre Stress**

Case (1): Transfer of Pre–stress

Let f_{u1} and f_{u2} be the extreme fibre stresses in concrete at transfer of pre–stress.

$$f_{u1} = \frac{F_o}{A} + \frac{F_o e y_1}{I} + \frac{M_G y_1}{1} = \left(\frac{441}{0.08}\right) + \left(\frac{441 \times (-0.3025) \times (0.2375)}{2.754 \times 10^{-3}}\right) + \left(\frac{24 \times 0.2375}{2.754 \times 10^{-3}}\right)$$

$$f_{u1} = -3922\text{kN/m}^2 = -3.922\text{N/mm}^2$$

$$f_{u2} = \frac{F_o}{A} + \frac{F_o e y_2}{I} + \frac{M_G y_2}{1} = \left(\frac{441}{0.08}\right) + \left(\frac{441 \times (-0.3025) \times (-0.3625)}{2.754 \times 10^{-3}}\right) + \left(\frac{24 \times (-0.3625)}{2.754 \times 10^{-3}}\right)$$

$$f_{u2} = 19.914\text{N/mm}^2$$

Case (2): Working Load Condition

Let fl_1 and fl_2 be the extreme fibre stresses in concrete at working load.

$$F = \eta F_o = 0.85 \times 441 = 374.85\text{kN} = 375\text{kN}$$

$$fl_1 = \frac{F}{A} + \frac{F e y_1}{I} + \frac{M_T y_1}{I} = \left(\frac{375}{0.08}\right) + \left(\frac{375 \times (-0.3025) \times 0.2375}{2.754 \times 10^{-3}}\right) + \left(\frac{149 \times 0.2375}{2.754 \times 10^{-3}}\right)$$

$$fl_1 = 7.755\text{N/mm}^2, \ fl_2 = 0$$

Problem 2.3

A simply supported beam 60mm×120mm in section has a span of 3m. If carries two point loads of 2kN each at 1/3 span points in addition to its own weight. Determine the initial pre–stressing force and its eccentricity so that the tensile stress in concrete will be limited to 1.5N/mm² and 1N/mm² at transfer and at working load respectively. Assume the loss ration as 80%.

Solution:

$f_{u1} = 1.5\text{N/mm}^2 = -1500\text{kN/m}^2$, $fl_2 = 1.0\text{N/mm}^2 = -1000\text{kN/m}^2$, $F_o = ?$, $e = ?$

$A = 60\text{mm} \times 120\text{mm} = 7.2 \times 10^{-3}\text{m}^2$, Self–weight $= g = 7.2 \times 10^{-3} \times 24 = 0.1728\text{kN/m}$

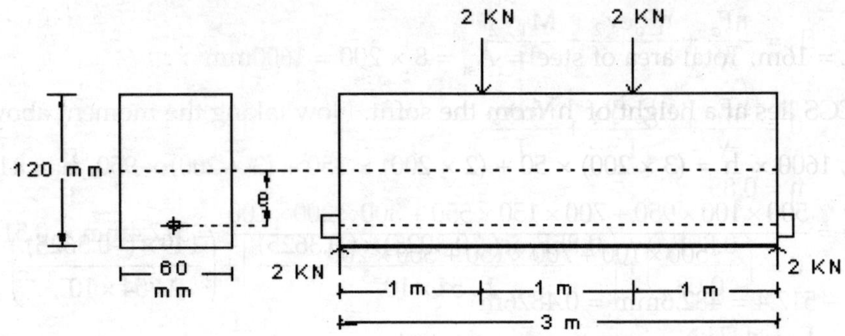

Fig. 2.8 Cross and Longitudinal Section of Simply Supported Beam

$$M_G = \frac{gL^2}{8} = \frac{0.1728 \times 3^2}{8} = 0.1944\text{kNm},\ M_L = 2 \times 1 = 2\text{kNm},\ y_1 = y_2 = 60\text{mm} = 0.06\text{m}$$

$$M_T = M_G + M_L = 0.1944 + 2 = 2.1944\text{kNm}$$

$$\text{Moment of Inertia} = I = \frac{b \times d^3}{12} = \frac{60 \times 120^3}{12} = 8.64 \times 10^{-6}\text{m}^4,\ \eta = 0.80$$

$$f_{u1} = \frac{F_o}{A} + \frac{F_o e y_1}{1} + \frac{M_G y_1}{I}, -1500 = \left(\frac{F_o}{7.2 \times 10^{-3}}\right) + \left(\frac{F_o e \times 0.06}{8.64 \times 10^{-6}}\right) + \left(\frac{0.1944 \times 0.06}{8.64 \times 10^{-6}}\right) \quad ...(1)$$

$$fl_2 = \frac{\eta F_o}{A} + \frac{\eta F_o e y_2}{I} + \frac{M_T y_2}{I}, -1000 = \frac{0.8F_o}{7.2 \times 10^{-3}} + \frac{0.8F_o e(-0.06)}{8.64 \times 10^{-6}} + \frac{2.1944 \times (-0.06)}{8.64 \times 10^{-6}} \quad ...(2)$$

Solving (1) and (2), $F_o = 53.81\text{kN}$, $e = 0.027\text{m}$

Problem 2.4

Fig. 2.9 shows the mid–span section of simply supported girder of span 16m. It is pre–stressed with 8 cables, each of area 200mm^2. Determine the central concentrated load required to nullify the bottom fibre stress, effective pre–stress in tensioned steel is 800N/mm^2.

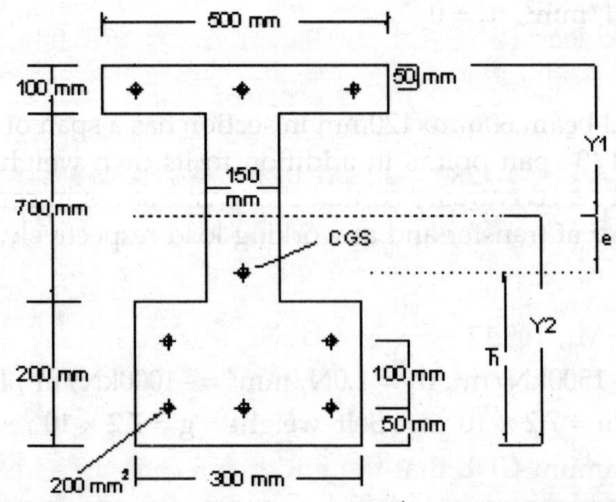

Fig. 2.9 Mid–span Section of Simply Supported Girder

Solution:

Span = L = 16m, Total area of steel = A_{st} = 8 × 200 = 1600mm^2

Let the CGS lies at a height of $\bar{h}$ from the soffit. Now taking the moment above the soffit

We have, 1600 × $\bar{h}$ = (3 × 200) × 50 + (2 × 200) × 150 + (3 × 200) × 950, $\bar{h}$ = 412.5mm

$$\bar{y} = \frac{\Sigma ay}{A} = \frac{500 \times 100 \times 950 + 700 \times 150 \times 550 + 300 \times 200 \times 100}{500 \times 100 + 700 \times 150 + 300 \times 200} = 517.4\text{mm} = 0.5147\text{m} = y_2$$

y_1= 1000 – 517.4 = 482.6mm = 0.4826m

∴ Eccentricity, e = $y_2 - \bar{h}$ = 517.4 – 412.5 = 104.9mm = 0.1049m

$$I_{XX} = I_{XG} + AY^2 = \frac{bd^3}{12} + A(\bar{y} - y)^2$$

$$I_{xx} = \frac{500 \times 100^3}{12} + 500 \times 100 \times (517.4 - 950)^2 + \frac{150 \times 700^3}{12} + 150 \times 700 \times (517.4 - 550)^2$$

$$\frac{300 \times 200^3}{12} + 300 \times 200 \times (517.4 - 100)^2 = 0.0244\text{m}^4$$

A = 500 × 100 + 150 × 700 + 300 × 200 = 215000mm^2 = 0.215m^2

Self–weight = g = 0.215×24 = 5.16kN/m

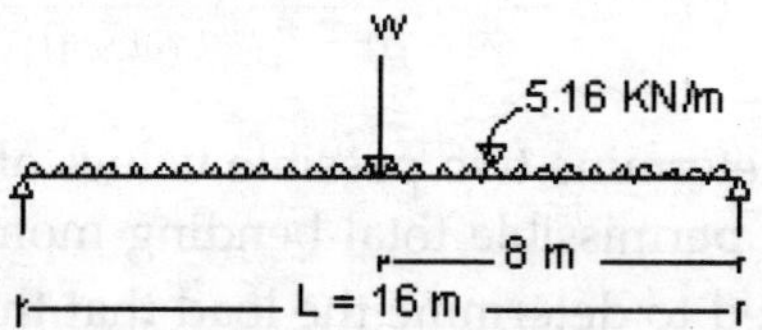

Fig. 2.10 Simply Supported Beam

$$M_G = \frac{gL^2}{8} = \frac{5.16 \times 16^2}{8} = 165.12\text{kNm}$$

Central concentrated load (W) = ?, f_{se} = 800N/mm^2, A_{st} = 1600mm^2

F = $f_{se} \times A_{st}$ = 800 × 1600 = 1280000N = 1280kN

Given, fl_2 = 0

$$fl_2 = \frac{F}{A} + \frac{Fey_2}{I} + \frac{M_T y_2}{I}, 0 = \frac{1280}{0.215} + \frac{1280(-0.1049) \times (-0.5174)}{0.0244} + \frac{M_T \times (-0.5174)}{0.0244}$$

M_T = 415.3kNm

We have $M_G + M_L = M_T$, $165.12 + \frac{WL}{4} = 415.3$, ∴ $165.12 + \frac{W \times 16}{4} = 415.3$, W = 62.56kN

Problem 2.5

Determine the maximum UDL that the girder can carry in addition to its own weight. The mid–span of which is shown in Fig. 2.11. Compressive and tensile stress in concrete are limited to $15N/mm^2$ and $1.5N/mm^2$ respectively. The girder has simply supported span of 15m and effective pre–stress in the tensioned steel is $800N/mm^2$.

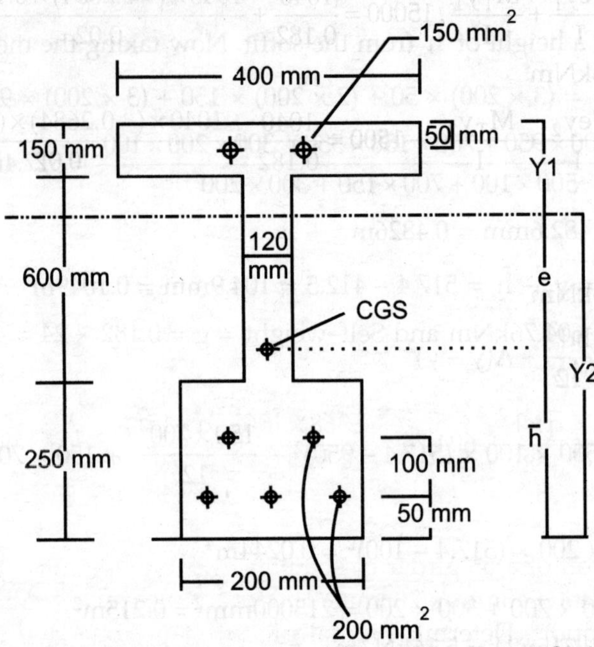

Fig. 2.11 Mid–span Section of Simply Supported Girder

Solution:

On the basis of fl_1 and fl_2 determine two possible values of M_T. Hence select the lower of these two values as maximum permissible total bending moment.

Note: Whenever we required to determine the load that the member can carry, it become obvious that the given permissible stress refer to loaded condition.

Total Area of steel = $A_{st} = 2 \times 150 + 5 \times 200 = 1300mm^2$

$$1300 \times \bar{h} = 2 \times 150 \times 950 + 2 \times 200 \times 150 + 3 \times 200 \times 50, \; \bar{h} = 288.46mm$$

$$\bar{y} = \frac{\Sigma ay}{A} = \frac{400 \times 150 \times 925 + 600 \times 120 \times 550 + 250 \times 200 \times 125}{400 \times 150 + 600 \times 120 + 250 \times 200} = 556.86mm$$

$$= 0.5568m = y_2$$

$$y_1 = 150 + 600 + 250 - 556.86 = 443.14mm = 0.4431m$$

$$\therefore \quad \text{Eccentricity, } e = y_2 - \bar{h} = 556.86 - 288.46 = 268.4 \text{ mm} = 0.2684m$$

$$I_{XX} = I_{XG} + AY^2 = \frac{bd^3}{12} + A(\bar{y} - y)^2 = 0.02m^4$$

$$fl_1 = 15N/mm^2 = 15000kN/m^2$$

$$fl_2 = -1.5N/mm^2 = -1500kN/m^2$$

$$F = f_{se} \times A_{st} = 800 \times 1300 = 104000 \text{ N} = 1040kN$$

$$A = 150 \times 400 + 120 \times 600 + 200 \times 250 = 0.182m^2$$

$$fl_1 = \frac{F}{A} + \frac{Fey_1}{I} + \frac{M_T y_1}{I}, 15000 = \frac{1040}{0.182} + \frac{1040 \times (-0.2684) \times 0.4431}{0.02} + \frac{M_T \times 0.4431}{0.02}$$

$$M_T = 698.26kNm$$

$$fl_2 = \frac{F}{A} + \frac{Fey_2}{1} + \frac{M_T y_2}{I}, -1500 = \frac{1040}{0.182} + \frac{1040 \times (-0.2684) \times (-0.4431)}{0.02} + \frac{M_T \times (-0.4431)}{0.02}$$

$$M_T = 604.76kNm$$

Hence, adopt $M_T = 604.76kNm$ and Self–weight = g = 0.182 × 24 = 4.368kN/m

$$M_T = M_G + M_L,$$

$$M_T = \frac{g \times L^2}{8} + \frac{q \times L^2}{8}, 604.76 = \frac{4.368 \times 15^2}{8} + \frac{q \times 15^2}{8}, q = 17.13kN/m$$

Problem 2.6

The mid–span section of a 55 girder of span 5m is T–section. The flange of the section is 200mm × 50mm and web is 40mm × 200mm. The beam carries two point loads of 10kN each at 1.5m from either support. Determine initial pre–stressing force and e, if the tensile stress in concrete limited to 1.5N/mm², 1.2N/mm² at transfer of pre–stress and working load condition. Assume loss ratio as 80%.

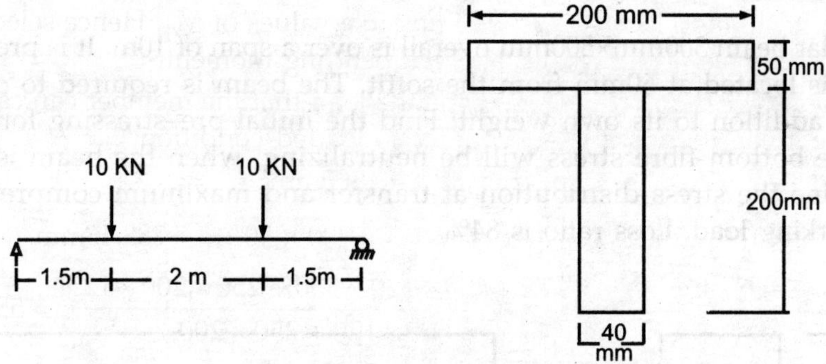

Fig. 2.12 Simply Supported Beam and Section of T–Girder

Solution:

Moment of inertia for T–Section $(I_{XX}) = I_{XG} + AY^2 = \frac{bd^3}{12} + A(\bar{y} - y)^2$

$$\bar{y} = \frac{\Sigma ay}{A} = \frac{200 \times 50 \times 225 + 40 \times 200 \times 100}{200 \times 50 + 40 \times 200} = 0.1694m = y_2$$

$$y_1 = 250 - 169.44 = 80.56mm = 0.0805m$$

$$I_{xx} = \frac{200\times50^3}{12} + 200\times50\times(169.44-225)^2 + \frac{40\times200^3}{12} + 40\times200\times(169.44-100)^2$$

$$I_{xx} = 9.81\times10^{-5}\text{m}^4$$

Area $= A = 200\times50 + 200\times40 = 0.018\text{m}^2$, Self–weight $= g = 0.018\times24 = 0.432\text{kN/m}$

$$M_G = \frac{g\times L^2}{8} = \frac{0.432\times5^2}{8} = 1.35\text{kNm}$$

$M_L = 10\times1.5 = 15\text{kNm}$, $M_T = M_G + M_L = 1.35 + 15 = 16.35\text{kNm}$

$f_{u1} = -1500\text{kN/mm}^2$, $fl_2 = -1200\text{kN/mm}^2$

$$f_{u1} = \frac{F_o}{A} + \frac{F_o e y_1}{I} + \frac{M_G y_1}{I},$$

$$-1500 = \left(\frac{F_o}{0.018}\right) + \left(\frac{F_o e\times0.0805}{9.81\times10^{-5}}\right) + \left(\frac{1.35\times0.0805}{9.81\times10^{-5}}\right) \quad ...(1)$$

$$fl_2 = \frac{\eta F_o}{A} + \frac{\eta F_o e y_2}{I} + \frac{M_T y_2}{I},$$

$$-1200 = \frac{0.8F_o}{0.018} + \frac{0.8F_o e(-0.1694)}{9.81\times10^{-5}} + \frac{16.35\times(-0.1694)}{9.81\times10^{-5}} \quad ...(2)$$

Solving (1) and (2), $F_o = 164\text{kN}$, $e = 86.9\text{m}$

Problem 2.7

A rectangular beam 300mm×500mm overall is over a span of 10m. It is pre–stressed with a force which is located at 60mm from the soffit. The beam is required to carry a load of 11.40kN/m in addition to its own weight. Find the initial pre–stressing force required to ensure that the bottom fibre stress will be neutralizing, when the beam is fully loaded. Hence determine the stress distribution at transfer and maximum compressive stress in concrete at working load. Loss ratio is 84%.

Solution:

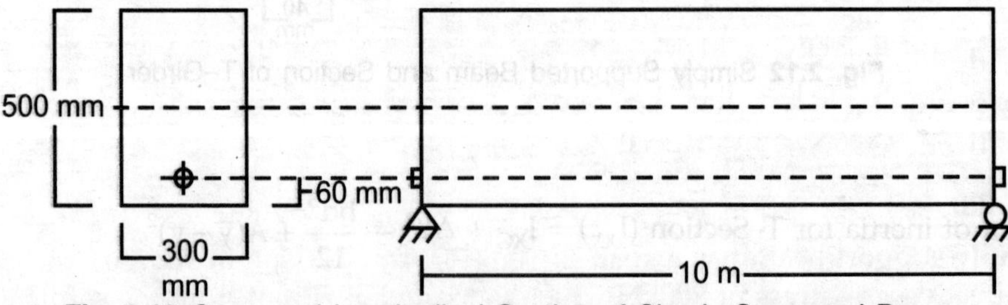

Fig. 2.13 Cross and Longitudinal Section of Simply Supported Beam

$y_1 = 250\text{mm} = 0.25\text{m}$, $y_2 = -250\text{mm} = -0.25\text{m}$, $e = 250–60 = -190\text{mm} = -0.190\text{m}$

$I = \frac{0.3 \times 0.5^3}{12} = 0.003125m^4$, $A = 0.3 \times 0.5 = 0.15m^2$, $\eta = 0.84$

Self–weight = g = 0.15 × 24 = 3.6kN/m, Super imposed load = q = 11.40kN/m

Bending moment due to self–weight, $M_G = \frac{gL^2}{8} = \frac{3.6 \times 10^2}{8} = 45kNm$

Bending moment due to super imposed load, $M_L = \frac{qL^2}{8} = \frac{11.40 \times 10^2}{8} = 142.5kNm$

$$M_T = M_G + M_L = 45 + 142.5 = 187.5kNm$$

Under the fully loaded condition to nullify the bottom fibre stress the required initial pre–stressing force is calculated as follows. Make $fl_2 = 0$.

$$fl_2 = \frac{\eta F_o}{A} + \frac{\eta F_o e y_2}{I} + \frac{M_T y_2}{I}$$

$$0 = \frac{\eta F_o}{A} + \frac{\eta F_o e y_2}{I} + \frac{M_T y_2}{I},$$

$$0 = \left(\frac{0.84F_o}{0.15}\right) + \left(\frac{0.84F_o \times (-0.19) \times (-0.25)}{0.003125}\right) + \left(\frac{187.5 \times (-0.25)}{0.003125}\right)$$

$$F_o = 816.63kN$$

• *At Transfer of Pre-stress*

$$f_{u1} = \frac{F_o}{A} + \frac{F_o e y_1}{I} + \frac{M_G y_1}{I} = \left(\frac{816.63}{0.15}\right) + \left(\frac{816.63 \times (-0.190) \times (0.25)}{0.003125}\right) + \left(\frac{45 \times 0.25}{0.003125}\right)$$

$$f_{u1} = -3.36N/mm^2$$

$$f_{u2} = \frac{F_o}{A} + \frac{F_o e y_2}{I} + \frac{M_G y_2}{I} = \left(\frac{816.63}{0.15}\right) + \left(\frac{816.63 \times (0.190) \times (0.25)}{0.003125}\right) + \left(\frac{45 \times (-0.25)}{0.003125}\right)$$

$$f_{u2} = 14.25N/mm^2$$

• *At Working Load Condition*

$$F = F_o = 0.84 \times 816.63 = 685.9kN$$

$$fl_1 = \frac{F}{A} + \frac{F e y_1}{I} + \frac{M_T y_1}{I} = \left(\frac{685.9}{0.15}\right) + \left(\frac{685.9 \times (-0.190) \times 0.25}{0.003125}\right) + \left(\frac{187.5 \times 0.25}{0.003125}\right)$$

$fl_1 = 9.146N/mm^2$, $fl_2 = 0$

Problem 2.8

A simply support beam has a span of 10m, 300mm×500mm in section. It is pre–stressed with an effective pre–stress of 800kN which is located at 50mm from the soffit. The beam is required to carry two point load of equal magnitude at 3m from either end in addition to its own weight, if the compressive and tensile stresses in concrete are not exceed to 12N/mm² and 1N/mm² respectively.

Solution:

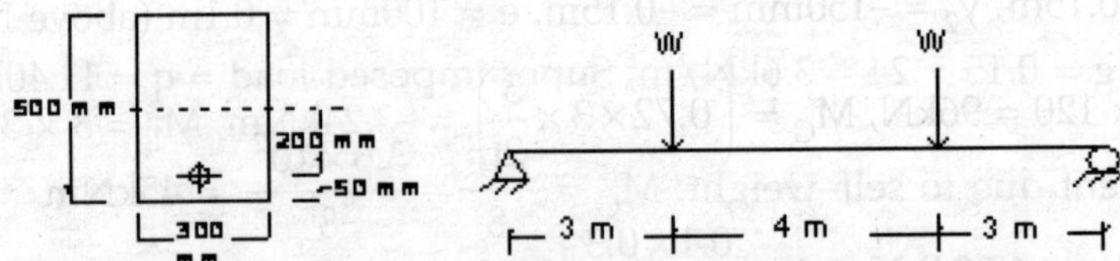

Fig. 2.14 Cross–section of Beam and Simply Supported Beam

y_1 = 250mm = 0.25m, y_2 = –250mm = –0.25m, e = –200mm = –0.20m, F = 800kN,

$$I = \frac{0.3\times0.5^3}{12} = 0.003125\text{m}^4, A = 0.3 \times 0.5 = 0.15\text{m}^2$$

$$fl_1 = 12\text{N/mm}^2 = 12000\text{kN/m}^2,$$

$$fl_2 = -1\text{N/mm}^2 = -1000\text{kN/m}^2$$

$$fl_1 = \frac{F}{A}+\frac{Fey_1}{I}+\frac{M_Ty_1}{I}, 12000 = \frac{800}{0.15}+\frac{800\times(-0.20)\times0.25}{0.003125}+\frac{M_T\times0.25}{0.003125}$$

$$M_T = 243.375\text{kNm}$$

$$fl_2 = \frac{F}{A}+\frac{Fey_2}{I}+\frac{M_Ty_2}{I}, -1000 = \frac{800}{0.15}+\frac{800\times(-0.20)\times(-0.25)}{0.003125}+\frac{M_T\times(-0.25)}{0.003125}$$

$$M_T = 239.162\text{kNm (Adopt)}$$

$$\text{Self–weight} = g = 0.15 \times 24 = 3.6\text{kN/m}, M_G = \frac{gL^2}{g} = \frac{3.6\times10^2}{8} = 45\text{kNm}$$

$$M_T = M_G + M_L, \therefore 239.162 = 45 + M_L, 239.162 = 45 + \frac{w\times10}{3}, \therefore w = 58\text{kN}$$

Problem 2.9

A cantilever beam of span 3m is required to carry a point load of 3kN at its free end in addition to its own weight. The beam is 100mm × 300mm in section and is pre–stressed with initial pre–stressing force 120kN which is located at 50mm from the top. Assume the loss ratio as 80%. Determine

(a) Stress distribution at transfer of pre–stress
(b) Stress distribution at working load.

Solution:

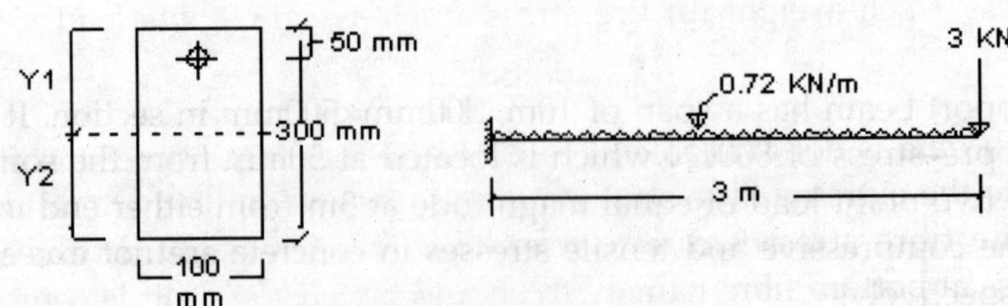

Fig. 2.15 Cross–section of Beam and Simply Supported Beam

Area = A = 0.3 × 0.1 = 0.03m², F_o = 120kN, Self–weight = g = 0.03 × 24 = 0.72kN/m, y_1 = 150mm = 0.15m, y_2 = –150mm = –0.15m, e = 100mm = 0.1m (above NA),

$$F = \eta F_o = 0.80 \times 120 = 96\text{kN},\ M_G = \left(0.72\times 3\times \frac{3}{2}\right) = -3.24\text{kNm},\ M_L = 3\times 3 = -9\text{kNm},$$

$$M_T = -(9+3.24) = -12.24\text{kNm},\ I = \frac{0.1\times 0.3^3}{12} = 225\times 10^6\text{mm}^4 = 2.25\times 10^{-4}\text{mm}^4$$

Case (1): At the Transfer of Pre–stress

Let f_{u1} and f_{u2} be the extreme fibre stresses in concrete at transfer of pre–stress.

$$f_{u1} = \frac{F_o}{A} + \frac{F_o e y_1}{1} + \frac{M_G y_1}{I} = \left(\frac{120}{0.03}\right) + \left(\frac{120\times 0.1\times 0.15}{2.25\times 10^{-4}}\right) + \left(\frac{-3.24\times 0.15}{2.25\times 10^{-4}}\right)$$

$$f_{u1} = 9840\text{kN/m}^2 = 9.8\text{N/mm}^2$$

$$f_{u2} = \frac{F_o}{A} + \frac{F_o e y_2}{1} + \frac{M_G y_2}{I} = \left(\frac{120}{0.03}\right) + \left(\frac{120\times 0.1\times (-0.15)}{2.25\times 10^{-4}}\right) + \left(\frac{-3.24\times (-0.15)}{2.25\times 10^{-4}}\right)$$

$$f_{u2} = -1840\text{kN/m}^2 = -1.840\text{N/mm}^2$$

Case (2): The Working Load Condition

Let fl_1 and fl_2 be the extreme fibre stresses in concrete at working load.

$$fl_1 = \frac{F}{A} + \frac{Fey_1}{I} + \frac{M_T y_1}{I} = \left(\frac{96}{0.03}\right) + \left(\frac{96\times 0.1\times 0.15}{2.25\times 10^{-4}}\right) + \left(\frac{(-12.24)\times 0.15}{2.25\times 10^{-4}}\right)$$

$$fl_1 = 1440\text{kN/m}^2 = 1.44\text{N/mm}^2$$

$$fl_2 = \frac{F}{A} + \frac{Fey_2}{I} + \frac{M_T y_2}{I} = \left(\frac{96}{0.03}\right) + \left(\frac{96\times 0.1\times (-0.15)}{2.25\times 10^{-4}}\right) + \left(\frac{-12.24\times (-0.15)}{2.25\times 10^{-4}}\right)$$

$$fl_2 = 4960\text{kN/m}^2 = 4.96\text{N/mm}^2$$

Problem 2.10

The cross section of a pathway is shown in Fig. 2.16. It is simply supported over a span of 10m and carries a live load of 5kN/m². It is pre–stressed with 4 cables of each area 150mm². Effective pre–stress in the cable is 850MPa. Determine the extreme fibre stresses in concrete due to (a) Pre–stress + self–weight (b) Pre–stress + self–weight + live load.

Solution:

Note: If enough data is not available to compute both extreme fibre stresses. We can use F_o itself to compute the extreme fibre stresses. If η is not known then we can use F itself for computing extreme fibre stresses due to pre–stress + self–weight. In the working load condition, the stress at bottom fibre namely fl_2 should be tension or little compression in the case of simply supported girder.

Span = L = 10m, Total area of steel = A_{st} = 4 × 150 = 600mm²

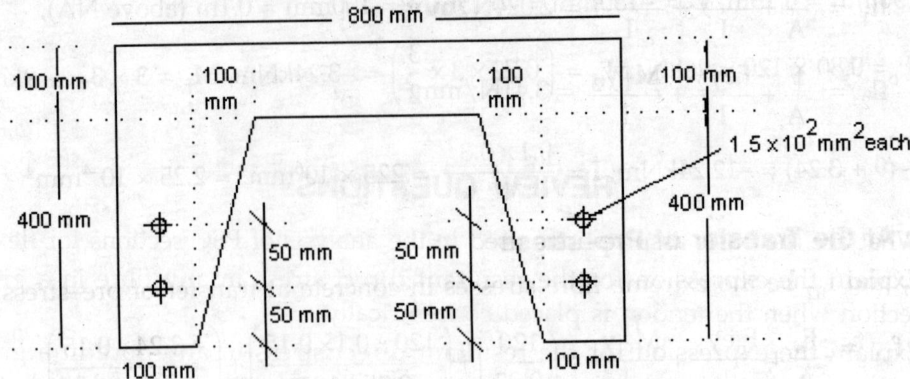

Fig. 2.16 Cross–section of a Pathway

Let the CGS lies at a height of from the soffit. Now taking the moment above the soffit

We have, $600 \times \bar{h} = 2 \times 150 \times 50 + 2 \times 15 \times 100$,

$\bar{h} = 75$mm, Area of section = 0.2m²,

Dead load = g = 0.2 × 24 = 4.8kN/m, Live load = q = 5 × 0.8 = 4kN/m

Bending moment due to self–weight, $M_G = \frac{gL^2}{8} = \frac{4.8 \times 10^2}{8} = 60\text{kNm}$

Bending moment due to super imposed load, $M_L = \frac{qL^2}{8} = \frac{4 \times 10^2}{8} = 50\text{kNm}$

$$M_T = M_G + M_L = 60 + 50 = 110\text{kNm}$$

$$\bar{y} = \quad = 186.6\text{mm} = 0.1866\text{m} = y_1,$$

$$y_2 = 186.6 - 500 = -313.4 = -0.3134\text{m}$$

$\therefore$ Eccentricity, $e = y_2 + \bar{h} = -313.4 + 75 = -238.4\text{mm} = -0.2384\text{m}$

$$I_{XX} = I_{XG} + AY^2 = \frac{bd^3}{12} + A(\bar{y} - y)^2 = 4.0975 \times 10^{-3}\text{m}^4,$$

$f_{se} = 850\text{N/mm}^2$, $A_{st} = 600\text{mm}^2$, $F = f_{se} \times A_{st} = 850 \times 600 = 510 \times 10^3\ \text{N} = 510\text{kN}$

- Extreme fibre stresses in concrete due to (a) Pre–stress + Self–weight

$$f_{u1} = \frac{F_o}{A} + \frac{F_o e y_1}{I} + \frac{M_G y_1}{I} = -0.254\text{N/mm}^2$$

$$f_{u2} = \frac{F_o}{A} + \frac{F_o e y_2}{I} + \frac{M_G y_2}{I} = 7.26\text{N/mm}^2$$

- Extreme fibre stresses in concrete due to (b) Pre–stress + Self–weight + live load

$$fl_1 = \frac{F}{A} + \frac{Fey_1}{I} + \frac{M_T y_1}{I} = 2.024\text{N/mm}^2$$

$$fl_2 = \frac{F}{A} + \frac{Fey_2}{I} + \frac{M_T y_2}{I} = 3.41\text{N/mm}^2$$

REVIEW QUESTIONS

2.1 What are the basic assumptions used in the analysis of PSC sections for flexure?

2.2 Explain the expression for the resultant direct stress in any fibre in a given PSC section when the tendon is placed concentrically.

2.3 Explain the expression for the resultant direct stress in any fibre in a given PSC section when the tendon is placed eccentrically w.r.t. the centroid of the concrete section.

2.4 Explain the extreme fibre stresses in concrete for the following two cases.
(a) Transfer of pre–stress (b) Working load condition

2.5 Explain loss ratio or loss factor with respect to analysis of sections for flexure.

EXERCISE PROBLEMS

2.1 A rectangular concrete beam, 100mm wide and 250mm deep spanning over 8m is pre–stressed by a straight cable carrying an effective pre–stressing force of 250kN located at an eccentricity of 40mm. The supports a live load of 1.2kN/m.

a. Calculate the resultant stress distribution for the central cross–section of the beam. The density of the concrete is 24kN/m^3.

b. Find the magnitude of the pre–stressing force with an eccentricity of 40mm which can balance the stresses due to dead and live loads at the bottom fibre of the central cross–section of the beam.

[Ans: a. Stress at top = 14.2N/mm^2 (Compression), Stress at the bottom = 5.8N/mm^2 (Compression); b. Pre–stressing force = 170kN]

2.2 A pre–stressed concrete beam supports a live load of 4kN/m over a simply supported span of 8m. The beam has an I–section with an overall depth of 400mm. The thickness of flanges and web are 60mm and 80mm respectively. The width of the flanges is 200mm. The beam is to be pre–stressed by a pre–stressing force of 235kN at a suitable eccentricity such that the resultant stresses at the soffit of the beam at the center of span are zero.

a. Find the eccentricity required for the force.

b. If the tendon is concentric, what should be the magnitude of the pre–stressing force for the resultant stress to be zero at the bottom fibre of the central span section?

[Ans: a. e = 84mm, b. Pre–stressing force = 450kN]

2.3 A pre–stressing concrete beam, 200mm wide and 300mm deep, is used over an effective span of 6m to support an imposed load of 4kN/m. The density of concrete is 24kN/m^3. At the quarter span section of the beam, find the magnitude of:

a. The concentric pre–stressing force necessary for zero fibre stress at the soffit when the beam is fully loaded; and

b. The eccentric pre–stressing force located 100mm from the bottom of the beam which would nullify the bottom fibre stress due to loading.

[Ans: a. 367.2kN; b. 183.6kN]

2.4 A concrete beam of symmetrical I–section spanning 8m has flange width and thickness of 200mm and 60mm respectively. The overall depth of the beam is 400mm. the thickness of the web is 80mm. the beam is pre–stressed by a parabolic cable with an eccentricity of 15mm at the centre zero at the supports with an effective force of 100kN. The live load on the beam is 2kN/m. Draw the stress distribution diagram at the central section for:

a. Pre–stress + self–weight (Density of concrete = 24kN/m^3); and

b. Pre–stress + self–weight + live load.

[Ans: a. 0.7N/mm^2 (Compression) at top and 3.6N/mm^2 (Compression); b. 7.4N/mm^2 (Compression) at top and –0.2N/mm^2 (Tension) at bottom]

2.5 A concrete beam with a double overhang has the middle span equal to 10m and the equal overhang on either side is 2.5m. Determine the profile of the pre–stressing cable with an effective force of 250kN which can balance a uniformly distributed load of 8kN/m on the beam, which includes the self–weight of the beam.

Sketch the cable profile making the eccentricity of cable at support and mid–span.

[Ans: e (support) = 100mm; e (centre of span) = 300mm]

2.6 A concrete beam of rectangular section having a width of 300mm and depth 500mm is pre–stressed by a cable carrying a force of 750kN at an eccentricity of 100mm. If the beam supports a live load of 20kN/m over an effective span of 7m, estimate the resultant stress at the top and bottom fibres at mid–span section due to the effect of pre–stress, dead and live loads. Assume unit weight of concrete as 24kN/m^3.

[Ans: Stress at top = 10.56N/mm^2 (Compression), Stress at the bottom = 2.94N/mm^2 (Compression)]

2.7 A pre–stressed concrete beam, 120mm wide by 300mm deep, is pre–stressed by a cable which has an eccentricity of 100mm at the centre of span section. The span of the beam is 6m. If the beam supports two concentrated loads of 10kN each at one third spans points. Assume unit weight of concrete as 24kN/m^3. Determine the magnitude of the pre–stressing force in the cable for the following cases:

a. Considering live loads but neglecting self–weight of the beam; and

b. Considering both self–weight of beam and live load.

[Ans: a. 200kN; b. 238.88kN]

OBJECTIVE QUESTIONS

2.1 The analysis of stresses developed in a pre–stressed concrete structural element is based on the assumptions

a. Concrete is homogeneous elastic material

b. Within the range of working stresses, both the concrete and steel behave elastically, notwithstanding the small amount of creep which occurs in both the materials under sustained loading and

c. The plane section before bending is assumed to remain plane even after bending, which implies a linear strain across the depth of the member

d. All of above

2.2 The minimum cube strength required for concrete to be used in PSC construction is

a. $30N/mm^2$ b. $40N/mm^2$ c. $45N/mm^2$ d. $50N/mm^2$

2.3 The permissible tension in steel is of order

a. $1000N/mm^2$ b. $1200N/mm^2$ c. $1400N/mm^2$ d. $1600N/mm^2$

2.4 Loss ratio or Loss factor is defined as

a. the ratio of the effective pre–stressing force (F) to the initial pre–stressing force (F_o)

b. the ratio of the initial pre–stressing force (F_o) to the effective pre–stressing force (F)

c. the product of the effective pre–stressing force (F) to the initial pre–stressing force (F_o)

d. None of the above

2.5 Expression for the resultant direct stress in any fibre in a given PSC section when the tendon is placed concentrically

a. $f = \frac{F_o}{A} + \frac{MI}{y}$ b. $f = \frac{F_o}{A} - \frac{MI}{y}$

c. $f = \frac{F_o}{A} + \frac{My}{I}$ d. None of above

2.6 If the loss of pre–stress is 16% then, loss ratio = η =

a. 0.94 b. 1.16 c. 0.84 d. 1.06

2.7 Expression for the resultant direct stress in any fibre in a given PSC section when the tendon is placed eccentrically w.r.t. the centroid of the concrete section

a. $f = \frac{F_o}{A} + \frac{F_o ey}{I} + \frac{My}{I}$ b. $f = \frac{F_o}{A} - \frac{F_o ey}{I} + \frac{My}{I}$

c. $f = \frac{F_o}{A} + \frac{F_o ey}{I} - \frac{My}{I}$ d. None of above

2.8 Radius of gyration can be computed as ...., where, I is the moment of inertia and A is the area of the section

a. $\sqrt{I/A}$ b. $\sqrt{A/I}$ c. $\sqrt{(I-A)/A}$ d. $\sqrt{A/(I-A)}$

2.9 Expression for resultant stress when along with moment due to pre–stressing force, if external moment M due to the weight of the beam, is also applied then,......

a. $f=\frac{F_o}{A}\left(1-\frac{ey}{K^2}\right)+\frac{My}{I}$ b. $f=\frac{F_o}{A}\left(1+\frac{ey}{K^2}\right)+\frac{My}{I}$

c. $f=\frac{F_o}{A}\left(1+\frac{ey}{K^2}\right)-\frac{My}{I}$ d. $f=\frac{F_o}{A}\left(1-\frac{ey}{K^2}\right)-\frac{My}{I}$

2.10 The resultant stresses due to pre–stressing moment and external moment (due to self–weight of beam and due to imposed loads) at the top and bottom fibres are given by

a. $f_{sup}=\left(\frac{F_o}{A}-\frac{F_oe}{Z_t}\right)+\frac{M_G}{Z_t}+\frac{M_L}{Z_t}$ and $f_{inf}=\left(\frac{F_o}{A}+\frac{F_oe}{Z_b}\right)+\frac{M_G}{Z_b}-\frac{M_L}{Z_b}$

b. $f_{sup}=\left(\frac{F_o}{A}+\frac{F_oe}{Z_t}\right)+\frac{M_G}{Z_t}+\frac{M_L}{Z_t}$ and $f_{inf}=\left(\frac{F_o}{A}+\frac{F_oe}{Z_b}\right)-\frac{M_G}{Z_b}-\frac{M_L}{Z_b}$

c. $f_{sup}=\left(\frac{F_o}{A}-\frac{F_oe}{Z_t}\right)+\frac{M_G}{Z_t}+\frac{M_L}{Z_t}$ and $f_{inf}=\left(\frac{F_o}{A}+\frac{F_oe}{Z_b}\right)-\frac{M_G}{Z_b}-\frac{M_L}{Z_b}$

d. $f_{sup}=\left(\frac{F_o}{A}-\frac{F_oe}{Z_t}\right)+\frac{M_G}{Z_t}+\frac{M_L}{Z_t}$ and $f_{inf}=\left(\frac{F_o}{A}+\frac{F_oe}{Z_b}\right)-\frac{M_G}{Z_b}+\frac{M_L}{Z_b}$

Answers

2.1 d	2.2 a	2.3 a	2.4 a	2.5 c
2.6 c	2.7 a	2.8 a	2.9 b	2.10 c

❑

CHAPTER 3

Losses of Pre–stress

3.1 INTRODUCTION

In pre–stressed concrete applications, the most important variable is the pre–stressing force. In the early days, it was meant that the pre–stressing force does not stay constant, but reduces with time. Even during pre–stressing of the tendons and the transfer of pre–stress to the concrete member, there is a drop of the pre–stressing force from the recorded value in the jack gauge. The various reductions of the pre–stressing force are termed as the losses of pre–stress.

The losses are broadly bifurcated into two groups, immediate and time–dependent. The immediate losses occur during pre–stressing of the tendons and the transfer of pre–stress to the concrete member. The time–dependent losses occur during the service life of the pre–stressed member. The losses due to elastic shortening of the member, friction at the tendon–concrete interface and slip of the anchorage are the immediate losses. The losses due to the shrinkage and creep of the concrete and relaxation of the steel are the time–dependent losses. The causes of the various losses in pre–stress are shown in the following Fig. 3.1.

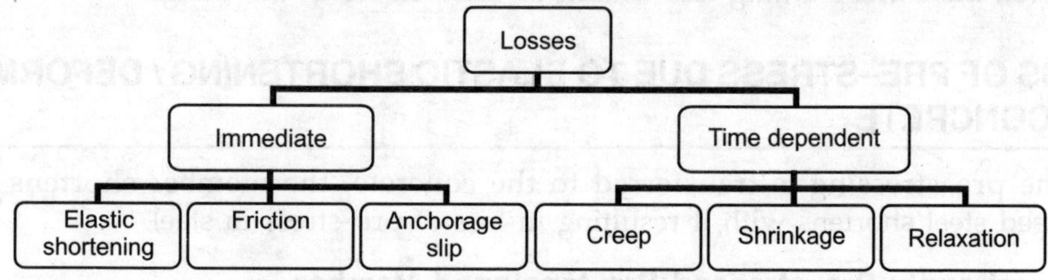

Fig. 3.1 Causes of the Various Losses in Pre–stress

According to IS: 1343–2012, while assessing the stresses in concreteand steel during tensioning operations and later in service, due regard shall be paid to all losses and variations in stress resulting from creep of concrete, shrinkage of concrete, relaxation of steel, the shortening (elastic deformation) of concrete at transfer, and friction and slip of anchorage. Unless otherwise determined by actual tests, allowance for these losses shall be made in accordance with the values specified under Sections 18.5.2.1 to 18.5.2.6.

In computing the losses in pre–stress when un–tensioned reinforcementis present, the effect of the tensile stresses developed by the un–tensioned reinforcement due to shrinkage and creep shall be considered.

3.2 NATURE OF LOSSES OF PRE–STRESS

The initial pre–stress in concrete undergoes a gradual reduction with time from the stage of transfer due to various causes. This is generally referred to as 'loss of pre–stress'. A reasonably good estimate of the magnitude of loss of pre–stress is necessary from the point of view of design. The different types of losses encountered in the pre–tensioning and post–tensioning systems are compiled in Table 3.1.

Table 3.1 Types of Losses of Pre–stresses

Sl. No.	Pre–tensioning	Sl. No.	Post–tensioning
1	Elastic deformation of concrete	1	No loss due to elastic deformation if all the wires are simultaneously tensioned. If the wires are successively tensioned, there will be loss of pre–stress due to elastic deformation of concrete
2	Relaxation of stress in steel	2	Relaxation of stress in steel
3	Shrinkage of concrete	3	Shrinkage of concrete
4	Creep of concrete	4	Creep of concrete
		5	Friction
		6	Anchorage slip

In addition to the above, there may be losses of pre–stress due to sudden changes in temperature, especially in steam curing of pre–tensioned units. The rise in temperature causes a partial transfer of pre–stress (due to elongation of tendons between adjacent units in the long–line process) which may be cause a large amount of creep if the concrete is not properly cured. If there is a possibility of a change of temperature between the times of tensioning and transfer, the corresponding loss should be allowed for in the design.

3.3 LOSS OF PRE–STRESS DUE TO ELASTIC SHORTENING / DEFORMATION OF CONCRETE

As the pre–stressing is transferred to the concrete, the member shortens and the pre–stressed steel shortens with it resulting in loss of pre–stress in steel.

3.3.1 In–uniformly Pre–stressed Pre–tensioned Member

If P_i is the initial pre–stressing force, stress in concrete,

$$f_c = P_i/A \qquad \text{...(3.1)}$$

Strain in concrete, $e_c = f_c/E_c$ ...(3.2)

Reduction in steel strain equals e_c

Hence, Loss of stress in steel $= e_c\,E_s$ ...(3.3)

$$= f_c\,E_s/E_c$$

$$= m\,f_c \qquad \text{...(3.4)}$$

Where, m is the modular ratio between steel and concrete.

3.3.2 In–uniformly Pre–stressed Post–tensioned Member

If there is only single tendon or cable in a post–tensioned member or when all available tendons are pulled simultaneously. The concrete shortens as the tendon is jacked against the concrete. Since the force in the tendon is measured after the elastic shortening of the concrete has taken place no loss in pre–stress due to that shortening need be accounted for. If there is more than one tendon and the tendons are stressed in succession, the shortening of the concrete increases as each cable is tightened against it, and the loss of pre–stress due to elastic shortening differs in the tendon. Each tendon or group of wires, as it is tensioned, does not lose pre–stress due to elastic deformation of the concrete it imposes on the concrete. But each tendon or group is subjected to the strain imposed on the concrete by all tendons or groups tensioned subsequently. Thus, the tendon that is first tensioned would suffer the maximum amount of loss due to the shortening of concrete and the tendon that tensioned last, will not suffer any loss due to the shortening of the concrete. The computation of such losses may be quite complicated. Therefore, the average loss of stress in steel due to elastic shortening of concrete may be calculated as follows when the number of tensioning operations is large.

If f_c is the stress in concrete due to initial pre–stressing force at the centroid of the steel, the strain in concrete is f_c / E_c and average loss of strain in steel is taken as $0.5f_c / E_c$.

Hence, Loss of stress in steel due to elastic shortening of concrete = $0.5f_c\, E_s / E_c$

$$= 0.5mf_c \qquad \ldots(3.5)$$

When the number of tensioning operations is few, the loss in first operation can be calculated and half of the value is taken as the loss in all cables.

In actual practice, either the following two methods are used to counteract the loss of pre–stress due to elastic shortening of the concrete in post–tensioned members.

1. All tendons are stressed to the specified initial pre–stress and the average loss is allowed for in the design.
2. All tendons are stressed to a value above the specified initial pre–stress by the magnitude of the average loss and then, while designing, the loss due to the elastic shortening of the concrete is not to be considered again.

If the loss of pre–stress due to this cause is not significant, the first method is followed. If the steel can withstand some over–tensioning, and if a high effective pre–stress is desired, the second method can be adopted.

3.3.3 Eccentrically Pre–stressed Pre–tensioned Member

In a pre–tensioned beam the steel is bonded to the concrete before pre–stress is transferred. After the transfer of pre–stress, the beam bends upward and the wires shorten because of that bending. Shortening of the wires due to bending is not considered as a loss of as a loss of pre–stress, just as the eventual lengthening under load is not considered again in pre–stress. Bending of beam does not change the pre–stress at its ends.

In this case, it is assumed that all the tendons transfer the pre–stress simultaneously and therefore they suffer an equal loss of pre–stress. Here loss of pre–stress due to elastic shortening of concrete = mf_c. Where m is modular ratio, f_c is compression stress in concrete fibers which are situated at the level of centre of gravity of steel (CGS), further

$$f_c = \frac{F_o}{A} + \frac{F_o ey}{I} = \frac{F_o}{A} + \frac{F_o e^2}{I} \qquad \ldots(3.6)$$

Where,

F_o = Initial pre–stressing force in the tendons

e = Eccentricity of CGS

A = Total cross sectional area of given section

I = Moment of Inertia of the given section about the centroidal axis of bending

3.3.4 Eccentrically Pre–stressed Post–tensioned Member

Before the tendon is bonded to the concrete, bending of the member will affect the pre–stress in the tendon. If friction between tendon and the surrounding duct is neglected, any stain in the tendon will be uniformly modified. After the tendon is bonded to the concrete, any further loading of the beam, causes stress in tendon near the mid–span to change quite a bit, but that at the end sections does not change at all since there is no change in bending moment at end sections. Hence the pre–stress, as mentioned at the ends of the beam in its initial condition, remains constant. Therefore, the value which should be allotted f_c in the equation loss = 0.5mf_c is the average concrete stress at the level of the centroidal line of steel in the beam's initial condition immediately after pre–stress is transferred. For practical purposes, the average value of the stress at end section, quarter span section and mid–span section may be adopted. Fortunately, the loss due to this source is not more than 2 to 3% and for all practical purposes can be approximately estimated and allowed for.

3.4 LOSS OF PRE–STRESS DUE TO FRICTION

In post–tensioning, there is movement of the greater part of the tendon relative to the surrounding material during the tensioning operation. Therefore where a tendon is in contact with its surrounding concrete or ducting material there is friction which imposes a reduction on the pre–stressing force and a consequent loss in the pre–stress. This friction loss occurs in two ways.

(a) Friction loss due to variation in the duct from its intended path. This is termed the 'length' or 'wobble' or 'wave' effect.

(b) Friction loss is due to curvature of the cable path, termed the 'curvature' effect.

The wobble friction loss occurs in every tendon whether straight or curved. The curvature friction loss occurs only when a tendon is curved. (Friction also occurs during tensioning itself and in the anchorage components. This too affects the pre–stress in steel and must be taken into account using the specifications given by the manufacturers.

3.4.1 Friction Loss due to 'Wave' and 'Curvature' Effects

For straight and moderately curved structures, with curved and straight cables, the value of pre–stressing force P at a distance x metres from tensioning end and acting in the direction of the tangent to the curve of the cable shall be calculated as below. (Refer Fig. 3.2)

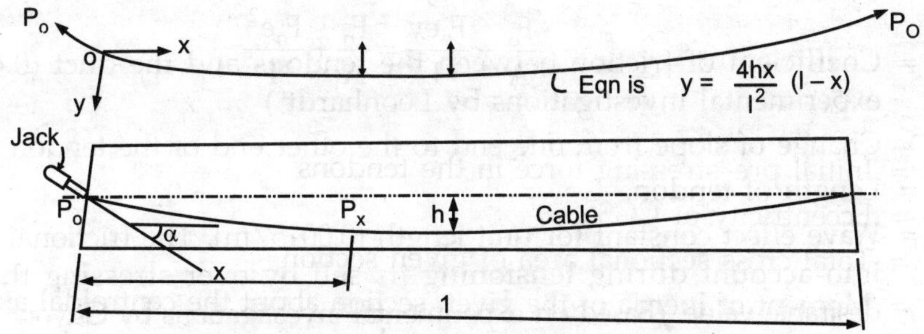

Fig. 3.2 Loss of Pre–stress due to Friction

$$P = P_o e^{-(\mu\alpha + kx)} \quad ...(3.7)$$

Where,

P_o = Pre–stressing force in steel at the tensioning end acting in the direction of the tangent to the curve of the cable.

α = Curvature angle in Radian through which the tangent to the cable profile at the section considered has turned in relation to that tangent at the tensioning end.

e = 2.7183 (base of the natural logarithm).

μ = Coefficient of friction in curve.
= 0.55 for steel moving on smooth concrete
= 0.30 for steel moving on steel fixed to duct
= 0.25 for steel moving on lead

k = Coefficient for wave effect varying from 15×10^{-4} to 50×10^{-4} per metre.

Note 1: Expansion of the equation for P for small values of (μα + kx) may be $P = P_o(1 - \mu\alpha - kx)$.

Note 2: In circular construction such as tanks where circumferential tendons are tensioned by jacks, value of μ for calculating friction be taken as:

μ = 0.45 for steel moving on smooth concrete
= 0.25 for steel moving on steel bearer fixed to concrete
= 0.10 for steel moving on steel rollers.

Note 3: The effect of reverse friction shall be taken into consideration in such cases where the initial tension applied to a pre–stressing tendon is partially released and action of friction in the reverse direction causes an alteration in the distribution of stress along the length of the tendon.

Frictional loss comes into play only in the case of post–tensioned tendons due to relative movement between the tendon and the duct. The internal surface of the duct does not possess the desired smoothness. On the other hand the internal surface of the duct has a sort of wavy surface which is responsible for offering the frictional resistance, as the length of duct increases, the frictional resistance also increases. Further the ducts are usually curved and this curvature gives rise additional frictional resistance. In this case percentage loss of pre–stress due to friction.

$$\text{(Expressed as a decimal)} = \mu\theta + kl \quad ...(3.8)$$

Where,

μ = Coefficient of friction between the tendons and the duct (0.4) (Based on experimental investigations by Leonhardt[6])

θ = Change of slope from one end to the other end of the tendon in radian.

l = Length of tendon.

k = Wave effect constant for unit length (0.0015/m). The frictional loss is taken into account during tensioning its self by over stressing the wires to a desirable value. (Based on experimental investigations by Guyon[2] and Cooley[3])

Note: Since the depth of the members is small compared with the length, the projected length of tendon measured along the axis of the member can be used for computing frictional losses.

3.4.2 Counteracting Friction Losses

Friction losses can be counteracted wholly or impart by using one or three of the following three methods.

(a) By jacking both ends simultaneously. This has no effect on the losses at mid–span but does keep losses at all other sections as low as possible and less than the losses at the mid–span sections.

(b) By overstressing tendons during the tensioning operation. The amount of over–tension required for overcoming friction is not cumulative over that required for overcoming anchorage losses or for minimizing creep in steel. It is sufficient to take the greatest of the three required values and over–tension for that amount. When the frictional loss is a high percentage of the pre–stress, it cannot be totally overcome because, at the time of initial tensioning, the maximum tensile stress immediately behind the anchorages shall not exceed 80% of the ultimate tensile strength of the wire or bar or strand (as per IS: 1343–2012 code).

(c) In post–tensioned members in the tendons are to remain unbounded, friction may be reduced by the use of a suitable lubricant. Lubricants should never be used if the steel is to be bonded to the concrete by grouting.

3.5 LOSS OF PRE–STRESS DUE TO SLIP OF ANCHORAGE COMPONENTS

In post–tensioning, after a tendon is tensioned, the jack is released and pre–stress is transferred to the concrete by bearing of the anchorage. When this occurs, the anchorage components which are subjected to force deform, and friction wedges, often used, slip slightly before the wires are firmly gripped. The amount of slippage and deformation depends on the type of wedge and the deformation depends on the type of wedge and the stress in the wires. It varies from 1 to 6mm. the values for particular system may be obtained from the suppliers.

If δ_a is the amount by which the anchorage slips and deforms, then

$$\text{The loss in steel strain} = \delta_a / l \qquad \text{...(3.9)}$$

Where, l = Length of tendon (which may take equal to the length of the member).

$$\text{Loss in steel stress} = \delta_a E_s / l \qquad \text{...(3.10)}$$

From this it should be noted that the loss of pre–stress due to this cause is considerably influenced by the length of tendons, being high when a tendon is short and small when a tendon is long. This loss should be allowed for in the design, when the length of wire is less than 10m. When the length is more, the wires may be over stretched by a pre–determined value to take care of this loss. When preparing specifications and drawings, it should be made clear which losses should be allowed for at the site by increasing the initial tensioning stress, and that the specified pre–stressing force is that which is to be applied when pre–stress allowance has been made for this additional losses.

Any loss of pre–stress which may occur due to slip of wires during anchoring or due to the strain of anchorage shall be allowed for in the design. Loss due to slip in anchorage is of special importance with short members and the necessary additional elongation should be provided for at the time of tensioning to compensate for this loss.

3.6 LOSS OF PRE–STRESS DUE TO CREEP OF CONCRETE

As there is shortening of concrete member due to creep, the tendon also get shortened thereby losing some of its pre–stress. Loss of strain in steel is equal to the creep strain in concrete at the level of steel.

The loss of pre–stress due to creep of concrete under load shall be determined for all the permanently applied loads including the pre–stress. The creep loss due to live load stresses, erection stresses and other stresses of short duration may be ignored. The loss of pre–stress due to creep of concrete is obtained as the product of the modulus of elasticity of the pre–stressing steel (see 4.5.3) and the ultimate creep strain of the concrete fibre (see 5.2.5.1) integrated along the line of centre of gravity of the pre–stressing steel over its entire length (IS: 1343–2012).

Creep coefficient = Creep strain / Elastic strain

Hence, Creep strain C = Creep coefficient × Elastic strain

$$C = C_c \times f_c / E_c \qquad ...(3.11)$$

Where, f_c = average stress in concrete at the level of centroidal line of steel.

Loss of stress in steel due to creep of concrete

$$= (C_c - 1) \times f_c \times E_s / E_c \qquad ...(3.12)$$

Loss of stress in steel due to creep of concrete

$$= (C_c - 1) \times f_c \times m \qquad ...(3.13)$$

Where, C_c = creep coefficient which varies from 2 to 3.

The total creep strain during any specific period shall be assumed for all practical purposes, to be the creep strain due to sustained stress equal to the average of the stresses at the beginning and end of the period.

3.7 LOSS OF PRE–STRESS DUE TO SHRINKAGE OF CONCRETE

Shrinkage of concrete means the contraction of concrete on drying. As shrinkage strain occurs in the concrete of a member, its pre–stressing steel also suffers the same strain thereby losing stress. This loss of pre–stress is termed the shrinkage loss. If e_{sh} is the shrinkage strain in concrete, then loss of pre–stress in steel = $e_{sh}\,E_s$. Shrinkage strain of concrete may usually take as 2×10^{-4} cm/cm. E_s is Young's modulus of elasticity of steel usually 2.1×10^5 N/mm^2.

The loss of pre–stress due to shrinkage of concrete shall be the product of the modulus of elasticity of steel (see Section 4.5.3) and the shrinkage strain of concrete (see Section 5.2.4.1) of IS: 1343–2012.

3.8 LOSS OF PRE–STRESS DUE TO RELAXATION OF STEEL

Relaxation of steel is defined as decrease of stress in steel at constant strain. The relaxation losses in pre–stressing steels vary with type of steel, initial pre–stress, age, and temperature and, therefore, shall be determined from experiments. When experimental values are not available, the relaxation losses may be assumed as given in Table 3.2.

Table 3.2 Relaxation Losses for Pre–stressing Steel at 1000 Hours at 27°C

Initial stress	Relaxation loss (MPa)
$0.5f_p$	0
$0.6f_p$	35
$0.7f_p$	70
$0.8f_p$	90

Where, f_p is the characteristic strength of pre–stressing steel.

For tendons at higher temperatures or subjected to large lateral loads, greater relaxation losses as specified by the engineer–in–charge shall be allowed for. No reduction in the value of the relaxation losses should be made for a tendon with a load equal to or greater than the relevant jacking force that has been applied for a short time prior to the anchoring of the tendon.

3.9 TOTAL LOSSES ALLOWED IN DESIGN

It is normal practice in the design of pre–stressed concrete members to assume the total loss of stress as a percentage of the initial stress and provide for this in the design computations. Since the loss of pre–stress depends on several factors, it is difficult to generalize the exact amount of total loss of pre–stress. However, typical values of the total losses of the stress that could be encountered under normal conditions of the work as recommended by Lin as shown in Table 3.3 and an exhaustive treatment of losses due to creep, shrinkage and relaxation is presented by Neville.

Table 3.3 Total Losses of Stress under Normal Conditions of the Work by Lin

Type of Loss	Percentage loss of stress	
	Pre–tensioning	Post–tensioning
Elastic shortening and bending of concrete	4	1
Creep of concrete	6	5
Shrinkage of concrete	7	6
Creep in steel	8	8
Total	25	20

In these recommendations it is assumed that temporary overstressing is done to reduce relaxation and to compensate for friction and anchorage losses.

If f_{pe} = Effective stress in tendons after losses

f_{pi} = Stress in tendon at transfer

η = Reduction factor for loss of pre-stress

$$\eta = \left(\frac{f_{pe}}{f_{pi}}\right) \quad \ldots(3.14)$$

The value of is generally taken as 0.75 for pre-tensioning and 0.80 for post-tensioned members.

3.10 SOLVED EXAMPLES

Problem 3.1

A rectangular pre-tensioned beam 180mm × 300mm in section is pre-stressed with two cables each of area 150mm² located at 50mm from soffit. Initial pre-stress in tendon is 1000N/mm². Determine the percentage loss of pre-stress due to concrete assume the following. Shrinkage strain of concrete = 2×10^{-4}mm/mm, ultimate creeps strain of concrete= 30×10^{-6}mm/mm/N/mm², modulus of elasticity of steel = 2.1×10^5N/mm², modulus of elasticity of concrete = 0.35×10^5N/mm².

Solution:

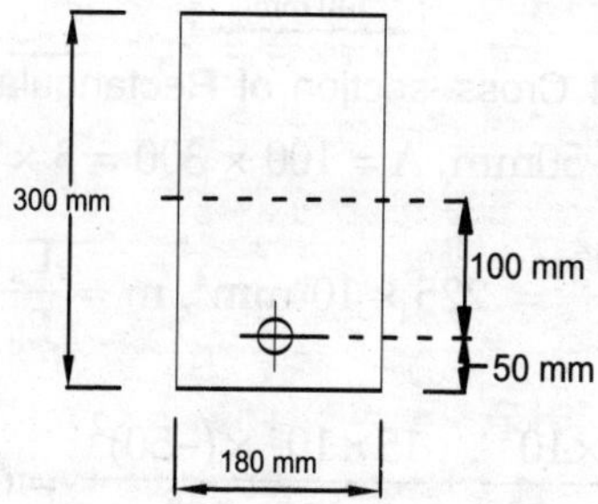

Fig. 3.3 Cross-section of Rectangular Beam

$$m = \frac{E_s}{E_c} = \frac{2.1\times10^5}{0.35\times10^5} = 6,\ f_{si} = 1000\text{N/mm}^2,\ A_{st} = 2 \times 150 = 300\text{mm}^2$$

$$F_o = A_{st} \times f_{si} = 300 \times 1000 = 30000\text{N} = 300\text{kN},$$

$$I = \frac{bd^3}{12} = \frac{180\times300^3}{12} = 405 \times 10^6\text{mm}^4$$

$$f_c = \frac{F_o}{A} + \frac{F_o e^2}{I} = \frac{300000}{300\times180} + \frac{300000\times(-100)^2}{405\times10^6} = 12.96\text{N/mm}^2$$

Loss due to elastic shortening = $mf_c = 6 \times 12.96 = 77.77\text{N/mm}^2$

$\delta_c = Cf_c = 30 \times 10^{-6} \times 12.96 = 3.88 \times 10^{-4}$

Loss due to creep = $(C_c - 1)\, mf_c = \delta_c\, E_s = 3.88 \times 10^{-4} \times 2.1 \times 10^5 = 81.48\text{N/mm}^2$

Loss due to shrinkage of concrete = $\delta_s \times E_s = 2 \times 10^{-4} \times 2.1 \times 10^5 = 42\text{N/mm}^2$

$\therefore$ Total loss of pre–stress in tensional steel = $77.77 + 81.48 + 42 = 201.25\text{N/mm}^2$

$$\therefore \text{ Loss of pre–stress} = \frac{201.25 \times 100}{1000} = 20.125\%.$$

Problem 3.2

A pre–tensioned concrete beam 100mm wide and 300mm deep is pre–stressed by straight wires carrying an initial force of 15×10^4N at an eccentricity of 50mm. The modulus of elasticity of steel and concrete are $210 \times 10^3\text{N/mm}^2$ and $35 \times 10^3\text{N/mm}^2$ respectively. Estimate the % loss of stress in steel due to elastic deformation of concrete if the area of steel wires is 188mm^2.

Solution:

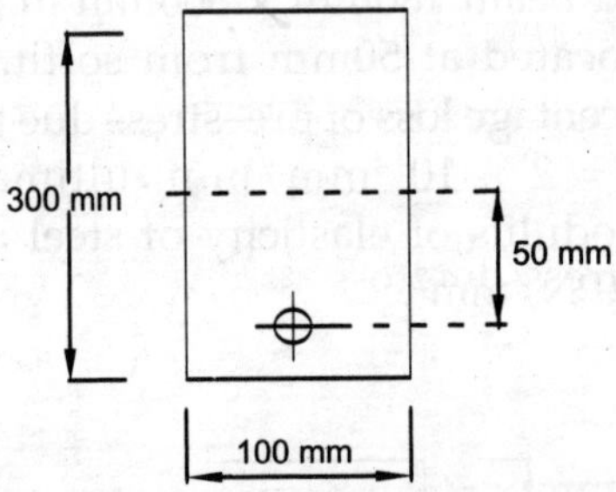

Fig. 3.4 Cross–section of Rectangular Beam

$F_o = 15 \times 10^4$ N, $e = -50$mm, $A = 100 \times 300 = 3 \times 10^4\text{mm}^2$, $A_{st} = 188\text{mm}^2$,

$$I = \frac{bd^3}{12} = \frac{100 \times 300^2}{12} = 225 \times 10^6\text{mm}^4,\ m = \frac{E_s}{E_c} = \frac{210 \times 10^3}{35 \times 10^3} = 6,$$

$$f_c = \frac{F_o}{A} + \frac{F_o e^2}{I} = \frac{15 \times 10^4}{100 \times 300} + \frac{15 \times 10^4 \times (-50)^2}{225 \times 10^6} = 6.6\text{N/mm}^2$$

$$\text{But, } F_o = f_{si} \times A_{st},\ f_{si} = \frac{F_o}{A_{st}} = \frac{15 \times 10^4}{188} = 797.87\text{N/mm}^2 \text{ Say } 800\text{N/mm}^2$$

Loss of stress due to elastic deformation of concrete = $mf_c = 6 \times 6.6 = 39.96 \approx 40\text{N/mm}^2$

$$\therefore \text{ \% Loss of stress in steel} = \frac{40 \times 100}{800} = 5\%$$

Problem 3.3

A pre–tensioned beam 200mm × 500mm in section is pre–stressed with 6 cables each of area 150mm^2. The cables are being symmetrically distributed about the centroidal axis. Determine the percentage loss of pre–stress due to concrete. Assume the following:

Initial pre–stress in tendon = $1000N/mm^2$, shrinkage strain of concrete = 2×10^{-4}mm/mm, creep coefficient of concrete = 2.5, $E_s = 2.1 \times 10^5 N/mm^2$, $E_c = 0.35 \times 10^5 N/mm^2$.

Solution:

Since the cables are placed symmetrically over the cross–section i.e., CGS co insides with CGC. Hence, e = 0, i.e., concentrically pre–stressed member.

$$A_{st} = 6 \times 150 = 900mm^2, F_o = f_{si} \times A_{st} = 900 \times 1000 = 9 \times 10^5 N$$

$$m = \frac{E_s}{E_c} = \frac{2.1\times10^5}{0.35\times10^5} = 6, I = \frac{bd^3}{12} = \frac{200\times500^3}{12} = 2.083 \times 10^9 mm^4,$$

$$f_c = \frac{F_o}{A} + \frac{F_o e^2}{I} = \frac{9\times10^5}{200\times500} + \frac{9\times10^5\times(0)^2}{2.083\times10^9} = 9N/mm^2$$

Since the beam is pre–tensioned

Loss due to elastic shortening of concrete = $mf_c = 6 \times 9 = 54N/mm^2$

Loss due to shrinkage of concrete = $\delta_s E_s = 2 \times 10^{-4} \times 2.1 \times 10^5 = 42 N/mm^2$

Loss due to creep of concrete = $(C_c-1)\, mf_c = (2.5-1) \times 6 \times 9 = 81N/mm^2$

Total loss = $177N/mm^2$

$$\therefore \text{Percentage loss of pre–stress due to concrete} = \frac{177\times100}{1000} = 17.7\%.$$

Problem 3.4

A post–tensioned beam 300mm × 600mm in section is pre–stressed with four cables each of $250mm^2$. The cables are being located at 100mm from the soffit. Determine the % loss of pre–stress due to concrete. Assume,

$f_{si} = 900N/mm^2$, shrinkage strain of concrete = 1.5×10^{-4}mm/mm, ultimate creep strain of concrete = 40×10^{-6}mm/mm/N/mm², m = 6, $E_s = 2.1 \times 10^5 N/mm^2$, $f_c = 11.66N/mm^2$.

Solution:

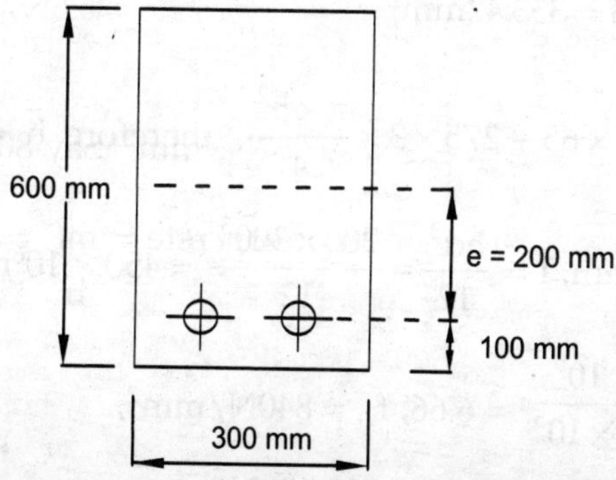

Fig. 3.5 Cross–section of Rectangular Beam

Loss due to elastic shrinking = $1/2\ m\ f_c = 1/2 \times 6 \times 11.66 = 35N/mm^2$

Due to shrinkage of concrete = $\delta_s E_s = 1.5 \times 10^{-4} \times 2.1 \times 10^5 = 31.5 N/mm^2$

$\delta_c = Cf_c = 40 \times 10^{-6} \times 11.66 = 4.664 \times 10^{-4}$

Loss due to creep of concrete = $\delta_c E_s = 4.664 \times 10^{-4} \times 2.1 \times 10^5 = 97.944 N/mm^2$

Total loss = $35 + 31.5 + 97.944 = 164.44 N/mm^2$

$\therefore$ Percentage loss of pre–stress due to concrete = $\dfrac{164.44 \times 100}{900} = 18.27\% \approx 18.3\%$.

Problem 3.5

A rectangular pre–stressed concrete beam 300mm deep and 200 mm wide is pre–stressed by means of 15 nos. of 5 mm dia wires located 65mm from the bottom of the beam and 5mm dia wires 3 in number located at 25mm from the top of the beam. If the wires are initially tensioned to a stress of $840 N/mm^2$, calculate the percentage loss of stress in steel immediately after transfer allowing for the loss of stress due to elastic deformation of concrete only. $E_s = 2.1 \times 10^5 N/mm^2$. $E_c = 0.315 \times 10^5 N/mm^2$.

Solution:

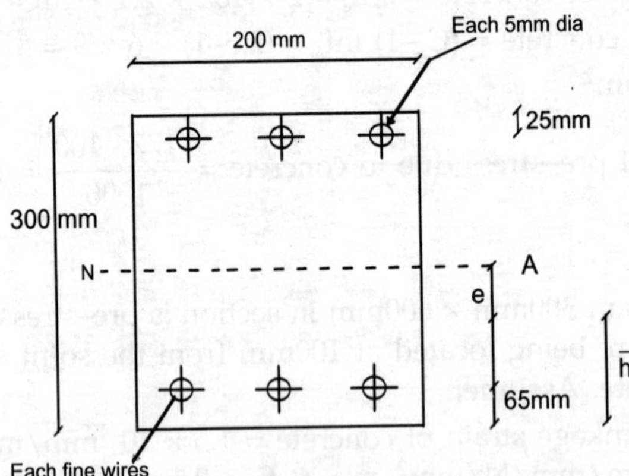

Fig. 3.6 Cross–section of Rectangular Beam

$$A_{st} = \frac{\pi \times 5^2}{4} \times 18 = 353.42 mm^2$$

$$\bar{h} \times A_{st} = \frac{\pi \times 5^2}{4} \times 15 \times 65 + 275 \times 3 \times \frac{\pi \times 5^2}{4}, \text{ therefore } \bar{h} = 100mm$$

$$\text{So, } e = \frac{300}{2} - 100 = 50mm, \; I = \frac{bd^3}{12} = \frac{200 \times 300^3}{12} = 450 \times 10^6 mm^4,$$

$$m = \frac{E_s}{E_c} = \frac{2.1 \times 10^5}{0.315 \times 10^5} = 6.66, \; f_{si} = 840 N/mm^2,$$

$$F_o = f_{si} \times A_{st} = 840 \times 353.42 = 296.88 kN$$

$$f_c \text{ (top wires)} = \frac{F_o}{A} + \frac{F_o e y_t}{I} = \frac{296.88 \times 10^3}{200 \times 300} + \frac{296.88 \times 10^3 \times (-50)\,(125)}{450 \times 10^6} = 0.833 N/mm^2$$

$$f_c \text{ (bottom wires)} = \frac{F_o}{A} + \frac{F_o e y_b}{I} = \frac{296.88 \times 10^3}{200 \times 300} + \frac{296.88 \times 10^3 \times (-50)(-85)}{450 \times 10^6} = 7.751 \text{N/mm}^2$$

Loss of stress in wires at top = mf_c = 6.66 × 0.833 = 5.547N/mm²

Loss of stress in wires at bottom = mf_c = 6.66 × 7.751 = 51.621N/mm²

$$\text{Percentage loss of stress for top} = \frac{5.547 \times 100}{840} = 0.66\%$$

$$\text{Percentage loss of stress for bottom} = \frac{51.621 \times 100}{840} = 6.14\%.$$

Problem 3.6

A pre–tension beam with a rectangular cross–section 100mm wide by 250mm deep is pre–stressed by 10 numbers of 5 mm dia wires with an initial pre–stress of 1050N/mm². Two wires are located at 25mm from top and 4 wires each at 25mm and 50mm from the soffit of the beam. Calculate the percentage loss of stress in each layer of wires due to elastic deformation of concrete. Assume m = 6.

Solution:

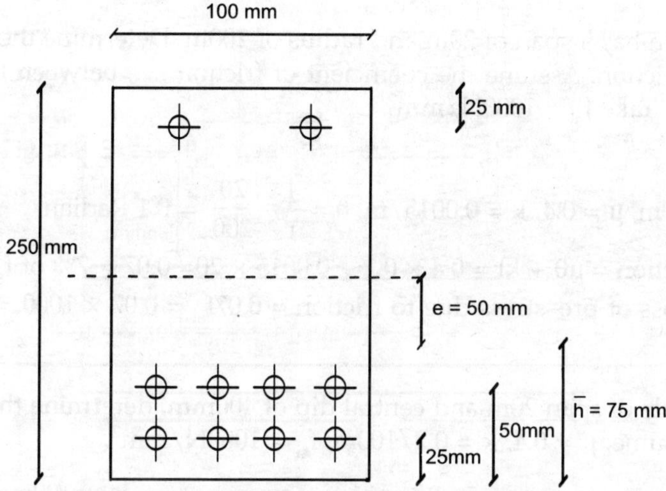

Fig. 3.7 Cross–section of Rectangular Beam

$$\text{Area of each wire} = \frac{\pi \times 5^2}{4} = 19.63\text{mm}^2, \; I = \frac{bd^3}{12} = \frac{100 \times 250^3}{12} = 130.20 \times 10^6 \text{mm}^4$$

So, area of 10 numbers = 10 × 19.63 = 196.3mm²

Taking the moment about the bottom of soffit of the beam,

$\bar{h} \times A_{st}$ = 4 × 19.63 × 25 + 4 × 19.63 × 50 + 2 × 19.63 × 225, $\bar{h}$ = 75mm

f_{si} = 1050N/mm² ; $F_o = f_{si} \times A_{st}$ = 1050 × 196.34 = 206.157kN

$$f_c \text{ (top wires)} = \frac{F_o}{A} + \frac{F_o e y_1}{I} = \frac{206.157 \times 10^3}{100 \times 250} + \frac{206.157 \times 10^3 \times (-50)(100)}{130.20 \times 10^6} = 0.329 \text{N/mm}^2$$

Loss of stress in wire at top = $mf_c = 6 \times 0.329 = 1.97 N/mm^2$.

$$\therefore \text{ Percentage loss of stress} = \frac{1.97 \times 100}{1050} = 0.188\%$$

$$f_c \text{ (bottom wires)} = \frac{F_o}{A} + \frac{F_o e y_2}{I} = \frac{206.157 \times 10^3}{100 \times 250} + \frac{206.157 \times 10^3 \times (-50)(-100)}{130.20 \times 10^6}$$

$$= 16.163 N/mm^2$$

Loss of stress in wires at bottom most = $mf_c = 6 \times 16.163 = 96.97 N/mm^2$

$$\therefore \text{ Percentage loss of stress for bottom} = \frac{96.97 \times 100}{1050} = 9.236\%$$

$$f_c \text{ (bottom wires)} = \frac{F_o}{A} + \frac{F_o e y_2}{I} = \frac{206.157 \times 10^3}{100 \times 250} + \frac{206.157 \times 10^3 \times (-50)(-75)}{130.20 \times 10^6} = 14.19 N/mm^2$$

Loss of stress in second level of bottom wires = $mf_c = 6 \times 14.19 = 85.14 N/mm^2$

$$\% \text{ loss of stress} = \frac{85.14 \times 100}{1050} = 8.1\%.$$

Problem 3.7

A curved tendon has a span of 20m and radius of 200m. Determine the percentage loss of pre–stress due to friction. Assume the coefficient of friction i.e., between tendon and duct is 0.4, k = 0.15/100m. Take $f_{si} = 1000 N/mm^2$.

Solution:

$$l = 20m,\ r = 200m,\ \mu = 0.4,\ k = 0.0015/m,\ \theta = \frac{l}{r} = \frac{20}{200} = 0.1 \text{ Radian}$$

Loss due to friction = $\mu\theta + kl = 0.4 \times 0.1 + 0.0015 \times 20 = 0.07$ = 7% of f_{si}

So the actual loss of pre–stress due to friction = $0.07 f_{si} = 0.07 \times 1000 = 70 N/mm^2$.

Problem 3.8

A parabolic cable of span 20m and central dip of 400mm, determine the loss of pre–stress due to friction. Assume, $\mu = 0.4$, k = 0.2/100m, $f_{si} = 1000 N/mm^2$.

Solution:

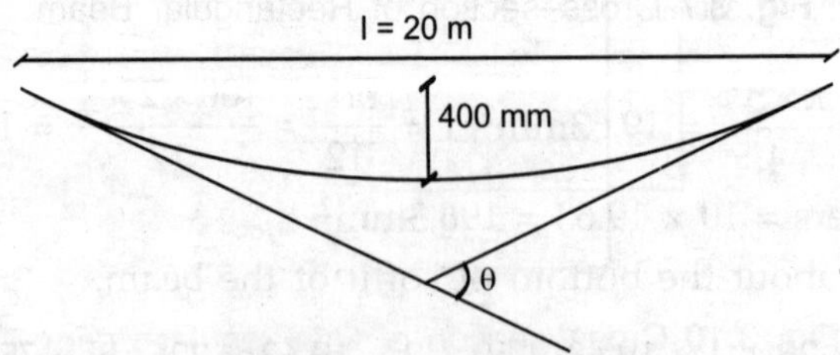

Fig. 3.8 Parabolic Cable

$$\theta = \frac{8r}{l} = \frac{8 \times 400}{20000} = 0.16 \text{ Radian}$$

Loss due to friction = $\mu\theta + kl = 0.4 \times 0.16 + 0.002 \times 20 = 0.104 = 10.4\%$ of f_{si}

Actual loss of pre–stress due to friction = $104 N/mm^2$.

Problem 3.9

A cable has a span of 16m and is cured at the ends at a slope of 1:20. Determine the loss of pre–stress due to friction. $\mu = 0.4$, $k = 0.15/100m$. Take $f_{si} = 1000 N/mm^2$.

Solution:

$\tan \alpha_A = \alpha_A = 1/20$ Radian, $\therefore\ \theta = \alpha_A + \alpha_B = 2 \times 1/20 = 0.1$ Radian

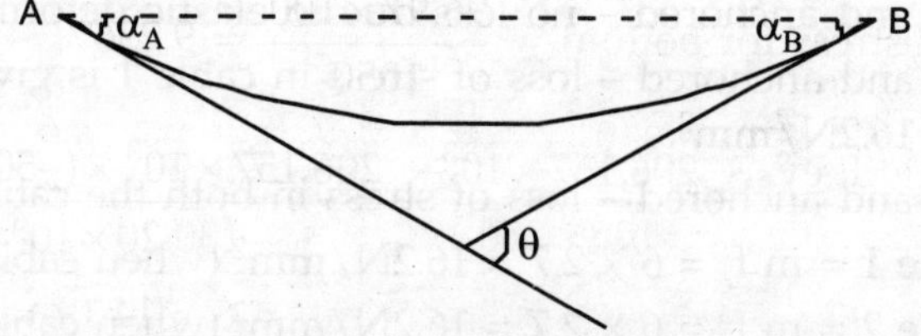

Fig. 3.9 Cable Having Slope of 1:20

Loss due to friction = $\mu\theta + kl = 0.4 \times 0.1 + 0.0015 \times 16 = 0.064 = 6.4\%$ of f_{si}.

Actual loss of pre–stress due to friction = $64 N/mm^2$.

Problem 3.10

A post–tensioned concrete beam 100mm wide and 300mm deep, is pre–stressed by three cables, each with a cross sectional area of $50 mm^2$ and with an initial stress of $1200 N/mm^2$. All the three cables are straight and located 50mm from the soffit of the beam. If the modular ratio is 6, calculate the loss of stress in the three cables due to elastic deformation of concrete for only the following cases.

(a) Simultaneous tensioning and anchoring of all the three cables.

(b) Successive tensioning of the three cables one at a time.

Solution:

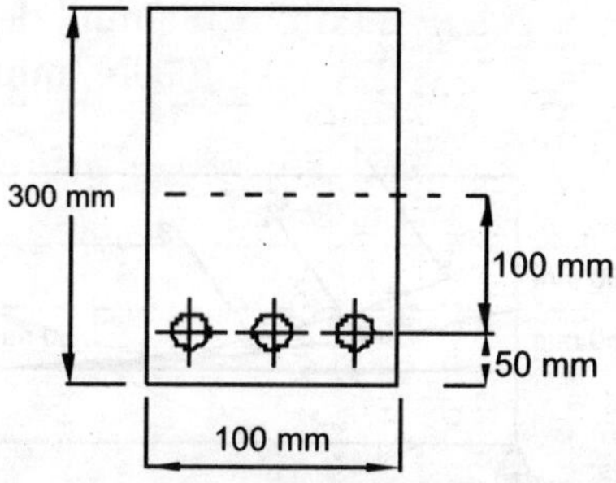

Fig. 3.10 Cross–section of Rectangular Beam

$f_{si} = 1200 N/mm^2$, F_o in each cable = $f_{si} \times A_{st} = 1200 \times 50 = 60 \times 10^3 N$

$$I = \frac{bd^3}{12} = \frac{100 \times 300^3}{12} = 225 \times 10^6 mm^4$$

$$f_c \text{ at level steel (each)} = \frac{F_o}{A} + \frac{F_o ey}{I} = \frac{60 \times 10^3}{100 \times 300} + \frac{60 \times 10^3 \times (-100)(-100)}{225 \times 10^6} = 4.66 N/mm^2$$

Case (a): Under simultaneous tensioning and anchoring of all the three cables there will be no loss due to elastic deformation of concrete.

Case (b): When the cables are succession tensioned

Cable 1 is tensioned and anchored – no loss due to elastic deformation.

Cable 2 is tensioned and anchored – loss of stress in cable 1 is given by, loss of stress in cable 1 = m f_c= 6 × 2.7 = 16.2N/mm².

Cable 3 is tensioned and anchored – loss of stress in both the cables 1 and 2 given by

Loss of stress in cable 1 = m f_c = 6 × 2.7 = 16.2N/mm² (when cable 3 is tensioned).

Loss of stress in cable 2 = m f_c = 6 × 2.7 = 16.2N/mm² (when cable 3 is tensioned).

Therefore total loss of stress due to elastic deformation of concrete in,

Cable 1 = 16.2 + 16.2 = 32.4N/mm². Cable 2 = 16.2N/mm². Cable 3 = 0.

So, average loss of stress considering all the 3 cables = 16.2N/mm².

Problem 3.11

A post–tensioned beam 100mm wide and 300mm deep spanning over 10m is stressed by successive tensioning and anchoring of three cables 1, 2 and 3 respectively. The cross sectional area of each cable is 200mm² and the initial stress in the cable is 1200N/mm², m = 6. The first cable is parabolic with an eccentricity of 50mm below the centroidal axis at the centre of span and 50mm above the centroidal axis at the support section. The second cable is parabolic with zero eccentricity at supports and an eccentricity 50mm at the centre of the span. The third cable is straight with a uniform eccentricity of 50mm below the centroidal axis. Estimate the percentage loss of stress in each of the cable if they are success only tensioned and anchored.

Solution:

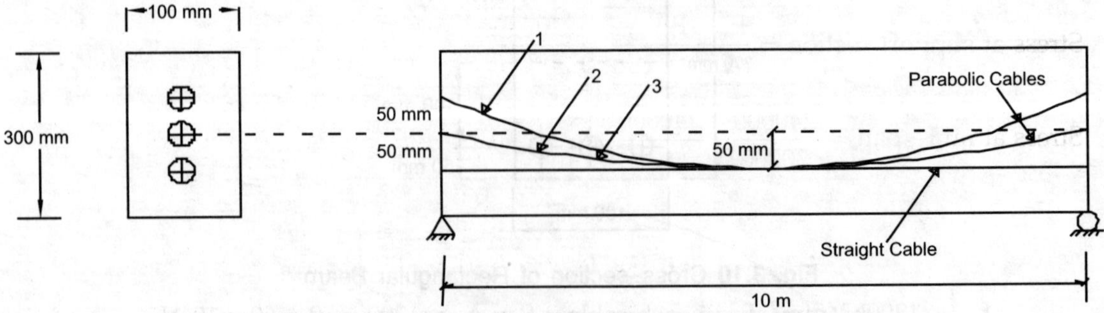

Fig. 3.11 Cross and Longitudinal Section of Rectangular Beam

Force in each cable, $F_o = f_{si} \times A_{st} = 1200 \times 200 = 240000N$, $A = 300 \times 100 = 30000mm^2$

$$m = 6, I = \frac{bd^3}{12} = \frac{100 \times 300^3}{12} = 225 \times 10^6 mm^4$$

When cable 1 tensioned and anchored – no loss of stress due to elastic deformation of concrete.

When cable 2 is tensioned and anchored, stress at the level of cable 1 is given by

$$\text{Stress at support section} = \frac{F_o}{A} + \frac{F_o ey}{I} = \frac{F_o}{A} + \frac{F_o y_1 y_2}{I}$$

$$= \frac{240000}{30000} + \frac{240000 \times 50 \times (0)}{225 \times 10^6} = 8.0N/mm^2$$

$$\text{Stress at centre of span} = \frac{F_o}{A} + \frac{F_o y_1 y_2}{I}$$

$$= \frac{240000}{30000} + \frac{240000 \times (-50) \times (-50)}{225 \times 10^6} = 10.7N/mm^2$$

Fig. 3.12 Stress Variation Diagram

Stress distribution at level of cable 1 when cable 2 is tensioned.

Average stress in concrete = $8.0 + 2/3 \times 2.7 = 9.8N/mm^2$.

Therefore loss of stress in cable 1 = $mf_c = 6 \times 9.8 = 58.8N/mm^2$.

When cable 3 is tensioned and anchored, stress distribution at the level of cable 1 and cable 2 and the average stress and the loss of stress is obtained as follows.

Cable 1:

$$\text{Stress at support section} = \frac{F_o}{A} + \frac{F_o y_1 y_3}{I} = \frac{240000}{30000} + \frac{240000 \times (50) \times (-50)}{225 \times 10^6} = 5.3N/mm^2$$

$$\text{Stress at mid–span} = \frac{240000}{30000} + \frac{240000 \times (50) \times (50)}{225 \times 10^6} = 10.7N/mm^2$$

Fig. 3.13 Stress Variation Diagram

So average stress = 5.3 + 2/3 × 5.4 = 8.9 N/mm²

Therefore, loss of stress in cable 1 = 6 × 8.9 = 53.4N/mm².

Cable 2:

$$\text{Stress at support section} = \frac{F_o}{A} + \frac{F_o y_2 y_3}{I} = \frac{240000}{30000} + \frac{240000 \times (0) \times (-50)}{225 \times 10^6} = 8\text{N/mm}^2$$

$$\text{Stress at mid–span} = \frac{240000}{30000} + \frac{240000 \times (50) \times (50)}{225 \times 10^6} = 10.7\text{N/mm}^2$$

So, average stress = 8 + 2/3 × 2.7 = 9.8 N/mm²

Loss of stress in cable 2 = 6 × 9.8 = 58.8N/mm²

For cable 1: Total loss = 58.8 + 53.4 = 112.2N/mm² = 9.34%.

For cable 2: Total loss = 58.8N/mm² = 4.9%.

For cable 3: Total loss = Zero percentage.

Problem 3.12

A post–tensioned beam is shown in Fig. 3.14 is pre–stressed with cable which consists of 24 parallel wires each of 7mm dia. The wires have to be tensioned two at a time from one end to have a value f_1. So as to over–come friction, then release to a value f_2, so that immediately after anchoring an initial stress of 1000N/mm² would be obtained. Hence compute the final design stress. Assume the following, creep of concrete = 2.2, m = 6, shrinkage strain of concrete = 1.5×10^{-4}, relaxation of steel = 3%, anchorage slip = 2mm/end, μ = 0.4, k = 0.15/100m, $E_s = 2.1 \times 10^5$N/mm².

Solution:

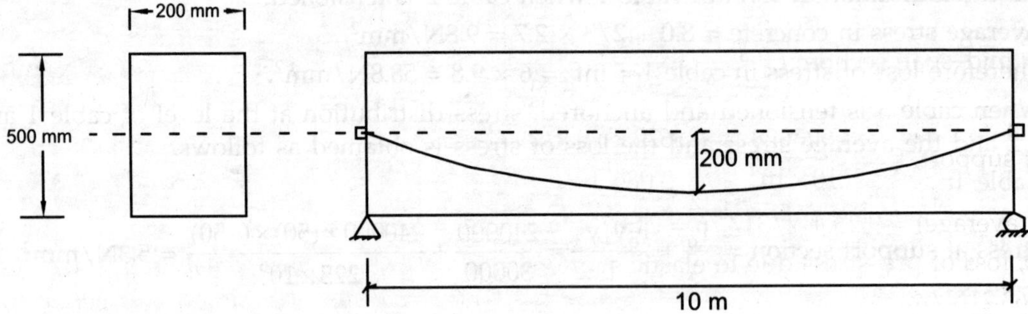

Fig. 3.14 Cross and Longitudinal Section of Rectangular Beam

$$l = 10\text{m},\ r = 200\text{m},\ \mu = 0.4,\ k = 0.0015/\text{m},\ \theta = 2\left(\frac{4r}{l}\right) = 2\left(\frac{4 \times 200}{10000}\right) = 0.16 \text{ Radian}$$

$$f_{si} = 1000\text{N/mm}^2$$

Loss of pre–stress due to friction = μθ + kl = 0.4 × 0.16 + 0.0015 × 10 = 0.079= 7.9% of f_{si}

Normally the loss of pre–stress due to friction is expressed as percentage of f_{si}. However, when the wires are initially stressed to a value of f_1 it is greater than f_{si}. The effective stress available in tendon after frictional loss = 92.1% of f_1. This should be equal to designed initial pre–stress f_{si}; i.e., $0.921f_1 = f_{si}$.

Therefore, $f_1 = \frac{f_{si}}{0.921} = \frac{1000}{0.921} = 1086\text{N/mm}^2$

Anchorage slip, $\delta_a = 2\text{mm}$

Loss of pre–stress due to anchorage slip = $\frac{\delta_a}{l} \times E_s = \frac{2}{10000} \times 2 \times 10^5 = 42\text{N/mm}^2$

Thus $f_2 = f_{si} + 42 = 1000 + 42 = 1042\text{N/mm}^2$

Hence, the wires have to be initially tensioned to a value $f_1 = 1086\text{N/mm}^2$ and then released to a value $f_2 = 1042\text{N/mm}^2$. So that an initial stress of 1000N/mm^2 would be obtained immediately after anchoring at the jacking end.

To compute the final design stress or effective pre–stress available in tensioned steel since the frictional loss and anchorage slip have been taken into account during tensioning itself. We have to consider the loss of pre–stress due to remaining factors.

$A = 200 \times 500 = 10 \times 10^4\text{mm}^2$, $I = \frac{bd^3}{12} = \frac{200 \times 500^3}{12} = 2.08 \times 10^9\text{mm}^4$,

$A_{st} = 24 \times \frac{\pi \times 7^2}{4} = 923.62\text{mm}^2$, $f_{si} = 1000\text{N/mm}^2$, $F_o = 1000 \times 923.62 = 923628\text{N}$,

$C_c = 2.2$, $m = 6$, $E_s = 2.1 \times 10^5\text{N/mm}^2$, $\delta_s = 1.5 \times 10^{-4}$

Let f_{c1} and f_{c2} represents the compressive stresses in concrete fibres laying at the level of CGC at the mid–span section at the support.

At mid–span section, $f_{c1} = \frac{F_o}{A} + \frac{F_o ey}{I} = \frac{923628}{10 \times 10^4} + \frac{923628 \times (-200) \times (-200)}{2.08 \times 10^9} = 26.99\text{N/mm}^2$

At support, $f_{c2} = \frac{923628}{10 \times 10^4} + \frac{923628 \times (0) \times (0)}{2.08 \times 10^9} = 9.23\text{N/mm}^2$.

f_c (average) = 9.23 + 2/317.76 = 21.07N/mm^2.

So, loss of pre–stress due to elastic shortening of concrete = ½ mf_c = ½ × 6 × 21.07 = 63.21N/mm^2.

Loss of pre–stress due to creep = $(C_c - 1)\, mf_c = (2.2 - 1) \times 6 \times 21.07 = 151.70\text{N/mm}^2$.

Loss of pre–stress due to shrinkage of concrete = $\delta_s E_s = 1.5 \times 10^{-4} \times 2.1 \times 10^5 = 31.5\text{N/mm}^2$.

Loss of pre–stress due to relaxation of steel= 3% = 0.03 of $f_{ci} = 0.03 \times 1000 = 30\text{N/mm}^2$.

∴ Total loss of pre–stress = 63.21 + 151.70 + 31.5 + 30 = 276.41N/mm^2.

∴ Effective pre–stress available in tensioned steel = 1000–276.4 = 723.59N/mm^2.

Note: In the above example it is given that the wires are tensioned from one end. However if the wires are tensioned from both the ends, then for computing the loss due to anchorage slip, we must consider one half of the length of the tendon. While computing the frictional loss we must consider the change of slope from one end up to the centre of the cable.

REVIEW QUESTIONS

3.1 List the various types of loss of pre–stress in pre–tensioned and post–tensioned members.
3.2 Explain the various causes of loss in pre–stress.
3.3 Explain the nature of losses of pre–stress.
3.4 Discuss losses of pre–stress according to IS: 1343–2012.
3.5 How do you compute the loss of stress due to elastic shortening of concrete?
3.6 "Post–tensioned members do not suffer the loss of pre–stress due to elastic shortening". Why?
3.7 How do you compute the loss of stress due to elastic deformation of concrete in post–tensioned members with several cables which are successively tensioned?
3.8 How do you compute friction loss due to 'wave' and 'curvature' effects?
3.9 Explain counteracting friction losses.
3.10 What is anchorage slip? Discuss loss of pre–stress due to slip of anchorage components.
3.11 How do you compute loss of pre–stress due to creep of concrete?
3.12 Discuss loss of pre–stress due to shrinkage of concrete as per IS: 1343–2012.
3.13 What is relaxation of stress in steel? How do it account for it in pre–stressed members? Explain the provisions made in IS: 1343 for relaxation loss.
3.14 "The Indian Standard Code IS: 1343 specifies different strains for pre–tensioned and post–tensioned members". Explain with reasons?

EXERCISE PROBLEMS

3.1 A pre–tensioned beam of rectangular cross–section, 150mm wide and 300mm deep, is pre–stressed by 8–7 mm wires located 100mm from the soffit of the beam. If the wires are initially tensioned to a stress of 1100N/mm^2. Calculate their stress at transfer and the effective stress after all losses, given the following data:

	Up to time of transfer	**Total**
Relaxation of steel	35N/mm^2	70N/mm^2
Shrinkage of concrete	100×10^{-6}	300×10^{-6}
Creep coefficient	——	1.6

E_s = 210kN/mm^2, E_c = 31.5kN/mm^2

[Ans: Stress in steel at transfer = 977.5N/mm^2 and stress in steel after all losses = 793.6N/mm^2]

3.2 A pre–stressed concrete pile of cross section, 250mm × 250mm contains 60 pre–tensioned wires each of 2mm diameter, uniformly distributed over the section. The wires are initially tensioned on the pre–stressing bed with a total force of 300kN. If E_s = 210kN/mm² and E_c = 32kN/mm², calculate the respective stresses in steel and concrete immediately after the transfer of pre–stress assuming that up to this point, the only loss of stress is that due to elastic shortening.

If the concrete undergo a further shortening due to a shrinkage of 200×10^{-6} per unit length, while there is a relaxation of 5% of steel stress due to creep of steel, find the greatest tensile stress which can occur in a pile, 20m long when lifted at two points 4m from each end. Assume creep coefficient as 1.6.

[Ans: Final stress in steel = 1389.3N/mm², stress in concrete = 4.18N/mm² and tensile stress in concrete = –0.42N/mm²]

3.3 A post–tensioned cable of a beam 10m long is initially tensioned to a stress of 1000N/mm² at one end. If the tendons are curved so that the slope is 1 in 24 at each end with an area of 600mm². Calculate the loss of pre–stress due to friction for the following data: Coefficient of friction between the duct and cable = 0.55, friction coefficient for wave effect = 0.0015/m, during anchoring, if there is a slip of 3mm at the jacking end; calculate the final force in the cable and the percentage loss of pre–stress due to friction and slip.

[Ans: Loss of stress due to friction = 60N/mm², final force in the cable = 526.2N and percentage loss of pre–stress due to friction and slip = –0.42N/mm²]

3.4 A pre–tensioned beam 250mm wide and 300mm deep is pre–stressed by 12 wires each 7mm diameter initially stressed to 1200N/mm² with their centroids located 100mm from the soffit. Estimate the final percentage loss of stress due to elastic deformation, creep, shrinkage and relaxation using IS: 1343–2012 and the following data: Relaxation of steel stress = 90N/mm², E_s = 210kN/mm², E_c = 35kN/mm², creep coefficient = 1.6 and residual shrinkage strain = 3×10^{-4}.

[Ans: Loss due to elastic deformation = 54N/mm², loss due to relaxation = 60N/mm², loss due to shrinkage = 63N/mm² loss of stress in steel due to creep of concrete = 86N/mm²]

3.5 In a post–tensioned beam of length 12m, a cable is laid symmetrically with its central 6m length and the two straight end portions sloping up at an angle with horizontal whose tangent is equal to 0.075. The cable is tensioned by jacking at one end and anchored at the remote end of the beam. At the jacking end the measured stress is 1040N/mm². The 'wobble' coefficient 'k' may be assumed as 0.004/m. Calculate the stress in the cable at the remote end and the two points where the alignment of the cable changes. Assume the coefficient of friction between the cable and duct is 0.40. What is the percentage loss of pre–stress between the jacking end and the anchoring end?

[Ans: Stress in the cable at the remote end = 928N/mm², stress at the two points where the alignment of the cable changes = 44N/mm², percentage loss of pre–stress = 10.8%]

3.6 A post–tensioned pre–stressed concrete beam of span length 10m has a rectangular section 300mm × 800mm. The beam is pre–stressed by a parabolic cable concentric at the support and with an eccentricity of 250mm at the centre of the span. The cross sectional area of high tensile wires in the cable is 500mm². The wires are stressed by using a jack at the left end so that the initial force in the cable at the right end is 250kN. Using the following data, calculate the total loss of stress in the wires.
Coefficient of friction for curvature effect = 0.55,
Friction coefficient for wave effect = 0.003/m,
Anchorage slip at the jacking end = 3mm, Relaxation of steel stress = 4%,
Shrinkage of concrete = 0.0002, Creep coefficient = 2.2,
Modulus of elasticity of steel = 210kN/mm²,
Modulus of elasticity of concrete = 35kN/mm²,
Modulus ratio = m = 6. **[Ans: Total loss of stress in steel = 235N/mm²]**

3.7 A rectangular beam of 180mm wide by 400mm deep is simply supported over a span of 8m and is reinforced with 3 wires of 8mm diameter. The wires are located at the constant eccentricity of 80mm and are subjected to an initial stress of 1200N/mm². Calculate percentage loss of pre–stress in the wires if the beam is (a) pre–tensioned and (b) post–tensioned. Relaxation of steel stress = 6%, E_s = 210kN/mm², modular ratio = 6, slip at anchorage = 0.8mm and friction coefficient = 0.002/m. Adopt shrinkage and creep coefficients from the IS: 1343–2012 specifications.

[Ans: Percentage loss of stress in the wires when the beam is (a) pre–tensioned = 16.05% and (b) post–tensioned = 15.80%]

3.8 A concrete beam of span 10m is post–tensioned by a cable with an initial stress of 1200N/mm² from on end. The cable is concentric at supports and has an eccentricity of 100mm at the centre of the span. Coefficient of friction due to curvature effect is 0.5 and friction coefficient for wobble effect = 0.003/m. Compute loss of stress in the tendons assuming the cable to have (a) Circular profile (b) Parabolic profile.

[Ans: Loss of stress in the tendons when the cable to have (a) Circular profile = 84N/mm² (b) Parabolic profile = 84N/mm²]

3.9 A pre–stressed concrete girder of 30m span is pre–stressed by a parabolic cable concentric at supports and having an eccentricity of 850mm at the centre of the span. The effective stress in the tendon is 1200N/mm². The coefficient of friction between the tendon and the cable is 0.5 and the friction coefficient for the wave effect is 0.0015/m. if the anchorage slip is 5mm; compute the loss of stress in tendons due to friction and anchorage slip. Assume E_s = 210kN/mm².

[Ans: Loss of stress in tendons due to friction and anchorage slip = 225N/mm²]

3.10 A pre–stressed concrete girder of 30m span is pre–stressed by a parabolic cable concentric at supports and having an eccentricity of 1000mm at the centre of span. The stress in the wire is 1200N/mm², ultimate shrinkage strain = 0.0002, E_s = 210kN/mm², coefficient of friction = 0.5, anchorage slip = 8mm. Estimate the percentage loss of stress in the wires.

[Ans: Percentage loss of stress in wires = 17%]

3.11 A pre–stressed concrete beam of section 300mm wide and 600mm deep is pre–stressed by 5 No.s of 8mm high tensile wires at 500mm from top and 15 No.s of 8mm high tensile wires at 83mm from the soffit. The stress in the wire is 1200N/mm^2. Estimate the loss of stress in the wires due to elastic shortening. Assume modular ratio = m = 6.

[Ans: Loss of stress in wiresdue to elastic shortening, a. Top wires = 6.66N/mm^2 and b. Bottom wires = 68.88N/mm^2]

3.12 A concrete beam is post–tensioned by a cable carrying an initial stress of 1200N/mm^2. The slip at the jacking end is 8mm. E_s = 210kN/mm^2. Estimate the percentage loss of stress due to anchorage slip only if the length of the beam is (a) 10m and (b) 50m.

[Ans: Percentage loss of stress due to anchorage slip, For 10m length beam = 14% and for 50m length beam = 2.8%]

OBJECTIVE QUESTIONS

3.1 The immediate type of loss of pre–stress is
- a. friction
- b. creep
- c. shrinkage
- d. relaxation

3.2 The time dependent type of loss of pre–stress is
- a. friction
- b. elastic shortening
- c. anchorage slip
- d. relaxation

3.3 Relaxation of stress in steel the type of loss of pre–stress for
- a. pre–tensioning
- b. post–tensioning
- c. Both a and b
- d. Neither a nor b

3.4 Shrinkage of concrete is the type of loss of pre–stress for
- a. pre–tensioning
- b. post–tensioning
- c. Both a and b
- d. Neither a nor b

3.5 Creep of concrete is the type of loss of pre–stress for
- a. pre–tensioning
- b. post–tensioning
- c. Both a and b
- d. Neither a nor b

3.6 Friction is the type of loss of pre–stress for
- a. pre–tensioning
- b. post–tensioning
- c. Both a and b
- d. Neither a nor b

3.7 Anchorage slip is the type of loss of pre–stress for
- a. pre–tensioning
- b. post–tensioning
- c. Both a and b
- d. Neither a nor b

3.8 Modular ratio (m) is given by
- a. E_s/E_c
- b. E_c/E_s
- c. $E_s \times E_c$
- d. None of the above

3.9 Loss of stress in steel due to elastic shortening of concrete for in–uniformly pre–stressed pre–tensioned member is given by

a. $0.25\ mf_c$ b. $0.50\ mf_c$ c. $0.75\ mf_c$ d. mf_c

3.10 Loss of stress in steel due to elastic shortening of concrete for in–uniformly pre–stressed post–tensioned member is given by

a. $0.25\ mf_c$ b. $0.50\ mf_c$ c. $0.75\ mf_c$ d. mf_c

3.11 As per IS: 1343–2012, friction loss due to 'wave' and 'curvature' effects is given by

a. $P = P_o\, e^{-(\mu\alpha + kx)}$ b. $P = P_o\, e^{(\mu\alpha + kx)}$

c. $P = P_o\, e^{-(\mu\alpha - kx)}$ d. $P = P_o\, e^{(\mu\alpha - kx)}$

3.12 Loss of stress in steel due to anchorage slip is given by

a. $\delta_a E_s l$ b. $E_s / l\, \delta_a$ c. $\delta_a / l\, E_s$ d. $\delta_a E_s / l$

3.13 Loss of stress in steel due to creep of concrete is given by

a. $(C_c - 1)\, f_c\, m$ b. $C_c f_c m$ c. $(1 - C_c)\, f_c\, m$ d. $f_c\, m$

3.14 Creep coefficient which varies from

a. 1 to 2 b. 2 to 3 c. 3 to 3.5 d. 3.5 to 4

3.15 Loss of stress in steel due to shrinkage of concrete is given by

a. $e_{sh} E_c$ b. e_{sh} / E_s c. $e_{sh} E_s$ d. e_{sh}/E_c

3.16 Shrinkage strain of concrete may usually take as

a. 2×10^{-4}cm/cm b. 2.5×10^{-4}cm/cm

c. 3×10^{-4}cm/cm d. 3.5×10^{-4}cm/cm

3.17 Young's modulus of elasticity of steel usually taken as

a. 2.1×10^{5}N/mm^2 b. 2.1×10^{6}N/mm^2

c. 2.1×10^{7}N/mm^2 d. 2.1×10^{8}N/mm^2

3.18 Relaxation loss for $0.7f_p$ characteristic strength of pre–stressing steel is given by

a. 70MPa b. 90MPa c. 0MPa d. 35MPa

3.19 Relaxation loss for $0.5f_p$ characteristic strength of pre–stressing steel is given by

a. 70MPa b. 90MPa c. 0MPa d. 35MPa

3.20 Relaxation loss for $0.6f_p$ characteristic strength of pre–stressing steel is given by

a. 70MPa b. 90MPa c. 0MPa d. 35MPa

3.21 Relaxation loss for $0.8f_p$ characteristic strength of pre–stressing steel is given by

a. 70MPa b. 90MPa c. 0MPa d. 35MPa

3.22 According to IS: 1343–2012, loss of pre–stress due to relaxation of steel at 27°C is

a. 1000 hours b. 2000 hours c. 3000 hours d. 4000 hours

3.23 According to IS: 1343–2012, creep loss of pre–stress due to .... may be ignored.

a. live load stresses b. erection stresses

c. other stresses of short duration d. all of the above

3.24 According to IS: 1343–2012, anchorage slip may vary from

a. 1 to 6mm b. 1 to 2mm

c. 2 to 3mm d. 3 to 4mm

3.25 In circular construction such as tanks where circumferential tendons are tensioned by jacks, value of μ for calculating friction be taken as……… for steel moving on smooth concrete

a. 0.25 b. 0.35 c. 0.45 d. 0.55

3.26 Coefficient for wave effect varying from

a. 15×10^{-4} to 50×10^{-4} per metre b. 25×10^{-4} to 40×10^{-4} per metre
c. 10×10^{-4} to 35×10^{-4} per metre d. 5×10^{-4} to 20×10^{-4} per metre

3.27 The main types of frictional losses are due to ………..

a. curvature effects b. wave effects
c. counteraction d. All of the above

3.28 As per T. Y. Lin, the total loss of pre–stress for post–tensioned members is ……… %

a. 15 b. 20 c. 25 d. 30

3.29 As per T. Y. Lin, the total loss of pre–stress for pre–tensioned members is ……… %

a. 15 b. 20 c. 25 d. 30

3.30 According to Lin, If f_{pe} = effective stress in tendons after losses, f_{pi}= stress in tendon at transfer, what is the formula for (reduction factor for loss of pre–stress)

a. $\left(\frac{f_{pe}}{f_{pi}}\right)$ b. $\left(\frac{f_{pi}}{f_{pe}}\right)$ c. $\left(\frac{f_{pe}-f_{pi}}{f_{pi}}\right)$ d. $\left(\frac{f_{pe}}{f_{pi}-f_{pe}}\right)$

3.31 According to Lin, the value of (reduction factor for loss of pre–stress) for pre–tensioning is ………..

a. 0.70 b. 0.75 c. 0.80 d. 0.85

3.32 According to Lin, the value of (reduction factor for loss of pre–stress) for post–tensioning is ………..

a. 0.70 b. 0.75 c. 0.80 d. 0.85

Answers

3.1 a	3.2 d	3.3 c	3.4 c	3.5 c
3.6 b	3.7 b	3.8 a	3.9 d	3.10 b
3.11 a	3.12 d	3.13 a	3.14 b	3.15 c
3.16 a	3.17 a	3.18 a	3.19 c	3.20 d
3.21 b	3.22 a	3.23 d	3.24 a	3.25 c
3.26 a	3.27 d	3.28 b	3.29 c	3.30 a
3.31 b	3.32 c			

❑

CHAPTER

4

Analysis and Design of Sections for Shear

4.1 ANALYSIS OF SECTIONS FOR SHEAR

4.1.1 Introduction

The analysis of reinforced concrete and pre–stressed concrete members for shear is more difficult compared to the analyses for axial load or flexure.

The analysis for axial load and flexure are based on the following principles of mechanics.

1. Equilibrium of internal and external forces
2. Compatibility of strains in concrete and steel
3. Constitutive relationships of materials.

The conventional analysis for shear is based on equilibrium of forces by a simple equation. The compatibility of strains is not considered. The constitutive relationships (relating stress and strain) of the materials, concrete or steel, are not used. The strength of each material corresponds to the ultimate strength.

The strength of concrete under shear although based on test results, is empirical in nature.

Shear stresses generate in beams due to bending or twisting. The two types of shear stress are called flexural shear stress and torsional shear stress, respectively.

To understand flexural shear stress, the behavior of a simply supported beam under uniformly distributed load, without pre–stressing, will be explained first. The effect of pre–stressing force will be subsequently introduced. The presentation will be in the following sequence. The design of shear reinforcement in PSC beams is carried out by conventional methods employed or in RCC.

1. Stresses in an un–cracked (homogenous) beam.
2. Types of cracks generated due to the combination of flexure and shear.
3. Components of shear resistance and the modes of failure.
4. Effect of pre–stressing force.

4.1.1.1 Stresses in an Un–cracked Beam

The following Fig. 4.1 shows the variations of shear and moment along the span of a simply supported beam under a uniformly distributed load. The variations of normal stress and shear stress along the depth of a section of the beam are also shown.

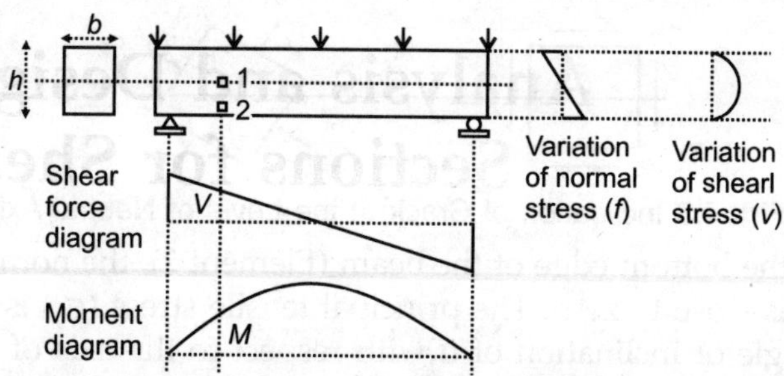

Fig. 4.1 Variation of Forces and Stresses in Simply Supported Beam

Under a general loading, the shear force and the moment vary along the length. The normal stress and the shear stress vary along the length, as well as along the depth. The combination of the normal and shear stresses generate a two–dimensional stress field at a point. At any point in the beam, the state of two–dimensional stresses can be expressed in terms of the principal stresses. The Mohr's circle of stress is helpful to understand the state of stress.

Before cracking, the stress carried by steel is negligible. When the principal tensile stress exceeds the cracking stress, the concrete cracks and there is redistribution of stresses between concrete and steel. For a point on the neutral axis (Element 1), the shear stress is maximum and the normal stress is zero. The principal tensile stress (σ_1) is inclined at 45° to the neutral axis. The following Fig. 4.2 shows the state of in–plane stresses.

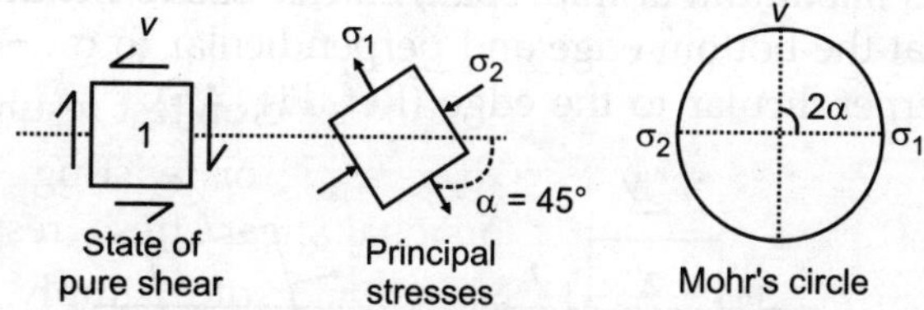

Fig. 4.2 State of Stresses at a Point on the Neutral Axis of a Beam

At the level of neutral axis, the normal stress is zero and the shear stress is maximum. An element at that level is under pure shear. A state of pure shear can be conceived as a state of biaxial tensile–compressive stresses. These principle stresses are inclined at 45° with respect to the axis of the beam. It is necessary to study the principle stresses to understand the cracking of concrete. The Mohr's circle is a representation of the state of in–plane stresses on surfaces of various inclinations passing through a point. The horizontal and vertical axes represent the normal and shear stresses, respectively. For a state of pure shear, the centre of the Mohr's circle coincides with the origin of the axes. It is expected that the reader is familiar with these concepts from a course in strength of materials.

Since the shear force is maximum near the supports, cracks due to shear occur near the supports. The cracks are formed around the neutral axis and perpendicular to the principal tensile stress (σ_1). The cracks are thus inclined at 45° to the axis of the beam. The following Fig. 4.3 shows the inclination of the cracks forming at the neutral axis.

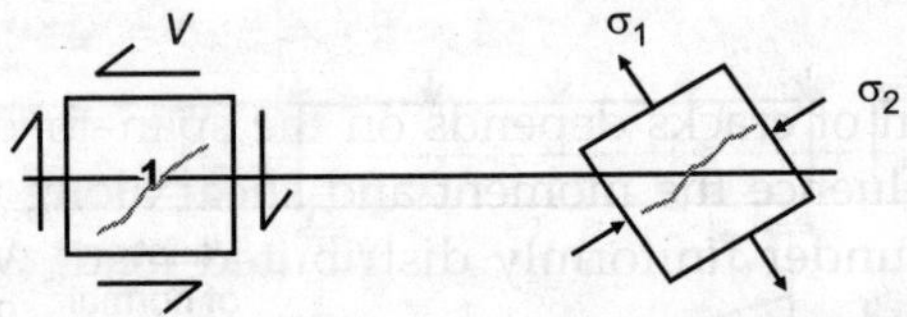

Fig. 4.3 Inclination of Crack at the Level of Neutral Axis

For a point near the bottom edge of the beam (Element 2), the normal stress is maximum and the shear stress is close to zero. The principal tensile stress (σ_1) is almost parallel to the bottom edge. The angle of inclination of σ_1 with respect to the axis of the beam (σ) is much smaller than 45°. The following Fig. 4.4 shows the state of in plane stresses.

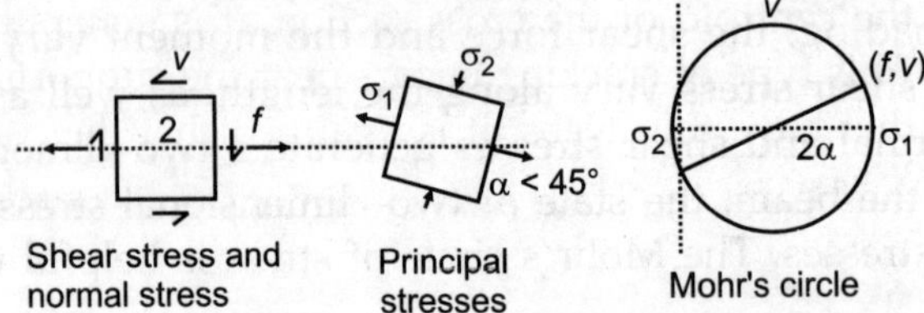

Fig. 4.4 State of Stresses at a Point Close to the Edge under Tension

Adjacent to the bottom edge (edge under tension), the tensile stress due to flexure is maximum and the shear stress is zero. The state of stress is nearly uni–axial tensile stress. The principal compressive stress is negligible. The Mohr's circle is shifted towards the axis of principal tensile stress.

Since the moment is maximum at mid–span, cracks due to flexure occur near mid–span. The cracks are formed at the bottom edge and perpendicular to σ_1. Since σ_1 is parallel to the edge, the cracks are perpendicular to the edge (Ref. Fig. 4.5).

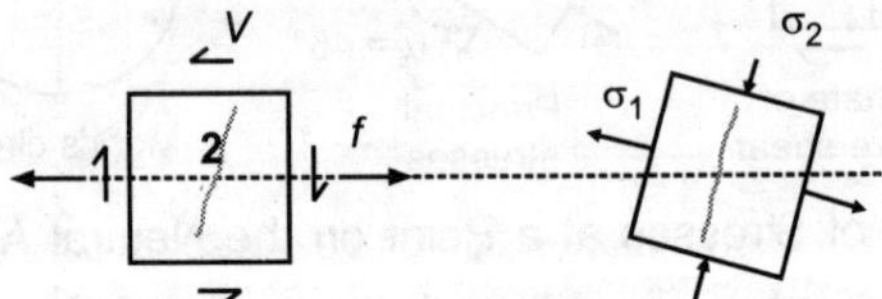

Fig. 4.5 Inclination of Crack Close to the Edge under Tension

The previous concepts can be used to develop the principal stress trajectories. The following Fig. 4.6 shows the trajectories for a simply supported beam under a uniformly distributed load. The crack pattern can be predicted from these trajectories.

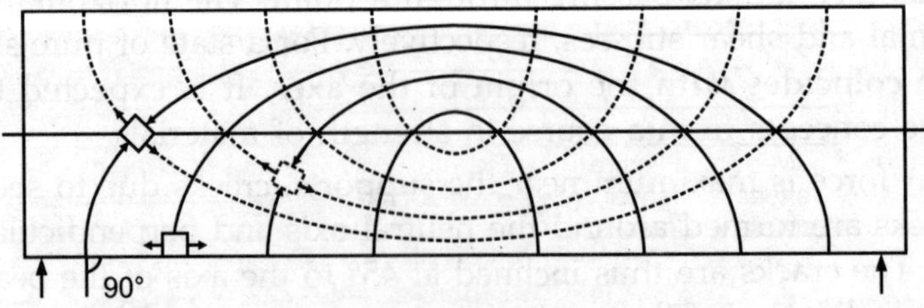

Fig. 4.6 Principle Stress Trajectories

4.1.1.2 Types of Cracks

The types and formation of cracks depends on the span–to–depth ratio of the beam and loading. These variables influence the moment and shear along the length of the beam. For a simply supported beam under uniformly distributed load, without pre–stressing, three types of cracks are identified.

- **Flexural Cracks:** These cracks form at the bottom near the mid–span and propagate upwards.
- **Web Shear Cracks:** These cracks form near the neutral axis close to the support and propagate inclined to the beam axis. This type of cracking is associated with principal tensile stresses where there are no flexural cracks. Since the maximum principal tensile stress occurs at the centroid of the cross section, it is generally sufficient to check at this point only. (This type of cracking occurs near the supports of highly pre–stressed concrete beams that have thin webs. It is assumed that the concrete cracks when the maximum principal tensile stress exceeds a value $f_t = 0.29\sqrt{f_{ck}}$).
- **Flexure Shear Cracks:** These cracks form at the bottom due to flexure and propagate due to both flexure and shear. This type of cracking is associated with flexural cracking and it may adversely affect the capacity of a member. (In addition, principal tensile stresses along a potential crack may aggravate the cracking). These flexural shear cracks begin as flexural cracks extending approximately vertically into the member. With the combination of flexural and shear cracks at the top of a flexural crack, the inclined flexural shear crack forms.

In the following Fig. 4.7, the formation of cracks for a beam with large span–to–depth ratioand uniformly distributed loading is shown.

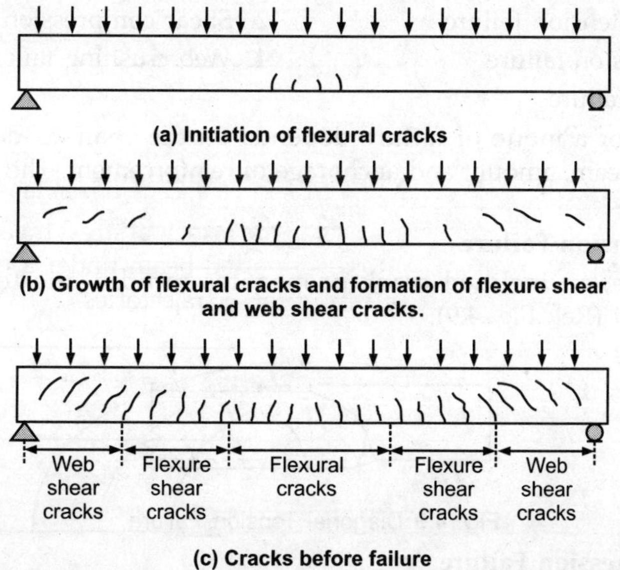

Fig. 4.7 Formation of Cracks in a Reinforced Concrete Beam

4.1.1.3 Components of Shear Resistance

The components of shear resistance are studied based on the internal forces at a flexure shear crack. The forces are shown in the following Fig. 4.8.

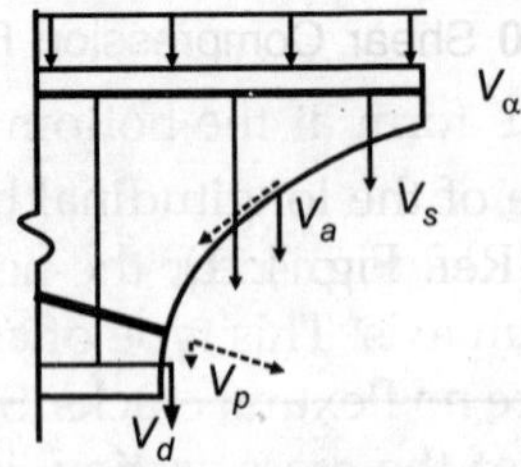

Fig. 4.8 Internal Forces at a Flexure Shear Crack

The notations in the previous Fig. 4.8 are as follows.

V_{cz} = Shear carried by un–cracked concrete
V_a = Shear resistance due to aggregate interlock
V_d = Shear resistance due to dowel action
V_s = Shear carried by stirrups
V_p = Vertical component of pre–stressing force in inclined tendons

The magnitude and the relative value of each component change with increasing load.

4.1.1.4 Modes of Failure

For beams with low span–to–depth ratio or inadequate shear reinforcement, the failure can be due to shear. A failure due to shear is sudden as compared to a failure due to flexure. The following five modes of failure due to shear are identified.

1. Diagonal tension failure
2. Shear compression failure
3. Shear tension failure
4. Web crushing failure
5. Arch rib failure

The occurrence of a mode of failure depends on the span–to–depth ratio, loading, cross–section of the beam, amount and anchorage of reinforcement. The modes of failure are explained next.

❖ **Diagonal Tension Failure**

In this mode, an inclined crack propagates rapidly due to inadequate shear reinforcement (Ref. Fig. 4.9).

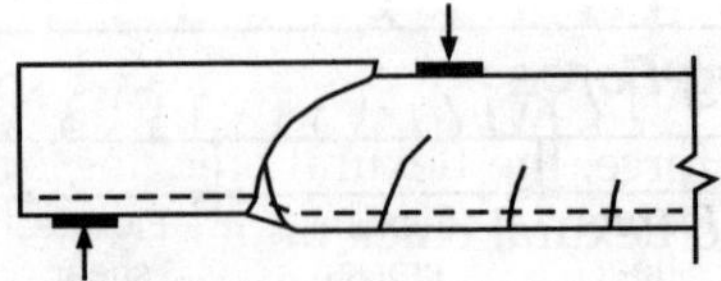

Fig. 4.9 Diagonal Tension Failure

❖ **Shear Compression Failure**

There is crushing of the concrete near the compression flange above the tip of the inclined crack (Ref. Fig. 4.10).

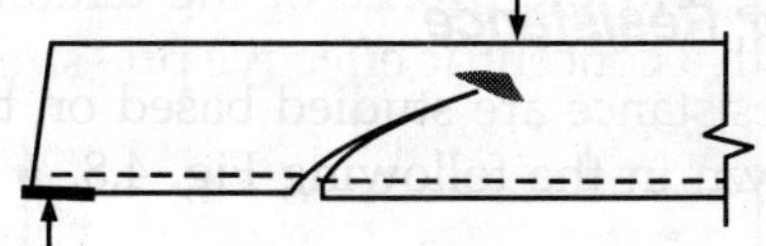

Fig. 4.10 Shear Compression Failure

❖ **Shear Tension Failure**

Due to inadequate anchorage of the longitudinal bars, the diagonal cracks propagate horizontally along the bars (Ref. Fig. 4.11).

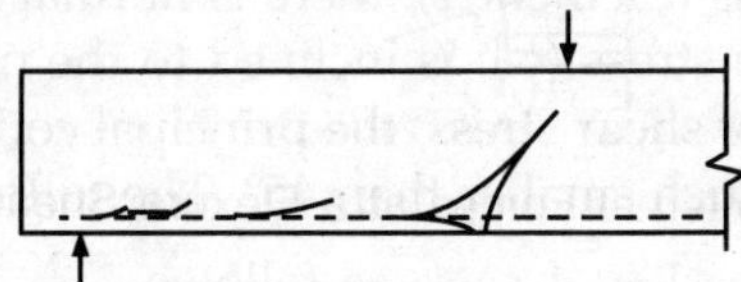

Fig. 4.11 Shear Tension Failure

❖ **Web Crushing Failure**

The concrete in the web crushes due to inadequate web thickness (Ref. Fig. 4.12).

Fig. 4.12 Web Crushing Failure

❖ **Arch Rib Failure**

For deep beams, the web may buckle and subsequently crush. There can be anchorage failure or failure of the bearing (Ref. Fig. 4.13).

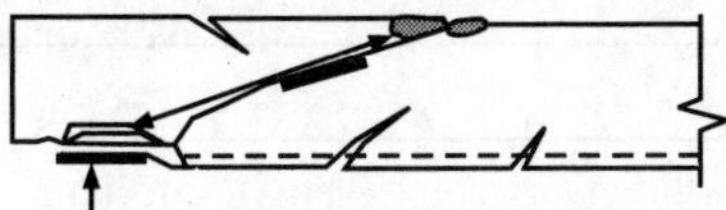

Fig. 4.13 Arch Rib Failure

The objective of design for shear is to avoid shear failure. The beam should fail in flexure at its ultimate flexural strength. Hence, each mode of failure is addressed in the design for shear. The design involves not only the design of the stirrups, but also limiting the average shear stress in concrete, providing adequate thickness of the web and adequate development length of the longitudinal bars.

4.1.1.5 Effect of Pre–stressing Force

In presence of pre–stressing force, the flexural cracking occurs at a higher load. For Type 1 and Type 2 sections, there is no flexural crack under service loads. This is evident from the typical moment versus curvature curve for a pre–stressed section. In presence of pre–stressing force, the web shear cracks also generate under higher load.

With increase in the load beyond the cracking load, the cracks generate in a similar sequence. But, the inclinations of the flexure shear and web shear cracks are reduced depending on the amount of pre–stressing and the profile of the tendon. The effect of pre–stressing force is explained for a beam with a concentric effective pre–stressing force (P_e) (Ref. Fig. 4.14).

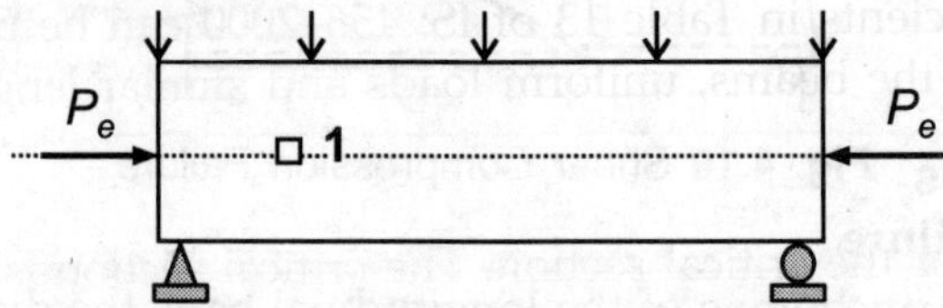

Fig. 4.14 Simply Supported Beam under Concentric Pre–stress and Uniformly Distributed Loads

For a point at the neutral axis (Element 1), there is normal stress due to the pre–stressing force ($-f_{pe}$). The principal tensile stress (σ_1) is inclined to the neutral axis at an angle greater than 45°. With the combination of shear stress, the principal compressive stress (σ_2) is inclined to the neutral axis at an angle much smaller than 45°. The following Fig. 4.15 shows the state of in–plane stresses.

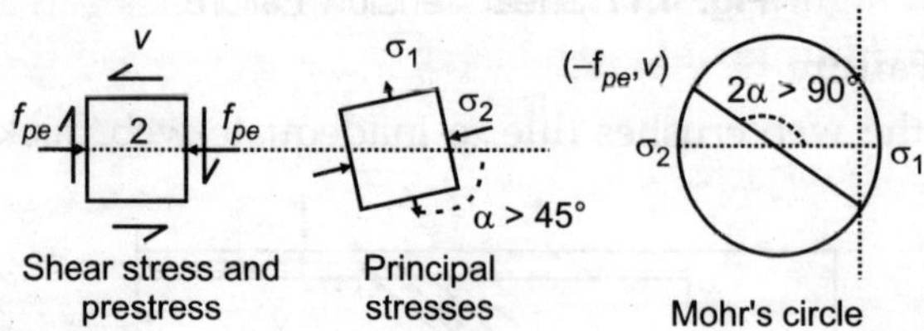

Fig. 4.15 State of Stresses at a Point on the Neutral Axis for a Pre–stressed Beam

The formation of cracks for a pre–stressed beam with large span–to–depth ratio and uniformly distributed loading is shown. This Fig. 4.16 can be compared with that for a reinforced concrete beam.

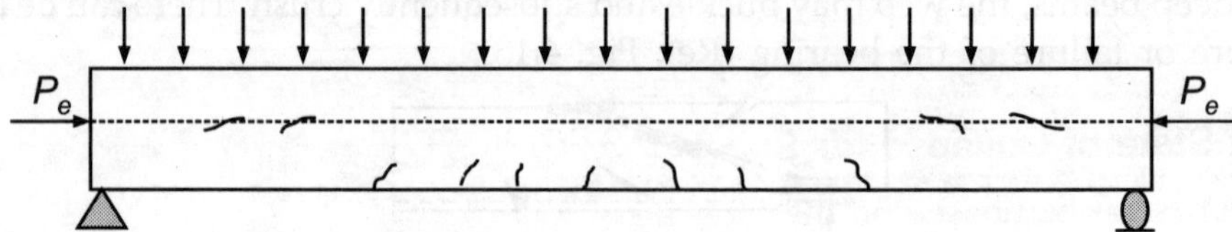

Fig. 4.16 Formation of Cracks in a Pre–stressed Beam

After cracking, in presence of pre–stressing force, the length and crack width of a diagonal crack are low. Thus, the aggregate interlock and zone of concrete under compression are larger as compared to a non–pre–stressed beam under the same load. Hence, the shear strength of concrete (V_c) increases in presence of pre–stressing force. This is accounted for in the expression of V_c.

4.2 DESIGN OF SECTIONS FOR SHEAR

The design of shear reinforcement in PSC beams is carried out by conventional methods employed as in RCC.

4.2.1 Calculation of Shear Demand

The objective of design is to provide ultimate resistance for shear (V_{uR}) greater than the shear demand under ultimate loads (V_u). For simply supported pre–stressed beams, the maximum shear near the support is given by the beam theory. For continuous pre–stressed beams, a rigorous analysis can be done by the moment distribution method.

Else, the shear coefficients in Table 13 of IS: 456–2000 can be used under conditions of uniform cross–section of the beams, uniform loads and similar lengths of span.

4.2.2 Design of Stirrups

The design is done for the critical section. The critical section is defined in Clause 22.6.2 of IS: 456–2000. In general cases, the face of the support is considered as the critical section. When the reaction at the support introduces compression at the end of the beam, the critical section can be selected at a distance effective depth from the face of the support.

The effective depth is selected as the greater of d_p or d_s.

d_p = Depth of CGS from the extreme compression fiber

d_s = Depth of centroid of non–pre–stressed steel.

Since the CGS is at a higher location near the support, the effective depth will be equal to d_s. To vary the spacing of stirrups along the span, other sections may be selected for design. Usually the following scheme is selected for beams under uniform load.

1. Close spacing for quarter of the span adjacent to the supports.
2. Wide spacing for half of the span at the middle.

For large beams, more variation of spacing may be selected. The following Fig. 4.17 shows the typical variation of spacing of stirrups. The span is represented by L.

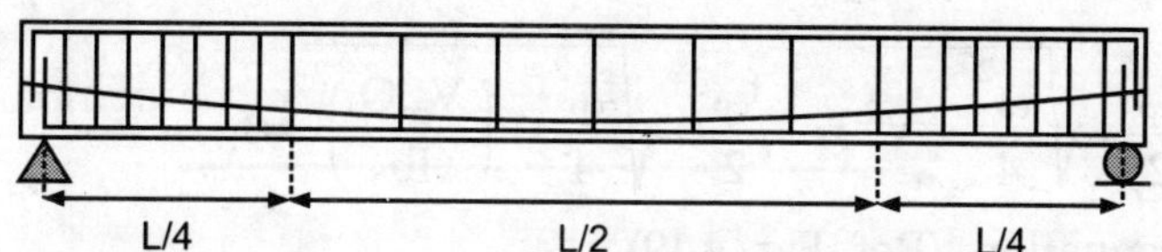

Fig. 4.17 Typical Variation of Spacing of Stirrups

4.2.3 Limit State of Collapse for Shear

The shear is studied based on the capacity of a section which is the limit state of collapse. The capacity (or ultimate resistance) of a section (V_{uR}) consists of a concrete contribution (V_c) and the stirrup contribution (V_S).

$$V_{uR} = V_c + V_S \qquad ...(4.1)$$

V_c includes V_{cz} (contribution from un–cracked concrete), V_a (aggregate interlock) and V_d (dowel action).

The value of V_c depends on whether the section is cracked due to flexure. Section 22.4 of IS: 1343–2012 gives two expressions of V_c, one for cracked section and the other for un–cracked section. Usually, the expression for the un–cracked section will govern near the support. The expression for the cracked section will govern near the mid–span. Of course, both the expressions need to be evaluated at a particular section. The lower value obtained from the two expressions is selected.

For un–cracked sections, $V_c = V_{co} = 0.67bD\sqrt{f_t^2 + 0.8f_{cp}f_t}$ ...(4.2)

V_{co} is the shear causing web shear cracking at CGC.

In the above expression,

b = Breadth of the section = b_w, breadth of the web for flanged sections

D = Total depth of the section (h)

f_t = Tensile strength of concrete = $0.24\sqrt{f_{ck}}$

f_{cp} = Compressive stress in concrete at CGC due to the pre–stress = P_e/A.

The value of f_{cp} is taken as positive (numeric value). Note that, a reduced effective pre–stress need to be considered in the transmission length (explained in Section 7.1 of IS: 1343–2012) region of a pre–tensioned beam.

The previous equation can be derived based on the expression of the principal tensile stress (σ_1) at CGC (Ref. Fig. 4.18).

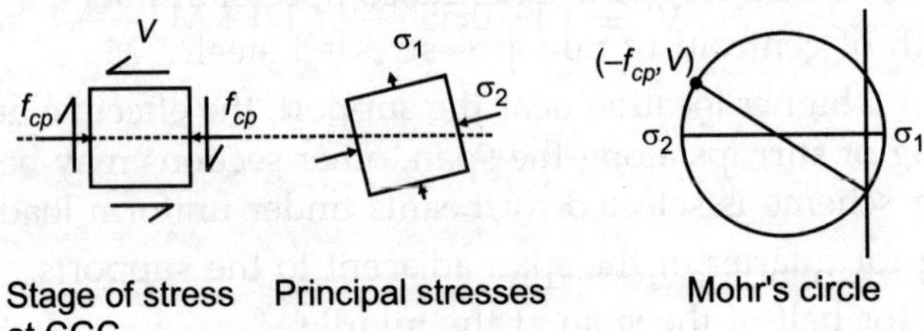

Fig. 4.18 State of Stresses at a Point on the Neutral Axis for a Pre–stressed Beam

The principal tensile stress is equated to the direct tensile strength of concrete (f_t).

$$\sigma_1 = -\frac{f_{cp}}{2}+\sqrt{\frac{f_{cp}^{\ 2}}{4}+V^2} = -\frac{f_{cp}}{2}+\sqrt{\frac{f_{cp}^{\ 2}}{4}+\left(\frac{V_{co}Q}{Ib}\right)^2} = f_t \qquad ...(4.3)$$

In the previous equation, (Ref. Fig. 4.19)

I = Gross moment of inertia

$Q = A_t\bar{y}$ ...(4.4)

A_t = Area of section above CGC

$\bar{y}$ = Vertical distance of centroid of A_t from CGC.

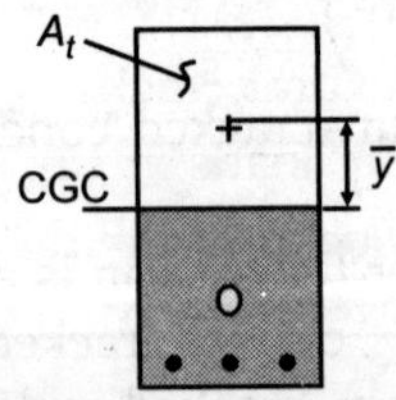

Fig. 4.19 Cross–section of a Beam Showing the Variables for Calculating Shear Stress in the Web

Transposing the terms,

$$V_{co} = \frac{Ib}{Q}\sqrt{f_t^2 + f_{cp}f_t} \rightarrow 0.67bD\sqrt{f_t^{\ 2} + 0.8f_{cp}f_t} \qquad ...(4.5)$$

The term 0.67bD represents Ib/Q for the section. It is exact for a rectangular section and conservative for other sections.

To be conservative, only 80% of the pre–stressing force is considered in the term $0.8f_{cp}$. For a flanged section, when the CGC is in the flange, the intersection of web and flange is considered to be the critical location. The expression of V_{co} is modified by substituting $0.8f_{cp}$ with 0.8 × (the stress in concrete at the level of the intersection of web and flange). In presence of inclined tendons or vertical pre–stress, the vertical component of the pre–stressing force (V_p) can be added to V_{co}.

$$V_c \rightarrow V_{co} + V_p = 0.67bD\sqrt{f_t^2 + 0.8f_{cp}f_t} + V_p \qquad \text{...(4.6)}$$

For cracked sections,

$$V_c = V_{cr} \qquad \text{...(4.7)}$$

$$V_c = \left(1 - 0.55\frac{f_{pe}}{f_{pk}}\right)\tau_c bd + M_o\frac{V_u}{M_u} \qquad \text{...(4.8)}$$

$$\geq 0.1bd\sqrt{f_{ck}} \qquad \text{...(4.9)}$$

V_{cr} is the shear corresponding to flexure shear cracking. The term $\left(1 - 0.55\frac{f_{pe}}{f_{pk}}\right)\tau_c bd$ is the additional shear that changes a flexural crack to a flexure shear crack.

The notations in the previous equation are as follows.

f_{pe} = Effective pre–stress in the tendon after all losses $\leq 0.6f_{pk}$

f_{pk} = Characteristic strength of pre–stressing steel

τ_c = Ultimate shear stress capacity of concrete, obtained from Table 6 of IS: 1343–2012.

It is given for values of A_p/bd, where, d is the depth of CGS. The values are plotted in the next Fig. 4.20.

b = Breadth of the section

= b_w, breadth of the web for flanged sections

d = Distance from the extreme compression fibre to the centroid of the tendons at the section considered

M_o = Moment initiating a flexural crack

M_u = Moment due to ultimate loads at the design section

V_u = Shear due to ultimate loads at the design section.

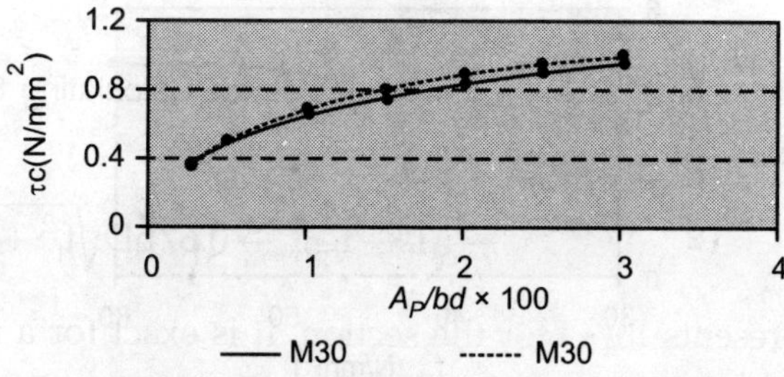

Fig. 4.20 Variation of Shear Strength of Concrete

The term $M_o \dfrac{V_u}{M_u}$ is the shear corresponding to the moment M_o, that decompresses (nullifies the effect of pre–stress) the tension face and initiates a flexural crack. The expression of M_o is given below.

$$M_o = 0.8 f_{pt} \frac{I}{y} \quad \text{...(4.10)}$$

In the above expression,

f_{pt} = Magnitude of the compressive stress in concrete at the level of CGS due to pre–stress only.

An equal amount of tensile stress is required to decompress the concrete at the level of CGS. The corresponding moment is $f_{pt}\dfrac{I}{y}$.

In the expression of M_o,

I = Gross moment of inertia

y = Depth of the CGS from CGC.

The factor 0.8 implies that M_o is estimated to be 80% of the moment that decompresses the concrete at the level of CGS. Since the concrete is cracked and the inclination of tendon is small away from the supports, any vertical component of the pre–stressing force is not added to V_{cr}.

4.2.4 Maximum Permissible Shear Stress

To check the crushing of concrete in shear compression failure, the shear stress is limited to a maximum value ($\tau_{c,max}$). The value of $\tau_{c,max}$ depends on the grade of concrete (Ref. Fig. 4.21) and is given in Table 7 of IS: 1343–2012.

$$\frac{V_u}{bd_t} \le t_{c,max} \quad \text{...(4.11)}$$

In the previous expression,

d_t = Greater of d_p or d_s

d_p = Depth of CGS from the extreme compression fiber

d_s = Depth of centroid of regular steel

V_u = Shear force at a section due to ultimate loads.

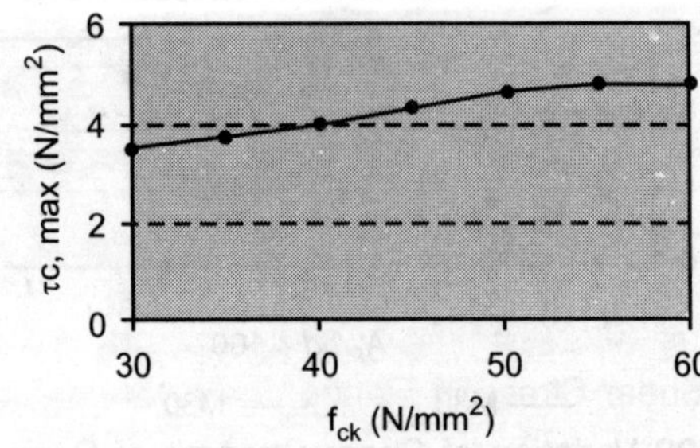

Fig. 4.21 Variation of Maximum Permissible Shear Stress in Concrete[8]

4.2.5 Design of Transverse Reinforcement

When the shear demand (V_u) exceeds the shear capacity of concrete (V_c), transverse reinforcements in the form of stirrups are required. The stirrups resist the propagation of diagonal cracks, thus checking diagonal tension failure and shear tension failure.

The stirrups resist a failure due to shear by several ways. The functions of stirrups are listed below.

1. Stirrups resist part of the applied shear.
2. They restrict the growth of diagonal cracks.
3. The stirrups counteract widening of the diagonal cracks, thus maintaining aggregate interlock to a certain extent.
4. The splitting of concrete cover is restrained by the stirrups, by reducing dowel forces in the longitudinal bars.

After cracking, the beam is viewed as a plane truss. The top chord and the diagonals are made of concrete struts. The bottom chord and the verticals are made of steel reinforcement ties. Based on this truss analogy, for the ultimate limit state, the total area of the legs of the stirrups (A_{sv}) is given as follows.

$$\frac{A_{sv}}{S_v} = \frac{V_u - V_c}{0.87 f_y d_t} \qquad \text{...(4.12)}$$

The notations in the above equation are explained.

S_v = Spacing of the stirrups

d_t = Greater of d_p or d_s

d_p = Depth of CGS from the extreme compression fiber

d_s = Depth of centroid of non–pre–stressed steel

f_y = Yield stress of the stirrups

The grade of steel for stirrups should be restricted to Fe 415 or lower.

4.2.6 Design of Stirrups for Flanges

For flanged sections, although the web carries the vertical shear stress, there is shear stress in the flanges due to the effect of shear lag. Horizontal reinforcement in the form of single leg or closed stirrups is provided in the flanges. The following Fig. 4.22 shows the shear stress in the flange at the face of the web.

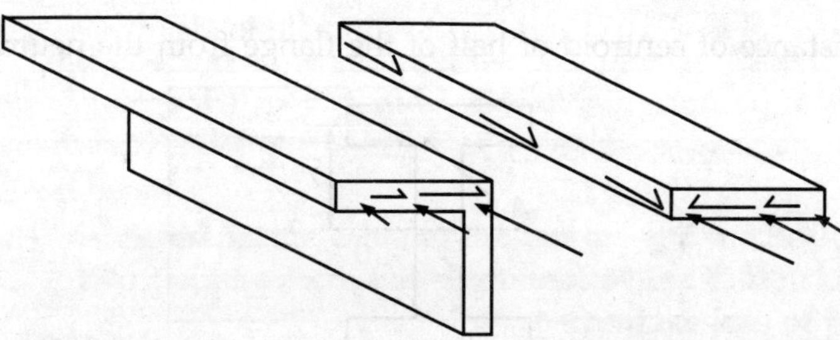

Fig. 4.22 Shear Stress in Flange due to Shear Lag Effect

The horizontal reinforcement is calculated based on the shear force in the flange. The relevant quantities for the calculation based on an elastic analysis are as follows.

1. Shear flow (shear stress × width)
2. Variation of shear stress in a flange (τ_f)
3. Shear forces in flanges (V_f).
4. Ultimate vertical shear force (V_u)

The following Fig. 4.23 shows the above quantities for an I–section (with flanges of constant widths).

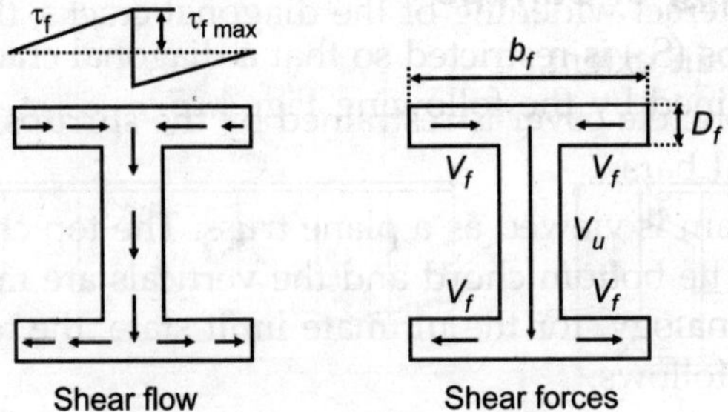

Fig. 4.23 Shear Flow and Shear Forces in an I–Section

The design shear force in a flange is given as follows.

$$V_f = \frac{\tau_{f,max}}{2}\frac{b_f}{2}D_f \qquad ...(4.13)$$

Here,

b_f = Breadth of the flange

D_f = Depth of the flange

$\tau_{f,\,max}$ = Maximum shear stress in the flange.

The maximum shear stress in the flange is given by an expression similar to that for the shear stress in web (Ref. Fig. 4.24).

$$\tau_{f,max} = \frac{V_u A_1 \bar{y}}{ID_f} \qquad ...(4.14)$$

Here, V_u = Ultimate vertical shear force

I = Moment of inertia of the section

A_1 = Area of half of the flange

$\bar{y}$ = Distance of centroid of half of the flange from the neutral axis at CGC.

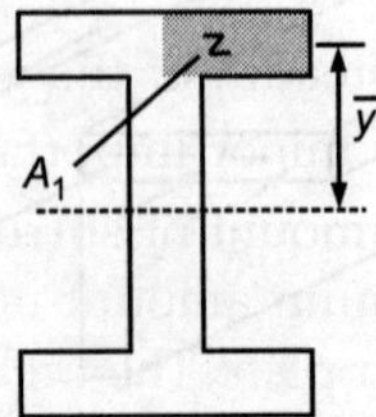

Fig. 4.24 Cross–section of a Beam Showing the Variables for Calculating Shear Stress in the Flange

The amount of horizontal reinforcement in the flange (A_{svf}) is calculated from V_f.

$$A_{svf} = \frac{V_f}{0.87f_y} \quad ...(4.15)$$

The yield stress of the reinforcement is denoted as f_y.

4.2.7 Detailing Requirements

The detailing requirements for the stirrups in IS: 1343–2012 are briefly mentioned.

4.2.7.1 Maximum Spacing of Stirrups

The spacing of stirrups (S_v) is restricted so that a diagonal crack is intercepted by at least one stirrup. This is explained by the following Fig. 4.25.

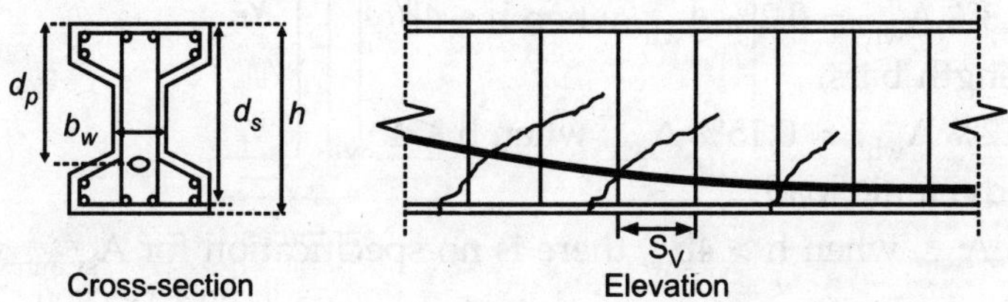

Fig. 4.25 Cross–section and Elevation of a Beam Showing Stirrups

As per Clause 22.4.3.2, the maximum spacing is $0.75d_t$ or $4b_w$, whichever is smaller. When V_u is larger than $1.8V_c$, the maximum spacing is $0.5d_t$.

The variables are as follows.

b_w = Breadth of web, d_t = Greater of d_p or d_s,
d_p = Depth of CGS from the extreme compression fiber,
d_s = Depth of centroid of non–pre–stressed steel,
V_u = Shear force at a section due to ultimate loads,
V_c = Shear capacity of concrete.

4.2.7.2 Minimum Amount of Stirrups

A minimum amount of stirrups is necessary to restrict the growth of diagonal cracks and subsequent shear failure. For $V_u < V_c$, minimum amount of transverse reinforcement is provided based on the following equation.

$$\frac{A_{sv}}{bS_v} = \frac{0.4}{0.87f_y} \quad ...(4.16)$$

b = Breadth of the section = b_w, breadth of the web for flanged sections.

If $V_u < 0.5V_c$ and the member is of minor importance, stirrups may not be provided.

Another provision for minimum amount of stirrups ($A_{sv, min}$) is given by Clause 18.6.3.2 for beams with thin webs. The minimum amount of stirrups is given in terms of A_{wh}, the horizontal sectional area of the web in plan. The area is shown in the following Fig. 4.26.

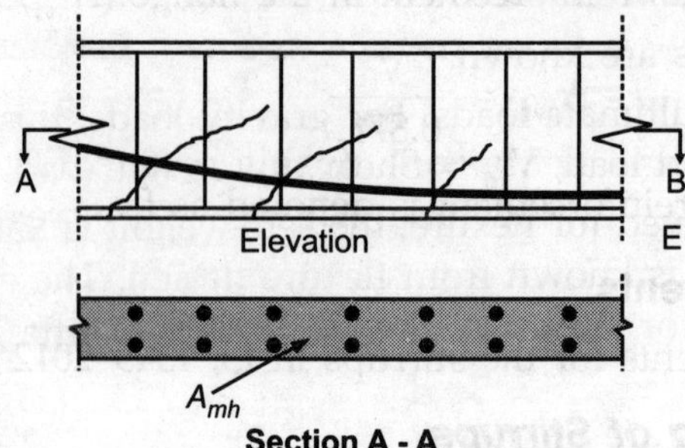

Fig. 4.26 Elevation and Horizontal Section of Beam Showing Stirrups

In presence of dynamic load,

$$A_{sv,min} = 0.3\% A_{wh}, = 0.2\% A_{wh}, \text{ when } h \le 4b_w$$

With high strength bars,

$$A_{sv,min} = 0.2\% A_{wh}, = 0.15\% A_{wh}, \text{ when } h \le 4b_w$$

In absence of dynamic load,

$A_{sv,min}$ = 0.1% A_{wh}, when $h > 4b_w$, there is no specification for $A_{sv,min}$ when $h \le 4b_w$.

4.2.7.3 Anchorage of Stirrups

The stirrups should be anchored to develop the yield stress in the vertical legs.

1. The stirrups should be bent close to the compression and tension surfaces, satisfying the minimum cover.
2. Each bend of the stirrups should be around a longitudinal bar. The diameter of the longitudinal bar should not be less than the diameter of stirrups.
3. The ends of the stirrups should be anchored by standard hooks.
4. There should not be any bend in a re–entrant corner. In a re–entrant corner, the stirrup under tension has the possibility to straighten, thus breaking the cover concrete.

The following Fig. 4.27 explains the requirement of avoiding the bend of a stirrup at a re–entrant corner.

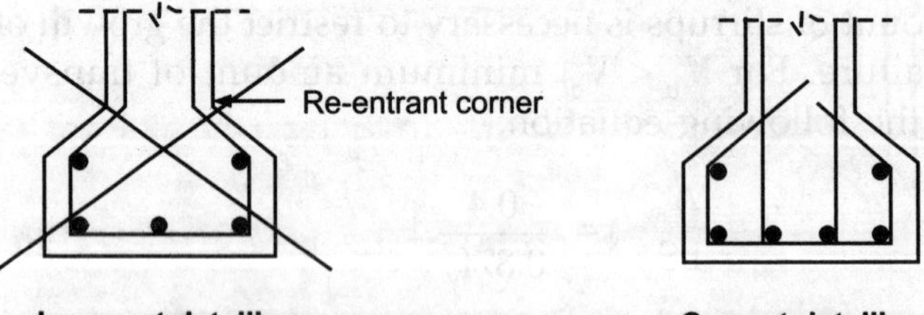

Fig. 4.27 Cross–section of the Bottom Flange of a Beam Showing Stirrups

4.2.7.4 Minimum Thickness (Breadth) of Web

To check web crushing failure, The Indian Roads Congress Code IRC: 18–2000 specifies a minimum thickness of the web for T–sections (Clause 9.3.1.1). The minimum thickness is 200mm plus diameter of the duct hole.

4.2.8 Design Steps

The following quantities are known.

V_u = Factored shear at ultimate loads. For gravity loads, this is calculated from V_{DL} and V_{LL}, V_{DL} = Shear due to dead load, V_{LL} = Shear due to live load.

After a member is designed for flexure, the self–weight is known. It is included as dead load. The grade of concrete is known from flexure design. The grade of steel for stirrups is selected before the design for shear. As per IS: 1343–2012, the grade of steel is limited to Fe 415.

The following quantities are unknown.

V_c = Shear carried by concrete, A_{sv} = Total area of the legs of stirrups within a distance S_v S_v = Spacing of stirrups.

The steps for designing stirrups along the length of a beam are given below.

1. Calculate the shear demand V_u at the critical location.
2. Check $(V_u/bd_t) < \tau_{c,max}$. If it is not satisfied, increase the depth or breadth of the section. Here, b is the breadth of the web (b_w) and d_t is larger of d_p and d_s.
3. Calculate the shear capacity of concrete V_c from the lower of V_{co} and V_{cr}. In presence of inclined tendons or vertical pre–stress, the vertical component of the pre–stressing force (V_p) can be added to V_{co}.
4. Calculate the requirement of shear reinforcement through A_{sv}/S_v. Compare the value with the minimum requirement.
5. Calculate the maximum spacing and round it off to a multiple of 5mm.
6. Calculate the size and number of legs of the stirrups based on the amount required, type of section and space to accommodate.

Repeat the calculations for other locations of the beam, if the spacing of stirrups needs to be varied.

4.2.9 Diagonal Tension or Principal Tension

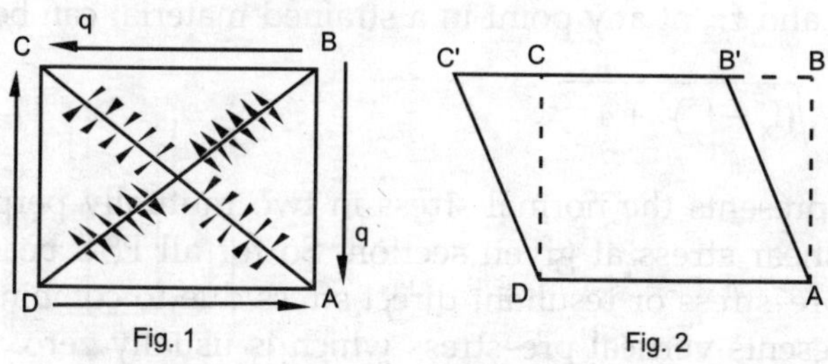

Fig. 4.28 Cubical Element ABCD Subjected All Round Shear Stress and its Deformations

Let us consider a cubical element ABCD subjected all round sharing stress of magnitude 'q' as shown in Fig. 1 of Fig. 4.28. As a result this element gets deformed as shown in Fig. 2 of Fig. 4.28. Here tensile and compressive stresses of equal magnitude are developed parallel to the diagonal AC and BD respectively. The tensile stress so developed known as diagonal or principal tension. It has been observed, that failure section occur due to the action of pure shear. As such it is usual to determine the principal tension due the combined action of shear and flexure. If this principal tension exceeds the permissible value it becomes necessary to provide shear reinforcement. Wherever the principal tension is within permissible limit normal shear reinforcement is usually provided.

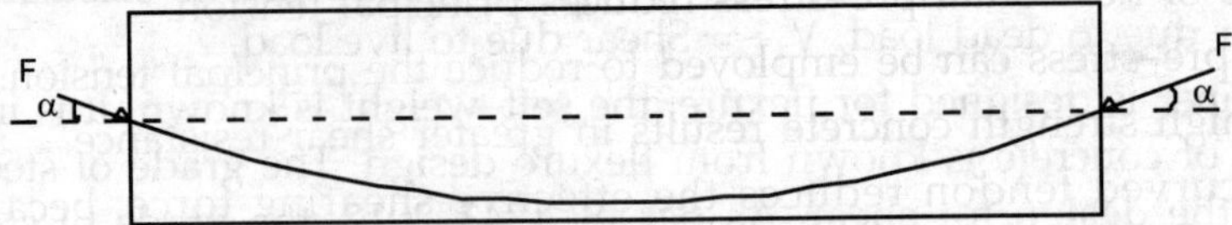

Fig. 4.29 Beam with Tendon

When a curved tendon is used in a beam vertical component of the effective pre–stressing force reduces the shear forces across the given section due to the dead and live load (Ref. Fig. 4.29). Further to b inclination of the tendon force F with the horizontal at the given section. Vertical component of F = FSinα = Fα. Where α is in radian.

Thus the effective shear force across any given section can be computed from the relation

$$P = V - F\alpha \qquad \ldots(4.17)$$

Whenever the longitudinal profile of the tendon is not known, we can always assume straight tendon, so that α = 0 at every section.

Further, the horizontal pre–stress along any fibre can be computed from the relation,

$$f_x = \frac{F}{A} + \frac{Fey}{I} + \frac{M_T y}{I} \qquad \ldots(4.18)$$

Where, F = Effective pre–stress in the tendon, A = Total cross sectional area of the given section, e = Eccentricity of the CGS, M_T = BM across the given section due to all–inclusive load, y = Distance of any fibre from the NA.

Note: Whenever the section is not specified we must always consider the most critical section, at which the shear force is maximum. Thus for computing principal tension, we must consider the support section in the case of simply supported beam and cantilever, as less otherwise specified. Further, we must compute F and q along the fibre which lie at the NA. This is necessary because shear stress is maximum along the NA.

Principal stress f_{n1}, and f_{n2} at any point in a strained material can be computed from the relation $f_n = \frac{f_x + f_y}{2} \pm \frac{1}{2}\sqrt{(f_x - f_y)^2 + 4\tau^2}$ ...(4.19)

Where f_x and f_y represents the normal stress in two mutually perpendicular directions and τ represents the shear stress at given section. So for all PSC beams are concerned f_x represents horizontal pre–stress or resultant direct stress due to combined action of bending and pre–stress. f_y represents vertical pre–stress which is usually zero.

Then the expression for the principal stresses takes the form,

$$f_{n1} = \frac{f_x}{2} + \frac{1}{2}\sqrt{(f_x)^2 + 4\tau^2} \text{ and } f_{n2} = \frac{f_x}{2} - \frac{1}{2}\sqrt{(f_x)^2 + 4\tau^2} \qquad \ldots(4.20)$$

f_{n1} represents principal compressive stress, f_{n2} represents principal tension or diagonal tension due to the combined action of shear and flexure. If this principal tension exceeds the permissible value, then it becomes necessary to provide shear reinforcement.

The strength of concrete subjected to pure shear being nearly twice that in tension.

4.2.10 Factors Responsible for Reducing Principal Tension or Methods of Improving Shear Resistance in PSC Beam

1. Presence of horizontal pre–stress reduces principal tension.
2. Vertical pre–stress can be employed to reduce the principal tension.
3. Use of high strength concrete results in greater shear resistance.
4. Use of curved tendon reduces the effective shearing force, because the vertical component of the tendon force acts in a direction opposite to that of the shear due to loads.

Note: In a rectangular section, the max. shearing stress along the CA is (Ref. Fig. 4.30),

$$\tau_{max} = \tau_{CA} = \frac{PA\bar{y}}{Ib} = \frac{P\times\left(\frac{bd}{2}\right)\times\frac{d}{4}}{\frac{bd^3}{12}\times b} = \frac{3}{2}\left(\frac{P}{3d}\right),\ \tau_{max} = 1.5\tau_{av} \qquad \ldots(4.21)$$

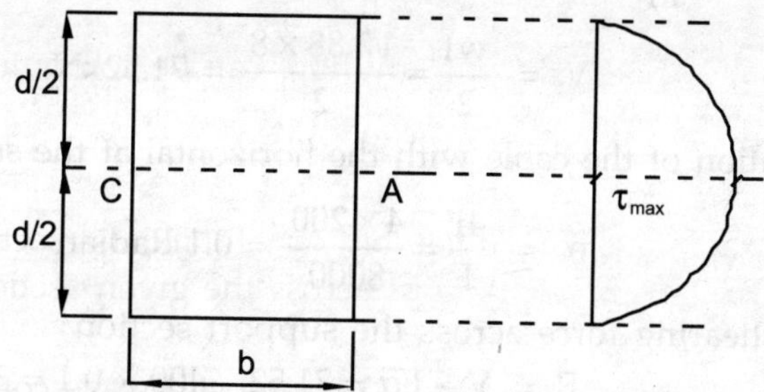

Fig. 4.30 Determination of Maximum Shear Stress

Where, Average stress $= \tau_{av} = \dfrac{P}{bd}$ ...(4.22)

4.3 SOLVED EXAMPLES

Problem 4.1

A rectangular beam 300mm × 400mm in section of simply supported over a span of 8m. It carries an imposed load of 15kN/m and is pre–stressed with a parabolic cable which has maximum eccentricity of 150mm at mid–span. At the supports the cable lies 50mm above CGC. Effective pre–stress in the cable is 400kN. Determine (a) Maximum principal tension (b) Magnitude of principal tension wherever horizontal pre–stress is absent (c) To find vertical pre–stress required, so as to reduce the principal tension to zero.

Solution:

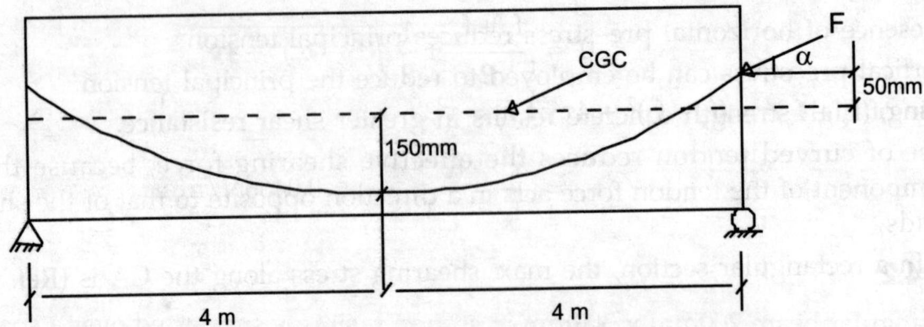

Fig. 4.31 Longitudinal Section of Rectangular Beam

A = 0.3 × 0.4 = 0.12m², Self-weight = g = 0.12 × 24 = 2.88kN/m, F = 400kN

Imposed load = q = 15kN/m, Total load = W = 17.88kN/m

(a) Let us consider the support section, here maximum vertical shear force exists

$$V = \frac{wl}{2} = \frac{17.88 \times 8}{2} = 71.52\text{kN}$$

Let 'α' be the inclination of the cable with the horizontal at the support section

$$\therefore \quad \alpha = \frac{4r}{l} = \frac{4 \times 200}{8000} = 0.1 \text{ Radian}$$

Therefore effective shearing force across the support section

$$P = V - F\alpha = 71.52 - 400 \times 0.1 = 31.52\text{kN}$$

At support section e = +50mm, $M_T = 0$, y = 0

$$f_x = \frac{F}{A} + \frac{Fey}{I} + \frac{M_T y}{I} = \frac{F}{A} = \frac{400 \times 10^3}{300 \times 400} = 3.33\text{N/mm}^2$$

Considering fibres along the central axis, $\tau = 1.5 \times \frac{31.52 \times 10^3}{300 \times 400} = 0.394\text{N/mm}^2$

Thus the principal tension is given by

$$f_{n2} = \frac{f_x}{2} - \frac{1}{2}\sqrt{(f_x)^2 + 4\tau^2}$$

$$= \frac{3.33}{2} - \frac{1}{2}\sqrt{(3.33)^2 + 4(0.394)^2} = -0.045\text{N/mm}^2$$

(b) When the horizontal pre-stress, $f_x = 0$. Here the principal tension is given by

$$f_{n2} = -\frac{1}{2}\sqrt{4\tau^2} = -\frac{1}{2}\sqrt{4(0.394)^2} = -0.394\text{N/mm}^2$$

(c) To find the vertical pre-stress required so as to reduce the principal tension to zero. Let f_y be the vertical pre-stress required, so that the principal tension $f_{n2} = 0$.

Now $$f_{n2} = \frac{f_x + f_y}{2} \pm \frac{1}{2}\sqrt{(f_x - f_y)^2 + 4\tau^2},$$

$$0 = \frac{f_x + f_y}{2} \pm \frac{1}{2}\sqrt{(f_x - f_y)^2 + 4\tau^2}$$

Squaring, $(f_x + f_y)^2 = (f_x - f_y)^2 + 4\tau^2$, $f_x f_y = \tau^2$,

$$f_y = \frac{\tau^2}{f_x} = \frac{(0.394)^2}{3.33} = 0.0466\text{N/mm}^2.$$

Problem 4.2

A rectangular beam 200mm × 300mm in section is simply supported over a span of 5m. It is pre–stressed with a parabolic cable which has the maximum eccentricity 100mm at mid–span section and a minimum eccentricity of 40mm at the support. Determine the effective pre–stressing force required in the cable so that the shear stress across the support section will be zero. Hence determine the principal stresses. Assume that the beam caries an all–inclusive load of 20kN/m.

Solution:

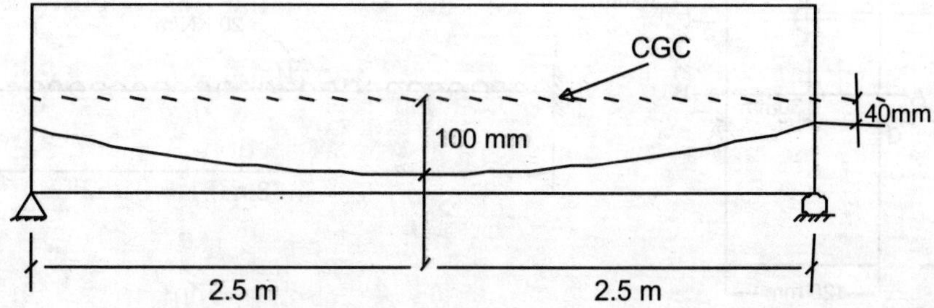

Fig. 4.32 Longitudinal Section of Rectangular Beam

Let α be inclination of the cable with the horizontal at the support section.

$$\alpha = \frac{4r}{l} = \frac{4 \times 60}{5000} = 0.048 \text{ Radian}$$

Vertical shear force across the support section.

$$V = \frac{wl}{2} = \frac{20 \times 5}{2} = 50\text{kN} = 50 \times 10^3\text{N}$$

It is given that the shear stress across the support section i.e., should be zero. Hence effective shear force, P = 0, P = V – Fα, 0 = 50 – F(0.048), F = 1041.66kN

• ***Principal Stresses***

Horizontal pre–stress along the centroidal axis is given by

$$f_x = \frac{F}{A} + \frac{Fey}{I} + \frac{M_T y}{I} \quad \text{since } y = 0,\ f_x = \frac{F}{A} = \frac{1041.66 \times 10^3}{200 \times 300} = 17.36\text{N/mm}^2$$

Here $\tau = 0$ Thus the principal stresses are given by

$$f_{n1} = \frac{f_x}{2} + \frac{1}{2}\sqrt{(f_x)^2 + 4\tau^2} = \frac{17.36}{2} + \frac{1}{2}\sqrt{(17.36)^2} = 17.36\text{N/mm}^2$$

$$f_{n2} = \frac{f_x}{2} - \frac{1}{2}\sqrt{(f_x)^2 + 4\tau^2} = 0$$

Note: The principal tension has the greatest value when $f_x = 0$. Further the principal tension reduces to zero when $\tau = 0$.

Problem 4.3

A cantilever beam of span 3m is 120mm × 300mm in section. It carries an all–inclusive UDL of 20kN/m together with a point load of 5kN at its free end. Determine the principal tension at 30mm above the centroidal fibre and at the support section. The beam is pre–stressed with an effective pre–stress of 360kN which is located at 50mm from the top edge.

Solution:

Since the longitudinal profile of the cable is not given, we shall assume a straight cable in which case eccentricity remains constant throughout the beam.

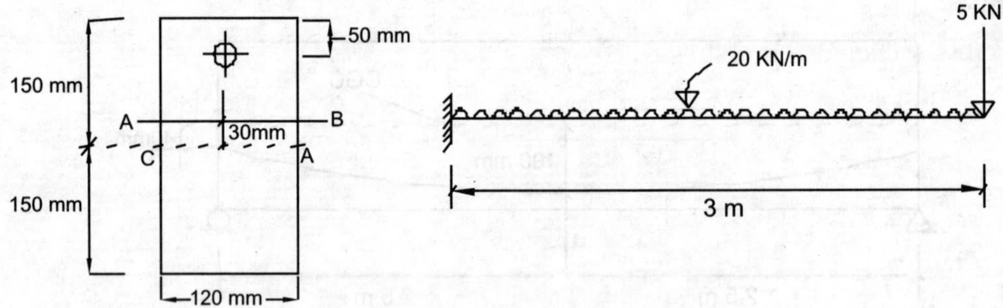

Fig. 4.33 Cross–section of Cantilever Beam

Bending moment at support section= –5 × 3 – 20 × 3 × 3/2 = –105kNm = M_T

$$A = 0.12 \times 0.3 = 0.036\text{m}^2,$$

$$I = \frac{b \times d^3}{12} = \frac{0.12 \times 0.3^3}{12} = 2.7 \times 10^{-4}\text{m}^4$$

Vertical shear force at the support section

$$V = 20 \times 3 + 5 = 65\text{kN},\ F = 360\text{kN},\ \alpha = 0$$

Effective shear force across the support section

$$P = V - F = V = 65\text{kN}$$

Horizontal pre–stress along AB is given by

$$f_x = \frac{F}{A} + \frac{Fey}{I} + \frac{M_T y}{I}$$

$$= \frac{360}{0.036} + \frac{360 \times 0.1 \times 0.03}{2.7 \times 10^{-4}} + \frac{-105 \times 0.03}{2.7 \times 10^{-4}} = 2.33\text{N/mm}^2$$

$$\bar{y} = 30 + \frac{150}{2} = 90\text{mm} = 0.09\text{m}$$

Shear stress along AB is given by

$$\tau = \frac{PA\bar{y}}{Ib} = \frac{65 \times (0.12 \times 0.3) \times 0.09}{2.7 \times 10^{-4} \times 0.12} = 6500\text{kN/m}^2$$

$$= 6.5\text{N/mm}^2$$

Thus the principal tension is given by

$$f_{n2} = \frac{f_x}{2} - \frac{1}{2}\sqrt{(f_x)^2 + 4\tau^2}$$

$$= \frac{2.33}{2} - \frac{1}{2}\sqrt{(2.33)^2 + 4(6.5)^2} = -5.43\text{N/mm}^2.$$

Note: In the case of flanged section, we must determine the principal tension along the centroidal fibre and also along the fibres laying at the junction of web and at the bottom flange unless otherwise specified. Having determined the principal tension at the above two values greater of these two values should be selected as the maximum principal tension.

Problem 4.4

The cross–section of a pre–stressed concrete beam T–section with the following dimensions.

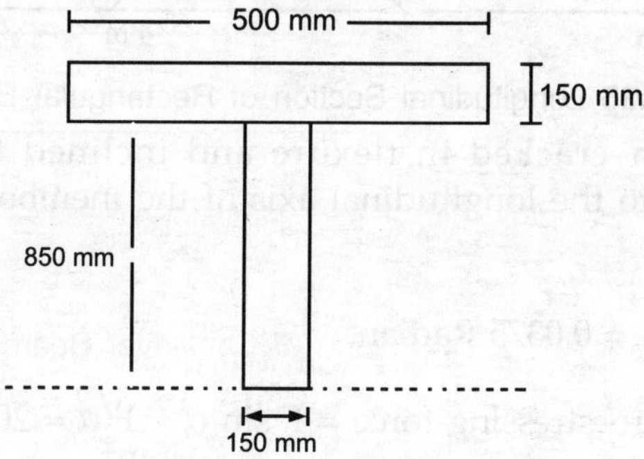

Fig. 4.34 Cross–section of a Pre–stressed Concrete T–Beam

f_{ck}= 35N/mm², effective pre–stress in tendon = 1000N/mm², area of tendons = 1500mm², calculate the shear resistance of the section shown in Fig. 4.34, if it is un–cracked in flexure.

Solution:

$$A = 202500\text{mm}^2,\ f_{sc} = 1000\text{N/mm}^2,\ A_{st} = 1500\text{mm}^2$$

Shear resistance of a section, if it is un–cracked in flexure

$$V_{co} = 0.67bD\sqrt{f_t^2 + 0.8f_{cp}f_t}$$

$$f_t = 0.24\sqrt{f_{ck}} = 0.24\sqrt{35} = 1.42\text{N/mm}^2$$

f_{cp} = Compressive stress at centroidal axis due to pre–stress

$$= \frac{F}{A} = \frac{f_{se}A_{st}}{A} = \frac{1000 \times 1500}{202500} = 7.4\text{N/mm}^2$$

$$V_{co} = 0.67 \times 150 \times 1000\sqrt{(1.42)^2 + 0.8 \times 7.4 \times 1.42} = 325\text{kN}.$$

Problem 4.5

A pre–stressed concrete beam of rectangular section 150mm × 300mm is pre–stressed by a parabolic cable carrying an effective pre–stress of 200kN. The cable has an eccentricity of 10mm at the centre of span and 25mm at the support. The beam supports UDL of 15kN/m including its self–weight. Determine the shear resistance of un–cracked section at the support. $f_{ck} = 45\text{N/mm}^2$.

Solution:

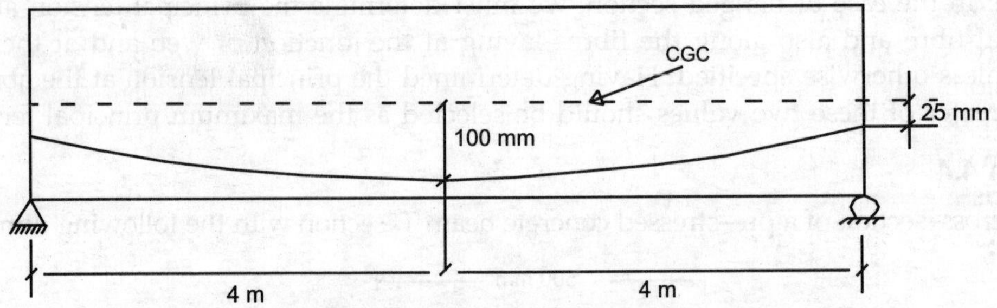

Fig. 4.35 Longitudinal Section of Rectangular Beam

Note: For a section un–cracked in flexure and inclined tends, the component of pre–stressing force normal to the longitudinal axis of the member should be added to web shear V_{co}

$$\alpha = \frac{4r}{l} = \frac{4 \times 75}{8000} = 0.0375 \text{ Radian}$$

Vertical component of pre–stressing force = F sin α = F α = 200 × 0.0375 = 7.5kN

$$V_{co} = 0.67bD\sqrt{f_t^2 + 0.8f_{cp}f_t} + F\,\alpha,\ f_{cp} = \frac{200 \times 1000}{150 \times 300} = 4.45\text{N/mm}^2$$

$$f_t = 0.24\sqrt{f_{ck}} = 0.24\sqrt{45} = 1.61\text{N/mm}^2$$

$$V_{co} = \frac{0.67 \times 150 \times 300\sqrt{(1.61)^2 + 0.8 \times 4.45 \times 1.61}}{1000} + 7.5 = 94.5\text{kN}$$

$$\text{Factored shear force} = \frac{wl}{2} \times 1.5$$

$$= \frac{15 \times 8}{2} \times 1.5 = 90\text{kN} < 94.5\text{kN}$$

Maximum permissible shear stress for M45 is 4.3N/mm² (As per Table 7 of IS: 1343)

So, maximum permissible shear force in the section = 4.3 × 150 × 300 = 193.5kN > 90kN.

Minimum shear reinforcement needs to be provided.

Let us provide 6mm, 2 legged mild steel shear stirrups

$$A_{sv} = 2 \times \frac{\pi \times 6^2}{4} = 56.5\text{mm}^2$$

$$\text{Spacing, } S_v = \frac{0.87 f_y A_{sv}}{0.4b} = \frac{0.87 \times 250 \times 56.5}{0.4 \times 150} = 205\text{mm}.$$

If the diameter of longitudinal reinforcement required to support the shear stirrups is 8mm and clear cover is 20mm. The depth d_t from the extreme compression fibre is given as $d_t = 300 - 20 - 8/2 = 276$mm.

Spacing of stirrups $S_V < 0.75d_t = 0.75 \times 276 = 207$mm, $< 4b = 4 \times 150 = 600$mm

Therefore, shear stirrups should be provided @ 200mm c/c.

Problem 4.6

The cross section of a pre–stressed beam is I–section as shown in Fig. 4.36. The area of high tension steel wires is 200mm². Determine the shear resistance of the section if it has developed flexural cracks. Assume the following data. f_{ck}= 40N/mm², ultimate moment at the section is 20kNm. Ultimate shear at the section is 25kN. Effective pre–stress is 1200N/mm². Characteristic strength of pre–stressing steel = 1500N/mm².

Solution:

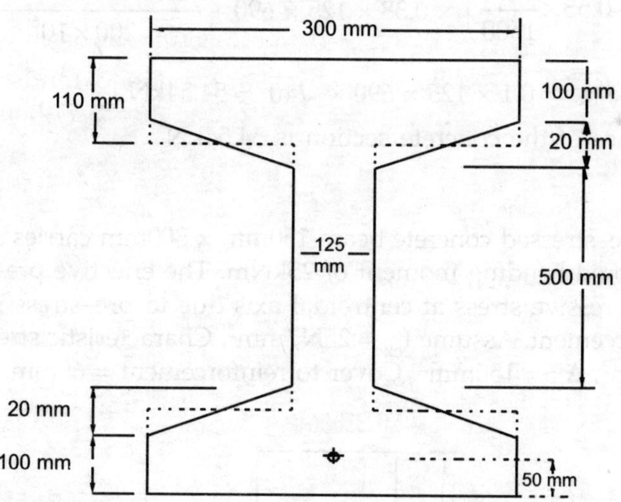

Fig. 4.36 Cross–section of a Pre–stressed I–Section Beam

Area of cross–section = 2 × 300 × 110 + 125 × 520 = 131000mm²

$$I = 2\left[\frac{300 \times 110^3}{12} + 300 \times 110 \times (370 - 55)^2\right] + \left[\frac{125 \times 520^3}{12} + 125 \times 520 \times (0)^2\right]$$

$$= 80.8 \times 10^8 \text{mm}^4$$

$$d = 740 - 50 = 690\text{mm}$$

Eccentricity of pre–stressing wires, e = 370 – 50 = 320mm

Effective pre–stressing force, $F = f_{se} A_{st}$ = 1200 × 200 = 240kN

Stress due to prestress only at depth 'd' at the level of tendon is

$$f_{pt} = \frac{F}{A} + \frac{Fey}{I} = \frac{240 \times 1000}{131000} + \frac{240 \times 1000 \times (-320)(-320)}{80.8 \times 10^8} = 4.87 \text{N/mm}^2$$

Moment necessary to produce zero stress in the concrete at the depth is given by

$$M_o = 0.8 f_{pt} \frac{I}{Y} = 0.8 \times 4.87 \times \frac{80.8 \times 10^8}{320} = 98.3 \times 10^6 \text{Nmm} = 98.3 \text{kNm}$$

[Effective pre–stress after all losses = f_{pe} = 1200N/mm² > $0.6 f_p$ = 0.6 × 1500 = 900N/mm²]

$$\% \text{ of pre–stressing steel} = \frac{100 \times 200}{125 \times 690} = 0.23\%$$

Therefore, shear strength of M40 concrete C_c = 0.38N/mm²

Shear Resistance of Section Cracked in Flexure

$$V_{cr} = \left(1 - 0.55 \frac{f_{pe}}{f_p}\right) C_c \, bd + M_o \frac{V}{M}$$

$$= \left(1 - 0.55 \times \frac{1200}{1500}\right) \times 0.38 \times 125 \times 690 + \frac{98.3 \times 10^6 \times 25 \times 10^3}{200 \times 10^6} = 30.641 \text{kN}$$

But $V_{cr} < 0.1 b_w d \sqrt{f_{ck}} = 0.1 \times 125 \times 690 \times \sqrt{40} = 54.54 \text{kN}$

∴ Shear resistance of the concrete section is 54.54kN.

Problem 4.7

A section of a pre–stressed concrete beam 150mm × 300mm carries a factored shear force of 110kN and a factored bending moment of 25kNm. The effective pre–stress after all losses is 600N/mm². Compressive stress at centroidal axis due to pre–stress is 6.5N/mm². Design suitable shear reinforcement. Assume f_{ck} = 25N/mm². Characteristic strength of pre–stressing steel f_p = 1600N/mm², A_{st} = 150mm². Cover to reinforcement = 60mm.

Solution:

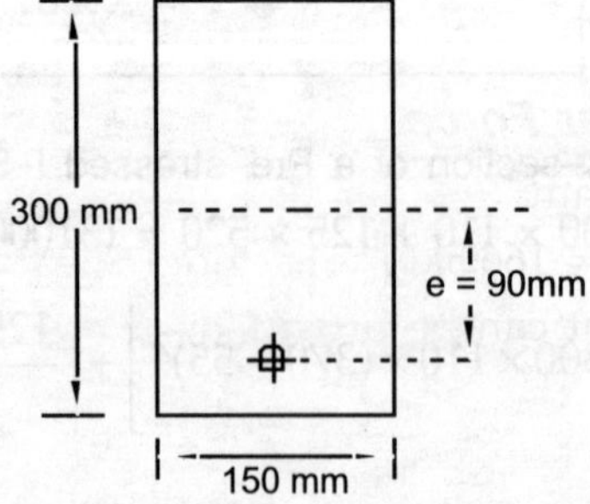

Fig. 4.37 Section of a Pre–stressed Concrete Beam

i. Shear Resistance of Concrete with Flexure Cracks

$$V_{cr} = \left(1 - 0.55 \times \frac{1200}{1500}\right) C_c\, bd + M_o \frac{V}{M} > 0.1bd\sqrt{f_{ck}}, \text{ Where, } M_o = 0.8f_{pt}\frac{I}{Y}$$

$$A = 150 \times 300 = 45000\text{mm}^2, I = \frac{150 \times 300^3}{12} = 337.5 \times 10^6\text{mm}^4, F = 150 \times 600 = 90000\text{N}$$

$$f_{pt} = \frac{F}{A} + \frac{Fey}{I} = \frac{90000}{45000} + \frac{90000 \times (-90)(-90)}{337.5 \times 10^6} = 4.16\text{N/mm}^2$$

$$M_o = 0.8f_{pt}\frac{I}{Y} = 0.8 \times 4.16 \times \frac{337.5 \times 10^6}{90} = 12.48 \times 10^6\text{Nmm}$$

$$\% \text{ of pre–stressing of steel} = \frac{100 \times 150}{150 \times 240} = 0.42\%$$

Therefore, shear strength of M35 concrete $C_c = 0.46\text{N/mm}^2$

$$V_{cr} = \left(1 - 0.55\frac{f_{pe}}{f_p}\right) C_c\, bd + M_o \frac{V}{M}$$

$$= \left(1 - 0.55 \times \frac{600}{1600}\right) \times 0.46 \times 150 \times 300 + \frac{12.48 \times 10^6 \times 110 \times 10^3}{25 \times 10^6} = 71.34\text{kN}$$

But $V_{cr} < 0.1bd\sqrt{f_{ck}} = 0.1 \times 150 \times 300 \times \sqrt{35} = 21.30\text{kN}$, $\therefore V_{cr} = 71.34\text{kN}$

ii. Shear Resistance of Concrete without Flexure Cracks

$$V_{co} = 0.67bD\sqrt{f_t^2 + 0.8f_{cp}f_t}$$

$$f_t = 0.24\sqrt{f_{ck}} = 0.24\sqrt{35} = 1.42\text{N/mm}^2$$

$$f_{cp} = 6.5\text{N/mm}^2$$

$$V_{co} = \frac{0.67 \times 150 \times 300\sqrt{(1.42)^2 + 0.8 \times 6.5 \times 1.42}}{1000} = 92.44\text{kN}$$

Shear strength of concrete is the lesser of V_{cr} and V_{co} that is shear strength of concrete $V_c = 71.34\text{kN}$.

iii. Maximum Permissible Shear Force

For M35 concrete, $\tau_c = 3.7\text{N/mm}^2$

$V_{max} = \tau_c bd = 3.7 \times 150 \times 300 = 166.5\text{kN} >$ Factored Shear force = 110kN

Therefore, shear reinforcement can be provided.

iv. Shear Reinforcement

Let us provide 6mm, 2 legged steel shear stirrups of Fe 415

$$A_{sv} = 2 \times \frac{\pi \times 6^2}{4} = 56.5\text{mm}^2$$

If the diameter of longitudinal reinforcement required to support the shear stirrups is 8mm and clear cover is 20mm.

So, $d_t = 300–8/2 – 20 = 276$mm

Spacing $$S_v = \frac{0.87 f_y d_t A_{sv}}{V - V_c}$$

$$= \frac{0.87 \times 415 \times 276 \times 56.5}{(110 - 71.34) \times 1000} = 145\text{mm}$$

Since $V > 1.8V_c$

Provide 6mm, 2 legged Fe 415 shear stirrups at 150mm c/c

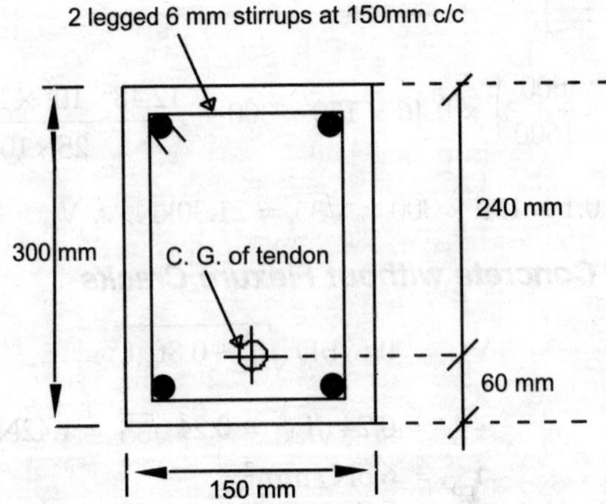

Fig. 4.38 Details of Shear Stirrups

Problem 4.8

A concrete beam of rectangular section, 200mm wide and 600mm deep, is pre–stressed by a parabolic cable located at an eccentricity of 100mm at mid–span and zero at the supports. If a beam has a span of 10m and carries a uniformly distributed live load of 4kN/m, find the effective force necessary in the cable for zero stress at the support section. For this condition calculate the principal stresses. The density of concrete is 24kN/m³.

Solution:

The longitudinal elevation and cross section of the beam is shown in Fig. 4.39.

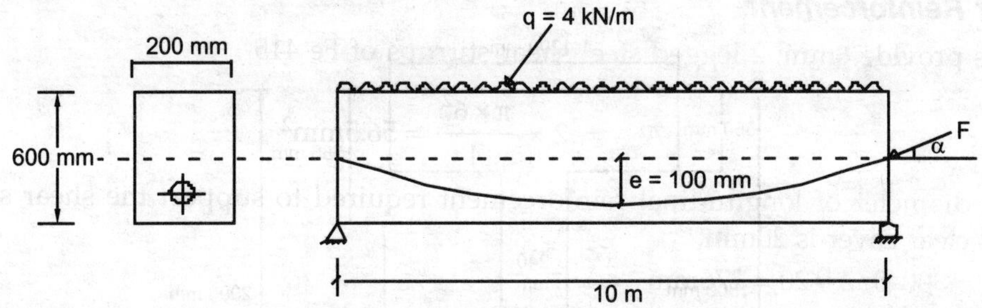

Fig. 4.39 Longitudinal Elevation and Cross–section of the Beam

$$\text{Self–weight} = (0.2 \times 0.6 \times 24) = 2.88\text{kN/m}$$

$$\text{Live load} = 4.00\text{kN/m}$$

$$\text{Total Load} = 6.88\text{kN/m} = w$$

$$V = \frac{wl}{2} = \frac{6.88 \times 10}{2} = 34.40\text{kN}$$

If P = Pre–stressing force,

$$\text{Slope of the cable} = \sin\alpha = \alpha = \frac{4r}{l} \times \frac{4 \times 0.1}{10} = 0.04 \text{ Radian}$$

$$P \sin\alpha = 34.40, \therefore P = 860\text{kN}$$

$$\text{Direct stress} = \frac{P}{A} = \frac{860 \times 10^3}{200 \times 600} = 7.2\text{N/mm}^2$$

Problem 4.9

An unsymmetrical I–section bridge girder has the following sectional properties:

Area of cross–section = 777×10^3 mm^2, second moment of area = 22×10^{10}mm^4, width and thickness of top flange = 1200 and 360 mm respectively, and thickness of web = 240mm. The centroid of the section is located at 580mm from the top. The girder is used over a span of 40m. The tendons with a cross section of 700mm^2 are parabolic with an eccentricity of 1220mm at the centre of span and zero at the supports. The effective pre–stress in the wires is 800N/mm^2. If the tensile strength of concrete is 4.5N/mm^2, estimate the ultimate shear resistance of the section, assuming failure to take place when the principal tensile stress reaches a value equal to the tensile strength of concrete. Overall depth is 2000mm.

Solution:

The un–symmetrical I–section of pre–stressed concrete beam is shown in Fig. 4.40.

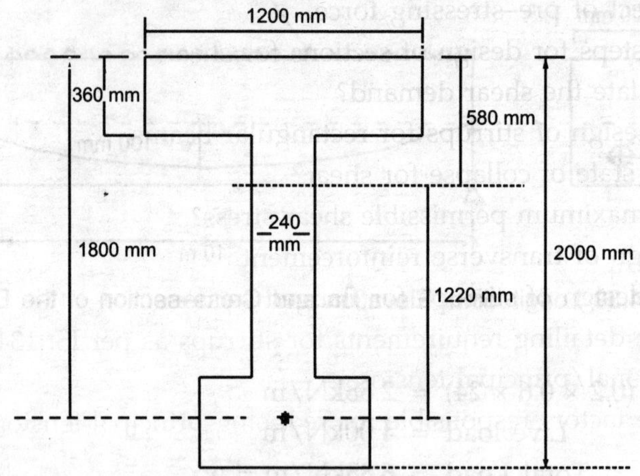

Fig. 4.40 Un–symmetrical I–section of Pre–stressed Concrete Beam

$$I = 22 \times 10^{10} mm^4, A = 777 \times 10^3 mm^2, f_t = 4.5 N/mm^2$$

$$S = A\bar{y} = [(1200 \times 360 \times 400) + (240 \times 220 \times 110)] = 17.78 \times 10^7 mm^3,$$

$$\therefore \quad e = (1220 - 580) = 640mm$$

$$V_{cw} = b\left(\frac{I}{S}\right)\sqrt{f_t^2 + f_{cp}f_t} + P \sin\alpha$$

$$\therefore \quad P = \frac{700 \times 800}{1000} = 560kN, f_{cp} = \frac{560 \times 10^3}{777 \times 10^3} = 0.72 N/mm^2$$

Slope of the cable at support = $\sin\alpha = \alpha = \frac{4r}{l} = \frac{4 \times 1.22}{40} = 0.122$ Radian

$$V_{cw} = 240\left(\frac{22 \times 10^{10}}{17.78 \times 10^7}\right)\sqrt{(4.5)^2 + (0.72 \times 4.5)} + (560 \times 10^3 \times 0.122) = 1507kN$$

According to IS: 1343 – 2012 code, shear strength is computed as

$$V_{co} = 0.67bD\sqrt{f_t^2 + 0.8f_{cp}f_t} + P \sin\alpha$$

$$V_{co} = (0.67 \times 240 \times 2000)\sqrt{(4.5)^2 + 0.48 \times 0.72 \times 4.5} + (560 \times 10^3 \times 0.122) = 1605kN$$

REVIEW QUESTIONS

4.1 Explain stresses in an un–cracked beam.

4.2 Distinguish the types of cracks and explain one by one.

4.3 What are the components of shear resistance?

4.4 Explain the modes of failure with the help of sketches.
4.5 Elaborate effect of pre–stressing force.
4.6 Mention the steps for design of sections for shear.
4.7 How to calculate the shear demand?
4.8 Explain the design of stirrups for rectangular beams.
4.9 What is limit state of collapse for shear?
4.10 What is the maximum permissible shear stress?
4.11 Explain design of transverse reinforcement.
4.12 Explain the design of stirrups for flanged sections.
4.13 What are the detailing requirements for stirrups as per IS: 1343–2012.
4.14 Explain diagonal/principal tension.
4.15 What are the factors responsible for reducing principal tension?

EXERCISE PROBLEMS

4.1 A pre–stressed concrete beam of rectangular section 300mm wide by 600mm deep is pre–stressed by two post–tensioned cables of area 600mm^2 each initially pre–stressed to 1600N/mm^2. The cables are located at a constant eccentricity of 100mm. The span of the beam is 10m. If f_{ck} = 40N/mm^2, estimate the ultimate shear resistance of support section un–cracked in flexure.

[Ans: Ultimate shear resistance of support section un–cracked in flexure = 475.164kN]

4.2 The support section of a pre–stressed concrete beam, 120mm wide and 250mm deep, is required to support an ultimate shear force of 60kN. The compressive pre–stress at the centroidal axis is 5N/mm^2. The characteristic strength of concrete is 40N/mm^2. The cover to the tension reinforcement is 50mm. If the characteristic tensile stress of steel in stirrups is 250N/mm^2, design suitable reinforcements at the section using IS: 1343 code specifications.

[Ans: 6mm diameter two legged stirrups at 180mm centres]

4.3 A double T–section used as bridge girder is made up of a flange 1200mm wide by 150mm thick. The ribs are 150mm wide by 1500mm deep. The sections is pre–stressed by high tensile wires of area 4700mm^2 located at 850mm from the soffit of the girder. The initial stress in the wires is 1200N/mm^2. If f_{ck} = 40N/mm^2, estimate the ultimate shear strength of the support section of the girder.

[Ans: Ultimate shear strength of double T–section at support section = 1203.117kN]

4.4 A pre–stressed concrete bridge girder of unsymmetrical I–section has an overall depth of 1300mm with a cross sectional area of 328500mm^2. Thickness of the web is 150mm, span of the girder is 25m. The girder is pre–stressed by a parabolic cable having an eccentricity of 650mm at the centroid and 285mm at the supports sections. The effective pre–stressing force in cable is 2070kN. If f_t = 1.6N/mm^2, estimate the ultimate shear resistance of the support section.

[Ans: Ultimate shear strength of support section = 546kN]

4.5 The support section of a pre–stressed concrete beam 100mm wide by 250mm deep is required to support an ultimate shear force of 60kN. The compressive pre–stress at centroid is 5N/mm^2, f_{ck} = 40N/mm^2, effective cover to reinforcement = 50mm. If f_y = 415N/mm^2, design suitable shear reinforcement in the section using IS: 1343 code.

[Ans: 6mm diameter two legged stirrups at 150mm centres]

OBJECTIVE QUESTIONS

4.1 The analysis for axial load and flexure are based on,

a. equilibrium of internal and external forces

b. compatibility of strains in concrete and steel

c. constitutive relationships of materials

d. All of the above

4.2 Concrete cracks when the maximum principal tensile stress exceeds a value of

a. $f_t = 0.29\sqrt{f_{ck}}$ b. $f_t = 0.39\sqrt{f_{ck}}$ c. $f_t = 0.49\sqrt{f_{ck}}$ d. $f_t = 0.59\sqrt{f_{ck}}$

4.3 Modes of failure due to shear are

1. Diagonal tension failure
2. Shear compression failure
3. Shear tension failure
4. Web crushing failure
5. Arch rib failure

a. 1, 2, 3, 5 b. 1, 3, 4, 5

c. 1, 2, 3, 4, 5 d. 1, 4, 5

4.4 As per IS: 1343–2012, the formula for V_c for un–cracked sections is

a. $0.67bD\sqrt{f_t^2 + 0.8f_{cp}f_t}$ b. $0.57bD\sqrt{f_t^2 + 0.8f_{cp}f_t}$

c. $0.47bD\sqrt{f_t^2 + 0.8f_{cp}f_t}$ d. $0.37bD\sqrt{f_t^2 + 0.8f_{cp}f_t}$

4.5 The principal tensile stress is equated to the direct tensile strength of concrete (f_t) as,

a. $f_t = -\frac{f_{cp}}{4} + \sqrt{\frac{f_{cp}}{4}^2 + \left(\frac{V_{co}Q}{Ib}\right)^2}$ b. $f_t = -\frac{f_{cp}}{2} + \sqrt{\frac{f_{cp}}{4}^2 + \left(\frac{V_{co}Q}{Ib}\right)^2}$

c. $f_t = -\frac{f_{cp}}{6} + \sqrt{\frac{f_{cp}}{4}^2 + \left(\frac{V_{co}Q}{Ib}\right)^2}$ d. $f_t = -\frac{f_{cp}}{2} + \sqrt{\frac{f_{cp}}{6}^2 + \left(\frac{V_{co}Q}{Ib}\right)^2}$

4.6 The value of V_c is ...., where f_{ck} = Characteristic strength of concrete.

a. $0.1bd\sqrt{f_{ck}}$ b. $0.2bd\sqrt{f_{ck}}$ c. $0.25bd\sqrt{f_{ck}}$ d. $0.3bd\sqrt{f_{ck}}$

4.7 As per IS: 1343–2012, the formula for V_c for un–cracked sections is

a. $V_c = \left(1 - 0.65\frac{f_{pe}}{f_{pk}}\right)\tau_c bd + M_o\frac{V_u}{M_u}$ b. $V_c = \left(1 - 0.55\frac{f_{pe}}{f_{pk}}\right)\tau_c bd + M_o\frac{V_u}{M_u}$

c. $V_c = \left(1 - 0.75\frac{f_{pe}}{f_{pk}}\right)\tau_c bd + M_o\frac{V_u}{M_u}$ d. $V_c = \left(1 - 0.85\frac{f_{pe}}{f_{pk}}\right)\tau_c bd + M_o\frac{V_u}{M_u}$

4.8 Maximum permissible shear stress is computed using the relation

a. $\frac{V_u}{bd_t} \geq \tau_{c,\,max}$ b. $\frac{V_u}{bd_t} \geq \tau_{c,\,max}$

c. $\frac{V_u}{bd_t} \geq \tau_{c,\,max}$ d. $\frac{V_u}{bd_t} \geq \tau_{c,\,max}$

4.9. The total area of the legs of the stirrups (A_{sv}) is given by,

a. $\frac{A_{sv}}{S_v} = \frac{V_u - V_c}{0.67f_y d_t}$ b. $\frac{A_{sv}}{S_v} = \frac{V_u - V_c}{0.67f_y d_t}$

c. $\frac{A_{sv}}{S_v} = \frac{V_u - V_c}{0.57f_y d_t}$ d. $\frac{A_{sv}}{S_v} = \frac{V_u - V_c}{0.87f_y d_t}$

4.10 The design shear force in a flange is given as

a. $V_f = \frac{\tau_{f,max}}{2}\frac{b_f}{4}D_f$ b. $V_f = \frac{\tau_{f,max}}{2}\frac{b_f}{2}D_f$

c. $V_f = \frac{\tau_{f,max}}{2}\frac{b_f}{6}D_f$ d. $V_f = \frac{\tau_{f,max}}{2}\frac{b_f}{3}D_f$

4.11 The maximum shear stress in the flange is given by an expression

a. $\tau_{f,max} = \frac{V_u A_1 \bar{y}}{ID_f}$ b. $\tau_{f,max} = \frac{V_u A_1 \bar{y}}{0.87ID_f}$

c. $\tau_{f,max} = \frac{V_u A_1 \bar{y}}{0.5ID_f}$ d. $\tau_{f,max} = \frac{V_u A_1 \bar{y}}{0.7ID_f}$

4.12 The amount of horizontal reinforcement in the flange (A_{svf}) is calculated from V_f as......................

a. $A_{svf} = \frac{V_f}{0.67f_y}$ b. $A_{svf} = \frac{V_f}{0.77f_y}$

c. $A_{svf} = \frac{V_f}{0.75f_y}$ d. $A_{svf} = \frac{V_f}{0.87f_y}$

4.13 The maximum spacing for shear stirrups is

a. $0.75d_t$ or $4b_w$, whichever is smaller b. $0.85d_t$ or $4b_w$, whichever is smaller

c. $0.75d_t$ or $5b_w$, whichever is smaller d. $0.75d_t$ or $6b_w$, whichever is smaller

4.14 When V_u is larger than $1.8V_c$, the maximum spacing is

a. $0.7d_t$ b. $0.6d_t$ c. $0.55d_t$ d. $0.5d_t$

4.15 The minimum thickness (breadth) of Web to check web crushing failure as per IRC: 18–2000 is
   a. 200mm plus diameter of the duct hole
   b. 200mm minus diameter of the duct hole
   c. 300mm plus diameter of the duct hole
   d. 300mm minus diameter of the duct hole

4.16 The maximum shearing stress is .... times the average stress
   a. 0.5 b. 1.5 c. 2.5 d. 3.5

Answers

4.1 d	4.2 a	4.3 c	4.4 a	4.5 b
4.6 a	4.7 b	4.8 c	4.9 d	4.10 b
4.11 a	4.12 d	4.13 a	4.14 d	4.15 a
4.16 c				

❑

CHAPTER

5

Deflection of Pre–stressed Concrete Members

5.1 INTRODUCTION

This chapter deals with the deflection of pre–stressed concrete members caused by application of pre–stressing and external forces. Both short time and long time deflections behaviour of pre–stressed concrete member have been considered. The deflection of structures on part thereof shall not adversely affect the appearance or efficiency of the structure or finishes. Hence the deflection shall generally be limited to the following. The final deflection, due to all loads including effects of temperature, creep and shrinkage and measured from the cast level of the supports of floors, roofs and all other horizontal members should not normally exceed span/250.

The deflection of a flexural member is calculated to satisfy a limit state of serviceability. Since a pre–stressed concrete member is smaller in depth than an equivalent reinforced concrete member, the deflection of a pre–stressed concrete member tends to be larger.

The total deflection is a resultant of the upward deflection due to pre–stressing force and downward deflection due to the gravity loads. Only the flexural deformation is considered and any shear deformation is neglected in the calculation of deflection. Shear deformation is included in members such as deep beams and wall type of structures.

The deflection of a member is calculated at least for two cases.

1. Short term deflection at transfer
2. Long term deflection under service loads

The short term deflection at transfer is due to the pre–stressing force (before long term losses) and self–weight. The effect of creep and shrinkage of concrete are not considered. The long term deflection under service loads is due to the effective pre–stressing force (after long term losses) and the gravity loads. The permanent components of the gravity loads are considered in the effect of creep. These components are dead load and sustained live load.

5.2 FACTORS INFLUENCING DEFLECTION

Following are various factors which have significant influence on deflection in PSC beams.

1. **Span:** As the span increases, the deflection also increases.
2. **Modulus of elasticity of concrete:** As the strength of concrete increases its modulus of elasticity also increases and deflection decreases.
3. **Superimposed load:** As the magnitude of super imposed load increases, the deflection also increases.

4. **Pre–stress:** As the magnitude of pre–stress increases the amount of deflection decreases.
5. **Moment of inertia:** As the moment of inertia of the cross section increases, the amount of deflection decreases.
6. **Creep and shrinkage of concrete:** These factors have significant influence on long term deflection, in particular as the creep of concrete increases, the deflection also increases.

5.3 LIMITS OF DEFLECTION

Clause 19.3.1 of IS: 1343–2012 specifies limits of deflection such that the efficiency of the structural element and the appearance of the finishes or partitions are not adversely affected. The limits of deflection are summarized next.

1. The total deflection due to all loads, including the effects of temperature, creep and shrinkage, should not exceed span / 250.
2. The deflection after erection of partitions or application of finishes, including the effects of temperature, creep and shrinkage, should not exceed span/350 or 20 mm, whichever is less.
3. If finishes are applied, total upward deflection due to pre–stressing force should not exceed span / 300.

5.4 SHORT TERM DEFLECTIONS AT TRANSFER

5.4.1 Mohr's Theorem

Short term or instantaneous deflections of pre–stressed members are governed by the bending moment distribution along the span and the flexural rigidity of the members.

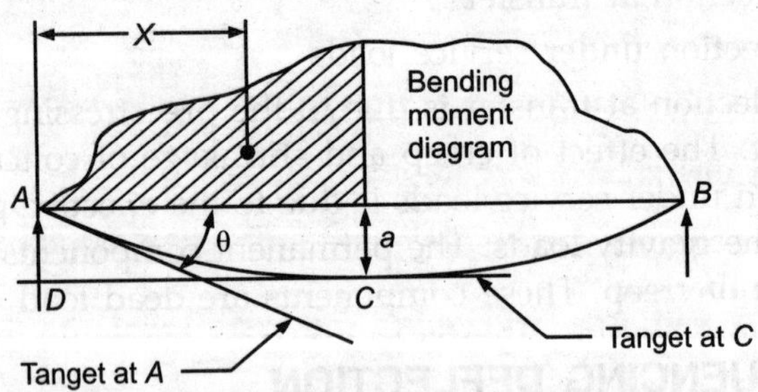

Fig. 5.1 Slope and Deflection of Beam

Mohr's moment area theorem are rapidly applicable for the estimation of deflections due to pre–stressing force, self–weight and imposed loads. Consider Fig. 5.1, in which beam AB subjected to a bending moment distribution due to pre–stressing force or self–weight or imposed loads. ACB is the centerline of the deformed structure under the system of given loads.

If θ = Slope of the elastic curve at A

AD = Intercept between the tangent at C and the vertical at A

δ = Deflection at the centre for symmetrically loaded, simply supported beam (since the tangent is horizontal for such cases)

A = Area of the BMD between A and C

x = Distance of the centroid of the BMD between A and C from the left support

EI = Flexural rigidity of the beam

Then according to Mohr's first theorem,

$$\text{Slope} = \frac{\text{Area of BMD}}{\text{Flexural Rigidity}}, \theta = \frac{A}{EI} \qquad \text{....(5.1)}$$

Mohr's second theorem states that

$$\text{Intercept, } \delta = \left(\frac{\text{Moment of the area of BMD}}{\text{Flexural Rigidity}}\right) = \left(\frac{Ax}{EI}\right) \qquad \text{...(5.2)}$$

The deflection of symmetrically loaded and simply supported beam, at the mid–span point is directly obtained from the second moment area theorem since the tangent is horizontal at this point. More complicated problems involving unsymmetrical loading may be solved by combining both the moment area theorems.

5.4.2 Deflection due to Gravity Loads

The methods of calculation of deflection are taught in a course on structural analysis. It is expected that the students are familiar with the methods. The methods include the following.

1. Double integration method
2. Moment–area method
3. Conjugate beam method
4. Principle of virtual work

Numerical solution schemes can be developed based on the above methods and executed in a computer. For members with prismatic cross–sections, common support conditions and subjected to conventional loading, the deflections are available in tables of text books on structural analysis.

The expressions of deflection (δ) for few cases are provided. Here,

I = Moment of inertia

E = Modulus of elasticity of concrete

5.4.2.1 Simply Supported Beams Carrying Central Concentrated Load

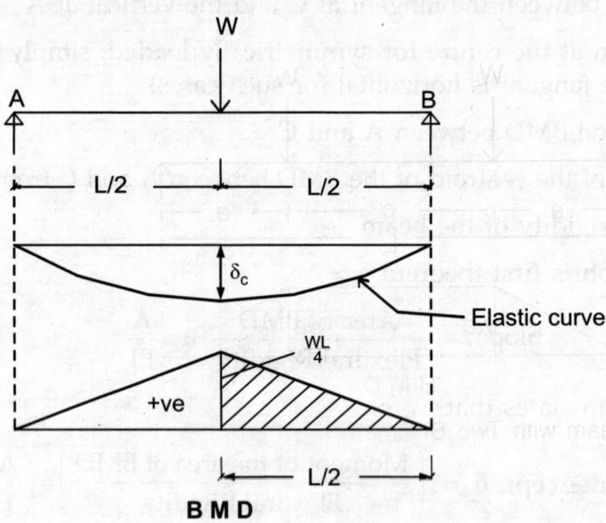

Fig. 5.2 Simply Supported Beam Carrying Central Concentrated Load

Referring Fig. 5.2, by Mohr's theorem, $\delta_c = \dfrac{Ax}{EI} = \dfrac{1}{EI}\left[\dfrac{1}{2}\times\dfrac{L}{2}\times\dfrac{wL}{4}\times\dfrac{2}{3}\times\dfrac{L}{2}\right]$

$$\delta_c = \frac{wL^3}{48EI} \qquad ...(5.3)$$

5.4.2.2 Simply Supported Beam Carrying UDL over the Entire Span

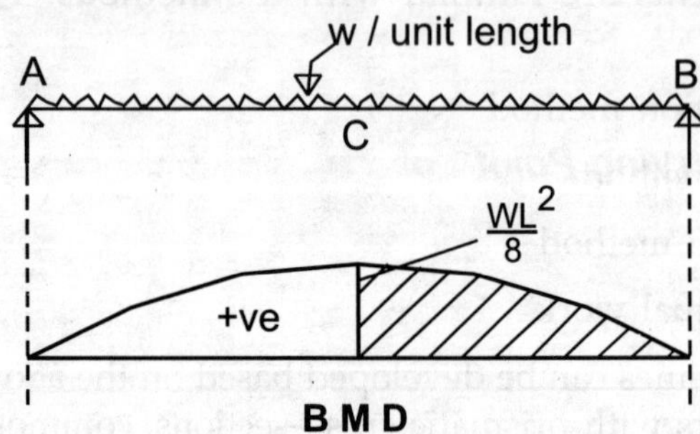

Fig. 5.3 Simply Supported Beam Carrying UDL over the Entire Span

Referring Fig. 5.3, $\delta_c = \dfrac{1}{EI}\left[\dfrac{2}{3}\times\dfrac{L}{2}\times\dfrac{wL^2}{8}\times\dfrac{5}{8}\times\dfrac{L}{2}\right]$

$$\delta_c = \frac{5wL^4}{384EI} \qquad ...(5.4)$$

5.4.2.3 Simply Supported Beam Carrying Two Symmetrically Placed Two Loads of Equal Magnitude

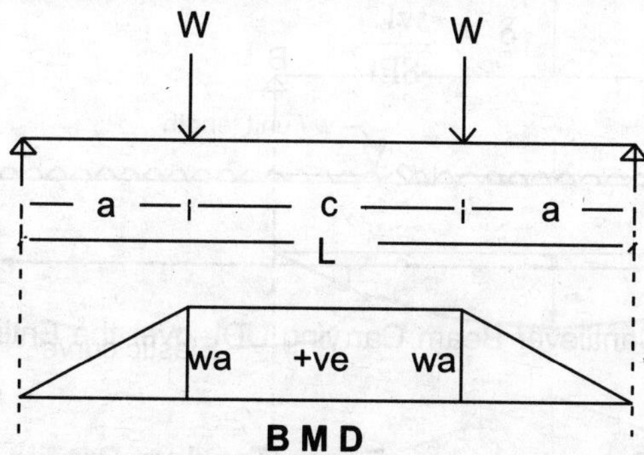

Fig. 5.4 Simply Supported Beam with Two Symmetrically Placed Two Loads of Equal Magnitude

Referring Fig. 5.4, $\delta_c = \frac{1}{EI}\left\{\left(\frac{1}{2}\times a\times wa\right)\frac{2}{3}a + wa\left(\frac{L}{2}-a\right)\times\left[a-\frac{1}{2}\right]\left(\frac{L}{2}-a\right)\right\}$

$$= \frac{1}{EI}\left[\frac{wa^3}{3}+\frac{wa}{2}\left(\frac{L}{2}-a\right)\left(\frac{L}{2}+a\right)\right]$$

$$= \frac{wa}{6EI}\left[2a^2+3\left(\frac{L^2}{4}-a^2\right)\right] = \frac{wa}{6EI}\left[\frac{8a^2+3L^2-12a^2}{4}\right]$$

$$\delta_c = wa\left[\frac{3L^2-4a^2}{24EI}\right] \qquad ...(5.5)$$

5.4.2.4 Cantilever Beam Carrying Point Load at the Free End

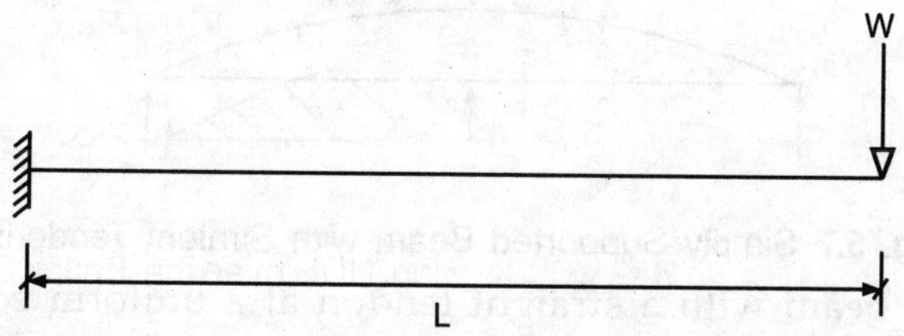

Fig. 5.5 Cantilever Beam Carrying Point Load at the Free End

Referring Fig. 5.5, deflection at the free end is,

$$\delta = \frac{wL^3}{3EI} \qquad ...(5.6)$$

5.4.2.5 Cantilever Beam Carrying UDL over the Entire Span

Referring Fig. 5.6, deflection at the free end is,

$$\delta = \frac{wL^4}{8EI} \qquad \text{...(5.7)}$$

Fig. 5.6 Cantilever Beam Carrying UDL over the Entire Span

5.4.3 Deflection due to Pre–stressing Force/Tendon Profile

In most of the cases of pre–stressed beams, tendons are located with eccentricities towards the soffit of the beams to counteract the sagging bending moments due to transverse loads. Consequently, the concrete beams deflects upwards (camber) on the application or transfer of pre–stress. Since the bending moment at every section is the product of the pre–stressing force and eccentricity, the tendon profile itself will represent the shape of the BMD. The deflection due to pre–stressing force is calculated by the load–balancing method.

5.4.3.1 Simply Supported Beam with Straight Tendons

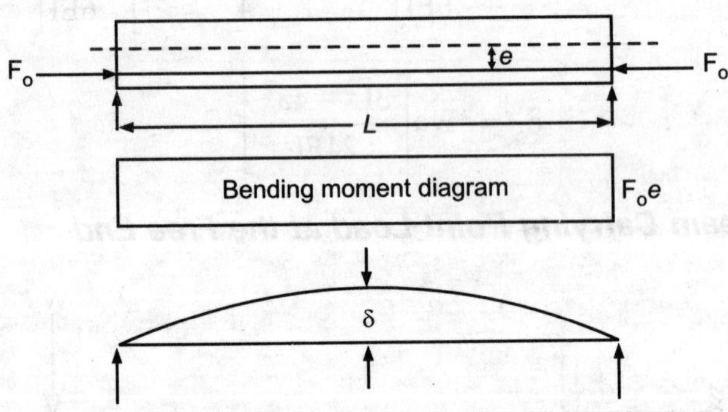

Fig. 5.7 Simply Supported Beam with Straight Tendons

Fig. 5.7 shows the beam with a straight tendon at a uniform eccentricity below the centroidal axis. If upward deflections are considered as negative and,

F_o = Effective pre–stressing force, e = Eccentricity, L = Length of the beam

$$\delta = -\frac{(F_o eL)(L/4)}{EI} = -\frac{F_o eL^2}{8EI} \qquad \text{...(5.8)}$$

5.4.3.2 Simply Supported Beam with Trapezoidal Tendons/Doubly Harped Cable

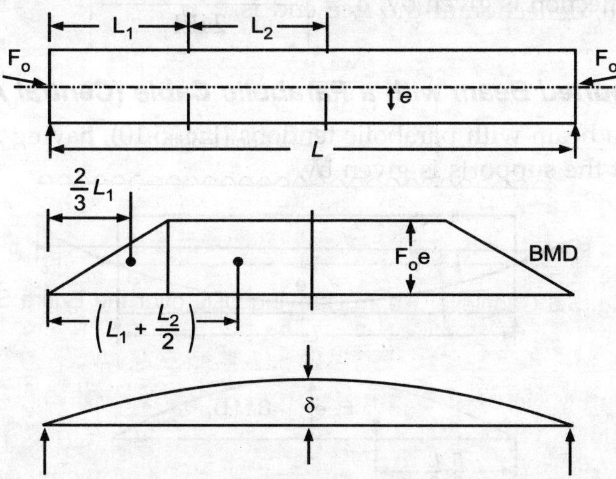

Fig. 5.8 Simply Supported Beam with Trapezoidal Tendons

A draped tendon with a trapezoidal profile is shown in Fig. 5.8. Considering the BMD, the deflection at the centre of the beam is obtained by taking the moment of the area of the BMD over the one half of the span. Thus,

$$\delta = -\frac{F_oe}{EI}[l_2(l_1 + l_2/2) + (l_1/2)(2/3l_1)],\ \delta = -\frac{F_oe}{EI}[2l_1^2 + 6l_1l_2 + 3l_2^2] \qquad ...(5.9)$$

OR

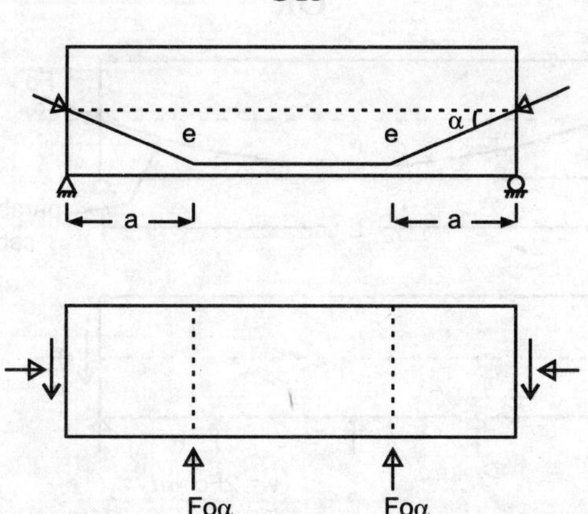

Fig. 5.9 Simply Supported Beam with Doubly Harped Cable

In this case the effect of bent up cable is reported by two upward concentrated load each of magnitude $F_o\alpha$ (Ref. Fig. 5.9). Where, $\alpha = \frac{e}{a}$. Therefore upward concentrated load is $w_{up} = \frac{F_oe}{a}$.

The maximum deflection is given by, $\delta_c = \left[\dfrac{W_{up}a(3L^2 - 4a^2)}{24EI}\right]$ ...(5.10)

5.4.3.3 Simply Supported Beam with a Parabolic Cable (Central Anchors)

The deflection of a beam with parabolic tendons (Fig. 5.10), having an eccentricity 'e', at the centre and zero at the supports is given by,

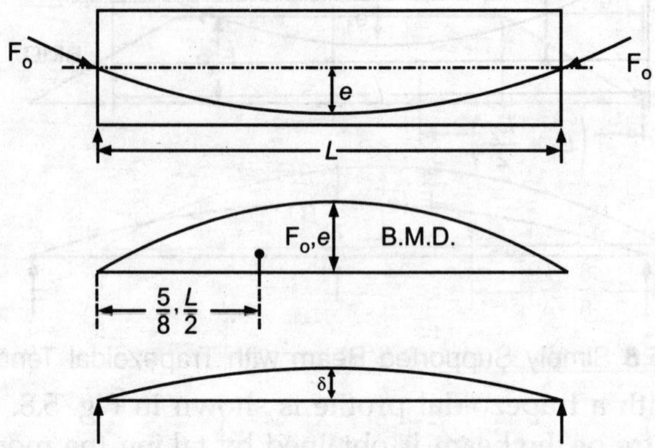

Fig. 5.10 Simply Supported Beam with a Parabolic Cable (Central Anchors)

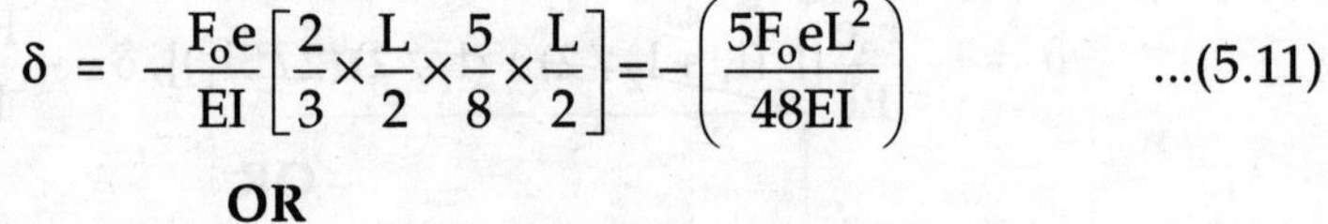

$$\delta = -\frac{F_o e}{EI}\left[\frac{2}{3}\times\frac{L}{2}\times\frac{5}{8}\times\frac{L}{2}\right] = -\left(\frac{5F_o eL^2}{48EI}\right) \quad ...(5.11)$$

OR

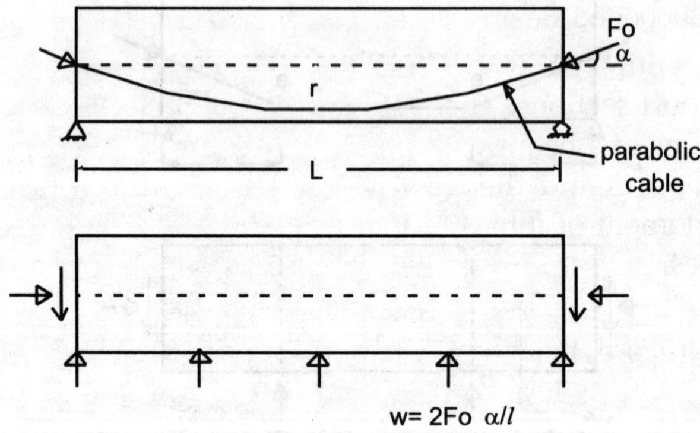

Fig. 5.11 Simply Supported Beam with a Parabolic Cable

Referring to Fig. 5.11, the effect of cable is reported by upward UDL of intensity $\left(\dfrac{2F_o\alpha}{L}\right)$.

Here $\alpha = \dfrac{4\gamma}{L}$ represents the inclination of the cable with the horizontal. Hence, $w_{up} = \dfrac{8F_o\gamma}{L^2}$.

Therefore the maximum deflection is given by, $\delta = \dfrac{5 w_{up} L^4}{384EI}$ ...(5.12)

5.4.3.4 Simply Supported Beam with a Parabolic Cable (Eccentric Anchors)

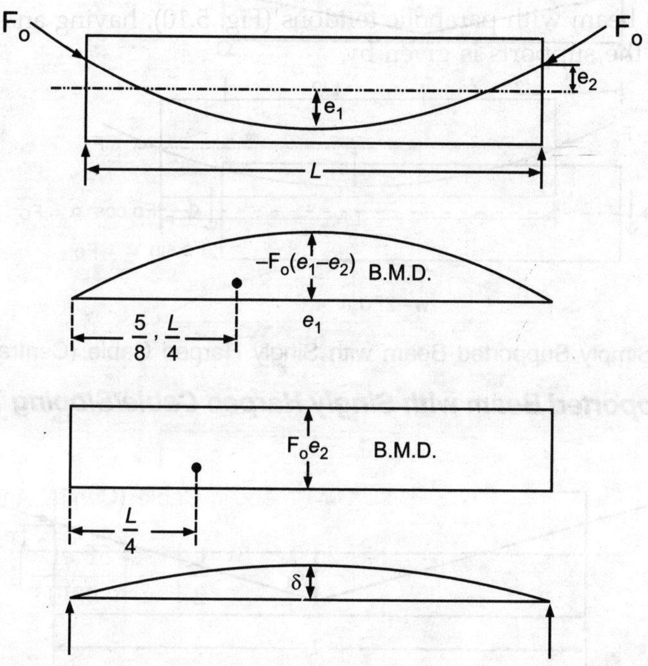

Fig. 5.12 Simply Supported Beam with a Parabolic Cable (Eccentric Anchors)

Fig. 5.12 shows a beam, with a parabolic tendon having an eccentricity e_1 at the centre of the beam and e_2 at the support sections. The resultant deflection at the centre is obtained by sum of the upward deflections of a beam with parabolic tendon of eccentricity $(e_1 + e_2)$ at the centre and zero at the supports and the downward deflection of a beam subjected to a uniform sagging bending moment of intensity $F_o e_2$ throughout the length. Consequently, the resultant deflection becomes,

$$\delta = \left[-\frac{5}{48}\frac{F_o L^2}{EI}(e_1 + e_2)\right] + \left[\frac{F_o e_2 L^2}{8EI}\right], \delta = \frac{F_o L^2}{48EI}(-5e_1 + e_2) \quad ...(5.13)$$

5.4.3.5 Simply Supported Beam with Singly Harped Cable/Sloping Tendon (Central Anchors)

Horizontal component of the pre–stressing force produce a column effect in the beam. However, deflection due to column effect in insignificant and therefore it usually ignored. The reaction to the downward component is represented by a central upward concentrated load of magnitude, $w_{up} = w = 2F_o\alpha$. Where, $\alpha = \dfrac{e}{L/2}$. Thus, the effect of bent cable shown in Fig. 5.13 is represented by a central concentrated load. Therefore deflection is given by,

$$\delta = \frac{w_{up} L^3}{48EI} \qquad ...(5.14)$$

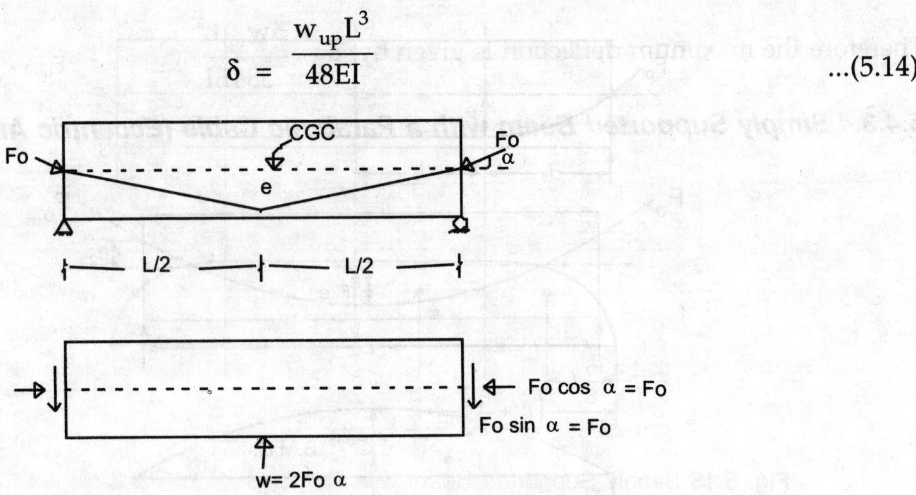

Fig. 5.13 Simply Supported Beam with Singly Harped Cable (Central Anchors)

5.4.3.6 Simply Supported Beam with Singly Harped Cable/Sloping Tendon (Eccentric Anchors)

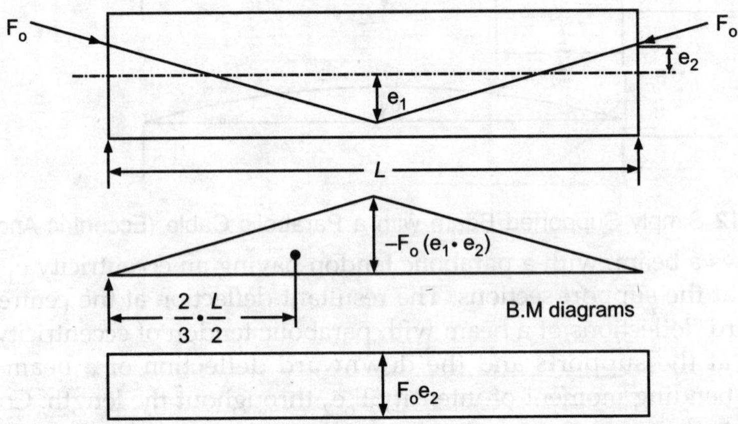

Fig. 5.14 Simply Supported Beam with Sloping Tendon (Eccentric Anchors)

From Fig. 5.14, the deflection computed in the similar way as discussed in section 5.4.3.4

$$\delta = \left[-\frac{F_o L^2}{12EI}(e_1 + e_2) \right] + \left[\frac{F_o e_2 L^2}{8EI} \right], \delta = \frac{F_o L^2}{24EI}(-2e_1 + e_2) \qquad ...(5.15)$$

5.4.3.7 Simply Supported Beam with Combination of Parabolic and Straight Tendons (Central Anchors)

Referring to Fig. 5.15, the deflection at the centre of the beam is calculated as follows,

$$\delta = -\frac{F_o e}{EI}[(2/3)l_1 (5/8)l_1 + l_2(l_1 + l_2/2)], \delta = -\frac{F_o e}{EI}[5l_1^2 + 12l_1 l_2 + 6l_2^2] \qquad ...(5.16)$$

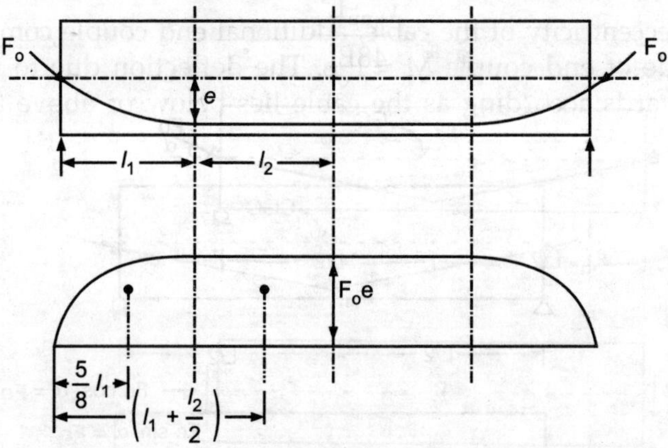

Fig. 5.15 Simply Supported Beam with Combination of Parabolic and Straight Tendons (Central Anchors)

5.4.3.8 Simply Supported Beam with Combination of Parabolic and Straight Tendons (Eccentric Anchors)

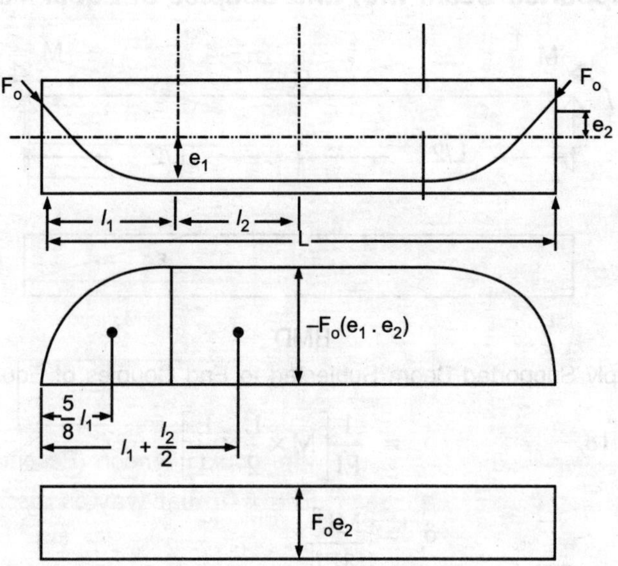

Fig. 5.16 Simply Supported Beam with Combination of Parabolic and Straight Tendons (Eccentric Anchors)

Referring to Fig. 5.16, the maximum central deflection is obtained as follows,

$$\delta = -\frac{F_o(e_1 + e_2)}{12EI}[5l_1^2 + 12l_1l_2 + 6l_2^2] + \left[\frac{F_oe_2L^2}{8EI}\right] \quad ...(5.17)$$

5.4.3.9 Simply Supported Beam with Cable Eccentric at the Ends

Due to the end eccentricity of the cable, additional end couple comes into picture at the ends. The magnitude of end couple $M = F_o e$. The deflection due to these end couples is upwards or downwards according as the cable lies below or above the CGC at the ends (Ref. Fig. 5.17).

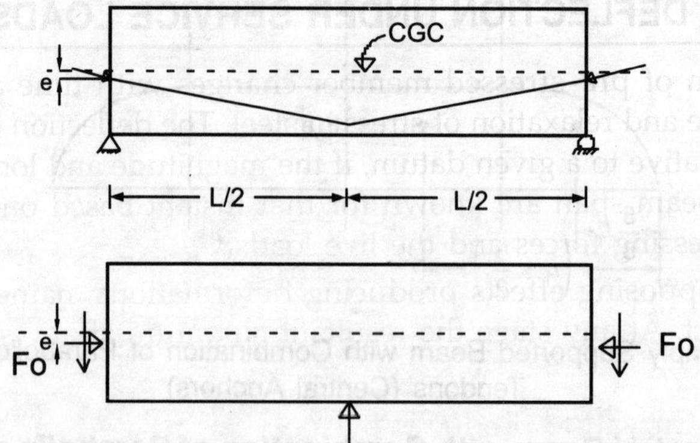

Fig. 5.17 Simply Supported Beam with Cable Eccentric at the Ends

5.4.3.10 Simply Supported Beam with End Couples of Equal Magnitude M

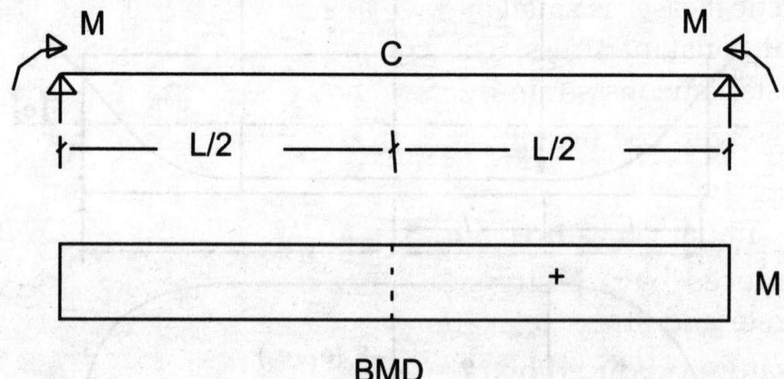

Fig. 5.18 Simply Supported Beam Subjected to End Couples of Equal Magnitude M

Referring Fig. 5.18, $$\delta_c = \frac{1}{EI}\left[M \times \frac{L}{2} \times \frac{L}{4}\right]$$

$$\delta_c = \frac{ML^2}{8EI} \qquad \text{...(5.18)}$$

5.4.4 Deflection due to Self–weight and Imposed Loads

At the time of transfer of pre–tress, the beam hogs up due to the effect of pre–stressing. At this stage, the self–weight of the beam induces downward deflections, which further increases due to effect of imposed load on the beam.

If g = Self–weight of the beam per m

q = Imposed load per m (uniformly distributed)

The downward deflection computed as $\delta = \frac{5(g+q)L^4}{384EI}$ ...(5.19)

Deflections due to concentrated live loads can be computed using Mohr's theorems as discussed in earlier sections.

5.5 LONG TERM DEFLECTION UNDER SERVICE LOADS

The deformation of pre–stressed member changes with time as a result of creep and shrinkage of concrete and relaxation of stress in steel. The deflection of pre–stressed members can be computed relative to a given datum, if the magnitude and longitudinal distribution of curvatures for the beam span are known for that instant based on the load history, which includes the pre–stressing forces and the live loads.

There are two opposing effects producing deformations, namely pre–stress effect and transverse load effect. At any stage the resultant curvature at any section is given by,

$$\phi_t = \phi_{mt} + \phi_{pt} \qquad ...(5.20)$$

Where, ϕ_{mt} = Curvature produced by transverse loads

ϕ_{pt} = Curvature produced by pre–stress

When maintained under sustained transverse loads, due to the effect of creep there will be increase in strain and consequently there is an increase of curvature. However the change in stress in the practical cases is small and we may assume that the creep of the concrete takes place practically at constant stress. The creep strain at any time t may be determined by multiplying the instantaneous strain by the quantity $(1+\phi)$.

Where, ϕ is the creep coefficient

Hence $\phi_{mt} = (1+\phi)\,\phi_i$ ...(5.21)

Where, ϕ_i = Initial curvature caused by application of transverse load.

Curvature produced by pre–stress (ϕ_{pt}) depends on the combined effect of creep and shrinkage of concrete and stress relaxation of steel.

The creep curvature produced by pre–stress is determined by assuming that, the creep is produced based on the average pre–stress which acts on the given time.

Let P_i = Initial pre–stress, P_t = Pre–stress at time t,

$\Delta P_i = P_i - P_t$ = Loss of pre–stress due to creep, shrinkage and stress relaxation.

The curvature due to pre–stress at time t is given by

$$\phi_{pt} = -\frac{P_i e}{EI}\left\{1-\frac{\Delta P_i}{P_i}+\left(1-\frac{\Delta P_i}{2P_i}\right)\phi\right\} \qquad ...(5.22)$$

Let δ_{pt} = Initial deflection due to transverse loads

δ_{ip} = Initial deflection due to pre–stress

Now the total long term deflection after time t (Neville's formula) is given by,

$$\delta_t = \delta_{il}(1-\phi) - \delta_{ip}\left\{1-\frac{\Delta P_i}{P_i}+\left(1-\frac{\Delta P_i}{2P_i}\right)\phi\right\} \qquad ...(5.23)$$

Note: In the above expression negative sign indicates upward deflection.

Simplified approximate method suggested by Lin, for determining long term deflection is given by, the initial deflection due to pre–stress and transverse loads is modified due to the loss of pre–stress which causes a decrease in deflection and the effect of creep which increase the deflection and it is suggested to adopt a reduced modulus which involves the creep coefficient. The long term deflection is given by,

$$\delta_t = \left(\delta_{il} - \delta_{ip} \times \frac{P_t}{P_i}\right)(1+\phi) \quad \text{...(5.24)}$$

Eq. 5.24 termed as Lin's formula.

5.6 CAMBER IN PSC BEAM

The initial reversed curvature in PSC beam is usually referred as camber. Following are the various advantages of camber.

1. Load bearing capacity of the structure increases.
2. Size of the member can be reduced and this inter reduces dead load.
3. Deflection under working load reduced to a minimum.
4. Camber adds to aesthetic beauty of the structure.

The major disadvantage of the excessive camber lies in the fact, it results in uncomfortable driving qualities.

5.7 SOLVED EXAMPLES

Problem 5.1

A simply supported beam 300mm × 600mm in section is pre–stressed with a bent cable which has maximum eccentricity of 200mm at mid–span and minimum eccentricity of 50mm at the supports. Initial pre–stress in the tendon is 800kN and loss of pre–stress is 20%. Determine maximum deflection at transfer of pre–stress and maximum deflection under the working load of 2kN/m. Assume M40 concrete and span 10m.

Solution:

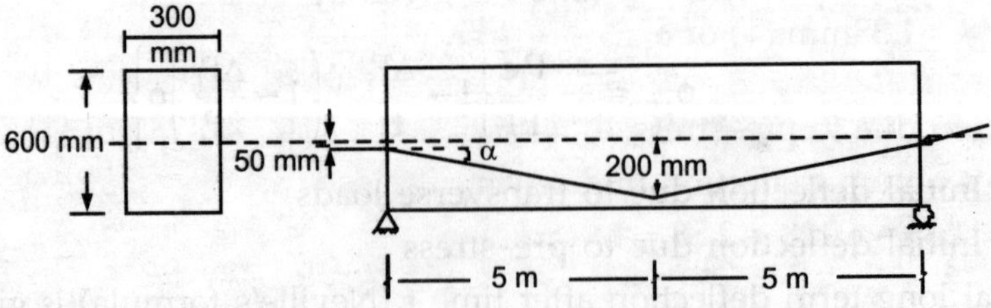

Fig. 5.19 Simply Supported Beam with Bent Cable

$A = 300 \times 600 = 180000\text{mm}^2$, $I = \dfrac{300 \times 600^3}{12} = 5.4 \times 10^9\text{mm}^4$, Span = L = 10m = 10000mm

Unit weight = 24kN/m^3 of concrete, Self–weight = $0.3 \times 0.6 \times 24$
$= 4.32\text{kN/m} = 4.32\text{N/mm}$

$E = E_c = 5700\sqrt{f_{ck}} = 5700\sqrt{40} = 36.049 \times 10^3\text{N/mm}^2$, $\tan\alpha = \alpha = \dfrac{150}{5000} = 0.03$ Radian

$F_o = 800\text{kN} = 800 \times 10^3\text{N}$,

Central upward concentrated load due to bent cable = $w = 2F_o\alpha$
$= 2 \times 800 \times 10^3 \times 0.03 = 48000\text{N}$

Due to end eccentricity, couple at each end
$M = F_o e = 800 \times 10^3 \times 50 = 40 \times 10^6\text{N/mm}$.

- ***Maximum Deflection at Transfer of Pre–stress***

Central deflection due to self–weight (δ_1)

$$\delta_1 = \frac{5wL^4}{384EI} = \frac{5 \times 4.32(10000)^4}{384 \times 36.049 \times 10^3 \times 5.4 \times 10^9} = 2.88\text{mm} (\downarrow)$$

Central deflection due to upward concentrated load (δ_2)

$$\delta_2 = \frac{wL^3}{48EI} = \frac{48000(10000)^3}{48 \times 36.049 \times 10^3 \times 5.4 \times 10^9} = 5.1\text{mm} (\uparrow)$$

Central deflection due to end couple (δ_3)

$$\delta_3 = \frac{ML^2}{8EI} = \frac{40 \times 10^6 (10000)^2}{8 \times 36.049 \times 10^3 \times 5.4 \times 10^9} = 2.56\text{mm} (\uparrow)$$

Resultant central deflection due to pre–stress
$\delta_4 = \delta_2 + \delta_3 = 5.1 + 2.56 = 7.66\text{mm} (\uparrow)$

Resultant max. deflection at transfer of pre–stress = $\delta_4 - \delta_1 = 7.66 - 2.88 = 4.79\text{mm} (\uparrow)$

- ***Maximum Deflection under Working Load***

$$\delta_5 = \frac{2}{4.32} \times 2.88 = 1.33\text{mm} (\downarrow) \text{ or } \delta_5 = \frac{5wL^4}{384EI} = \frac{5 \times 2 \times (10000)^4}{384 \times 36.049 \times 10^3 \times 5.4 \times 10^9} = 1.33\text{mm} (\uparrow)$$

Central deflection due to pre–stress, $\delta_6 = \eta\delta_4 = 0.8 \times 7.66 = 6.128\text{mm} (\uparrow)$

∴ Resultant maximum deflection under working load
$= \delta_5 - \delta_6 + \delta_1 = 1.33 - 6.128 + 2.88 = 1.918\text{mm} (\uparrow)$

Problem 5.2

A simply supported beam 300mm × 800mm in section has a span of 12m and is pre–stressed with a parabolic cable which has maximum eccentricity of 300mm at mid–span section. At the supports, the cable lies 60mm above. Initial pre–stress in cable is 1500kN.

Determine (1) Maximum deflection at transfer of pre–stress (2) Maximum deflection under working load which consists of two point load of 50kN each placed at one third points. Assume M40 concrete and loss ratio as 80%

Solution:

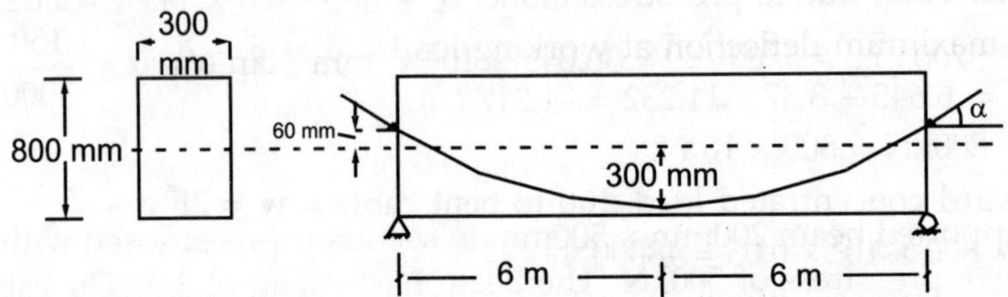

Fig. 5.20 Simply Supported Beam with Parabolic Cable

$$I = \frac{300 \times 800^3}{12} = 1.28 \times 10^{10}\text{mm}^4;\ A = 300 \times 800 = 0.24\text{m}^2,\ \alpha = \frac{4r}{L} = \frac{4 \times 360}{12000} = 0.12 \text{ Radian}$$

Self–weight = 0.24 × 24.0 = 5.76kN/m = 5.76N/mm, F_o = 1500kN, Span = L = 12m

$$E = E_c = 5700\sqrt{f_{ck}} = 5700\sqrt{40} = 36.049 \times 10^3\text{N/mm}^2$$

Intensity of upward UDL due to parabolic cable

$$= \left(\frac{2F_o\alpha}{L}\right) = \left(\frac{2 \times 150 \times 0.12}{12}\right) = 30\text{kN/m} = 30\text{N/mm}$$

Magnitude of end couple M = F_oe = 1500 × 0.06 = 90kNm = 90 × 10^6Nmm

i. Maximum Deflection at Transfer of Pre–stress

Central deflection due to self–weight

$$\delta_1 = \frac{5wL^4}{384EI} = \frac{5 \times 5.76 \times (12000)^4}{384 \times 36.049 \times 10^3 \times 1.28 \times 10^{10}} = 3.37\text{mm } (\downarrow)$$

Central deflection due to upward UDL

$$\delta_2 = \frac{5wL^4}{384EI} = \frac{5 \times 30 \times (12000)^4}{384 \times 36.049 \times 10^3 \times 1.28 \times 10^{10}} = 17.55\text{mm } (\uparrow) \text{ or}$$

$$\delta_2 = \frac{30}{5.76} = 17.55\text{mm } (\uparrow)$$

Central deflection due to end couples

$$\delta_3 = \frac{ML^2}{8EI} = \frac{90 \times 10^6 (12000)^2}{8 \times 36.049 \times 10^3 \times 1.28 \times 10^{10}} = 3.51\text{mm } (\downarrow)$$

Resultant deflection due to pre–stress alone

$$\delta_4 = \delta_2 - \delta_3 = 17.55 - 3.51 = 14.04\text{mm } (\downarrow)$$

Resultant max. deflection at transfer of pre–stress = $\delta_4 - \delta_1$ = 14.04 – 3.37 = 10.67mm (↑)

ii. Maximum Deflection at Working Load

$$\delta_5 = \left[\frac{wa(3L^2 - 4a^2)}{24EI}\right] = \left[\frac{50000 \times 4000\,(3 \times 12000^2 - 4 \times 4000^2)}{24 \times 36.049 \times 10^3 \times 1.28 \times 10^{10}}\right] = 6.64\text{mm} (\downarrow)$$

Central deflection due to pre–stress alone, $\delta_6 = \eta\delta_4 = 0.8 \times 14.04 = 11.232\text{mm} (\uparrow)$

Resultant maximum deflection at working load $= \delta_5 + \delta_1 - \delta_6$

$= 6.645 + 3.37 - 11.232 = -1.217 (\uparrow)$

Problem 5.3

Simply supported beam 200mm × 500mm in section is pre–stressed with a parabolic cable with an effective pre–stress of 500kN. The beam has a span of 8m. The cable has maximum eccentricity of 200mm at mid–span and zero eccentricity at support.

Determine (a) Maximum deflection due to pre–stress + self–weight (b) Central concentrated load required to nullify this deflection (c) The UDL required to cause a central downward deflection of being the span. Assume M40 concrete.

Solution:

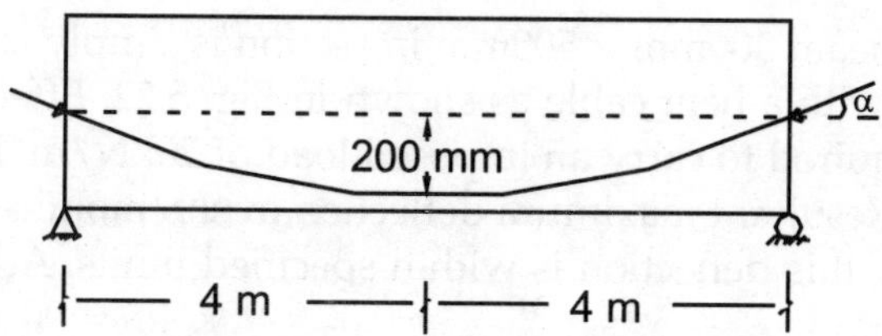

Fig. 5.21 Simply Supported Beam with Parabolic Cable

$F = 500\text{kN}$ $A = 0.2 \times 0.5 = 0.1\text{m}^2$, Self–weight $= 0.1 \times 24 = 2.4\text{kN/m} = 2.4\text{N/mm}$

$$I = \frac{200 \times 500^3}{12} = 2.08 \times 10^9\,\text{mm}^4;\ \alpha = \frac{4r}{L} = \frac{4 \times 200}{8000} = 0.1 \text{ Radian, Span} = L = 8\text{m}$$

$$E = E_c = 5700\sqrt{f_{ck}} = 5700\sqrt{40} = 36.049 \times 10^3\text{N/mm}^2$$

$$\text{Intensity of upward UDL due to parabolic cable} = \left(\frac{2F_o\alpha}{L}\right) = \left(\frac{2 \times 500 \times 0.11}{8}\right) = 12.5\text{N/mm}$$

(a) Maximum Deflection due to Pre–stress + Self–weight

Central deflection due to self–weight

$$\delta_1 = \frac{5wL^4}{384EI} = \frac{5 \times 2.4 \times (8000)^4}{384 \times 36.049 \times 10^3 \times 2.08 \times 10^9} = 1.70\text{mm}$$

Central deflection due to upward UDL, $\delta_2 = \dfrac{12.5}{2.4} = 8.85 \text{ mm} (\uparrow)$

∴ Resultant maximum deflection due to pre–stress + self–weight

$= \delta_2 - \delta_1 = 8.85 - 1.70 = 7.15\text{mm}$

(b) Central Concentrated Load Required for Nullifying this Deflection

Let 'w' be the central concentrated load required to nullify the above deflection

$$\delta = \frac{wL^3}{48EI}; 7.15 = \frac{w \times 8000^3}{48 \times 36.049 \times 10^3 \times 2.08 \times 10^9}; w = 50261N = 50.26kN$$

(c) UDL Required, to Cause a Central Downward Deflection of L/300, 'L' Being Span

Let 'w' N/mm be the UDL required to cause a central downward deflection of L/300

Now L/300 = 8000/300 = 26.66mm, since there is initial upward deflection of 7.15mm

∴ Actual central downward deflection = 7.15 + 26.66 = 33.81mm

$$\text{Now } \delta = \frac{5wL^4}{384EI}; 33.81 = \frac{5 \times w \times 8000^4}{384 \times 36.049 \times 10^3 \times 2.08 \times 10^9}; w = 47.74N/mm$$

$$\text{or } \delta = \frac{w}{2.4} \times 1.70; \therefore w = \frac{33.81 \times 2.4}{1.7} = 47.74N/mm$$

Problem 5.4

A simply supported beam 300mm × 500mm in section is simply supported over a span of 10m and is pre–stressed with a bent cable as shown in Fig. 5.22. Effective pre–stress in cable is 600kN. The beam is required to carry an imposed load of 20kN/m. Determine (a) Resultant maximum deflection (b) Resultant maximum deflection over 6 months taking creep coefficient as 2.2. Also state whether this deflection is within specified limits. Adopt M40 concrete. Loss of pre–stress = 20%.

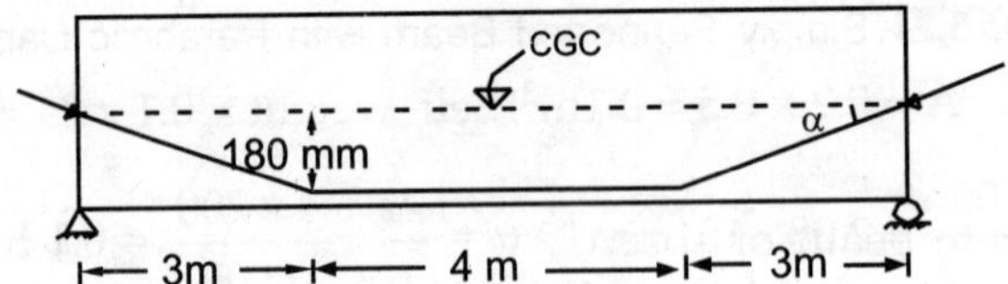

Fig. 5.22 Simply Supported Beam with Doubly Harped Cable

Solution:

$$F = 600kN, \alpha = \frac{180}{3000} = 0.06 \text{ Radian}, A = 0.3 \times 0.5 = 0.15m^2, \text{Span} = L = 10m$$

$$I = \frac{300 \times 500^3}{12} = 3.125 \times 10^9 mm^4,$$

$$E = E_c = 5700\sqrt{f_{ck}} = 5700\sqrt{40} = 36.049 \times 10^3 N/mm^2$$

Magnitude of each of the two upward concentrated load due to the bent cables

$$w = F\alpha = 600 \times 0.06 = 36kN$$

Self–weight = 0.15 × 24 = 3.6kN/m = 3.6N/mm, Live load = 20kN/m = 20N/mm

(a) Resultant Maximum Deflection

Central deflection due to self–weight

$$\delta_1 = \frac{5wL^4}{384EI} = \frac{5\times3.6\times(10000)^4}{384\times36.049\times10^3\times3.125\times10^9} = 4.16\text{mm} (\downarrow)$$

Central deflection due to upward point load

$$\delta_2 = \left[\frac{wa\,(3L^2-4a^2)}{24EI}\right] = \left[\frac{36000\times3000\,(3\times10000^2-4\times3000^2)}{24\times36.049\times10^3\times3.125\times10^9}\right] = 10.548\text{mm} (\uparrow)$$

Central deflection due to live load

$$\delta_3 = \frac{20}{3.6}\times4.16 = 23.11\text{mm} (\downarrow)$$

Resultant maximum deflection at working load

$$\delta_4 = \delta_1+\delta_3-\delta_2 = 4.16+23.11-10.548 = 16.72\text{mm}$$

Note: The long term deflection at any instant is approximately determined by multiplying the resultant maximum deflection under working load with the corresponding creep coefficient.

(b) Resultant Maximum Deflection Over 6 Months Taking Creep Coefficient As 2.2

Resultant maximum deflection after 6 months = $C_c \times \delta_4 = 2.2 \times 16.72 = 36.79$mm

Permissible maximum deflection as per IS code = $\frac{L}{250} = \frac{10000}{250} = 40\text{mm} > 36.79\text{mm}$

Thus the long term deflection does not exceed the permissible deflection and therefore the beam is safe.

Problem 5.5

A pre–stress concrete beam of span 10m is of rectangular section, 120mm wide and 300mm deep and is pre–stressed by a parabolic cable, the initial pre–stressing force being 280kN. The eccentricity of the cable at the centre is 20mm and the cable concentric at the supports. The beam carries a live load of 2.20kN/m. Calculate the short time deflection at the centre of the span. Take E = 40kN/mm² and creep coefficient $\phi = 2.0$. Loss of pre–stress = 18% of the initial stress after a duration of 6 months. Find the long term deflection at the centre. Assume that the beam is subjected to dead load and live load simultaneously when the pre–stress is applied.

Solution:

$$P_i = 280\text{kN},\ I = \frac{120\times300^3}{12} = 2.7\times10^8\text{mm}^4,\ e = 50\text{mm}$$

Dead load = 0.12 × 0.30 × 24 = 0.864kN/m,

Total load = 0.864 + 2.200 = 3.064kN/m = 0.003064kN/mm

Loss of pre–stress = $\Delta P_i = 0.2P_i$

i. Short Term Deflection

Upward deflection due to pre–stress = $\left(\frac{5F_oeL^2}{48EI}\right) = \left(\frac{5\times 280\times 50\times 10000^2}{48\times 40\times 2.7\times 10^8}\right)$ = 36.94mm

Downward deflection due to DL and LL = $\left(\frac{5\times 0.003064\times 10000^4}{384\times 40\times 2.7\times 10^8}\right)$ = 13.50mm

∴ Net deflection = 36.24 – 13.50 = 23.44mm (Downward)

ii. Long Term Deflection

δ_{pt} = Initial deflection due to transverse loads = 36.94mm

δ_{ip} = Initial deflection due to pre–stress = 13.50mm

By Neville's formula, long term deflection

$$\delta_t = \delta_{il}(1+\phi) - \delta_{ip}\left\{1-\frac{\Delta P_i}{P_i}+\left(1-\frac{\Delta P_i}{2P_i}\right)\phi\right\}$$

$$\delta_t = 36.94(1+2) - 13.50\left\{1-\frac{0.2P_i}{P_i}+\left(1-\frac{0.2P_i}{2P_i}\right)2\right\}$$

$$= 110.82 - 13.50\,(1 - 0.2 + 1.8) = 110.82 - 35.10 = 75.72\text{mm (Downward)}$$

By Lin's formula

$$\delta_t = \left(\delta_{il} - \delta_{ip}.\frac{P_t}{P_i}\right)(1+\phi)$$

$$= (36.94 - 13.50 \times 0.8)(1 + 2) = 78.42\text{mm (Downward)}$$

5.8 PRESSURE LINE

At any section of a beam combined effect of the pre–stressing force and the externally applied load will result in a distribution of concrete stresses that can be resolved into a single force. The locus of the points of application of this force in any beam is called the pressure line. Pressure lines are very useful in understanding the load carrying mechanism in pre–stressed concrete.

5.9 SOLVED EXAMPLES

Problem 5.6

Compute and draw the location of pressure line for a rectangular beam 250mm wide by 300mm deep that is pre–stressed with a force of 500kN at a constant eccentricity of 75mm. The beam is supporting a 50kN concentrated load at the middle of span of 4m. Neglect dead load of beam.

Solution:

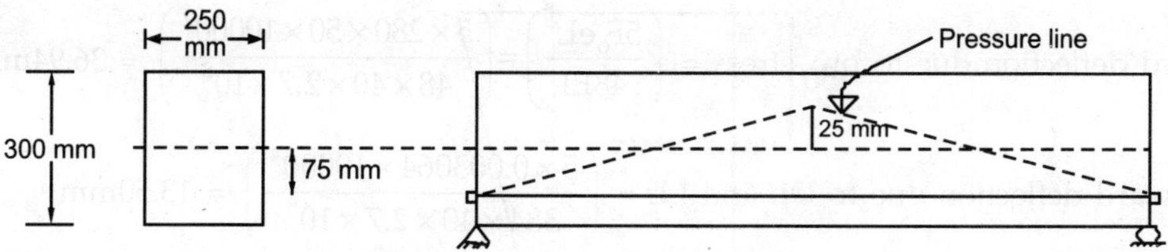

Fig. 5.23 Simply Supported Beam

Given: $A = 300 \times 250 = 75000\text{mm}^2$, $F = 500\text{kN}$, $I = \dfrac{250 \times 300^3}{12} = 562.5 \times 10^6\text{mm}^4$, $L = 4\text{m}$

Stress at Support

$$f_{(\text{at bottom})} = \frac{F}{A} + \frac{Fey}{I} = \frac{500 \times 10^3}{75000} + \frac{500 \times 10^3 (-75)(-150)}{562.5 \times 10^6} = 16.66\text{N/mm}^2$$

$$f_{(\text{at top})} = \frac{F}{A} + \frac{Fey}{I} = \frac{500 \times 10^3}{75000} + \frac{500 \times 10^3 (-75)(150)}{562.5 \times 10^6} = -3.33\text{N/mm}^2$$

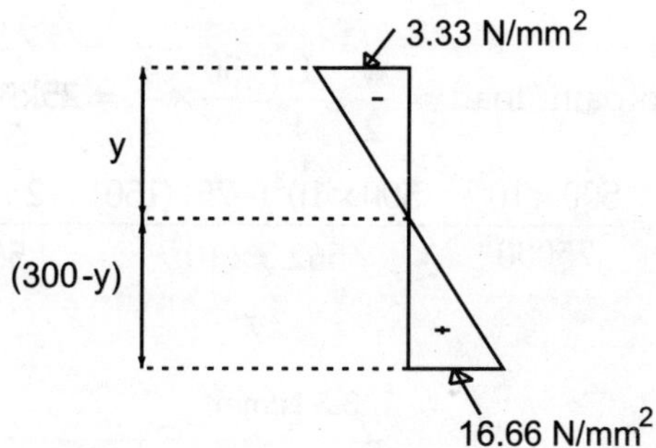

Fig. 5.24 Stress Distribution at Supports

$$\frac{3.33}{16.66} = \frac{y}{300 - y};\ \therefore\ y = 50\text{mm}$$

BM is zero at the support of the beam. Therefore the pressure line is at e = 75mm.

$$\text{Maximum BM at Centre} = \frac{wL}{4} = \frac{50 \times 4}{4} = 50\text{kNm} = M$$

Stress Distribution at Centre of Span

$$f_{(\text{at top})} = \frac{F}{A} + \frac{Fey_1}{I} + \frac{My_1}{I} = \frac{500 \times 10^3}{75000} + \frac{500 \times 10^3 (-75)(150)}{562.5 \times 10^6} + \frac{50 \times 10^6 (150)}{562.5 \times 10^6} = 10\text{N/mm}^2$$

$$f_{(\text{at bottom})} = 3.3\text{N/mm}^2$$

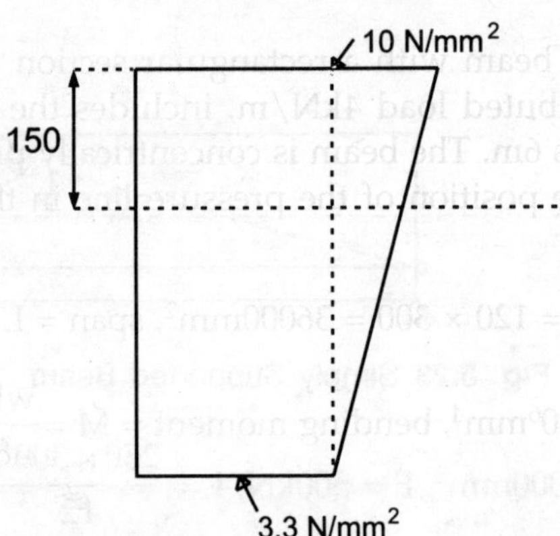

Fig. 5.25 Stress Distribution at Mid–span

Position of resultant thrust 'C' from top edge of stress diagram

$$y = \frac{3.33 \times 300 \times 150 + \frac{1}{2} \times 6.66 \times 300 \times \frac{1}{3} \times 300}{3.3 \times 300 + \frac{1}{2} \times 6.66 \times 300} = 124.44\text{mm say } 125\text{mm}$$

Stresses at Quarter Span

$$\text{BM at quarter span due to paint load} = \frac{w}{2} \times \frac{L}{4} = \frac{50}{2} \times \frac{4}{4} = 25\text{kNm}$$

$$f_{(\text{at top})} = \frac{F}{A} + \frac{Fey_1}{I} + \frac{My_1}{I} = \frac{500 \times 10^3}{75000} + \frac{500 \times 10^3 (-75)(150)}{562.5 \times 10^6} + \frac{25 \times 10^6 (150)}{562.5 \times 10^6} = 3.33\text{N/mm}^2$$

$$f_{(\text{at bottom})} = 10\text{N/mm}^2$$

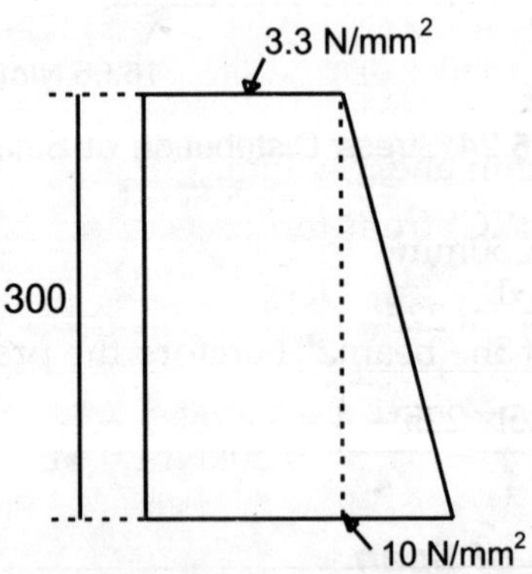

Fig. 5.26 Stress Distribution at Quarter Span

Position of resultant thrust 'C' from top of stress block

$$y = \frac{3.33 \times 300 \times 150 + \frac{1}{2} \times 6.66 \times 300 \times \frac{2}{3} \times 300}{3.3 \times 300 + \frac{1}{2} \times 6.66 \times 300} = 175\text{mm. Let, e} = 25\text{mm}$$

Problem 5.7

A pre–stresses concrete beam with a rectangular section 120mm wide by 300mm deep supports a uniformly distributed load 4kN/m, includes the self–weight of the beam. The effective span of the beam is 6m. The beam is concentrically pre–stressed by a cable carrying a force of 180kN. Locate the position of the pressure line in the beam.

Solution:

F = 180kN, e = 0, area = 120 × 300 = 36000mm², span = L = 6m,

$$I = \frac{120 \times 300^3}{12} = 270 \times 10^6 mm^4, \text{ bending moment} = M = \frac{wL^2}{8} = \frac{4 \times 6^2}{8} = 18\text{kNm, at center}$$

Stress at Centre

$$f_{(at\ top)} = \frac{F}{A} + \frac{Fey_1}{I} + \frac{My_1}{I} = \frac{180 \times 10^3}{36000} + 0 + \frac{18 \times 10^6 (150)}{270 \times 10^6} = 15\text{N/mm}^2$$

$$f_{(at\ bottom)} = -5\text{N/mm}^2$$

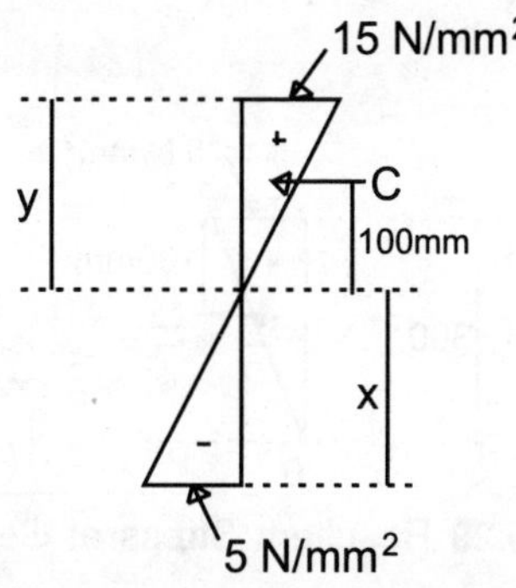

Fig. 5.27 Stress Distribution at Mid–span

$$\frac{5}{15} = \frac{x}{y}; y = \frac{15x}{5} = \frac{x}{y}; y = \frac{5}{15}(300 - y) \quad ..(1), \quad \text{But, } x + y = 300 \quad ...(2)$$

Put (1) in (2) we get = 225mm and x = 75mm

Position of resultant thrust 'C' from top edge of stress block

$$y = \frac{\frac{1}{2} \times 15 \times 225 \times \frac{1}{3} \times 225 - \frac{1}{2} \times 5 \times 75 \times 274.5}{\frac{1}{2} \times 15 \times 225 - \frac{1}{2} \times 5 \times 75} = 50\text{mm.}$$

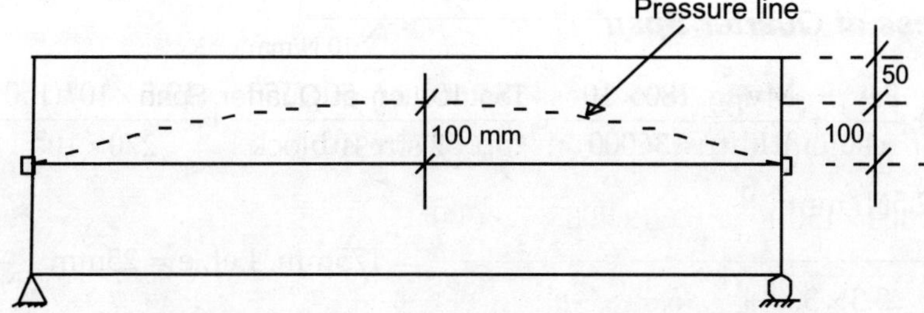

Fig. 5.28 Location of Pressure Line in PSC Beam

Problem 5.8

A pre–stressed concrete beam of section 120mm wide by 300mm deep is used over an effective span of 6m to support a uniformly distributed load 4kN/m which includes the self–weight of beam. The beam is pre–stressed by a straight cable carrying a force of 180kN, located at an eccentricity of 50mm (Below NA). Determine the location of the thrust line in the beam and plot its position at quarter and central span sections.

Solution:

F = 180kN, e = – 50mm, A = 120 × 300 = 36000mm², I = 270 × 10⁶mm⁴,

$$\text{BM at centre} = \frac{4\times6^2}{8} = 18\text{kNm}$$

Resultant Stress at Centre

$$f_{(\text{at top})} = \frac{F}{A}+\frac{Fey_1}{I}+\frac{My_1}{I} = \frac{180\times10^3}{36000}+\frac{18\times10^6\times(-50)\,(150)}{270\times10^6}+\frac{18\times10^6(150)}{270\times10^6} = 10\text{N/mm}^2$$

$$f_{(\text{at bottom})} = 0$$

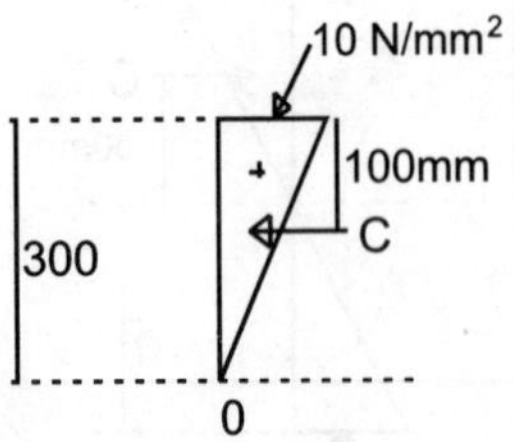

Fig. 5.29 Resultant Stress at Centre

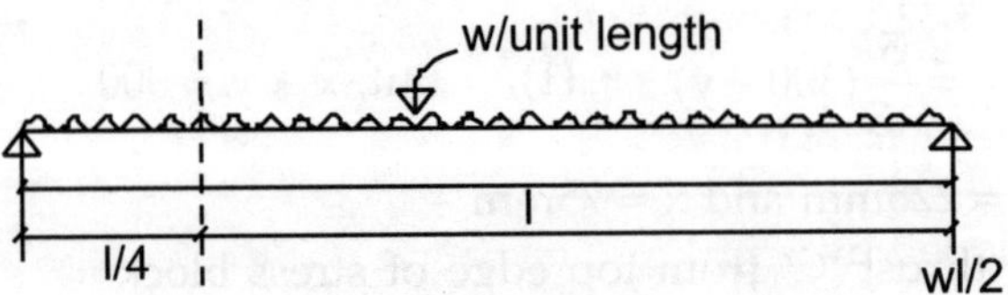

Fig. 5.30 Simply Supported Load Carrying UDL

$$\text{BM at quarter span} = \frac{wL}{2}\times\frac{L}{4}-w\times\frac{L}{4}\times\frac{L}{8}=\frac{wL^2}{8}-\frac{wL^2}{32}=\frac{3wL^2}{32}=\frac{3\times4\times6^2}{32} = 13.5\text{kNm}$$

Resultant Stress at Quarter Span

$$f_{(\text{at top})} = \frac{F}{A}+\frac{Fey_1}{I}+\frac{My_1}{I} = \frac{180\times10^3}{36000}+\frac{18\times10^6\times(-50)\,(150)}{270\times10^6}+\frac{13.5\times10^6(150)}{270\times10^6} = 7.5\text{N/mm}^2$$

$$f_{(\text{at bottom})} = 2.5\text{N/mm}^2$$

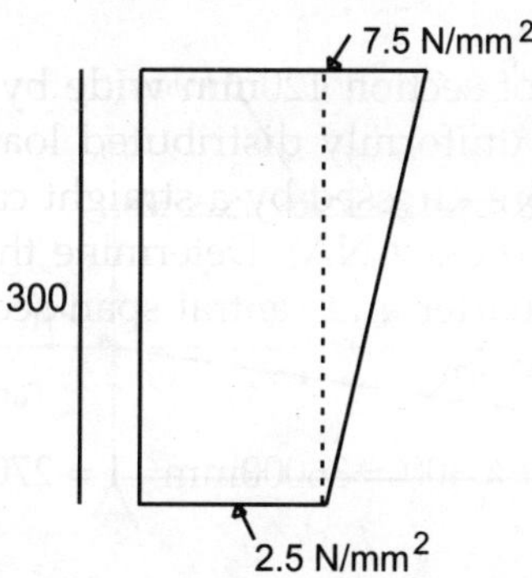

Fig. 5.31 Resultant Stress at Quarter Span

Position of the resultant thrust 'C' from top edge of stress block

$$y = \frac{2.5\times 300\times 150+\frac{1}{2}\times 5\times 300\times\frac{1}{3}\times 300}{300\times 2.5+\frac{1}{2}\times 5\times 300} = 125\text{mm}.$$

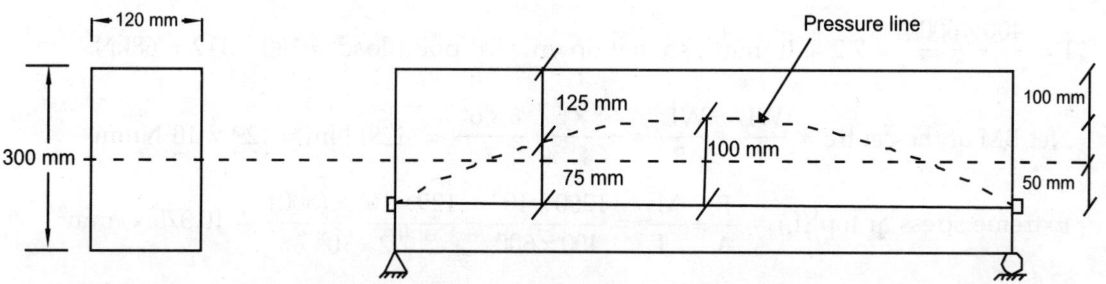

Fig. 5.32 Location of Pressure Line in PSC Beam

5.10 LOAD BALANCING METHOD

This concept was developed by T. Y. Lin. The beam is so designed that the upward forces transmitted by the cable exactly neutralize the external loading. The method of design is called load balancing method.

5.11 SOLVED EXAMPLES

Problem 5.9

A PSC beam is pre–stressed with a tendon bent as shown in Fig. 5.33. The external load on the beam consists of a concentrated load of 180kN at mid–span. If effective pre–stressing force is 1200kN, calculate the extreme stresses in concrete for the mid–span section. Let span = 6m.

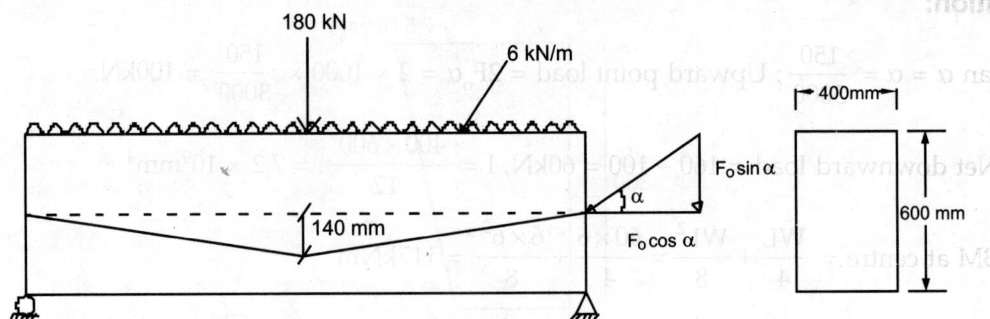

Fig. 5.33 PSC Beam is Pre–stressed with a Tendon Bent

Solution:

DL of Beam = 0.4 × 0.6 × 25 = 6kN/m, $\tan\alpha = \alpha = \dfrac{140}{3000} = \dfrac{7}{150}$

Upward point load transmitted by tendon at centre = $2F_o\alpha = 2 \times 1200 \times \dfrac{7}{150} = 112\text{kN}$

$I = \dfrac{400 \times 600^3}{12} = 7.2 \times 10^9 \text{mm}^4$, so, net downward point load = 180 – 112 = 68kN

Net BM at the centre = $\dfrac{WL}{4} + \dfrac{WL^2}{8} = \dfrac{68 \times 6}{4} + \dfrac{6 \times 6^2}{8} = 129\text{kNm} = 129 \times 10^6\text{Nmm}$

Extreme stress at top $(f_t) = \dfrac{F}{A} + \dfrac{My}{I} = \dfrac{1200 \times 10^3}{400 \times 600} + \dfrac{129 \times 10^6 \times (300)}{7.2 \times 10^9} = 10.375\text{N/mm}^2$

Similarly, stress at bottom $(f_b) = \dfrac{1200 \times 10^3}{400 \times 600} + \dfrac{129 \times 10^6 \times (-300)}{7.2 \times 10^9} = -0.375\text{N/mm}^2$

Problem 5.10

Fig. 5.34 shows a pre–stressed concrete beam provided with inclined tendon. The beam carries a point load of 160kN at the centre. If effective pre–stressing force is 1000kN, determine the stress distribution at mid–section of the beam. The DL of the beam is 6kN/m.

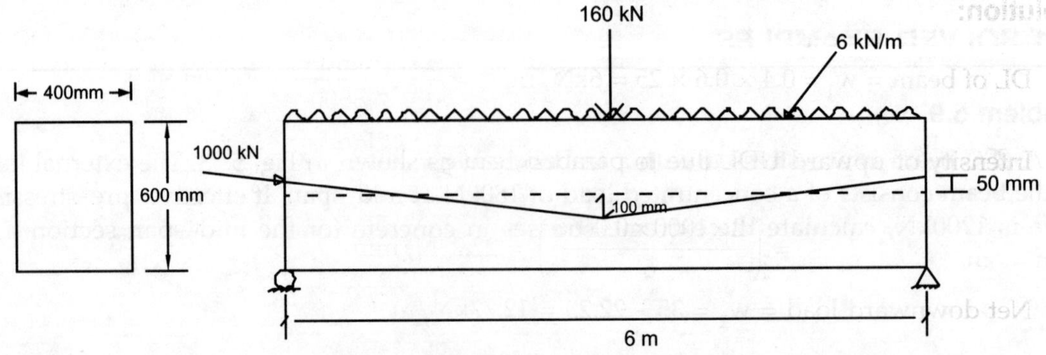

Fig. 5.34 Pre–stressed Concrete Beam Provided with Inclined Tendon

Solution:

$\tan\alpha = \alpha = \dfrac{150}{3000}$; Upward point load $= 2F_o\alpha = 2 \times 1000 \times \dfrac{150}{3000} = 100\text{kN}$

Net downward load $= 160 - 100 = 60\text{kN}$, $I = \dfrac{400 \times 600^3}{12} = 7.2 \times 10^9 \text{mm}^4$

BM at centre $= \dfrac{WL}{4} + \dfrac{WL^2}{8} = \dfrac{60 \times 6}{4} + \dfrac{6 \times 6^2}{8} = 117\text{kNm}$

BM due to eccentricity of the load at the end $= Fe = 1000 \times \dfrac{50}{1000} = 50\text{kNm}$ (Sagging)

Total BM $= 117 + 50 = 167\text{kNm}$

Extreme stress at top $(f_t) = \dfrac{F}{A} + \dfrac{My}{I} = \dfrac{1000 \times 10^3}{400 \times 600} + \dfrac{167 \times 10^6 \times (300)}{7.2 \times 10^9} = 11.13\text{N/mm}^2$

Similarly, stress at bottom $(f_b) = \dfrac{1000 \times 10^3}{400 \times 600} + \dfrac{167 \times 10^6 \times (-300)}{7.2 \times 10^9} = -2.79\text{N/mm}^2$

Problem 5.11

Fig. 5.35 shows a PSC beam provided with a tendon having a parabolic profile. It the external load on the beam is 35kN/m on the whole span. Calculate the extreme stresses for the mid–span section. The tendon carries a pre–stressing force of 1000kN.

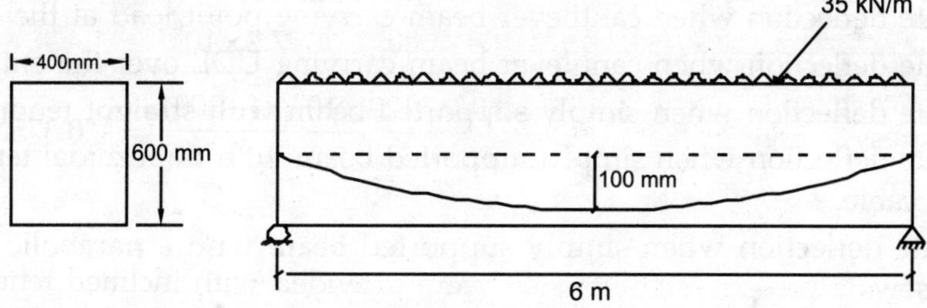

Fig. 5.35 PSC Beam Provided with a Tendon having a Parabolic Profile

Solution:

DL of beam $= w_1 = 0.4 \times 0.6 \times 25 = 6\text{kN/m}$, $\alpha = \dfrac{4\gamma}{L} = \dfrac{4 \times 100}{6000} = 0.0666$

Intensity of upward UDL due to parabolic cable $= \left(\dfrac{2F_o\alpha}{L}\right)$

$= \left(\dfrac{2 \times 1000 \times 0.0666}{6}\right) = 22.23\text{N/mm}$

Net downward load $= w_2 = 35 - 22.23 = 12.77\text{kN/m}$

$$I = \frac{400 \times 600^3}{12} = 7.2 \times 10^9 mm^4$$

$$\text{BM at centre} = M = \frac{W_1L^2}{8} + \frac{W_2L^2}{8} = \frac{6 \times 6^2}{8} + \frac{12.77 \times 6^2}{8} = 84.465 kNm$$

$$\text{Extreme stress at top } (f_t) = \frac{F}{A} + \frac{My}{I} = \frac{1000 \times 10^3}{400 \times 600} + \frac{84.465 \times 10^6 \times (300)}{7.2 \times 10^9} = 7.68 N/mm^2$$

$$\text{Similarly, stress at bottom } (f_b) = \frac{1000 \times 10^3}{400 \times 600} + \frac{84.465 \times 10^6 \times (-300)}{7.2 \times 10^9} = 0.647 N/mm^2$$

REVIEW QUESTIONS

5.1 What are the factors that influences deflection?

5.2 Mention the limits of deflection as per IS: 1343–2012.

5.3 Explain short term deflection and long term deflection.

5.4 Explain Mohr's method of calculating slope and deflection of beam.

5.5 What are all the methods used for the calculation of deflection under gravity loads?

5.6 Discuss deflection when simply supported beam carrying central concentrated load.

5.7 Discuss deflection when simply supported beam carrying UDL over the entire span.

5.8 Calculate deflection when simply supported beam carrying two symmetrically placed two loads of equal magnitude.

5.9 Calculate deflection when cantilever beam carrying point load at the free end.

5.10 Calculate deflection when cantilever beam carrying UDL over the entire span.

5.11 Calculate deflection when simply supported beam with straight tendons.

5.12 Calculate deflection when simply supported beam with trapezoidal tendons/doubly harped cable.

5.13 Calculate deflection when simply supported beam with a parabolic cable (central anchors).

5.14 Calculate deflection when simply supported beam with a parabolic cable (eccentric anchors).

5.15 Calculate deflection when simply supported beam with singly harped cable/sloping tendon (central anchors).

5.16 Calculate deflection when simply supported beam with singly harped cable/sloping tendon (eccentric anchors).

5.17 Calculate deflection when simply supported beam with combination of parabolic and straight tendons (central anchors).

5.18 Calculate deflection when simply supported beam with combination of parabolic and straight tendons (eccentric anchors).

5.19 Calculate deflection when simply supported beam with cable eccentric at the ends.

5.20 Calculate deflection when simply supported beam with end couples of equal magnitude M.

5.21 Explain deflection due to self–weight and imposed loads.

5.22 Explain Neville's formula for long term deflection.

5.23 Explain Lin's formula for long term deflection.

5.24 Elaborate camber in a PSC beam.

5.25 Explain pressure line.

5.26 Explain Lin's load balancing method.

EXERCISE PROBLEMS

5.1 A post–tensioned concrete beam of rectangular section is 250mm wide and 450mm deep. The beam is pre–stressed by two cables of area 510mm^2 each, which are initially pre–stressed to 1500N/mm^2. The eccentricity of the cables is 90mm throughout the length of the beam, the span of the beam being 8m. Ignoring all losses find the deflection at centre when the beam supports its own weight.

What would be the deflection at the centre when the beam carries an imposed load of 15kN/m^2 and there is 20% loss in pre–stress. Concrete weighs 24kN/m^3. Modulus of elasticity of concrete is 40kN/mm^2.

[Ans: Deflection at centre when beam supports its own weight = 14.52mm, deflection at centre when beam supports imposed load and loss of pre–stress = 0.84mm]

5.2 A pre–stressed concrete beam of span 9m has a section of area 40 × 10^{3}mm^3, the moment of inertia of the section being 1.70 × 10^{8}mm^4. The beam is pre–stressed by a parabolic cable providing a pre–stressing force of 240kN. The cable having an eccentricity of 50mm at the centre and has no eccentricity at the ends. Ignoring all the losses, find the deflection at the centre,

(i) When the beam carries its own weight and pre–stress.

(ii) When the beam carries its own weight, pre–stress and a superimposed load of 1.75kN/m.

Take weight of concrete equal to 24kN/m^3 and modulus of elasticity of concrete equal to 40kN/mm^2.

[Ans: (i) Deflection at centre when beam carries its own weight and pre–stress = 2.83mm, (ii) Deflection at centre when beam carries its own weight, pre–stress and a superimposed load = 19.15mm]

5.3 A pre–stressed concrete beam of I–section having top and bottom flange width = 200mm, flange thickness = 60mm, web thickness = 75mm, overall depth = 400mm and span = 8m. The cable carrying a pre–stressing force of 980kN. Calculate the initial deflection at mid–span due to pre–stress and the dead load of the beam. Modulus of elasticity of concrete may be taken as 38kN/mm^2. State also whether this deflection is within the permissible limit. Take weight of concrete equal to 24kN/m^3.

[Ans: Net upward deflection = 2.29mm, permissible deflection = 26.7mm, the deflection is within the permissible limits]

5.4 A pre–stressed concrete beam of rectangular section is 125mm wide and 250mm deep and has span of 6m. The beam is provided with a straight tendon at a uniform eccentricity of 40mm, the pre–stressing force being 190kN. Find the deflection at the centre.

(i) Under the action of pre–stress and dead load of the beam.

(ii) Under the action of pre–stress, dead load and live load of 3.75kN/m and including the effect of creep and shrinkage, take creep coefficient as 1.75. Compare the deflection with the permissible limits.

[Ans: Net upward deflection = 3.48mm, permissible deflection = 20mm, the deflection is within the permissible limits]

5.5 A concrete beam with a rectangular section 300mm wide and 500mm deep is pre–stressed by 2 post–tensioned cables of area 600mm^2 each. Initially stressed to 1600N/mm^2. The cables are located at a constant eccentricity of 100mm throughout the length of the beam having a span of 10m. The modulus of elasticity of steel and concrete is 210 and 38kN/mm^2 respectively.

a. Neglecting all losses, find the deflection at the centre of span when it is supporting its own weight.

b. Allowing for 20% loss in pre–stress, find the final deflection at the centre of the span when it carries an imposed load of 18kN/m. Take density of concrete = 24kN/m^3.

[Ans: a. Net upward deflection when it supports its own weight = 16.3mm, b. Deflection after loss of pre–stress and carries imposed load = 7.29mm]

5.6 A rectangular concrete beam of cross section 150mm and 300mm deep is simply supported over a span of 8m and is pre–stressed by means of a symmetric parabolic cable, at a distance of 75mm from the bottom of the beam at mid–span and 125mm from the top of the beam at supports. If the force in the cable is 350kN and the modulus of elasticity of concrete is 38kN/mm^2, calculate:

a. The deflection at mid–span when the beam is supporting its own weight

b. The concentrated load which must be applied at mid–span to restore it to the level of supports.

[Ans: a. Deflection due to pre–stress and self–weight = –8.52mm (Upward), b. Concentrated load required at centre of span = 9.9kN]

5.7 A concrete beam with a cross sectional area of 32 × 10^{3}mm^2 and radius of gyration = 72mm is pre–stressed by a parabolic cable carrying an effective stress of 1000N/mm^2. The span of the beam is 8m. The cable, composed of 6 wires of 7mm diameter, has an eccentricity of 50mm at the centre and zero at the supports. Neglecting all the losses, find the central deflection of the beam as follows:

a. Self–weight + pre–stress and

b. Self–weight + pre–stress + live load of 2kN/m.

Assume, E = 38kN/mm^2 and density of concrete = 24kN/m^3.

[Ans: a. Deflection at, self–weight + pre–stress = 5.7mm (Upwards) b. Deflection at, self–weight + pre–stress + live load = 11.2mm (Downwards)]

5.8 A concrete beam having a rectangular section 100mm wide and 300mm deep is pre–stressed by a parabolic cable carrying an initial force of 240kN. The cable has an eccentricity of 50mm at the centre span and is concentric at the supports. If the span of the beam is 10m and live load is 2kN/m, estimate the short term deflection at the centre of the span.

Assume, E = 38kN/mm^2 and creep coefficient = 2.0, loss of pre–stress = 20% of the initial stress after 6 months. Estimate the long term deflection at the centre of the span at this stage, assuming that the dead and live loads are simultaneously applied after the release of pre–stress.

[Ans: Short term deflection = 26.8mm (Downwards)
long term deflection = 89.1mm (Downwards)]

OBJECTIVE QUESTIONS

5.1 Which of the following factors affect deflection?

a. Span
b. Modulus of elasticity of concrete
c. Moment of inertia
d. All of the above

5.2 As per IS: 1343–2012, total deflection due to all loads, including the effects of temperature, creep and shrinkage, should not exceed

a. span/150
b. span/250
c. span/300
d. span/350

5.3 As per IS: 1343–2012, deflection after erection of partitions or application of finishes, including the effects of temperature, creep and shrinkage,

a. should not exceed span/350 or 20 mm, whichever is less
b. should not exceed span/250 or 20 mm, whichever is less
c. should exceed span/350 or 20 mm, whichever is less
d. should exceed span/250 or 20 mm, whichever is less

5.4 As per IS: 1343–2012, when finishes are applied, total upward deflection due to pre–stressing force

a. should exceed span/300
b. should not exceed span/250
c. should exceed span/250
d. should not exceed span/300

5.5 Deflection when simply supported beams carrying central concentrated load

a. $\delta_c = \frac{wL^3}{48EI}$
b. $\delta_c = \frac{wL^3}{24EI}$
c. $\delta_c = \frac{3wL^3}{48EI}$
d. $\delta_c = \frac{5wL^3}{48EI}$

5.6 What is the method for calculation of deflection due to gravity loads?

a. Moment–area method
b. Conjugate beam method
c. Principle of virtual work
d. All of the above

5.7 Deflection when simply supported beams carrying UDL over the entire span

a. $\delta_c = \frac{3wL^4}{384EI}$
b. $\delta_c = \frac{7wL^4}{384EI}$
c. $\delta_c = \frac{5wL^4}{384EI}$
d. $\delta_c = \frac{9wL^4}{384EI}$

5.8 Deflection when simply supported beam carrying two symmetrically placed two loads of equal magnitude

a. $\delta_c = wa\left[\dfrac{3L^2 + 4a^2}{24EI}\right]$ b. $\delta_c = wa\left[\dfrac{3L^2 - 4a^2}{24EI}\right]$

c. $\delta_c = wa\left[\dfrac{4L^2 - 3a^2}{24EI}\right]$ d. $\delta_c = wa\left[\dfrac{4L^2 + 3a^2}{24EI}\right]$

5.9 Deflection when cantilever beam carrying point load at the free end

a. $\delta = \dfrac{wL^3}{3EI}$ b. $\delta = \dfrac{wL^3}{2EI}$ c. $\delta = \dfrac{wL^3}{5EI}$ d. $\delta = \dfrac{2wL^3}{3EI}$

5.10 Deflection when cantilever beam carrying UDL over the entire span

a. $\delta = \dfrac{wL^4}{14EI}$ b. $\delta = \dfrac{wL^4}{6EI}$ c. $\delta = \dfrac{wL^3}{8EI}$ d. $\delta = \dfrac{wL^4}{8EI}$

5.11 Deflection when simply supported beam with straight tendons

a. $\delta = -\dfrac{F_o eL^2}{6EI}$ b. $\delta = -\dfrac{F_o eL^2}{8EI}$ c. $\delta = -\dfrac{F_o eL^2}{10EI}$ d. $\delta = -\dfrac{F_o eL^2}{12EI}$

5.12 Deflection when simply supported beam with trapezoidal tendons/doubly harped cable

a. $\delta = -\dfrac{F_o e}{10EI}[2l_1^2 + 6l_1l_2 + 3l_2^2]$ or $\delta = w_{up}a\left[\dfrac{3L^2 - 4a^2}{24EI}\right]$

b. $\delta = -\dfrac{F_o e}{6EI}[2l_1^2 + 6l_1l_2 + 3l_2^2]$ or $\delta = w_{up}a\left[\dfrac{3L^2 - 4a^2}{24EI}\right]$

c. $\delta = -\dfrac{F_o e}{12EI}[2l_1^2 + 6l_1l_2 + 3l_2^2]$ or $\delta = w_{up}a\left[\dfrac{3L^2 - 4a^2}{24EI}\right]$

d. $\delta = -\dfrac{F_o e}{14EI}[2l_1^2 + 6l_1l_2 + 3l_2^2]$ or $\delta = w_{up}a\left[\dfrac{3L^2 - 4a^2}{24EI}\right]$

5.13 Deflection when simply supported beam with a parabolic cable (central anchors)

a. $\delta = -\left(\dfrac{5F_o eL^2}{68EI}\right)$ or $\delta = \dfrac{5w_{up}L^4}{384EI}$ b. $\delta = -\left(\dfrac{F_o eL^2}{98EI}\right)$ or $\delta = \dfrac{5w_{up}L^4}{384EI}$

c. $\delta = -\left(\dfrac{5F_o eL^2}{48EI}\right)$ or $\delta = \dfrac{5w_{up}L^4}{384EI}$ d. $\delta = -\left(\dfrac{3F_o eL^2}{48EI}\right)$ or $\delta = \dfrac{5w_{up}L^4}{384EI}$

5.14 Deflection when simply supported beam with a parabolic cable (eccentric anchors)

a. $\delta=\frac{F_oL^2}{48EI}(-7e_1+e_2)$ b. $\delta=\frac{F_oL^2}{48EI}(-5e_1+e_2)$

c. $\delta=\frac{F_oL^2}{48EI}(-3e_1+e_2)$ d. $\delta=\frac{F_oL^2}{48EI}(-2e_1+e_2)$

5.15 Deflection when simply supported beam with singly harped cable/sloping tendon (central anchors)

a. $\delta=\frac{5w_{up}L^3}{48EI}$ b. $\delta=\frac{3w_{up}L^3}{48EI}$ c. $\delta=\frac{w_{up}L^3}{48EI}$ d. $\delta=\frac{7w_{up}L^3}{48EI}$

5.16 Deflection when simply supported beam with singly harped cable/sloping tendon (eccentric anchors)

a. $\delta=\frac{F_oL^2}{24EI}(-3e_1+e_2)$ b. $\delta=\frac{F_oL^2}{24EI}(-2e_1+e_2)$

c. $\delta=\frac{F_oL^2}{24EI}(-4e_1+e_2)$ d. $\delta=\frac{F_oL^2}{24EI}(-6e_1+e_2)$

5.17 Deflection when simply supported beam with combination of parabolic and straight tendons (central anchors)

a. $\delta=-\frac{F_oe}{EI}[5l_1^2+12l_1l_2+6l_2^2]$ b. $\delta=-\frac{2F_oe}{EI}[5l_1^2+12l_1l_2+6l_2^2]$

c. $\delta=-\frac{4F_oe}{EI}[5l_1^2+12l_1l_2+6l_2^2]$ d. $\delta=-\frac{6F_oe}{EI}[5l_1^2+12l_1l_2+6l_2^2]$

5.18 Deflection when simply supported beam with combination of parabolic and straight tendons (eccentric anchors)

a. $\delta=-\frac{F_o(e_1+e_2)}{12EI}[2l_1^2+12l_1l_2+6l_2^2]+\left[\frac{F_oe_2L^2}{8EI}\right]$

b. $\delta=-\frac{F_o(e_1+e_2)}{12EI}[7l_1^2+12l_1l_2+6l_2^2]+\left[\frac{F_oe_2L^2}{8EI}\right]$

c. $\delta=-\frac{F_o(e_1+e_2)}{12EI}[3l_1^2+12l_1l_2+6l_2^2]+\left[\frac{F_oe_2L^2}{8EI}\right]$

d $\delta=-\frac{F_o(e_1+e_2)}{12EI}[5l_1^2+12l_1l_2+6l_2^2]+\left[\frac{F_oe_2L^2}{8EI}\right]$

5.19 Deflection when simply supported beam with end couples of equal magnitude M

a. $\delta_c=\frac{5ML^2}{8EI}$ b. $\delta_c=\frac{3ML^2}{8EI}$ c. $\delta_c=\frac{ML^2}{8EI}$ d. $\delta_c=\frac{7ML^2}{8EI}$

5.20 Deflection due to self–weight and imposed loads is calculated as

a. $\delta_c = \frac{5(g+q)L^4}{384EI}$ b. $\delta = \frac{3(g+q)L^4}{384EI}$ c. $\delta = \frac{7(g+q)L^4}{384EI}$ d. $\delta = \frac{9(g+q)L^2}{384EI}$

5.21 Long term deflection is due to

a. creep b. shrinkage of concrete
c. relaxation of stress in steel d. All of the above

5.22 The creep strain at any time t may be determined by multiplying the instantaneous strain by the quantity .... where, ϕ is the creep coefficient

a. $(1 + 2\phi)$ b. $(1 + 3\phi)$ c. $(1 + \phi)$ d. $(3 + \phi)$

5.23 Total long term deflection after time t (Neville's formula) is given by .... negative sign indicates upward deflection

a. $\delta_t = \delta_{il}(1-\phi) - \delta_{ip}\left\{1-\frac{\Delta P_i}{P_i} + \left(1-\frac{\Delta P_i}{2P_i}\right)3\phi\right\}$

b. $\delta_t = \delta_{il}(1-\phi) - \delta_{ip}\left\{1-\frac{\Delta P_i}{P_i} + \left(1-\frac{\Delta P_i}{2P_i}\right)\phi\right\}$

c. $\delta_t = \delta_{il}(1-\phi) - \delta_{ip}\left\{1-\frac{\Delta P_i}{P_i} + \left(1-\frac{\Delta P_i}{2P_i}\right)2\phi\right\}$

d. $\delta_t = \delta_{il}(1-\phi) - \delta_{ip}\left\{1-\frac{\Delta P_i}{P_i} + \left(1-\frac{\Delta P_i}{2P_i}\right)5\phi\right\}$

5.24 Lin's formula for long term deflection calculation is..........

a. $\delta_t = \left(\delta_{ip} - \delta_{ip}.\frac{P_t}{P_i}\right)(1+5\phi)$ b. $\delta_t = \left(\delta_{ip} - \delta_{ip}.\frac{P_t}{P_i}\right)(1+3\phi)$

c. $\delta_t = \left(\delta_{il} - \delta_{ip}.\frac{P_t}{P_i}\right)(1+\phi)$ d. $\delta_t = \left(\delta_{il} - \delta_{ip}.\frac{P_t}{P_i}\right)(1+2\phi)$

Answers

5.1 d	5.2 b	5.3 a	5.4 d	5.5 a
5.6 d	5.7 c	5.8 b	5.9 a	5.10 d
5.11 b	5.12 b	5.13 c	5.14 b	5.15 c
5.16 b	5.17 a	5.18 d	5.19 c	5.20 a
5.21 d	5.22 c	5.23 b	5.24 c	

❑❑❑

CHAPTER 6

Ultimate Flexural Resistance

6.1 INTRODUCTION

The single most important design objective for a structure or a component of a structure is the provision of adequate strength. The consequences and costs of strength failures are large and therefore the probability of such failures must be very small.

The satisfaction of concrete and steel stress limits at service loads does not necessarily ensure adequate strength and does not provide a reliable indication of either the actual strength or the safety of a structural member. It is important to consider the non–linear behaviour of the member in the over–loaded range to ensure that it has an adequate structural capacity. Only by calculating the ultimate capacity of a member can a sufficient margin between the service load and the ultimate load be guaranteed.

In the 1950s and 1960s, there was a gradual swing away from the use of elastic stress calculations for the satisfaction of the design objective of adequate strength. The so–called ultimate strength design approach emerged as the most appropriate procedure. The ultimate strength of a cross–section in bending M_u is calculated from a rational and well established flexural strength theory, which involves consideration of the strength of both the concrete and the steel in the compressive and tensile parts of the cross–section. The prediction of ultimate flexural strength is described and illustrated in this chapter. When M_u is determined, the design requirements for the strength limit state may be checked and satisfied.

In addition to calculating the strength of a section, a measure of the ductility of each section must also be established. Ductility is an important objective in structural design. Ductile members undergo large deformations prior to failure, thereby providing warning of failure and allowing indeterminate structures to establish alternative load paths. In fact, it is only with adequate ductility that the predicted strength of indeterminate members and structures can beachieved in practice.

6.2 FLEXURAL BEHAVIOR AT OVERLOADS

The load at which collapse of a flexural member occurs is called the ultimate load. If the member has sustained large deformations prior to collapse, it is said to have ductile behaviour. If, on the other hand, it has only undergone small deformations prior to failure, the member is said to have brittle behaviour. There is no defined deformation or curvature which distinguishes ductile from brittle behaviour. Codes of practice, however, usually impose a ductility requirement by limiting the curvature at ultimate to some minimum value, thereby ensuring that significant deformation occurs in a flexural member prior to failure. Since beam failures that result from a breakdown of bond between the concrete and the steel reinforcement, or from excessive shear, or from failure of the anchorage zone tend to be brittle in nature, every attempt should be made to ensure that the region of maximum moment is the weakest link. The design philosophy should ensure therefore that a member does not fail before the required design moment capacity of the section is attained.

6.2.1 Types of Flexural Failures

When pre–stressed concrete members are subjected to bending loads, different types of flexural failures are possible at critical sections, depending upon the principal controlling parameters, such as the percentage of reinforcement in the section, degree of bond between tendon and concrete, compressive strength of concrete and ultimate tensile strength of tendons. In the post–cracking stage, the behavior of a pre–stressed concrete member is more akin to that of a reinforced concrete member and the theories used for estimating the flexural strength of reinforced concrete section may as well be used for pre–stressed concrete sections.

The various types of flexural failures encountered in pre–stressed concrete members are examined in the light of recommendations of various codes of practice.

6.2.1.1 Fracture of Steel in Tension

The sudden failure of pre–stressed member without any warning is generally due to the fracture of steel in the tension zone. This type of failure is imminent when the percentage of steel provided in the section is so low that when the concrete in the tension zone cracks, the steel is not in a position bear the additional tensile stress transferred to it by the cracked concrete. This type of failure can be prevented by providing a certain minimum percentage of steel in the cross-section.

The Indian standard code IS: 1343–2012 prescribes a minimum longitudinal reinforcement of 0.2% of cross sectional are in all cases except in the case of pre–tensioned units of small sections. When high yield strength deformed reinforcement used, the minimum steel percentage is reduced to 0.15%. The percentage of steel provided, both tensioned and un–tensioned together, should be sufficient so what when the concrete in the pre–compressed tensile zone cracks, the steel is in a position to bear the additional tensile stress transferred it by the cracking of the adjacent fibres of concrete, thereby preventing a sudden failure of the beam due to fracture of steel in tension.

6.2.1.2 Failure of Under Reinforced Sections

If the cross sections is provided with an amount of steel greater than the minimum prescribed as in earlier case, the failure is characterized by an excessive elongation of steel raises the neutral axis closure to the compression face at the critical section.

The member approaches failure due to the gradual reduction of the compression zone, exhibiting large deflection and cracks, which develops at the soffit and progress towards the compression face. When the area of the concrete in compression bone is insufficient to resist the resultant internal compressive force, the ultimate flexural failure of the member takes place through the crushing of concrete. Large deflections and wide cracks are the characteristic features of under reinforced sections at failure (Fig. 6.1). This type of behavior is desirable since there is considerable warning before the impending failure. As such it is common practice to design under reinforced sections which becomes more important in case of statically indeterminate structures. An upper limit on the maximum area of steel is generally prescribed in various codes for under reinforced sections.

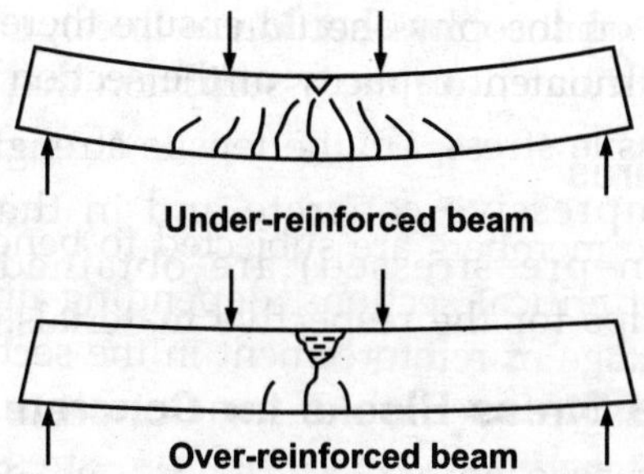

Fig. 6.1 Flexural Failure Modes of Pre–stressed Concrete Beams

6.2.1.3 Failure of Over Reinforced Sections

When the effective reinforcement index, which is expressed in terms of the percentage of reinforcement, the compressive strength of concrete and tensile strength of steel, exceeds a certain range of values, the section is said to be over reinforced. Generally, over reinforced concrete members fail by sudden crushing of concrete, the failure being characterized by small deflections and narrow cracks (Fig. 6.1). The area of steel being comparatively large, the stresses developed in steel at failure of the member may not reach the tensile strength and in many cases it may well be within the proof stress of tendons.

In structural concrete members, it is undesirable to have sudden failures without any warning in the form of excessive deflections and widespread cracks, and consequently the uses of over reinforced sections are discouraged. The amount of reinforcement used in practice should, preferably, not exceed that required for a balanced section, in this connection, most of the codes follow a conservative approach in formulating the evaluation procedures for flexural strength calculations of over reinforced sections.

The redistribution of moments in indeterminate structures depends upon the rotation capacities of the critical sections of the member under a given system of loads. The use of over reinforced sections in such structures curtails the rotation capacity of the sections, consequently affecting the ultimate load on the structure.

6.2.1.4 Other Modes of Failure

Pre–stressed concrete members subjected to transverse loads may fail in shear before their full flexural strength is attained, if they are not adequately designed for shear. Web shear cracks may develop if the principal stresses are excessive, and if thin webs are used, the failure may occur due to web crushing. In the case of pre–tensioned members, the failure the bond between the steel and the surrounding concrete is likely due to inadequate transmission lengths at the ends of the members. In post–tensioned members, anchorage failures may take place if the end block is not properly designed to resist the transverse tensile forces.

6.3 FLEXURAL STRENGTH THEORY

6.3.1 Assumptions

In the analysis of a cross–section to determine its ultimate bending strength M_u, the following assumptions are usually made:

(a) The variation of strain on the cross–section is linear, i.e, strains in the concrete and the bonded steel are calculated on the assumption that plane sections remain plane.

(b) Concrete carries no tensile stress, i.e, the tensile strength of the concrete is ignored.

(c) The stress in the compressive concrete and in the steel reinforcement (both pre–stressed and non–pre–stressed) are obtained from actual or idealized stress–strain relationships for the respective materials.

6.3.2 Idealized Compressive Stress Blocks for Concrete for Rectangular Sections

The Indian code method (IS: 1343–2012) for computing the flexural strength of rectangular section or T–sections in which the neutral axis (NA) lies within the flange is based on the rectangular and parabolic stress block as shown in Fig. 6.2.

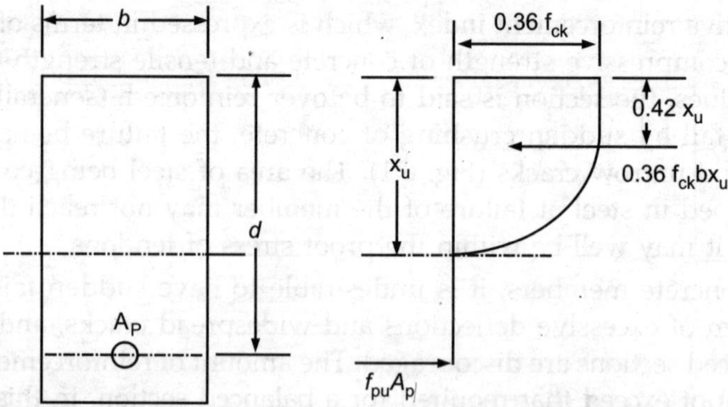

Fig. 6.2 Moment of Resistance of Rectangular Sections (IS: 1343–2012)

The moment of resistance of rectangular beam and flanged section, where the neutral axis lies within the flange, is expressed as $M_u = f_{pu} A_p (d - 0.42 x_u)$ ....(6.1)

Where, M_u = Ultimate moment of resistance of the section

f_{pu} = Tensile stress developed in tendon at the failure stage of the beam

A_p = Area of pre–stressing tendons

d = Effective depth

x_u = Neutral axis depth

The value of f_{pu} depends upon the effective reinforcement ratio = $\frac{A_p f_p}{bdf_{ck}}$ ...(6.2)

⇒ Ref. Table 11 and 12 of IS: 1343–2012 Page No. 59 and 60.

Where, f_p = Characteristic tensile strength of the pre–stressing steel

d = Effective depth

f_{pe} = Effective pre–stress in tendons after losses

f_{ck} = Characteristic strength of concrete

b = Width of beam in rectangular section or width of flange in T–section

For pre–tensioned and post–tensioned members with an effective bond between concrete and tendons, the values of f_{pu} and x_u are given in Table 6.1. The effective pre–stress f_{pe} after all losses should not be less $0.451f_p$. For post–tensioned rectangular beams with un–bonded tendon, the value of f_{pu} and x_u are influenced by the effective span to depth ratio, and their values for different span/depth ratios are given in Table 6.2.

6.3.3 Idealized Compressive Stress Blocks for Concrete for Flanged Sections

The ultimate moment resistance of flanged sections in which neutral axis falls outside the flange is computed by combining the moment of resistance of web and flange portions and considering the stress blocks as shown in Fig. 6.3.

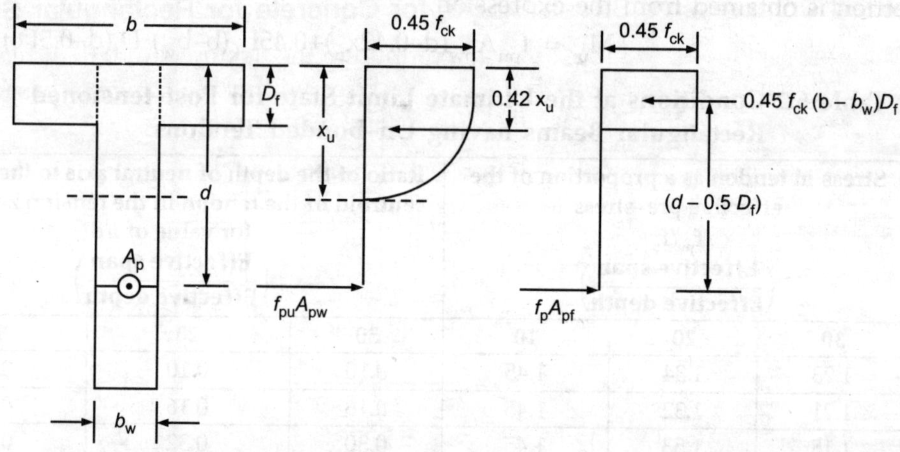

Fig. 6.3 Moment of Resistance of Flanged Sections (x_u>D_f) (IS: 1343–2012)

Table 6.1 Conditions at the Ultimate Limit State for Rectangular Beams with Pre–tensioned Tendons or with Post–tensioned Tendons having Effective Bond

$\frac{A_p f_p}{bdf_{ck}}$	Stress in tendon as a proportion of the design strength $f_{pu}/0.87f_p$		Ratio of the depth of neutral axis to that of the centroid of the tendon in the tension zone x_u/d	
	Pre–tensioning	Post–tensioning with effective bond	Pre–tensioning	Post–tensioning with effective bond
0.025	1.0	1.0	0.054	0.054
0.05	1.0	1.0	0.109	0.109
0.10	1.0	1.0	0.217	0.217
0.15	1.0	1.0	0.326	0.316
0.20	1.0	0.95	0.435	0.414
0.25	1.0	0.90	0.542	0.488
0.30	1.0	0.85	0.655	0.558
0.40	0.9	0.75	0.783	0.653

If A_{pw} = Area of pre–stressing steel for web
A_{pf} = Area of pre–stressing steel for flange
D_f = Thickness of flange

Then, $$A_p = (A_{pw} + A_{pf}) \quad ...(6.3)$$

But, $$A_{pf} = 0.45f_{ck}(b - b_w)\left(\frac{D_f}{f_p}\right) \quad ...(6.4)$$

After evaluating A_{pf}, the value of A_{pw} is obtained as
$$A_{pw} = (A_p - A_{pf}) \quad ...(6.5)$$

For the effective reinforcement ratio of $(A_{pw}f_p/b_w df_{ck})$, the corresponding values of $(f_{pu}/0.87f_p)$ and (x_u/d) are obtained from Table 6.1. The ultimate moment of resistance of the flanged section is obtained from the expression,
$$M_u = f_{pu}A_{pw}(d-0.42x_u)+0.45f_{ck}(b-b_w)\,D_f(d-0.5D_f) \quad ...(6.6)$$

Table 6.2 Conditions at the Ultimate Limit State for Post–tensioned Rectangular Beams having Un–bonded Tendons

$\frac{A_p f_p}{bdf_{ck}}$	Stress in tendon as a proportion of the effective pre–stress f_{pu}/f_p $\left(\frac{\text{Effective span}}{\text{Effective depth}}\right)$			Ratio of the depth of neutral axis to that of the centroid of the tendon in the tension zone x_u/d for value of l/d $\left(\frac{\text{Effective span}}{\text{Effective depth}}\right)$		
	30	20	10	30	20	10
0.025	1.23	1.34	1.45	0.10	0.10	0.10
0.05	1.21	1.32	1.45	0.16	0.16	0.18
0.10	1.18	1.53	1.45	0.30	0.32	0.36
0.15	1.14	1.20	1.36	0.44	0.46	0.52
0.20	1.11	1.16	1.27	0.56	0.58	0.64

6.4 SOLVED EXAMPLES

Problem 6.1

A pre–tensioned pre–stressed concrete beam having a rectangular section, 150mm wide and 350mm deep, has an effective cover of 50mm. If $f_{ck} = 40\text{N/mm}^2$, $f_p = 1600\text{N/mm}^2$ and the area of pre–stressing steel, $A_p = 461\text{mm}^2$. Calculate the ultimate strength of the section using code provisions.

Solution:

$f_{ck} = 40\text{N/mm}^2$ $\quad b = 150\text{mm}$ $\quad f_p = 1600\text{N/mm}^2$ $\quad d = 350–50 = 300\text{mm}$

$A_p = 461\text{mm}^2$.

The effective reinforcement ratio is given by,

$$\frac{A_p f_p}{bdf_{ck}} = \frac{461 \times 1600}{150 \times 300 \times 40} = 0.409 \text{ say } 0.41$$

Ref. Table No. 11 of Page No. 59 (IS: 1343–2012)

$$\frac{f_{pu}}{0.87f_p} = 0.9 \quad \therefore \ f_{pu} = 0.9 \times 0.87f_p = 0.9 \times 0.87 \times 1600 = 1252.8 \text{ say } 1253\text{N/mm}^2$$

and $\frac{x_u}{d} = 0.783, x_u = 0.783d = 0.783 \times 300 = 234.9\text{mm}$

$$\therefore \quad M_u = f_{pu}A_p(d - 0.42x_u) = 1253 \times 461\,(300 - 0.42 \times 234.9) = 116 \times 10^6\text{Nmm} = 116\text{kNm}.$$

Problem 6.2

A pre tensioned, T–section has a flange which is 300mm wide 200mm thick. The rib is 150mm wide by 350mm deep. The effective depth of the cross section is 500mm. Given $A_p = 200\text{mm}^2$, $f_{ck} = 50\text{N/mm}^2$, $f_p = 1600\text{N/mm}^2$. Estimate the ultimate moment capacity of the T–section using the Indian standard code regulations.

Solution:

$f_{ck} = 50\text{N/mm}^2$; $f_p = 1600\text{N/mm}^2$; $b = 300\text{mm}$; $d = 500\text{mm}$; $A_p = 200\text{mm}^2$. Assuming the neutral axis lies within the flange.

The value of b = 300mm for computations of effective reinforcement ratio.

$$\frac{A_p f_p}{bdf_{ck}} = \frac{200 \times 1600}{300 \times 500 \times 50} = 0.04$$

Ref. Table No. 11 of Page No. 59 (IS: 1343–2012)

$$\frac{f_{pu}}{0.87f_p} = 1.0 \quad \therefore \ f_{pu} = 1 \times 0.87f_p = 0.87 \times 1600 = 1392\text{N/mm}^2$$

and $\frac{x_u}{d} = 0.089, \therefore \ x_u = 0.089d = 0.089 \times 500 = 44.5\text{mm}.$

Hence the NA lies within the flange.

$\therefore \ M_u = f_{pu}A_p(d - 0.42x_u) = 1392 \times 200(500 - 0.42 \times 44.5) = 134 \times 10^6\text{Nmm} = 134\text{kNm}.$

Problem 6.3

A post–tensioned beam with un–bonded tendons is of rectangular section 400mm wide with an effective depth of 800mm. The cross sectional area of pre–stressing steel is 2840mm². The effective pre–stress in the steel after all losses is 900N/mm². The effective span of the beam is 16m. If $f_{ck} = 40\text{N/mm}^2$. Estimate ultimate moment of resistance of the section using IS: 1343–2012.

Solution:

Given:

$f_{ck} = 40\text{N/mm}^2$, $b = 400\text{mm}$, $d = 800\text{mm}$,

$f_{pe} = 900\text{N/mm}^2$, $A_p = 2840\text{mm}^2$, $L/d = 16000/800 = 20$

The effective reinforcement ratio is given by

$$\frac{A_p f_p}{bdf_{ck}} = \frac{2840 \times 900}{400 \times 800 \times 40} = 0.2,$$ Ref. Table No. 12 of Page No. 60 (IS: 1343–2012)

$$\frac{f_{pu}}{f_{pe}} = 1.16 \therefore f_{pu} = 1.16 \times 900 = 1044\text{N/mm}^2$$

and $\frac{x_u}{d} = 0.58, x_u = 0.58 \times 800 = 464\text{mm}.$

$\therefore M_u = f_{pu}A_p(d - 0.42x_u) = 1044 \times 2840(800 - 0.42 \times 464) = 1794 \times 10^6\text{Nmm} = 1794\text{kNm}.$

Problem 6.4

A pre–tensioned, T–section has a flange 1200mm wide and 150mm thick. The width and depth of the rib are 300mm and 1500mm respectively. The high tensile steel has an area of 4700mm² and is located at an effective depth of 1600mm. If the characteristic cube strength of the concrete and tensile strength of steel are 40N/mm² and 1600N/mm² respectively. Calculate the flexural strength of the T–section.

Solution:

$$A_p = 4700\text{mm}^2, f_{ck} = 40\text{N/mm}^2, D_f = 150\text{mm}, b = 1200\text{mm},$$
$$d = 1600\text{mm}, b_w = 300\text{mm}.$$
$$A_p = A_{pw} + A_{pf}$$
$$A_{pf} = 0.45f_{ck}(b - b_w)D_f/f_p = 0.45 \times 40(1200 - 300)(150/1600) = 1518\text{mm}^2$$
$$A_{pw} = A_p - A_{pf} = 4700 - 1518 = 3182\text{mm}^2$$
$$\frac{A_{pw}f_p}{b_w df_{ck}} = \frac{3182 \times 1600}{300 \times 1600 \times 40} = 0.265$$

Ref. Table No. 11 of Page No. 59 (IS: 1343–2012)

$$\frac{f_{pu}}{0.87f_p} = 1.0 \text{ and } \frac{x_u}{d} = 0.56, x_u = 0.56 \times 1600 = 896\text{mm, lies below the flange.}$$

$$\therefore \quad f_{pu} = 0.87 \times 1600 = 1392\text{N/mm}^2$$
$$M_u = f_{pu}A_{pw}(d - 0.42x_u) + 0.45f_{ck}(b - b_w)D_f(d - 0.5D_f)$$
$$= 1392 \times 3182(1600 - 0.42 \times 896) + 0.45 \times 40(1200 - 300)150(1600 - 0.5 \times 150)$$
$$= 10198 \times 10^6\text{Nmm} = 10198\text{kNm}.$$

Problem 6.5

A post–tensioned bridge girder with un–bonded tendon is of box section of overall dimension 1200mm wide by 1800mm deep with wall thickness of 150mm. The high tensile steel has an area of 4000mm² and is located at an effective depth of 1600mm. The effective pre–stress in steel after all losses is 1000N/mm² and the effective span of the girder is 24m. If f_{ck} = 40N/mm², f_p = 1600N/mm², estimate the ultimate flexural strength of the section.

Solution:

$$A_p = 4000\text{mm}^2,\ f_{ck} = 40\text{N/mm}^2,\ f_p = 1600\text{N/mm}^2,\ f_{pe} = 1000\text{N/mm}^2,\ b = 1200\text{mm},$$

$$d = 1600\text{mm},\ D_f = 150\text{mm},\ b_w = 300\text{mm}$$

$$A_p = A_{pw} + A_{pf}$$

$$A_{pf} = 0.45f_{ck}(b - b_w)\,D_f/f_p = 0.45 \times 40(1200 - 300)\,(150/1600) = 1518\text{mm}^2$$

$$A_{pw} = A_p - A_{pf} = 4000 - 1518 = 2482\text{mm}^2$$

$$\frac{A_{pw}f_p}{b_w d f_{ck}} = 0.129$$

Ref. Table No. 12 of Page No. 60 (IS: 1343–2012)

$$\frac{f_{pu}}{f_{pe}} = 1.310 \text{ and } \frac{x_u}{d} = 0.41,\ \therefore\ x_u = 0.41 \times 1600 = 656\text{mm, lies below the flange.}$$

$$\therefore\ f_{pu} = 1.31 \times 1000 = 1310\text{N/mm}^2$$

$$M_u = f_{pu}A_{pw}(d - 0.42x_u) + 0.45f_{ck}(b - b_w)\,D_f(d - 0.5D_f)$$

$$= 1310 \times 2482(1600 - 0.42 \times 656) + 0.45 \times 40(1200 - 300)\,150(1600 - 0.5 \times 150)$$

$$= 8011 \times 10^6\text{Nmm} = 8011\text{kNm.}$$

REVIEW QUESTIONS

6.1 Elaborate the flexural behavior at over loads.

6.2 What are the different types of flexural failures observed in concrete beams? Explain with sketches.

6.3 Explain flexural strength theory.

6.4 What are all the assumptions of flexural strength theory?

6.5 Explain fracture of steel in tension.

6.6 Explain failure of under reinforced sections.

6.7 Explain failure of over reinforced sections.

6.8 What are the difference in the types of stress blocks adopted in Indian code specifications regarding flexural strength computation?

6.9 Explain with sketches the method of estimating the ultimate flexural strength of rectangular pre–stressed concrete sections according to IS: 1343 specifications.

6.10 Explain with sketches the method of estimating the ultimate flexural strength of flanged pre–stressed concrete sections according to IS: 1343 specifications.

EXERCISE PROBLEMS

6.1 A post–tensioned bonded pre–stressed concrete beam is of rectangular section 200mm wide and 400mm deep. The beam is pre–stressed by 300mm² of high tensile steel located at an eccentricity of 100mm. the effective pre–stress after al losses is 800N/mm², the characteristic tensile strength of pre–stressing steel is 1600N/mm² and characteristic cube strength is 40N/mm^2. The ratio of yield to ultimate tensile stress of steel is 0.9. Calculate the ultimate moment capacity of the section using IS: 1343.

[Ans: Ultimate moment capacity of the section = 98.32kNm]

6.2 The cross–section of a symmetrical I–section pre–stressed beam is 300mm by overall depth of 750mm, with flanges web of 100mm thick. The beam is post–tensioned by cables containing 48 wires of 5mm diameter high tensile steel wires at an eccentricity of 250mm. The characteristic strength of concrete is 40MPa and the ultimate tensile strength of wires is 1700MPa. Using Indian standards, Determine the ultimate moment of the section.

[Ans: Ultimate moment of the section = 571kNm]

6.3 An unsymmetrical I–section has an overall depth of 2000mm. The top flange width and depth are 1200 and 300mm respectively, and the bottom flange width and depth are 750mm and 200mm respectively. The thickness of web is 300mm. The tendons having an cross sectional area of 7000mm^2 are located 200mm from the soffit. If the ultimate compressive strength of concrete and tensile strength of steel are 42 and 1750N/mm^2 respectively, and the tendons are effectively bonded to concrete, evaluate the flexural strength of the sections.

[Ans: Flexural strength of the section = 15366kNm]

6.4 A double T–section having a flange of 1200mm and 150mm thickness is pre–stressed by 4700mm^2 of high tensile steel is located at an effective depth of 1600mm. The ribs having an thickness of 150mm each. If the cube strength of concrete is 40MPa and tensile strength of steel is 1600MPa, determine the flexural strength of the double T–girder using IS: 1343 specifications.

[Ans: Flexural strength of the double T–girder = 9069kNm]

6.5 An unsymmetrical I–section bridge girder has the following section properties: width and depth of top flange = 1200 and 360mm respectively, thickness of web = 240mm, centroid of section is located at 580mm from the top, the girder is used over a span of 40m, and the tendons with a cross–section of 7000mm^2 are parabolic with an eccentricity of 1220mm at the centre of span and zero at the supports. Take f_{ck} = 45MPa and f_{pu} = 1700MPa, estimate the ultimate flexural strength of the center of span section.

[Ans: Ultimate flexural strength = 17416kNm]

6.6 A post–tensioned pre–stressed concrete T–beam with un–bonded tendons is made up of a flange of 300mm wide by 150mm thick and the width of the rib is 150mm. The effective depth of the section is 320mm. The beam is pre–stressed by 24 wires of each 5mm diametr having a characteristic strength of 1650MPa. The effective stress after all losses of 900MPa. If cube strength of concrete is 56MPa, evaluate the flexural strength of section using Indian code. Assume L/d =20

[Ans: Flexural strength = 185.2kNm]

6.7 A post–tensioned bonded T–section has a flange width of 800mm and thickness of 250mm. Thickness of the web is 200mm. The area of high tensile strength steel located at an effective depth of 1200mm is 4000mm^2. The characteristic strength of concrete and steel are respectively 40MPa and 1500MPa. The effective pre–stress after all losses is 900MPa and the ratio of yield stress to the tensile strength of steel is 0.90. Estimate the ultimate moment capacity of T–section using IS: 1343 recommendations.

[Ans: Ultimate flexural strength = 5716kNm]

OBJECTIVE QUESTIONS

6.1 Which among the following is not the type of flexural failure?

a. Fracture of steel in tension

b. Failure of under reinforced section

c. Failure of over reinforced section

d. Failure of balanced reinforced section

6.2 Which among the following is not the assumption of flexural strength theory?

a. The variation of strain on the cross–section is linear, i.e., strains in the concrete and the bonded steel are calculated on the assumption that plane sections remain plane.

b. Concrete carries no tensile stress, i.e., the tensile strength of the concrete is ignored.

c. Stress of concrete is directly proportional to steel reinforcement.

d. The stress in the compressive concrete and in the steel reinforcement (both pre–stressed and non–pre–stressed) are obtained from actual or idealized stress–strain relationships for the respective materials.

6.3 The moment of resistance of rectangular beam and flanged section, where the neutral axis lies within the flange, is expressed as......, where, M_u = Ultimate moment of resistance of the section, f_{pu} = Tensile stress developed in tendon at the failure stage of the beam, A_p = Area of pre–stressing tendons, d = Effective depth, x_u = Neutral axis depth.

a. $M_u = f_{pu}A_p(d - 0.42x_u)$ b. $M_u = f_{pu}A_p(d - 0.44x_u)$

c. $M_u = f_{pu}A_p(d - 0.46x_u)$ d. $M_u = f_{pu}A_p(d - 0.48x_u)$

6.4 The value of f_{pu} = Tensile stress developed in tendon at the failure stage of the beam is depends upon the effective reinforcement ratio........, Where, f_p = Characteristic tensile strength of the pre–stressing steel, d = Effective depth, f_{pe}= Effective pre–stress in tendons after losses, f_{ck} = Characteristic strength of concrete, b = Width of beam in rectangular section or width of flange in T–section.

a. $\left(\dfrac{A_p f_{ck}}{bdf_p}\right)$ b. $\left(\dfrac{A_p f_p b}{df_{ck}}\right)$ c. $\left(\dfrac{A_p f_p}{bdf_{ck}}\right)$ d. $\left(\dfrac{A_p f_p d}{bf_{ck}}\right)$

6.5 The effective pre–stress f_{pe} after all losses should not be less than ...., Where f_p = Characteristic tensile strength of the pre–stressing steel.

a. $0.415f_p$ b. $0.451f_p$ c. $0.431f_p$ d. $0.421f_p$

6.6 The ultimate moment of resistance of the flanged section when neutral axis lies in web, is obtained from the expression ...., where A_{pw} = Area of pre–stressing steel for web, D_f = Thickness of flange.

a. $M_u = f_{pu} A_{pw} (d - 0.45 x_u) + 0.42 f_{ck} (b - b_w) D_f (d - 0.5D_f)$
b. $M_u = f_{pu} A_{pw} (d - 0.42 x_u) + 0.5 f_{ck} (b - b_w) D_f (d - 0.45D_f)$
c. $M_u = f_{pu} A_{pw} (d - 0.5 x_u) + 0.42 f_{ck} (b - b_w) D_f (d - 0.45D_f)$
d. $M_u = f_{pu} A_{pw} (d - 0.42 x_u) + 0.45 f_{ck} (b - b_w) D_f (d - 0.5D_f)$

6.7 A_{pf}, area of pre–stressing steel in flange is obtained by the expression …………, Where f_p = Characteristic tensile strength of the pre–stressing steel.

a. $0.44 f_{ck}(b - b_w)\left(\frac{D_f}{f_p}\right)$ b. $0.45 f_{ck}(b - b_w)\left(\frac{D_f}{f_p}\right)$

c. $0.49 f_{ck}(b - b_w)\left(\frac{D_f}{f_p}\right)$ d. $0.48 f_{ck}(b - b_w)\left(\frac{D_f}{f_p}\right)$

6.8 What is the value of effective reinforcement ratio $\left(\frac{A_p f_p}{bdf_{ck}}\right)$ of rectangular section with post–tensioned tendons having effective bond for corresponding $(f_{pu}/0.87f_p) = 0.95$ and $(x_u/d) = 0.414$, as per IS: 1343–2012?

a. 0.10 b. 0.15 c. 0.20 d. 0.25

6.9 What is the value of effective reinforcement ratio $\left(\frac{A_p f_p}{bdf_{ck}}\right)$ of rectangular section with pre–tensioned tendons for corresponding $(f_{pu}/0.87f_p) = 0.9$ and $(x_u/d) = 0.783$, as per IS: 1343–2012?

a. 0.25 b. 0.30 c. 0.35 d. 0.40

6.10 What is the value of effective reinforcement ratio $\left(\frac{A_p f_p}{bdf_{ck}}\right)$ of rectangular section with post–tensioned un–bonded tendons for corresponding $(f_{pu}/0.87f_p) = 1.53$ and $(x_u/d) = 0.32$ for effective span/effective depth ratio of 20, as per IS: 1343–2012?

a. 0.05 b. 0.10 c. 0.15 d. 0.20

6.11 What is the value of (x_u/d) of rectangular section with post–tensioned tendons having effective bond for corresponding $(f_{pu}/0.87f_p) = 0.90$ and effective reinforcement ratio = 0.25, as per IS: 1343–2012?

a. 0.217 b. 0.316 c. 0.414 d. 0.488

6.12 What is the value of (x_u/d) of rectangular section with pre–tensioned tendons for corresponding $(f_{pu}/0.87f_p) = 1.0$ and effective reinforcement ratio $\left(\frac{A_p f_p}{bdf_{ck}}\right) = 0.15$, as per IS: 1343–2012?

a. 0.217 b. 0.326 c. 0.435 d. 0.542

6.13 What is the value of (x_u/d) of rectangular section with post–tensioned un–bonded tendons for corresponding $(f_{pu}/0.87f_p) = 1.18$ and effective reinforcement ratio $\left(\frac{A_p f_p}{bdf_{ck}}\right) = 0.10$ for effective span/effective depth ratio of 30, as per IS: 1343–2012?

a. 0.10 b. 0.16 c. 0.30 d. 0.44

6.14 The value of effective span/effective depth ratio of rectangular section with post–tensioned un–bonded tendons for corresponding $(f_{pu}/0.87f_p) = 1.45$, effective reinforcement ratio $\left(\frac{A_p f_p}{bdf_{ck}}\right) = 0.10$ and $(x_u/d) = 0.36$ as per IS: 1343–2012 is ....

a. 40 b. 30 c. 20 d. 10

6.15 The value of effective span/effective depth ratio of rectangular section with post–tensioned un–bonded tendons for corresponding $(f_{pu}/0.87f_p) = 1.14$, effective reinforcement ratio $\left(\frac{A_p f_p}{bdf_{ck}}\right) = 0.15$ and $(x_u/d) = 0.44$ as per IS: 1343–2012 is ....

a. 40 b. 30 c. 20 d. 10

Answers

6.1 d	6.2 c	6.3 a	6.4 c	6.5 b
6.6 d	6.7 b	6.8 c	6.9 d	6.10 b
6.11 d	6.12 b	6.13 c	6.14 d	6.15 b

❑

CHAPTER 7

Design of Pre–stress Concrete Sections for Flexure and Torsion

7.1 DESIGN OF PRE–STRESS CONCRETE SECTIONS FOR FLEXURE

7.1.1 Introduction

In design the aim is to arrive at a suitable section, amount of pre–stressing steel and cable path for the given span and loading. According to IS: 1343–2012 code, the design shall be based on limit state concepts. In this method of design, the structure can be designed to withstand safely all the loads liable to act in it throughout its life; it also satisfy the serviceability requirements such as limitations on deflection and cracking. The acceptable limit for the safety and serviceability requirement before failure occurs are called "limit states". The aim of the design is to achieve acceptable probabilities that the structure will not become unfit for the use for which it is intended, that is, that it will not reach a limit state. The four limit states to be considered in design are given below.

- **Limit State of Collapse**

The limit state of collapse of the structure or part of the structure could be assessed from rupture of one or more critical sections and from buckling due to elastic or plastic instability or overturning. The resistance due to bending, shear, torsion and axial loads at every section shall not be less than the value of these forces of moments induced at that section by the probable most unfavorable combination of loads on the structure multiplied by the appropriate partial safety factors. The partial safety factors for DL and LL is 1.5.

- **Limit State of Serviceability: Deflection**

The deflection of a structure or part there of shall not adversely affect the appearance or efficiency of the structures or finishes or partitions.

- **Limit State of Serviceability: Cracking**

Cracking of concrete shall not affect the appearance or durability of the structure, the criteria of the limit state of cracking for three types of pre–stressed concrete members shall be as follows.

(a) For type 1, no tensile stresses are permitted.

(b) For type 2, tensile stresses are allowed but no visible cracking is permissible.

(c) For type 3, cracking is allowed, but should not affect the appearance or durability of the structure; the acceptable limits of width of cracking vary between wide limits for different types of structure and environment and prediction of the absolute maximum is not possible.

The surface width of cracks should not in general exceed 0.1mm in members exposed to particularly aggressive environment and 0.2mm in all other members.

The flexural tensile stress at any section of the structure, both at transfer and under the most unfavorable combination of design loads, shall satisfy the criteria for the corresponding type of structure.

• Limit State of Serviceability: Maximum Compression

The compressive stresses both at transfer and under design loads at service condition shall be limited to the values given below for all types of structures.

❖ *Maximum Permissible Compressive Stress at Transfer (Fig. 7.1)*

Maximum permissible compressive stress due to bending and direct forces at the time of transfer of pre–stress shall be determined for the different grades of concrete from a straight line plot joining a point given by $0.54f_{ci}$ for concrete of grade M30 and a second point giving M60. f_{ci} being the cube strength of concrete at transfer (which is no case shall be less than half the corresponding characteristic compressive strength of concrete). These values apply to post–tensioned work.

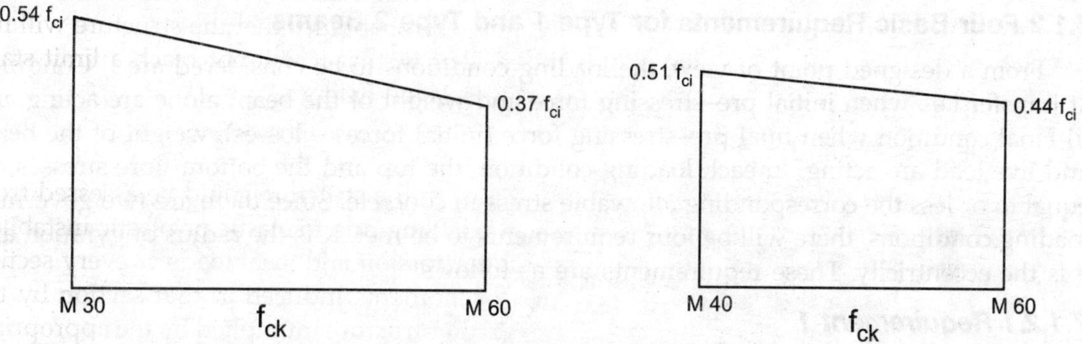

Fig. 7.1 Computation of Maximum Permissible Compressive Stressin Flexure at Transfer for (a) Post–tensioned Work and (B) Pre–tensioned Work

❖ *Maximum Permissible Compressive Stress under Service Conditions*

The maximum permissible compressive stress in concrete under the action of live load and effective pre–stressing force shall be determined by a straight line relation but different stress limits shall apply to the concrete of the structure depending on whether it falls in a part of the section where the compressive stresses are not likely to increase in service (Zone I) or in a part of the structure where the compressive stresses are likely to increase (Zone II).

For zone I, the straight line relation of permissible stress shall be determined by the straight line joining a point given by a permissible stress of $0.41f_{ck}$ for concrete of grade M30 to another point given by the permissible stress of $0.35f_{ck}$ for the concrete of grade M60.

For zone II, the determining points of the graph shall be reduced to $0.34f_{ck}$ and $0.27f_{ck}$ respectively. (Ref. Fig. 7.2)

Ex: For concrete of grade M40 in zone I, the maximum compressive stress in concrete

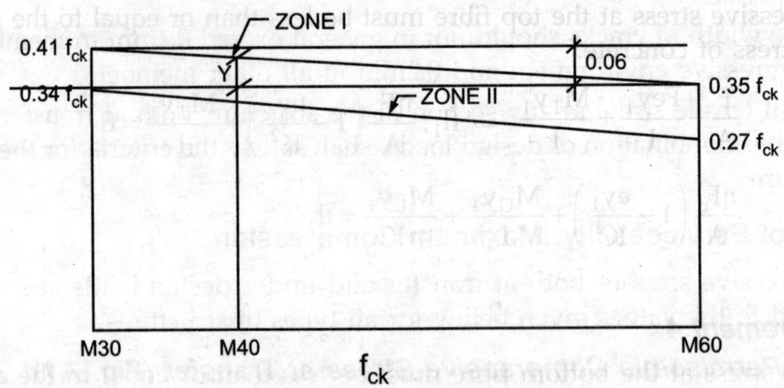

Fig. 7.2 Computation of Maximum Permissible Compressive Stress in Flexure due to Final Pre–stress

$$\frac{Y}{0.06} = \frac{20}{30} \quad \therefore\ Y = \frac{20 \times 0.06}{30} = 0.04$$

$\therefore$ For M40, $0.35f_{ck} + 0.04f_{ck} = 0.35 \times 40 + 0.04 \times 40 = 14 + 1.6 = 15.6$ N/mm^2.

7.1.2 Four Basic Requirements for Type 1 and Type 2 Beams

From a designed point of view, the loading conditions to be considered are i) Condition at transfer i.e., when initial pre–stressing force and weight of the beam alone are acting and ii) Final condition when final pre–stressing force (initial forces – losses), weight of the beam and live load are acting. In each loading condition, the top and the bottom fibre stresses are equal to or less the corresponding allowable stress in concrete. Since there are two governing loading conditions, there will be four requirements to be met. K is the radius of gyration and e is the eccentricity. These requirements are as follows.

7.1.2.1 Requirement 1

At transfer, the tensile stress at the top fibre must be less than or equal to the allowable tensile stress of the concrete at transfer. Algebraically, this requirement can be stated as follows.

$$\frac{F_o}{A} - \frac{F_o e y_1}{I} + \frac{M_G y_1}{I} = -f_{ul};\ \frac{F_o}{A} - \frac{F_o e y_1}{AK^2} + \frac{M_G y_1}{I} = -f_{u1},$$

$$\frac{F_o}{A}\left(1 - \frac{e y_1}{K^2}\right) + \frac{M_G y_1}{I} = -f_{u1} \qquad \text{...(7.1)}$$

7.1.2.2 Requirement 2

At transfer, the compressive stress at the bottom concrete fibre must be less than or equal to the allowable compressive stress of concrete at transfer.

$$\frac{F_o}{A} + \frac{F_o e y_2}{I} - \frac{M_G y_2}{I} = f_{u2};\ \frac{F_o}{A}\left(1 + \frac{e y_2}{K^2}\right) - \frac{M_G y_2}{I} = f_{u2} \qquad \text{...(7.2)}$$

7.1.2.3 Requirement 3

The compressive stress at the top fibre must be less than or equal to the allowable final compressive stress of concrete.

$$\frac{F}{A}-\frac{Fey_1}{I}+\frac{M_Ty_1}{I} = fl_1;\ \frac{\eta F_o}{A}\left(1-\frac{ey_1}{K^2}\right)+\frac{M_Ty_1}{I}=fl_1$$

$$\frac{\eta F_o}{A}\left(1-\frac{ey_1}{K^2}\right)+\frac{M_Gy_1}{I}+\frac{M_Ly_1}{I}=fl_1 \quad ...(7.3)$$

7.1.2.4 Requirement 4

The tensile stress at the bottom fibre must be less than or equal to the allowable final tensile stress of concrete.

$$\frac{F}{A}+\frac{Fey_2}{I}-\frac{M_Ty_2}{I} = -fl_2;\ \frac{\eta F_o}{A}\left(1+\frac{ey_2}{K^2}\right)-\frac{M_Ty_2}{I}=-fl_2$$

$$\frac{\eta F_o}{A}\left(1+\frac{ey_2}{K^2}\right)-\frac{M_Gy_2}{I}-\frac{M_Ly_2}{I} = -fl_2 \quad ...(7.4)$$

Now subtracting (7.3) from (7.1) we get

$$\frac{F_o}{A}\left(1-\frac{ey_1}{K^2}\right)+\frac{M_Gy_1}{I}-\frac{\eta F_o}{A}\left(1-\frac{ey_1}{K^2}\right)-\frac{M_Gy_1}{I}-\frac{M_Ly_1}{I}=-f_{u1}-fl_1;$$

$$\frac{F_o}{A}\left(1-\frac{ey_1}{K^2}\right)(1-\eta)-\frac{M_Ly_1}{I}=-f_{u1}-fl_1;\ \frac{F_o}{A}\left(1-\frac{ey_1}{K^2}\right)(1-\eta)+f_{u1}+fl_1=\frac{M_Ly_1}{I}$$

$$\therefore \quad \frac{I}{y_1} = \frac{M_L}{f_{u1}+fl_1+\frac{F_o}{A}\left(1-\frac{ey_1}{K^2}\right)(1-\eta)} \quad ...(7.5)$$

Similarly subtracting (7.4) from (7.2)

$$\frac{I}{y_1} = \frac{M_L}{f_{u2}+fl_2+\frac{F_o}{A}\left(1+\frac{ey_2}{K^2}\right)(1-\eta)} \quad ...(7.6)$$

In design practice the initial pre–stressing force F_o is assumed equal to $\frac{Af_{u2}}{2}$

So, $\frac{F_o}{A}=\frac{f_{u2}}{2}$, In many cases $\frac{ey_1}{K^2}$ and $\frac{ey_2}{K^2}$ are approximately equal to 2

Now substituting these values in (7.6)

$$\frac{I}{y_2}=\frac{M_L}{f_{u2}+fl_2-\frac{F_o}{A}\left(1+\frac{ey_2}{K^2}\right)(1-\eta)},\ \frac{I}{y_2}=\frac{M_L}{f_{u2}+fl_2-(1-\eta)\frac{F_o}{A}\left(1+\frac{ey_2}{K^2}\right)},$$

$$\frac{I}{y_2}=\frac{M_L}{f_{u2}+fl_2-(1-\eta)\frac{f_{u2}}{2}(1+2)},\ \frac{I}{y_2}=\frac{M_L}{f_{u2}+fl_2-3(1-\eta)\frac{f_{u2}}{2}},$$

$$\frac{I}{y_2}=\frac{M_L}{f_{u2}+fl_2-1.5f_{u2}+1.5f_{u2}\eta},\ \frac{I}{y_2}=\frac{M_L}{-0.5f_{u2}+fl_2+1.5f_{u2}\eta},$$

$$\frac{I}{y_2}=\frac{M_L}{fl_2+(1.5\eta-0.5)f_{u2}} \quad \text{...(7.7)}$$

7.1.3 Selection of Shapes of Cross–section

First a trial rectangular section shall be obtained using the equation of modulus of section.

$$\frac{I}{y_1}=\frac{I}{y_2}=\frac{BH^2}{6}=\frac{M_L}{fl_2+(1.5\eta-0.5)f_{u2}} \quad \text{...(7.8)}$$

The following Table 7.1 may be used in selecting the preliminary dimensions of equivalent I–section of rectangular beam of depth H and breadth B=0.6H

Table 7.1 Preliminary Dimensions of Equivalent I–Section of Rectangular Beam of Depth H and Breadth B = 0.6H

M_G/M_L	B1 (Top flange)	B2 (Bottom flange)	Total Depth
<0.3	0.5H	0.65H	1.05H
0.3 to 0.4	0.6H	0.6H	H
0.5	0.7H	0.4H	1.1H
0.7	0.75H	0.3H	1.15H
0.9	0.8H	0.2H	1.2H

7.1.3.1 Thickness of Flanges

The thickness of the top flange may be taken to vary between 0.1D to 0.15D. Thickness of the bottom flange should be such that at mid–span all the tendons are properly accommodate with appropriate cover and space in between and also the requirement of I/y_2 is satisfied. For preliminary design a thickness of 0.2D may be assumed.

7.1.3.2 Thickness of Web

For the first trial section, a thickness of D/36+55mm+exteranal diameter of sheath or duct about 0.125D to 0.15D may be adopted. The external dia of the sheath for 12ϕ7mm and 12ϕ5mm Freyssinet cables are 44 and 34mm respectively.

7.1.3.3 Un–tensioned Longitudinal Reinforcement

Clause 18.6.3.3 of IS: 1343 stipulate that a minimum longitudinal reinforcement of 0.2%

of the total concrete area shall be provided in all cases except in the case of pre–tensioned units of small sections. This reinforcement may be reduced to 0.15% in the case of high yield strength deformed reinforcements. When the depth of the web exceeds 500mm, longitudinal distribution reinforcement not less than 0.05% of the area of the web shall be provided in each face. The spacing of the individual bars of such reinforcement shall not exceed 200mm. All un–tensioned longitudinal reinforcement shall be restrained in the lateral direction.

7.1.3.4 Cover

In post–tensioned work, when cables and large sized bars are used, the minimum clear cover from sheathing/duct is at least 30mm or the size of the cable or bar whichever is greater. Where pre–stressed concrete members are located in aggressive environment, the cover shall be increased by 10mm.

7.1.3.5 Spacing

In the case of cables or large bars, the clear spacing (measured between sheathing/ducts, whichever used) shall not be less than the greater of the following a) 40mm b) maximum size of the cable or bar and c) 5mm+maximum size of aggregate.

7.1.4 Solved Examples

Problem 7.1

A post–tensioned pre–stressed concrete slab of effective span 12m is to be designed to carry an equivalent live load of 20kN/m. Adopting M45 grade of concrete and 7mm dia high tensile wires of ultimate strength 1520N/mm². Design the mid–span section of the slab assuming that at transfer, the cube strength of concrete is 35.6N/mm².

Solution:

❖ *Bending Moment due to Live Load*

Maximum bending moment due to live load = $M_L = \dfrac{qL^2}{8} = \dfrac{20 \times 12^2}{8} = 360\text{kNm}$

Assume $\eta = 0.8$

❖ *Permissible Stresses*

- At transfer

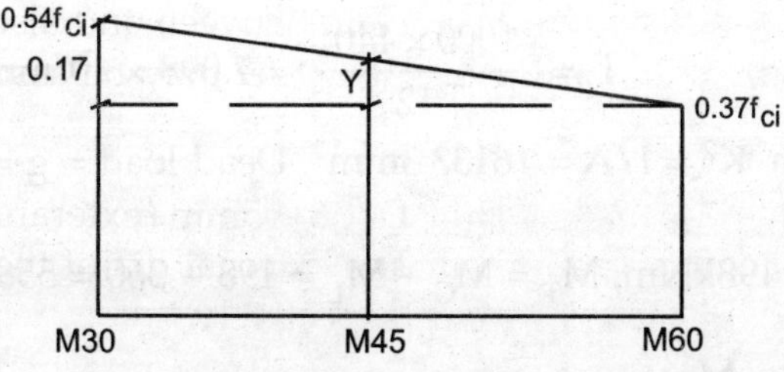

Fig. 7.3 Determination of Y

$$\frac{Y}{0.17} = \frac{15}{30} \therefore Y = \frac{15 \times 0.17}{30} = 0.085$$

$\therefore$ Permissible compressive stress at transfer $= f_{u2} = 0.37f_{ci} + 0.085f_{ci}$

$$= 0.455f_{ci} = 0.455 \times 35.6 = 16.2\text{N/mm}^2$$

- At working load

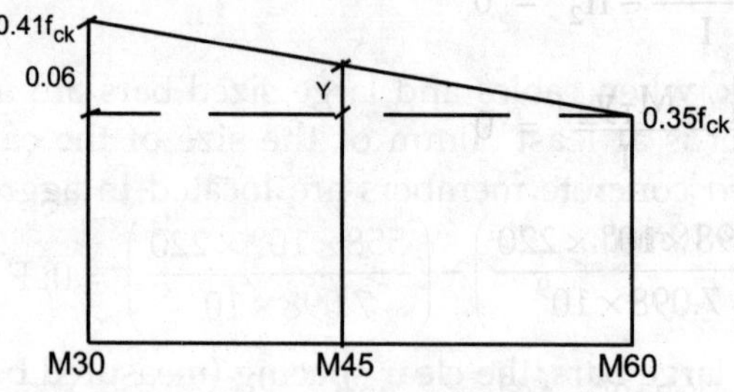

Fig. 7.4 Determination of Y

$$\frac{Y}{0.06} = \frac{15}{30} \therefore Y = \frac{15 \times 0.06}{30} = 0.03$$

$\therefore$ Permissible compressive stress at working $= 0.35f_{ck} + 0.03f_{ck}$

$$= 0.38f_{ck} = 0.38 \times 45 = 17.1\text{N/mm}^2$$

$$f_{u1} = fl_2 = 0 \text{ (For type 1 structure)}$$

❖ *Section Modulus Required for the Rectangular Section*

$$Z = \frac{BH^2}{6} = \frac{M_L}{fl_2 + (1.5\eta - 0.5)f_{u2}}$$

$$= \frac{M_L}{0 + (1.5 \times 0.8 - 0.5)16.2} = \frac{360 \times 10^6}{11.32}$$

$$\frac{BH^2}{6} = 31746032\text{mm}^3.\ H^2 = \frac{31746032 \times 6}{1000},$$

$$H = 436.43\text{mm Say } 440\text{mm}$$

$$A = 1000 \times 440 = 440000\text{mm}^2$$

$$I = \frac{1000 \times 440^3}{12} = 7.098 \times 10^9\text{mm}^4$$

$y_1 = y_2 = H/2 = 220\text{mm}$, $K^2 = I/A = 16133.3\text{mm}^2$. Dead load $= g = 0.44 \times 25 = 11\text{kN/m}^2$

$$M_G = \frac{gL^2}{8} = \frac{11 \times 12^2}{8} = 198\text{kNm},\ M_T = M_G + M_L = 198 + 360 = 558\text{kNm}$$

At transfer, $\frac{F_o}{A} - \frac{F_o e y_1}{I} + \frac{M_G y_1}{I} = -f_{u1}$ ......(1) and $\frac{F_o}{A} + \frac{F_o e y_2}{I} - \frac{M_G y_2}{I} = f_{u2}$ ...(2)

At working, $\frac{F}{A}+\frac{Fey_2}{I}-\frac{M_Ty_2}{I} = -fl_2$ ...(3), Take (1) × η + (3), and WKT, $F = \eta F_o$,

$$\frac{\eta F_o}{A}-\eta\frac{F_oey_1}{I}+\eta\frac{M_Gy_1}{I}-f_{u1} = 0$$

$$\frac{\eta F_o}{A}+\eta\frac{F_oey_2}{I}-\frac{M_Ty_2}{I}-fl_2 = 0$$

$$2\frac{\eta F_o}{A}+\eta\frac{M_Gy_1}{I}-\frac{M_Ty_2}{I} = 0$$

$$2\times\left(\frac{0.8F_o}{440000}\right)+0.8\times\left(\frac{198\times10^6\times220}{7.098\times10^9}\right)-\left(\frac{558\times10^6\times220}{7.098\times10^9}\right) = 0;\ F_o = 3405\text{kN}$$

From (1), $\left(\frac{3405\times10^3}{440000}\right)-\left(\frac{3405\times10^3\times e\times220}{7.098\times10^9}\right)+\left(\frac{198\times10^6\times220}{7.098\times10^9}\right) = 0;\ e = 132\text{mm}$

❖ *Check for Stresses*

- At transfer of pre–stress

$$f_{u1} = \frac{F_o}{A}-\frac{F_oey_1}{I}+\frac{M_Gy_1}{I}=\left(\frac{3405\times10^3}{440000}\right)-\left(\frac{3405\times10^3\times132\times220}{7.098\times10^9}\right)+\left(\frac{198\times10^6\times220}{7.098\times10^9}\right)$$

$f_{u1} = -0.056\text{N/mm}^2$ (Safe)

$$f_{u2} = \frac{F_o}{A}+\frac{F_oey_2}{I}-\frac{M_Gy_2}{I}=\left(\frac{3405\times10^3}{440000}\right)+\left(\frac{3405\times10^3\times132\times220}{7.098\times10^9}\right)-\left(\frac{198\times10^6\times220}{7.098\times10^9}\right)$$

$f_{u2} = 15.532\text{N/mm}^2 < 16.2\text{N/mm}^2$

- At working load

$$fl_1 = \frac{F}{A}-\frac{Fey_1}{I}+\frac{M_Ty_1}{I}=\left(\frac{0.8\times3405\times10^3}{440000}\right)-\left(\frac{0.8\times3405\times10^3\times132\times220}{7.098\times10^9}\right)+\left(\frac{558\times10^6\times220}{7.098\times10^9}\right)$$

$fl_1 = 12.34\text{N/mm}^2 < 17.1\text{N/mm}^2$

$$fl_2 = \frac{F}{A}+\frac{Fey_2}{I}-\frac{M_Ty_2}{I}=\left(\frac{0.8\times3405\times10^3}{440000}\right)+\left(\frac{0.8\times3405\times10^3\times132\times220}{7.098\times10^9}\right)-\left(\frac{558\times10^6\times220}{7.098\times10^9}\right)$$

$fl_2 = 0.040\text{N/mm}^2$ (Safe)

Assuming that, the wires are initially stretched to 1100N/mm². This must not de greater than 80% of ultimate strength.

$$\text{Required area of high tensile steel wires} = \frac{F_o}{1100} = \frac{3405\times10^3}{1100} = 3095.55\text{mm}^2$$

$$\text{Number of 7mm}\phi\text{ tensile wires} = \frac{3095.55}{\frac{\pi}{4}\times7^2} = 80\text{ No.s Say 84.}$$

Provide 7 Freyssinet cables each having 12 No.s of 7mmϕ.

Problem 7.2

Design a post–tensioned beam of span 20m for the live load of 15kN/m run. Adopt concrete of grade M40 and 5mm dia steel wires if characteristic strength of 1600N/mm². Design the beam as type 1 structure. Assume that the strength of concrete for transfer is 35N/mm² and wires can be initially stretched to 1200N/mm².

Solution:

❖ *Bending Moment due to Live Load*

$$\text{Maximum bending moment due to live load, } M_L = \frac{qL^2}{8} = \frac{15\times20^2}{8} = 750\text{kNm}$$

Assume $\eta = 0.8$

❖ *Permissible Stresses*

- At transfer

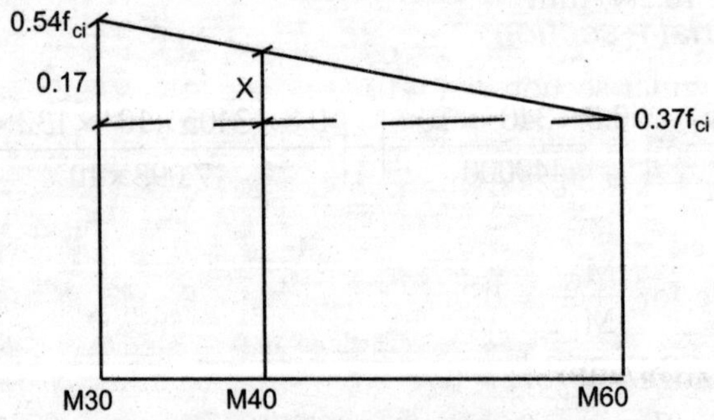

Fig. 7.5 Determination of X

$$\frac{X}{0.17} = \frac{20}{30} \quad \therefore X = \frac{20\times0.17}{30} = 0.1133$$

$$\therefore \text{Permissible compressive stress at transfer} = f_{u2} = 0.37f_{ci} + 0.1133f_{ci}$$

$$= 0.483f_{ci} = 0.483\times40 = 19.32\text{N/mm}^2$$

- At working load

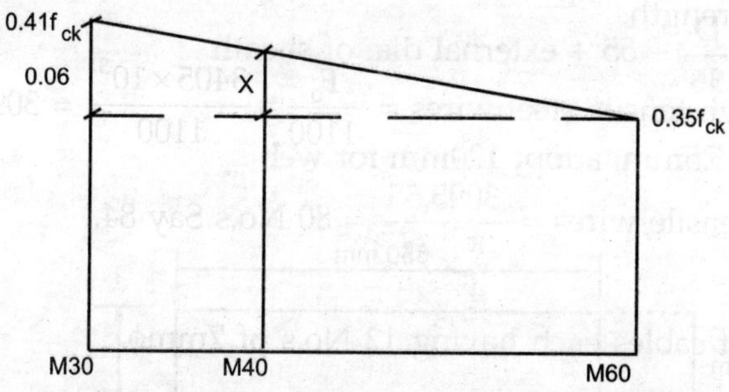

Fig. 7.6 Determination of X

$$\frac{X}{0.06} = \frac{20}{30} \therefore X = \frac{20 \times 0.06}{30} = 0.04$$

∴ Permissible compressive stress at working = $0.35f_{ck} + 0.04f_{ck}$

$= 0.39f_{ck} = 0.39 \times 40 = 15.6N/mm^2$

❖ *Section Modulus Required for the Rectangular Section*

$$Z = \frac{BH^2}{6} = \frac{M_L}{fl_2 + (1.5\eta - 0.5)f_{u2}}$$

The loss ratio generally lies in the range of 0.75 to 0.8 for pre–tensioned members and 0.8 to 0.85 for post–tensioned members.

Taking B = 0.6H and η = 0.8

$$\frac{0.6H \times H^2}{6} = \frac{M_L}{0 + (1.5 \times 0.8 - 0.5)16.9}, H = 859mm$$

❖ *Selection of First Trial I–section*

Self–weight of rectangular section = g = 0.6 × 0.859 × 0.859 × 25 = 11.068kN/m.

$$M_G = \frac{gL^2}{8} = \frac{11.068 \times 20^2}{8} = 553.4kNm, \frac{M_G}{M_L} = \frac{553.4}{750} = 0.738$$

Referring to the table for $\frac{M_G}{M_L} = 0.7$

Breadth of the top flange = 0.75H = 0.75 × 859 = 644.25mm Say 644mm

Breadth of the bottom flange = 0.3H = 0.3 × 859 = 257.7mm Say 258mm

Depth of section = 1.15H = 1.15 × 859 = 987.85 Say 988mm

Adopt Breadth of the top flange = 650mm

Breadth of the bottom flange = 260mm

Depth of section = 990mm

Thickness of the top flange = 0.125D = 0.125 × 990 = 123.75mm Say 130mm

Thickness of the bottom flange = 0.2D = 0.2 × 990 = 198mm Say 200mm

Thickness of web = $\frac{D}{36}$ + 55 + external dia. of sheath

$= \frac{990}{36} + 55 + 34 = 116.5\text{mm}$, adopt 120mm for web

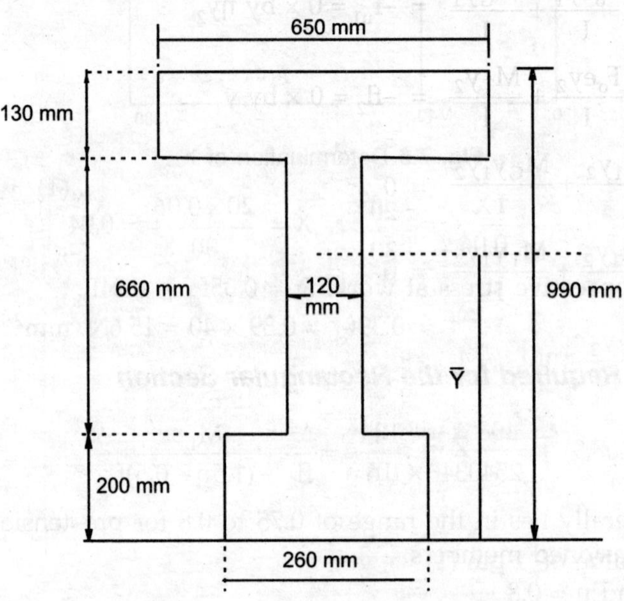

Fig. 7.7 Unsymmetrical I–section

❖ *Properties of Trial Section*

Area = A = 650 × 130 + 120 × 660 + 200 × 260 = 215700mm²

$$\bar{y} = \frac{\Sigma ay}{A} = \frac{260 \times 200 \times 100 + 660 \times 120 \times 530 + 650 \times 130 \times 925}{215700} = 581.1\text{mm} = y_2$$

So, y_2 = 581.1mm, y_1 = 990 – 581.1 = 408.9mm

$$I_{xx} = \frac{260 \times 200^3}{12} + 260 \times 200 \times (581.1 - 100)^2 + \frac{120 \times 660^3}{12} + 120 \times 660 \times (581.1 - 530)^2$$

$$+ \frac{650 \times 130^3}{12} + 650 \times 130 \times (581.1 - 925)^2 ; \quad I = 2540346 \times 10^4\text{mm}^4$$

$$K^2 = \frac{I}{A} = \frac{2540346 \times 10^4}{215700} = 117772\text{mm}^2$$

❖ *Calculation of Self–weight of Beam*

Self–weight of beam = g = 0.2157 × 25 = 5.3925kN/m

$$M_G = \frac{gL^2}{8} = \frac{5.3925 \times 20^2}{8} = 269.625\text{kNm}$$

❖ *Calculation of F_o and e*

$$f_{u1} = 0 \text{ and } fl_2 = 0$$

$$\frac{F_o}{A} + \frac{F_o e y_1}{I} + \frac{M_G y_1}{I} = -f_{u1} = 0 \times \text{ by } \eta y_2$$

$$\frac{\eta F_o}{A} + \eta\frac{F_o e y_2}{I} + \frac{M_T y_2}{I} = -fl_2 = 0 \times \text{ by } y_1$$

$$\frac{\eta F_o y_2}{A} + \eta\frac{F_o e y_1 y_2}{I} + \frac{M_G y_1 y_2}{I} = 0 \quad ...(1)$$

$$\frac{\eta F_o y_1}{A} + \eta\frac{F_o e y_1 y_2}{I} + \frac{M_T y_1 y_2}{I} = 0 \quad ...(2), \text{ Take (1)–(2) we get}$$

$$\frac{\eta F_o}{A}(y_2 - y_1) + \frac{y_1 y_2}{I}(\eta M_G - M_T) = 0$$

$$\frac{0.8F_o}{215700}(-581.1 - 408.9) + \frac{408.9(-581.1)}{2540346 \times 10^4}(0.8 \times 269.625 \times 10^6 - 1019.625 \times 10^6) = 0$$

$\therefore$ F_o= 2047kN

Substituting the value of F_o in (1) we get

$$\left(\frac{2047 \times 10^3}{215700}\right) - \left(\frac{2047 \times 10^3 \times e \times 408.9}{2540346 \times 10^4}\right) + \left(\frac{269.625 \times 10^6 \times 408.9}{2540346 \times 10^4}\right) = 0, \; e = 419.8\text{mm}$$

❖ *Check for Stresses*

- At transfer of pre–stress

$$f_{u1} = \frac{F_o}{A} - \frac{F_o e y_1}{I} + \frac{M_G y_1}{I} = \left(\frac{2047 \times 10^3}{215700}\right) - \left(\frac{2047 \times 10^3 \times 419.8 \times 408.9}{2540346 \times 10^4}\right) + \left(\frac{269.625 \times 10^6 \times 408.9}{2540346 \times 10^4}\right)$$

f_{u1} = –0.009N/mm² (Little tension)

$$f_{u2} = \frac{F_o}{A} + \frac{F_o e y_2}{I} - \frac{M_G y_2}{I} = \left(\frac{2047 \times 10^3}{215700}\right) + \left(\frac{2047 \times 10^3 \times 419.8 \times 581.1}{2540346 \times 10^4}\right) - \left(\frac{269.625 \times 10^6 \times 581.1}{2540346 \times 10^4}\right)$$

$f_{u2} = 17.138 N/mm^2 < 19.32 N/mm^2$

- At working load

$$fl_1 = \frac{F}{A} - \frac{Fey_1}{I} + \frac{M_T y_1}{I}$$

$$= \left(\frac{0.8 \times 2047 \times 10^3}{215700}\right) - \left(\frac{0.8 \times 2047 \times 10^3 \times 419.8 \times 408.9}{2540346 \times 10^4}\right) + \left(\frac{1019.625 \times 10^6 \times 408.9}{2540346 \times 10^4}\right)$$

$$fl_1 = 12.048 N/mm^2 < 15.6 N/mm^2$$

$$fl_2 = \frac{F}{A} + \frac{Fey_2}{I} - \frac{M_T y_2}{I}$$

$$= \left(\frac{0.8 \times 2047 \times 10^3}{215700}\right) + \left(\frac{0.8 \times 2047 \times 10^3 \times 419.8 \times 581.1}{2540346 \times 10^4}\right) - \left(\frac{1019.625 \times 10^6 \times 581.1}{2540346 \times 10^4}\right)$$

$$fl_2 = 0.057 N/mm^2 \text{ (Safe)}$$

Due to the loss of pre–stress, the stress will get reduced in due course. Hence the section seems to be adequate and feasible.

$$\therefore \text{ Area of high tensile steel wires} = \frac{F_o}{1200} = \frac{2047 \times 10^3}{1200} = 1706.86 mm^2$$

$$\text{Number of 5mm}\phi \text{ tensile wires} = \frac{1706.86}{\frac{\pi}{4} \times 5^2} = 87 \text{ No.s}$$

Provide 7 Freyssinet cables each having 12 No.s of 5mmϕ.

7.1.5 Stress Conditions

7.1.5.1 Minimum Section Modulus

Pre–stressed sections under the action of flexure should satisfy the limits specified for permissible stresses at the stage of transfer of pre–stress and at service loads. Expressions for minimum section moduli required, pre–stressing force and the corresponding eccentricity are developed using the four stress relationships established for the two extreme fibres of the section and considering the two critical combinations of pre–stress and moments. The general critical combinations considered are,

1. The maximum pre–stressing force at transfer together with the minimum moments sustained by the section, and
2. The minimum pre–stressing forces, after all losses in combination with the maximum design moment for the serviceability limit state.

Referring to the Fig. 7.8, the four fundamental conditions for stresses at transfer and service loads are as follows,

At transfer

Top fibres $\left(f_{sup} + \dfrac{M_G}{Z_t}\right) \geq f_{tt}$ ...(7.9)

Bottom fibres $\left(f_{inf} - \dfrac{M_G}{Z_b}\right) \leq f_{ct}$ ...(7.10)

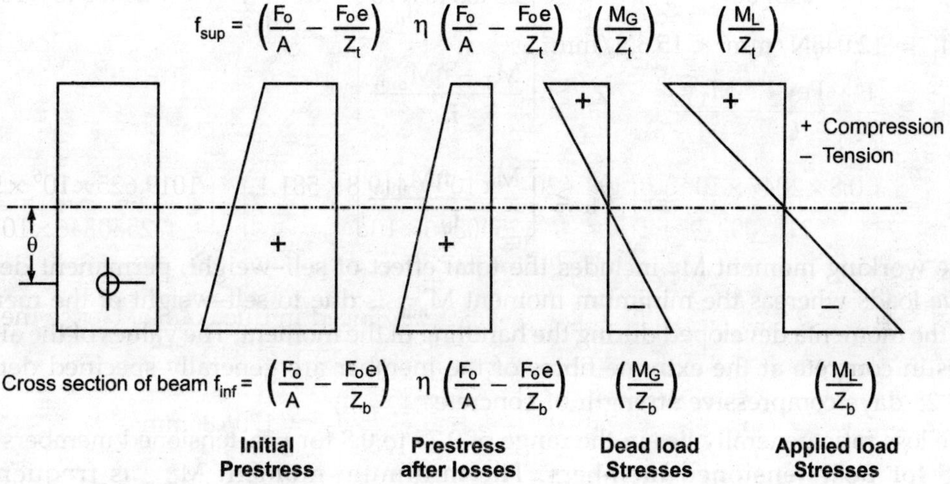

Fig. 7.8 Stresses due to Pre–stress, Dead and Applied Loads

At working load

Top fibres $\left(\eta f_{sup} + \dfrac{M_G}{Z_t} + \dfrac{M_L}{Z_t}\right) \leq f_{cw}$ ...(7.11)

Bottom fibres $\left(\eta f_{inf} - \dfrac{M_G}{Z_b} - \dfrac{M_L}{Z_b}\right) \geq f_{tw}$ ...(7.12)

From Eqs 7.9 and 7.11, we have, $\left(\dfrac{M_L + (1-\eta) M_G}{Z_t}\right) \leq (f_{cw} - \eta f_{tt}) \leq f_{tr}$ ...(7.13)

From Eqs 7.10 and 7.12, we have, $\left(\dfrac{M_L + (1-\eta) M_G}{Z_b}\right) \leq (\eta f_{ct} - f_{tw}) \leq f_{br}$ ...(7.14)

Where, f_{tr} and f_{br} are the ranges of stress at top and bottom fibres respectively. Hence the design formulae for the required section moduli are expressed as

$$Z_t \geq \left[\frac{M_L + (1-\eta) M_G}{f_{tr}}\right] \quad ...(7.15)$$

$$Z_b \geq \left[\frac{M_L + (1-\eta)\,M_G}{f_{br}}\right] \quad ...(7.16)$$

In case, where permanent dead loads in addition to the self–weight acts on the member, these equations are modified and used in the following form as shown below.

$$Z_t \geq \left[\frac{(M_L + M_G) - \eta M_{min}}{f_{tr}}\right] \quad ...(7.17)$$

$$Z_t \geq \left[\frac{M_T - \eta M_{min}}{f_{tr}}\right] \quad ...(7.18)$$

$$Z_b \geq \left[\frac{M_T - \eta M_{min}}{f_{br}}\right] \quad ...(7.19)$$

The working moment M_T includes the total effect of self–weight, permanent dead load and live loads whereas the minimum moment M_{min} is due to self–weight of the member or due to the momenta developed during the handling of the moment. The values of the allowable stresses in concrete at the extreme fibres of the member are generally specified depending on the 28 days compressive strength of concrete.

The loss ratio generally lies in the range of 0.75 to 0.8 for pre–tensioned members and 0.8 to 0.85 for post–tensioned members. The minimum moment M_{min} is frequently the self–weight moment of the beam, but a lower value lower value is possible during handling of pre–stressed members, such as lifting of a precast beam at points other than the supports.

It is important to be noted that the size of the concrete section is influenced both by the applied or live load moment and the minimum moment. However, in case of beams with large ratios of applied to minimum moments, the cross section is highly influenced by the applied moments and the self–weight may be considered to be counteracted by the pre–stress. When the minimum moment is of comparatively large in magnitude, it can no longer be completely counterbalanced by the pre–stressing force within the cross section. This situation is generally encountered in large span post–tensioned girders with considerable self–weight moments.

7.1.5.2 Pre–stressing Force

Generally, the sections selected are somewhat greater than the minimum indicated by the Eqs 7.15 to 7.19 and consequently the pre–stress can lie between the upper and the lower limit. Any value of pre–stress within these limits may be safely used without exceeding the permissible stresses at the extreme fibres. However, the minimum pre–stressing force requires will be obtained by selecting the maximum tensile pre–stress, indicated by Eq. 7.9 at the top fibres and the minimum compressive pre–stress, indicated by Eq. 7.12 corresponding to the bottom fibre. Rearranging these equations,

$$-f_{sup} \geq \left(f_{tt} - \frac{M_G}{Z_t}\right) \quad ...(7.20)$$

$$f_{inf} \geq \left[\frac{f_{tw}}{\eta} + \frac{(M_L + M_G)}{\eta Z_b}\right] \quad ...(7.21)$$

In these equations, Z_t and Z_b correspond to the actual values of the section selected. Eliminating e from the equations,

$$f_{sup} = \left(\frac{F_o}{A} - \frac{F_o e}{Z_t}\right) \text{ and } f_{inf} = \left(\frac{F_o}{A} + \frac{F_o e}{Z_b}\right)$$

We have the expression, after clubbing the above two equations, as shown below,

$$F_o = \frac{A(f_{inf}Z_b + f_{sup}Z_t)}{(Z_t + Z_b)} \quad ...(7.22)$$

Similarly eliminating F_o from the equations, the corresponding maximum eccentricity is given by

$$e = \frac{Z_t Z_b (f_{inf} - f_{sup})}{A(f_{inf}Z_b + f_{sup}Z_t)} \quad ...(7.23)$$

Where, f_{inf} and f_{sup} are computed from the Eqs 7.20 and 7.21.

7.1.5.3 Solved Examples

Problem 7.3

A post–tensioned pre–stressed beam of rectangular section 250mm wide is to be designed for an imposed load of 12kN/m, uniformly distributed on a span of 12m. The stress in the concrete must not exceed 17N/mm² in compression or 1.4N/mm² in tension at any time and the loss of pre–stress may be assumed to be 15%. Calculate,

1. The minimum possible depth of the beam and
2. For the section provided, the minimum pre–stressing force and the corresponding eccentricity.

Solution:

Imposed load = q =12kN/m, η = 0.85

Breadth of section, B = 250mm, $f_{ct} = f_{cw}$ = 17MPa

Overall depth of beam be 'H' mm, $f_{tt} = f_{tw}$ = –1.4MPa

$$\text{Live load moment} = M_L = \left(\frac{12 \times 12^2}{8}\right) = 216\text{kNm} = 216 \times 10^6\text{Nmm}$$

Self–weight of beam = Dead load = B × H × Density of concrete

$$= B \times H \times 24 = \frac{24BH}{10^6}\text{kN/m}$$

$$\text{Dead load moment} = M_G = \frac{24BH}{10^6} \times \frac{12^2}{8} = \frac{432BH}{10^6}\text{kNm} = 432BH\text{ Nmm}$$

Range of stress at bottom fibre, $f_{br} = (\eta f_{ct} - f_{tw}) = (0.85 \times 17 - (-1.4)) = 15.85$ MPa

1. Minimum section modulus is given by

$$Z_b = \left[\frac{M_L + (1-\eta)M_G}{f_{br}}\right], \frac{BH^2}{6} = \left[\frac{(216\times10^6) + (1-0.85)\,432BH}{15.85}\right]$$

Substituting B = 250mm and simplifying, $H^2 - 24.5\,H - 325000 = 0$

On solving, H = 580mm

2. For the section provided (B = 250mm and H = 580mm)

Area of section = A = B × H = 250 × 580 = 145 × 10^3mm²

$$Z_b = Z_t = \frac{BH^2}{6} = \frac{250\times580^2}{6} = 14\times10^6\text{mm}^3$$

$$\text{Self–weight moment} = M_G = 432BH \text{ Nmm}$$
$$= 432\times250\times580 = 625\times10^5\text{Nmm}$$

$$(M_G + M_L) = (2160 + 625)\times10^5\text{Nmm} = 2785\times10^5\text{Nmm}$$

$$f_{sup} = \left(f_{tt} - \frac{M_G}{Z_t}\right) = \left(-1.4 - \frac{625\times10^5}{14\times10^6}\right) = -5.9\text{N/mm}^2$$

$$f_{inf} = \left[\frac{f_{tw}}{\eta} + \frac{(M_L + M_G)}{\eta Z_b}\right] = \left[\frac{-1.4}{0.85} + \frac{2785\times10^5}{0.85\times14\times10^6}\right]$$
$$= 22\text{N/mm}^2$$

Minimum pre–stressing force is given by

$$F_o = \frac{A(f_{inf}Z_b + f_{sup}Z_t)}{(Z_t + Z_b)} = \frac{145\times10^3(22-5.9)\,14\times10^6}{(14+14)\times10^6}$$
$$= 1170\text{kN}$$

Corresponding eccentricity is given by

$$e = \frac{Z_t Z_b(f_{inf} - f_{sup})}{A(f_{inf}Z_b + f_{sup}Z_t)} = \frac{14\times10^6\times14\times10^6(22-(-5.9))}{145\times10^3(22-5.9)14\times10^6} = 167.5\text{mm}$$

Problem 7.4

A pre–tensioned beam, 80mm wide and 120m deep, is to be designed to support working loads of 4kN, each concentrated at the third points over the span of 3m. If the permissible stresses in tension are zero at transfer and 1.4N/mm² under working loads, design the number of 3mm wires and the corresponding eccentricity required at the mid–span section. Permissible tensile stress in wires is 1400MPa. The loss of pre–stress = 20% and density of concrete = 24kN/m³.

Solution:

Self–weight of the beam = (0.08 × 0.12 × 24) = 0.23kN/m

Section modulus = $Z_b = Z_t$ = (80 × 120²)/6 = 192 × 10³mm³

Self–weight moment = M_G = (0.23 × 3²)/8 = 0.26kNm

Live load moment = M_L = (4 × 1) = 4kNm

Permissible stresses, $f_{tt} = 0$, $f_{tw} = -1.4\text{N/mm}^2$

Required pre–stress at bottom and top fibres is given by

$$f_{inf} = \left[\frac{f_{tw}}{\eta} + \frac{(M_L + M_G)}{\eta Z_b}\right] = \left[\frac{-1.4}{0.8} + \frac{(4+0.26)\times 10^6}{0.8\times 192\times 10^3}\right] = 26\text{N/mm}^2$$

$$f_{sup} = \left(f_{tt} - \frac{M_G}{Z_t}\right) = \left(0 - \frac{0.26\times 10^6}{192\times 10^3}\right) = -1.35\text{N/mm}^2$$

Area of the section = 80 × 120 = 9600mm²

Minimum pre–stressing force is given by

$$F_o = \frac{A(f_{inf}Z_b + f_{sup}Z_t)}{(Z_t + Z_b)} = \frac{9600(26-1.35)192\times 10^3}{(192+192)\times 10^3} = 121\text{kN}$$

Corresponding eccentricity is given by

$$e = \frac{Z_t Z_b (f_{inf} - f_{sup})}{A(f_{inf}Z_b + f_{sup}Z_t)} = \frac{192\times 10^3 \times 192\times 10^3 (26-(-1.35))}{9600(26-1.35)192\times 10^3} = 22\text{mm}$$

Area of 3mm wires = $\frac{\pi}{4} \times 3^2 = 7.05\text{mm}^2$

Safe force in each wire = (7.05 × 1400) = 9780N

$$\therefore \text{ Number of wires} = \frac{121\times 10^3}{9780} = 12.3 \approx 13.$$

7.1.6 Limitations of Pre–stress in Long Spans

In the case of lightly loaded beams with long spans in which the minimum moment is relatively large, the theoretical value of eccentricity e, determined by Eq. 7.23 indicates the position of pre–stressing force. This force is either below the soffit of the section or too low for tendons to be accommodate with sufficient cover. In such cases, the theoretical eccentricity is reduced so that the pre–stressing force lies within the section at the lowest practicable position. Consequently, the magnitude of the pre–stressing force is to be increased so that the pre–stress at the bottom is unaltered (Fig. 7.9). There will be a reduction in the negative pre–stress at top at the transfer stage and an increase in the compressive stress at the top fibre under service loads. In the case of heavy girders associated with large self–weight moments, the minimum stress developed at transfer for the top fibre will be compressive in nature due to the limitations of the position of the pre–stressing force.

The minimum stress developed at transfer f_{mint} is computed from the below equations

$$\left(f_{sup} + \frac{M_{min}}{Z_t}\right) \geq f_{mint} \text{ and } \left(\eta f_{inf} - \frac{M_{min}}{Z_b} - \frac{M_L}{Z_b}\right) \geq f_{tw}$$

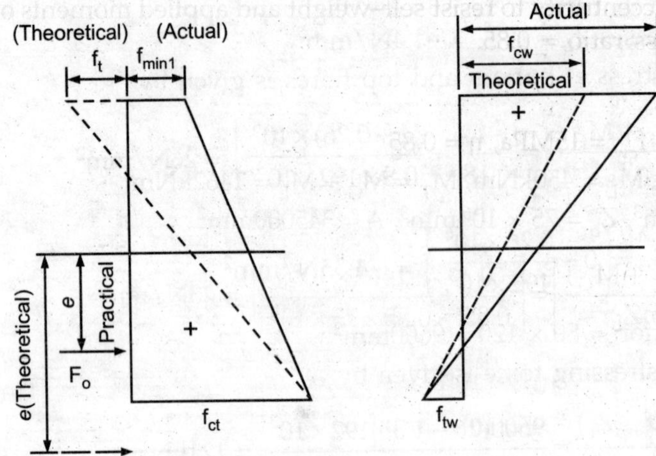

Fig. 7.9 Limitations of Pre–stress due to Large Minimum Moments

By substituting, $f_{sup} = \left(\frac{F_o}{A} - \frac{F_o e}{Z_t}\right)$ and $f_{inf} = \left(\frac{F_o}{A} + \frac{F_o e}{Z_b}\right)$ and eliminating F_o from the resulting equations we get,

$$f_{mint} = \left[\frac{M_{min}}{Z_t} + \frac{Z_b(Z_t - Ae)}{\eta Z_t(Z_t + Ae)}\left(\frac{M_T}{Z_b} + f_{tw}\right)\right] \quad ...(7.24)$$

The revised section modulus, obtained by using the value of f_{mint} in the range of stress f_{tr} and using the Eq. 7.15 to 7.19, should be checked, and if the chosen section satisfies the criteria, the magnitude of the pre–stressing force computed from Eq. 7.22.

Alternatively, the required pre–stressing force F_o acting at known eccentricity e, which can develop required pre–stress f_{inf} at the bottom fibre, is computed directly by the expression,

$$F_o = \left(\frac{A f_{inf} Z_b}{Z_b + Ae}\right) \quad ...(7.25)$$

Where, A and Z_b are the cross sectional area and the section modulus of the cross–section actually provided.

7.1.6.1 Solved Examples

Problem 7.5

An unsymmetrical I–section having the following properties is used for a bridge girder. The thickness of top and bottom flanges are respectively 200 and 250mm. The width of the top and bottom flanges are respectively 750 and 450mm. The thickness of the web is 150mm and overall depth is 1000mm and the area of section is 345000mm^2. Z_t = 95 × 10^{6}mm^3, Z_b = 75 × 10^6 mm^3 and the position of the centroid of the section is 440mm from the top. The permissible tensile and compressive stresses (at transfer and working loads) are not to exceed zero in tension and 15MPa in compression. Determine the pre–stressing force required and the corresponding eccentricity to resist self–weight and applied moments of 1012 and 450kNm respectively. The loss ratio = 0.85.

Solution:

$f_{tt} = f_{tw} = 0$, $f_{ct} = f_{cw} = 15\text{MPa}$, $\eta = 0.85$

$M_G = 1012\text{kNm}$, $M_L = 450\text{kNm}$, $M_T = M_G + M_L = 1462\text{kNm}$,

$Z_t = 95 \times 10^6\ \text{mm}^3$, $Z_b = 75 \times 10^6\ \text{mm}^3$, $A = 345000\text{mm}^2$

$$f_{inf} = \left[\frac{f_{tw}}{\eta} + \frac{(M_L + M_G)}{\eta Z_b}\right] = \left[\frac{0}{0.85} + \frac{1462\times10^6}{0.85\times75\times10^6}\right] = 23\text{N/mm}^2$$

$$f_{sup} = \left(f_{tt} - \frac{M_G}{Z_t}\right) = \left(0 - \frac{1012\times10^6}{95\times10^6}\right) = -10.6\text{N/mm}^2$$

Minimum pre–stressing force is given by

$$F_o = \frac{A(f_{inf}Z_b + f_{sup}Z_t)}{(Z_t + Z_b)} = \frac{345000(23\times75 - 10.6\times95)10^6}{(95+75)\times10^6} = 146 \times 10^4\text{N}$$

Corresponding eccentricity is given by

$$e = \frac{Z_tZ_b(f_{inf} - f_{sup})}{A(f_{inf}Z_b + f_{sup}Z_t)} = \frac{95\times10^6\times75\times10^6(23-(-10.6))}{345000(23\times75 - 10.6\times95)10^6} = 967\text{mm}$$

The maximum possible eccentricity (with suitable cover of 100mm) = (1000 – 440 – 100) = 460mm

Providing the maximum possible eccentricity e = 460mm, the pre–stressing force F_o, required to develop the pre–stress f_{inf}, is given by,

$$F_o = \left(\frac{Af_{inf}Z_b}{Z_b + Ae}\right) = \left(\frac{3450000\times23\times75\times10^6}{(75\times10^6)+(345000\times460)}\right) = 2547\text{kN}$$

7.1.7 Limiting Zone for Pre–stressing Force

The pre–stress along the length of the beam is generally adjusted by varying the eccentricity of the pre–stressing force. This practice generally used in the post–tensioned beams by using curved cables. In case of pre–tensioned beams, the tendons may be deflected by using deviating devices fixed to the mould before casting. After having once determined the magnitude of the pre–stressing force for the critical section, it is possible to fix the limiting zone for the force bounded by upper and lower limits expressed as a function of minimum and maximum moments, sectional properties, pre–stressing force and the permissible stresses in concrete at transfer and working loads.

The limiting zone is defined by the four equations obtained by combining the stress conditions of Eqs 7.9 to 7.12 and the pre–stress equations for the top and bottom fibres given by

$$f_{sup} = \left(\frac{F_o}{A} - \frac{F_o e}{Z_t}\right) \text{ and } f_{inf} = \left(\frac{F_o}{A} + \frac{F_o e}{Z_b}\right)$$

Combining these relations with Eqs 7.9 to 7.12, and noting that $M_G = M_{min}$, the following equations are obtained.

$$e \leq \frac{-Z_t f_{tt}}{F_o} + \frac{Z_t}{A} + \frac{M_{min}}{F_o} \qquad \text{...(7.26)}$$

$$e \leq \frac{Z_b f_{ct}}{F_o} - \frac{Z_b}{A} + \frac{M_{min}}{F_o} \qquad \text{...(7.27)}$$

$$e \geq \frac{-Z_t f_{cw}}{\eta F_o} + \frac{Z_t}{A} + \frac{M_T}{\eta F_o} \qquad \text{...(7.28)}$$

$$e \geq \frac{Z_b f_{tw}}{\eta F_o} - \frac{Z_b}{A} + \frac{M_T}{\eta F_o} \qquad \text{...(7.29)}$$

The curves represented by two of these four equations are given in Fig. 7.10. Positive eccentricities are plotted below the centroid of the section. The permissible tendon zone is controlled by only two of the above four equations. In case of prismatic members with a constant pre–stressing force, the permissible tendon zone is controlled by Eqs. 7.27 to 7.29.

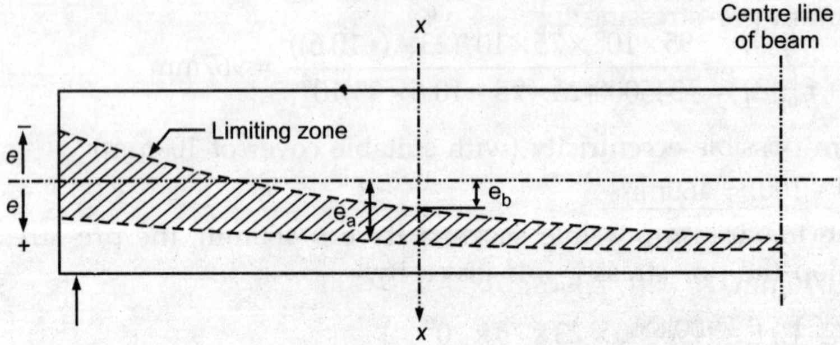

Fig. 7.10 Limiting Zone for Pre–stressing Force

7.1.7.1 Solved Examples

Problem 7.6

A pre–stressed beam has an unsymmetrical; I–section with an overall depth of 1840mm. the top and bottom flange widths are 1800 and 820mm respectively. The thickness of the top flange varies from 180mm at the ends and 430mm at the junction of the web, which is 180mm thick. The thickness of the bottom flange varies from 150mm at the ends to 450mm at the junction of the web. The beam is designed for the simply supported beam of span 40m. The permissible compressive stress at the transfer and working load is limited to 16MPa, while tensile stress at the transfer and working load is limited to zero and 1.4MPa respectively. The loss ratio is 0.80. Calculate

i. The permissible uniformly distributed imposed load
ii. The magnitude of pre–stressing force if at the mid–span section is located 130mm from the soffit and
iii. The vertical limits within which cable must lie at mid–span and support sections.

Solution:

Properties of I–section

$A = 1016500 mm^2$, $I = 4442 \times 10^8 mm^4$, $y_t = 700mm$ and $y_b = 1140mm$

$Z_t = 63 \times 10^7 mm^3$, $Z_b = 39 \times 10^7 mm^3$, $e = (1140 - 130) = 1010mm$

Self–weight $= g = (1.0165 \times 1 \times 24) = 24.4kN/m$, $M_G = (24.4 \times 40^2)/8 = 4880kNm$

$f_{tt} = 0$, $f_{tw} = -1.4MPa$, $f_{ct} = f_{cw} = 16MPa$, $\eta = 0.80$

i. If M_L is live load moment, the minimum section modulus required is given by

$$Z_b = \left[\frac{M_L + (1-\eta)M_G}{f_{br}}\right], \; f_{br} = (\eta f_{ct} - f_{tw}) = (0.8 \times 16 - (-1.4)) = 14.2MPa$$

$$\therefore \; 39 \times 10^7 = \left[\frac{M_L + (1-0.8)\,4880 \times 10^6}{14.2}\right].$$ On solving, $M_L = 4574 \times 10^6 Nmm = 4574kNm$

We have $M_L = (\text{Live load} \times L^2)/8$, $\therefore$ Live load $= q = 8M_L/L^2 = 8 \times 4574/40^2 = 22.87kN/m$

ii. The minimum pre–stressing force is obtained by the limiting stress conditions

$$\frac{F_o}{A}\left(1 + \frac{ey_b}{K^2}\right) - \frac{M_G}{Z_b} = 16.0$$

$$\frac{F_o}{A}\left(1 + \frac{ey_b}{K^2}\right) - \left(\frac{M_G + M_L}{Z_b}\right) = -1.4$$

Also $e = 1010mm$, $K^2 = (I/A) = 437000$ and $y_b = 1140mm$

On solving, $F_o = 7900kN$

iii. Substituting the values of limiting stresses, in Eqs. 7.27 and 7.29.

At the centre of span

$$e \le \frac{Z_b f_{ct}}{F_o} - \frac{Z_b}{A} + \frac{M_{min}}{F_o}$$

$$\le \frac{39 \times 10^7 \times 16}{7900 \times 10^3} - \frac{39 \times 10^7}{1.0165 \times 10^6} + \frac{4880 \times 10^6}{7900 \times 10^3} \le 1024mm$$

and $$e \ge \frac{Z_b f_{tw}}{\eta F_o} - \frac{Z_b}{A} + \frac{M_T}{\eta F_o}$$

$$\geq \frac{39\times10^7\times(-1.4)}{0.8\times7900\times10^3} - \frac{39\times10^7}{1.0165\times10^6} + \frac{(4880+4574)\times10^6}{0.8\times7900\times10^3} \geq 1026\text{mm}$$

support section

$$e \leq \frac{Z_b f_{ct}}{F_o} - \frac{Z_b}{A} \leq \frac{39\times10^7\times16}{7900\times10^3} - \frac{39\times10^7}{1.0165\times10^6} \leq (790-385) \leq 405\text{mm}$$

$$\text{and } e \geq \frac{Z_b f_{tw}}{\eta F_o} - \frac{Z_b}{A} \geq \frac{39\times10^7\times(-1.4)}{0.8\times7900\times10^3} - \frac{39\times10^7}{1.0165\times10^6} \geq (-86-385) \geq -471\text{mm}$$

7.1.8 Design of Sections for Limit State of Collapse in Flexure

In the design of pre–stressed concrete members for the limit state of collapse, the distribution of compressive stress in the section at the ultimate failure stage is considered and the effective depth required is estimated by equating total ultimate moment with the internal resisting couple.

According to IS: 1343–2012, the maximum ultimate moment of resistance of a rectangular section is given by

$$M_{ud} = 0.21 f_{ck} bd^2 \qquad \text{...(7.30)}$$

The dimensions based on this expression are the minimum values. It is often preferable to use a larger section, because it means a saving on the costly pre–stressing tendons.

In case of flanged sections, a reasonable proportion of h_f/d = 0.2 to 0.25 and b_w/d = 0.2 to 0.3, so that the ultimate moment varies from $0.08 f_{ck}\, bd^2$ to $0.12 f_{ck}\, bd^2$. Assuming a suitable value for the breadth of the compression face b, generally of the order of 0.5 h for T–sections and 0.6 to 0.8 h for I–sections, the effective and overall depth of the section is designed by providing suitable cover reinforcements.

The area of high tensile and un–tensioned reinforcement required to mobilize the desired flexural strength is computed using the force equilibrium at the limit state of collapse. However, the tensile reinforcement index should not be exceeding the specified permissible values. The limitation of reinforcement will result in under reinforced sections, ensuring progressive collapse of members.

7.1.8.1 Solved Examples

Problem 7.7

A pre–tensioned pre–stressed concrete beam of rectangular section is required to support a design ultimate moment of 100kNm. Design the section if f_{ck} = 50MPa and f_p = 1600MPa.

Solution:

If b and d are the breadth and effective depth of respectively, assuming the ratio (x_u/d) = 0.5,

We have, $$M_u = 0.36 f_{ck} b x_u (d - 0.42 x_u) = 0.14 f_{ck} bd^2$$

Assuming $b = 0.5d$, $d^3 = \left(\dfrac{M_u}{0.14 \times f_{ck} \times 0.5}\right) = \left(\dfrac{100 \times 10^6}{0.14 \times 50 \times 0.5}\right)$

Solving $d = 300$mm and hence $b = 150$mm

and $x_u = (0.5 \times 300) = 150$mm

From Table 11 of IS: 1343 – 2012, for $(x_u/d) = 0.5$, $f_{pu} = 0.87f_p$

$$A_p = \left[\frac{M_u}{0.87f_p(d - 0.42x_u)}\right] = \left[\frac{100 \times 10^6}{0.87 \times 1600(300 - 0.42 \times 150)}\right] = 300\text{mm}^2$$

Adopt a section, 150mm by 350mm deep, with 300mm² of high tensile steel located at an effective depth of 300mm.

Problem 7.8

A post–tensioned bonded beam of unsymmetrical I–section is required to support a design moment of 1200kNm. Determine the overall depth and thickness of compression flange required if $f_{ck} = 35$MPa and $f_p = 1500$MPa.

Solution:

For flanged section, $M_{ud} = 0.08f_{ck}bd^2$

Assuming $b = 0.5d$ and $b_w = 0.25b$, $d^3 = \left(\dfrac{M_u}{0.10 \times f_{ck} \times 0.5}\right) = \left(\dfrac{1200 \times 10^6}{0.10 \times 35 \times 0.5}\right)$

Selecting $d = 1000$mm and $b = 500$mm

Thickness of top flange $= h_t = 0.2d = (0.2 \times 1000) = 200$mm

Thickness of the web $= b_w = 0.25b = (0.25 \times 500) = 125$mm

Assuming the neutral axis depth, $x_u = h_t = 200$mm and $M_u = 0.87f_pA_p(d - 0.42x_u)$

$$A_p = \left[\frac{M_u}{0.87f_p(d - 0.42x_u)}\right] = \left[\frac{1200 \times 10^6}{0.87 \times 1500(1000 - 0.42 \times 200)}\right] = 1003\text{mm}^2$$

7.2 DESIGN OF PRE–STRESS CONCRETE SECTIONS FOR TORSION

7.2.1 Limit State of Collapse: Torsion

In general, where the torsional resistance or stiffness of members has not been taken into account in the analysis of the structure, no specific calculation for torsion will be necessary, adequate control in torsional cracking being provided by the required nominal shear reinforcement. Where the torsional resistance or stiffness of member is taken into account in the analysis, the member shall be designed for torsion.

7.2.2 Application of Design Rules for Torsion

As per IS: 1343–2012, the design rules applied to

a. Beams of solid rectangular cross–section (D>b),

b. Hollow rectangular beams with D>b and with a wall thickness, t ≥ b/4, and

c. T–beam and I–beams.

In all the three cases, the average intensity of pre–stress in the concrete shall be less than $0.3f_{ck}$.

This IS method is based on the design procedures of Rangan and Hall.[6] The space truss theory is suitable for evaluation of analysis of presence of large torsion.[7]

7.2.3 Longitudinal Reinforcement

As the combination of bending moment and twisting moment results in skew bending, the code gives recommendation recognizing this fact. As per the code, the longitudinal reinforcement shall be designed to resist an equivalent ultimate bending moment M_{e1}, given by,

$$M_{e1} = M + M_t \qquad ...(7.31)$$

Where, M = Applied ultimate bending moment at the cross–section acting in combination with ultimate torque T.

$$M_t = T\sqrt{1+\left(\frac{2D}{b}\right)} \qquad ...(7.32)$$

The sign of M_t, being the same as that of M

D = Overall depth of beam

b = Breadth of the member which for T and I beams shall be taken as the breadth of the web b_w.

If the numerical value of M is less than M_t, the member shall also be designed to withstand a moment M_{e2} given by,

$$M_{e2} = M_t - M \qquad ...(7.33)$$

The moment M_{e2} being taken as acting in the opposite sense to the moment M i.e., if M is sagging in nature, then M_{e2} is taken as hogging in nature.

In cases, when the numerical values of M is less than or equal to that of M_t, the beam shall be design to withstand an equivalent transverse bending moment M_{e3} (not acting simultaneously with M_{e1}), given by

$$M_{e3} = M_t\left(1+\frac{x_1}{2e}\right)^2\left(\frac{1+\dfrac{2b}{D}}{1+\dfrac{2D}{b}}\right) \qquad ...(7.34)$$

and acting about an axis at right angles to the axis of M, where x_1, is the smaller dimension of a closed hoop used as torsional shear reinforcement and e = T/V, where V is the ultimate shear force.

7.2.4 Transverse Reinforcement

In PSC beams, the horizontal pre–stress is beneficial in reducing the effective torsional moment and shear.

So, the reduced torsional moment carried by the concrete

$$T_{c1} = T_c \frac{e}{e+e_c} \quad ...(7.35)$$

Where,

$$T_c = 0.15b^2D\left(1-\frac{b}{3D}\right)\lambda_p\sqrt{f_{ck}} \quad ...(7.36)$$

$$e = \frac{T}{V}, e_c = \frac{T_c}{V_c}, \lambda_p = \sqrt{\left(1+\frac{12f_{cp}}{f_{ck}}\right)} \quad ...(7.37)$$

T_c = Torsional moment carried by concrete

b = Breadth of the member, which for T and I beams shall be taken as the breadth of the web b_w.

D = Overall depth of beam

f_{ck} = Characteristic compressive strength of concrete

T = Torsional moment applied to a cross section under ultimate load conditions

V = Shearing force at a cross section calculated for the specified ultimate load

V_c = Theoretical shear strength at a cross section. i.e., smaller of V_{co} and V_{cr} as per 22.4.1 and 22.4.2 of IS: 1343–2012.

$$\text{Shear force carried by the concrete} = V_{c1} = V_c \frac{e}{e+e_c} \quad ...(7.38)$$

The area of cross–section A_{sv} of the closed stirrup enclosing the corner longitudinal bars shall be taken as the larger of the following two values.

$$A_{sv} = \left(\frac{M_tS_v}{1.5b_1d_1f_y}\right) \text{ and } A_{sv} = A_v + 2A_t \quad ...(7.39)$$

Where,

$$A_v = \frac{(V-V_{c1})S_v}{0.87f_yd_1}, A_t = \frac{(T-T_{c1})S_v}{0.87f_yb_1d_1} \quad ...(7.40)$$

S_v = Spacing of stirrup reinforcement

b_1 = Centre to centre distance between corner bars in the direction of width

d_1 = Centre to centre distance between corner bars in the direction of depth

However, the maximum reinforcement should not be less than the given by,

$$\frac{A_{sv}}{bS_v} = \frac{0.4}{0.87f_y} \quad ...(7.41)$$

There shall be at least one longitudinal bar not less than 12mm in dia. in each corner of the stirrups.

7.2.5 Distribution of Torsional Reinforcement

When a member is designed for torsion, torsion reinforcement shall be provided as below:

a. All transverse reinforcement provided for torsion shall be in the form of closed stirrups perpendicular to the axis of the members.
b. The spacing, S_v, of the stirrups shall not exceed $(x_1 + y_1)/4$ or 200 mm whichever is smaller, where x_1 and y_1 are respectively short and long dimensions of the stirrup.
c. Each end of the bar forming the stirrup shall be anchored in accordance with IS: 456–2000.
d. Torsional reinforcement shall be continued to a distance not less than $(D + b_w)$ beyond the point at which it is no longer than theoretically required, where D is the overall depth and b_w is the effective width of the web of a flanged member.

7.2.6 Solved Examples

Problem 7.9

A post–tensioned bonded pre–stressed concrete beam of rectangular cross–section 400mm wide by 550mm deep, is subjected to a service load bending moment of 166.6kNm, torsional moment of 46.6kNm and shear force of 66.6kN. The section has an effective pre–stressing force determined from service load requirements, of magnitude 500kN at an eccentricity of 150mm, provided by 5 numbers of 12.5mm stress relieved strands of cross sectional area 506mm² with a an ultimate tensile strength of 1820N/mm². If the cube strength of concrete is 40N/mm², design asuitable longitudinal and transverse reinforcements in the beam using IS: 1343–2012 code recommendations based on the skew bending approach.

Solution:

Ultimate bending moment, $M = 1.5 \times 166.6 = 249.9$ Say 250kNm

Ultimate torsional moment, $T = 1.5 \times 46.6 = 69.9$ Say 70kNm

Ultimate shear force, $V = 1.5 \times 66.6 = 99.9$ Say 100kN

Area of pre–stressing strands, $A_p = 506\text{mm}^2$

Ultimate tensile strength of strands, $f_p = 1820\text{N/mm}^2$

$f_{ck} = 40\text{N/mm}^2$; $f_y = 250\text{N/mm}^2$

F = 500kN; e = 150mm; b = 400m; D = 550mm

Area of concrete section, $A = 22 \times 10^4\text{mm}^2$, $I = \dfrac{400 \times 550^3}{12} = 5.54 \times 10^9\text{mm}^4$

Effective stress in steel, $f_{se} = \dfrac{F}{A_{st}} = \dfrac{500 \times 10^3}{506} = 988.14$ Say $988\text{N/mm}^2 = f_{pe}$

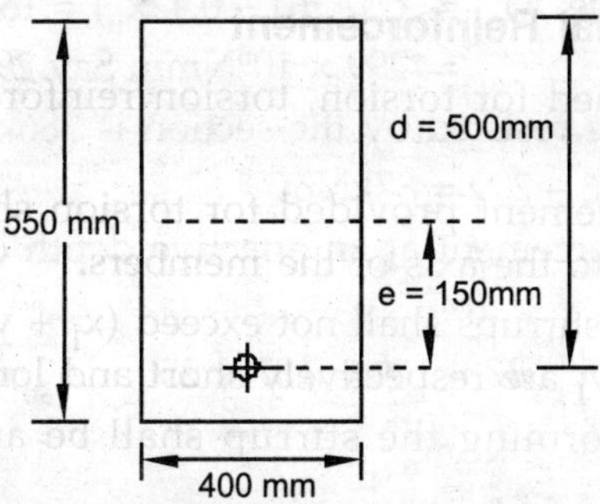

Fig. 7.11 Rectangular Cross–section

• ***Design of Longitudinal Reinforcement***

$M_{e1} = M + M_t$

Where, $M_t = T\sqrt{1+\left(\frac{2D}{b}\right)} = 70\sqrt{1+\left(\frac{2\times 550}{400}\right)}$ = 135.55 Say 136kNm

M_{e1} = 250 + 136 = 386kNm

The longitudinal steel in the section is designed for the ultimate moment of 386kNm.

The ultimate capacity of the section with pre–stressing steel is first determined and for the balance moment, mild steel reinforcement will be designed. Referring Page No. 59 and Table 11 of IS: 1343–2012.

f_p = 1820N/mm², b = 400mm, A_p = 506mm², d = 425mm, f_{ck} = 40N/mm²

$$\frac{A_p f_p}{f_{ck}bd} = \frac{1820\times 506}{40\times 400\times 425} = 0.135,\ \frac{f_{pu}}{0.87f_p} = 1,\quad \therefore\ f_{pu} = 1\times 0.87\times 1820 = 1583\text{N/mm}^2$$

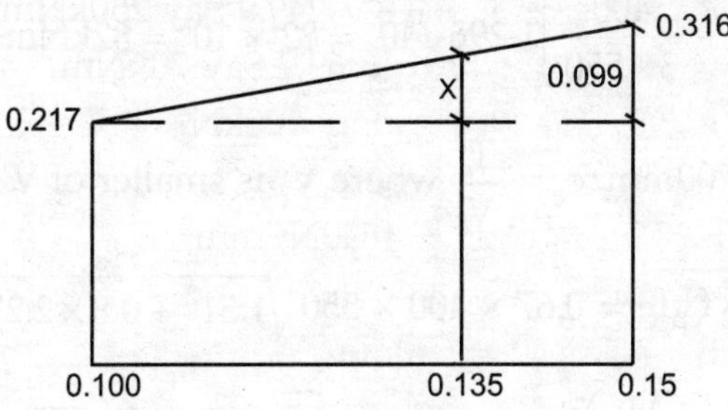

Fig. 7.12 Determination of X

$$\frac{X}{0.099} = \frac{0.035}{0.05},\quad \therefore\ X = 0.0693$$

At 0.135, 0.217 + 0.0693 = 0.2863 = $\frac{x_u}{d}$, x_u = 0.2863 × 425 = 121.67 Say 122mm

Ref. Page No. 59 of IS: 1343–2012,

Moment of resistance, $M_u = f_{pu}A_p(d - 0.42x_u) = 1583 \times 506(425 - 0.42 \times 122)$
$= 299 \times 10^6$Nmm Say 299kNm

But the total moment to be resisted by the section is 386kNm

∴ Balancing moment = 386 – 299 = 87kNm

Providing mild steel reinforcement at an effective depth of 500mm, we get

$$M_u = 0.87f_yA_{st}d\left(1-\left(\frac{A_{st}f_y}{bdf_{ck}}\right)\right), 87 \times 10^6 = 0.87 \times 250 \times A_{st} \times 500\left(1-\left(\frac{A_{st}\times 250}{400\times 500\times 40}\right)\right)$$

On solving, $A_{st} = 840\text{mm}^2$,

Provide 2 bars of 25mmϕ at an effective depth of 500mm

$A_{st\ provided} = 982\text{mm}^2$

- ***Design of Transverse Reinforcement***

The reduced torsional moment T_{c1} carried by the concrete is given by

$$T_{c1} = T_c\frac{e}{e+e_c}, \text{ Where, } T_c = 0.15b^2D\left(1-\frac{b}{3D}\right)\lambda_p\sqrt{f_{ck}}, \lambda_p = \sqrt{\left(1+\frac{12f_{cp}}{f_{ck}}\right)}$$

Average intensity of pre–stress at the centroid $= f_{cp} = \frac{F}{A} = \frac{500\times 10^3}{22\times 10^4} = 2.27\text{N/mm}^2$

Tensile strength of concrete $f_t = 0.27\sqrt{f_{ck}}$

$$\therefore \quad \lambda_p = \sqrt{\left(1+\frac{12\times 2.27}{40}\right)} = 1.296,$$

$$T_c = 0.15 \times 400^2 \times 550\left(1-\frac{400}{3\times 550}\right)1.296\sqrt{40} = 82 \times 10^6 = 82\text{kNm}$$

$$e = \frac{T}{V} = \frac{70}{100} = 0.7\text{m} = 700\text{mm}, e_c = \frac{T_c}{V_c} \text{ where } V_c \text{ is smaller of } V_{co} \text{ and } V_{cr}$$

$$V_{co} = 0.67bD\sqrt{f_t^2 + 0.8\times f_{cp}f_t} = 0.67 \times 400 \times 550\sqrt{1.51^2 + 0.8\times 2.27\times 1.51} = 330\text{kN}$$

$$V_{cr} = \left(1-0.55\frac{f_{pe}}{f_p}\right)\tau_c bd + \left(\frac{M_o}{M}\right)V, \text{ should not be less than } 0.1b_w\sqrt{f_{ck}}$$

Equivalent area of pre–stressing steel is obtained as

$$A_p = 506 + A_{st}\left(\frac{f_y}{f_p}\right) = 506 + 982\left(\frac{250}{1820}\right) = 640.89\text{mm}^2 \text{ Say } 640\text{mm}^2$$

$$\text{Then, } 100\left(\frac{A_p}{bd}\right) = 100\left(\frac{640}{400\times 425}\right) = 0.376$$

From the Table 6, Page No. 47 of IS: 1343–2012.

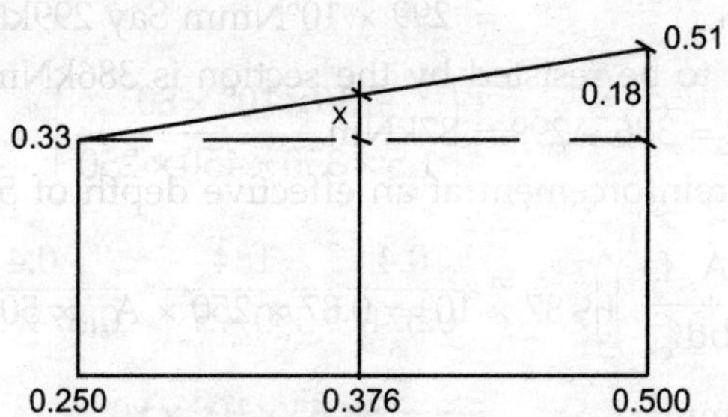

Fig. 7.13 Determination of X

$$\frac{X}{0.18} = \frac{0.126}{0.25}, \quad \therefore X = 0.090$$

So, $\tau_c = 0.420\text{N/mm}^2$, $M_o = 0.8 f_{pt} \frac{I}{y}$

$$f_{pt} = \frac{F}{A} + \frac{Fey}{I} = \frac{500 \times 10^3}{400 \times 550} + \frac{500 \times 10^3 (-150)(-275)}{5.54 \times 10^9} = 6\text{N/mm}^2$$

$$\text{So, } M_o = 0.8 \times 6 \times \frac{5.54 \times 10^9}{275} = 96.6\text{Nmm}$$

$$\text{So, } V_{cr} = \left(1 - 0.55 \times \frac{988}{1820}\right)(0.420 \times 400 \times 425) + \left(\frac{96.6 \times 10^6}{250 \times 10^6}\right) 100 \times 10^3 = 88.691\text{kN}$$

Smaller of V_{co} and V_{cr} is obtained as $V_c = 88.691\text{kN}$

$$\therefore \quad e_c = \frac{82}{88.691} = 0.924\text{m} = 924\text{mm}$$

$$\therefore \quad T_{c1} = T_c \frac{e}{e + e_c} = 82 \times \frac{150}{150 + 924} = 11.45\text{kNm}$$

$$V_{c1} = V_c \frac{e}{e + e_c} = 88.691 \times \frac{150}{150 + 924} = 12.387\text{kN}$$

$$V - V_{c1} = 100 - 12.387 = 87.613\text{kN}$$

$$T - T_{c1} = 70 - 11.45 = 58.55\text{kNm}$$

The cross–section A_{sv} of closed stirrups enclosing the longitudinal bars is taken as the larger of the following two values.

$$A_{sv} = \left(\frac{M_t S_v}{1.5 b_1 d_1 f_y}\right) \text{ and } A_{sv} = A_v + 2A_t$$

$$\text{Where, } A_v = \frac{(V - V_{c1})S_v}{0.87 f_y d_1}, \; A_t = \frac{(T - T_{c1})S_v}{0.87 f_y b_1 d_1}$$

In the present problem, b_1 = 350mm and d_1 = 450mm, f_y= 250N/mm²

Using a pitch of 80mm,

$$A_{sv} = \left(\frac{136\times10^6\times80}{1.5\times350\times450\times250}\right) = 184\text{mm}^2 \quad ...(1)$$

$$\frac{A_{sv}}{bS_v} = \frac{0.4}{0.87f_y}, \frac{184}{350\times S_v} = \frac{0.4}{0.87\times250}, S_v = 80\text{mm}$$

$$A_v = \frac{87.613\times10^3\times80}{0.87\times250\times450} = 72\text{mm}^2,\ A_t = \frac{58.55\times10^3\times80}{0.87\times250\times350\times450} = 136\text{mm}^2$$

$$A_{sv} = 72 + 2(136) = 344\text{mm}^2 \quad ...(2)$$

The larger of (1) and (2), A_{sv} = 344mm²

Provide 12mm diameter two legged stirrups at a spacing of 50mm which provides A_{sv} = 344mm²

• *Checking for Spacing of Stirrups*

X_1 = 400 – 1.5 × Dia. of stirrups = 400 – 1.5 × 12 = 382mm

Y_1 = 500 – 1.5 × Dia. of stirrups = 500 – 1.5 × 12 = 482mm

$$\therefore \frac{X_1 + Y_1}{4} = \frac{382+482}{4} = 216 > 80\text{mm}$$

REVIEW QUESTIONS

7.1 Explain limit state of collapse as per IS: 1343 code provisions.

7.2 Elaborate limit state of serviceability in deflection, cracking and maximum compression as per IS: 1343 code recommendations.

7.3 What are the four basic requirements for type 1 and type 2 pre–stress concrete beams.

7.4 How you are going to select the shapes of cross–section for design of pre–stress concrete sections for flexure.

7.5 Outline the IS: 1343 code provisions for calculation of thickness of flanges, web, cover, spacing and un–tensioned longitudinal reinforcement for design of pre–stress concrete sections for flexure.

7.6 Elaborate stress conditions with minimum section modulus and pre–stressing force

7.7 What are the limitations of pre–stress in long spans in design of pre–stress concrete sections for flexure?

7.8 Explain limiting zone for pre–stressing force in design of pre–stress concrete sections for flexure.

7.9 Outline design of sections for limit state of collapse in flexure.

7.10 Explain limit state of collapse in torsion as per Indian code provisions.

7.11 Where design rules for torsion are applicable?

7.12 Explain how to calculate longitudinal reinforcement for design of pre-stress concrete sections for torsion.

7.13 Elaborate how to calculate transverse reinforcement for design of pre-stress concrete sections for torsion.

7.14 How the distribution of torsional reinforcement takes place as per Indian code provisions for PSC members.

EXERCISE PROBLEMS

7.1 A pre–stressed concrete beam of rectangular section, 90mm by 180mm deep, is to be designed to support two imposed load of 3.5kN each located at one third points over a span of 3m. If there is to be no tensile stress in the concrete at transfer and working loads, calculate the minimum pre–stressing force and corresponding eccentricity. Take D_c = 24kN/m^3, Loss ratio = 0.80.

[Ans: F_o = 74.86kN, e = 35.6mm]

7.2 The pre–stressed concrete T–beam is to be designed to support an imposed load of 4.4kN/m over an effective span of 5m. The T–beam is made up of a flange 400mm deep and 40mm deep. The rib is of 100mm thick and 200mm deep. The stress in the concrete must not exceed 15MPa in compression and zero in tension at any stage. Check for the adequacy of the section provided, and calculate the minimum pre–stressing force and the corresponding eccentricity. Assume 20% pre–stress loss.

[Ans: Section is adequate, F_o = 181kN, e = 78mm]

7.3 A post–tensioned beam of span 15m and overall depth of 900mm, has a uniform symmetrical cross–section of area 2 × 10^{5}mm^2 and second moment of area 212 × 10^{8}mm^4. The pre–stress is provided by cable tensioned to a force of 1450kN at transfer. If the beam is to support an UDL of 21kN/m and the minimum load is that due to self–weight of the beam, calculate the vertical limits within which the cable must lie along the beam length. The permissible compressive stress at transfer and working load are 14 and 16.8MPa respectively. The tensile stresses at transfer and working load are zero and 1.75MPa respectively. Density of concrete = 24kN/m^3, η = 0.8.

[Ans: At mid–span: e_1 = 312mm, e_2 = 249mm

At support: e_1 = 219mm, e_2 = 376mm]

7.4 A pre–stressed I–section of minimum overall depth of 300mm, is required to have an ultimate flexural strength of 86kNm. Find (a) suitable minimum dimensions of the top flange and (b) the total number of 5mm wires required in the bottom flange. The cube strength of concrete is 60MPa and tensile strength of steel is 1600MPa.

[Ans: (a) 125mm, (b) 18 No.s]

7.5 Design a simply supported slab for a bridge deck using the following data: Span = 10m, permissible compressive strength in concrete at transfer = 16.5MPa, Type 1 member (No tensile stress at any stage), safe stress in steel = 950MPa, live load on slab is 10kN/m^2, loss of pre–stress = 18%. Design the pre–stressing force and eccentricity.

[Ans: F_o = 1112.5kN, e = 218.12mm, Spacing of cables = 500mm centres]

7.6 Design the pre–stressing force and the corresponding eccentricity for symmetrical I–section beam having flange width 260mm and thickness of flanges being 100mm. Thickness of web = 50mm, overall depth of the section = 520mm, span = 9m. The beam supports a uniformly distributed live load of 8kN/m. Assume compressive strength of concrete at transfer as 15MPa. Loss ratio = 0.8. No tensile stresses are permitted at any stages. If 5mm diameter high tensile wires initially stressed to 1200MPa are used, find the no. of wires.

[Ans: F_o = 376kN, e = 184mm, Number of wires = 16]

7.7 A post–tensioned pre–stressed concrete beam of rectangular section 250mm width by 580mm deep is to be designed to support an imposed load of 12kN/m uniformly distributed over a span of 12m. The stress in concrete must not exceed 17MPa in compression or 1.4MPa in tension at any stage and the loss of pre–stress is 15%. Design the pre–stressing force and the corresponding eccentricity.

[Ans: F_o = 1170kN, e = 167.5mm]

7.8 A post–tensioned pre–stressed concrete beam of rectangular section 400mm width by 1000mm deep spans over 15m and supports a live load of 25kN/m. If f_{ct} = 17MPa and $f_{tt} = f_{tw} = 0$, $\eta = 0.85$. Design the pre–stressing force and the corresponding eccentricity.

[Ans: F_o = 2624kN, e = 269.5mm]

7.9 The cross sectional area of an unsymmetrical pre–stressed I–section beam designed to carry a central point load on a simply supported beam of span 15m is of 194000mm^2. The second moment of area = 197 × 10^{8}mm^4. The overall depth of the section is 900mm, the centroid is located at 520mm from the soffit. The maximum permissible stresses are 14MPa in compression and zero at tension. The loss ratio is 0.8. Calculate

1. The breadth of a rectangular section, having the same depth designed for the same loading
2. The value of the point load
3. The saving in steel and concrete of the I–section compared to rectangular section
4. The maximum eccentricities of the cable at mid–span for the two sections.

[Ans: 1. Breadth of section = 280mm, 2. Point load = 106.66kN, 3. Saving in concrete and steel are 23% and 34.3% respectively 4. Maximum eccentricity for I and rectangular sections are 379mm and 250mm respectively]

7.10 A post–tensioned bonded pre–stressed concrete beam of rectangular section, 350mm by 700mm deep, is pre–stressed by an effective force of 180kN acting at an eccentricity of 190mm. At service load conditions, a section of the beam is subjected to a bending moment of 250kNm, a torsional moment of 100kNm and a transverse shear force of 100kN. If f_{ck} = 40MPa, f_y = 415MPa, f_p = 1600MPa, design the suitable longitudinal and transverse reinforcements in the section using IS: 1343 recommendations.

[Ans: Longitudinal reinforcements: 3 bars of 25mm diameter, Transverse reinforcement: 12mm diameter 2 legged stirrups at 100mm centres]

OBJECTIVE QUESTIONS

7.1 Among the below which is not a limit state?

a. Limit state of collapse b. Limit state of Creep
c. Limit state of shrinkage d. Limit state of Relaxation

7.2 Limit state of cracking for three types of pre–stressed concrete members is stated in below statements, find which of the following are correct?

1. For type 1, no tensile stresses are permitted
2. For type 2, tensile stresses are allowed but no visible cracking is permissible
3. For type 3, cracking is allowed, but should not affect the appearance or durability of the structure; the acceptable limits of width of cracking vary between wide limits for different types of structure and environment and prediction of the absolute maximum is not possible.

a. 1 b. 1, 2 c. 2, 3 d. 1, 2, 3

7.3 As per limit state of serviceability: cracking of IS: 1343–2012, the surface width of cracks should not in general exceed .... in case of PSC members exposed to particularly aggressive environment.

a. 0.10mm b. 0.15mm c. 0.20mm d. 0.25mm

7.4 As per IS: 1343, for M30 grade concrete, maximum permissible compressive stress due to bending and direct forces at the time of transfer of pre–stress for post–tensioned works shall be determined by.......

a. $0.51f_{ci}$ b. $0.54f_{ci}$ c. $0.44f_{ci}$ d. $0.37f_{ci}$

7.5 As per IS: 1343, for M60 grade concrete, maximum permissible compressive stress due to bending and direct forces at the time of transfer of pre–stress for post–tensioned works shall be determined by.......

a. $0.51f_{ci}$ b. $0.54f_{ci}$ c. $0.44f_{ci}$ d. $0.37f_{ci}$

7.6 As per IS: 1343, for M30 grade concrete, maximum permissible compressive stress due to bending and direct forces at the time of transfer of pre–stress for pre–tensioned works shall be determined by.......

a. $0.51f_{ci}$ b. $0.54f_{ci}$ c. $0.44f_{ci}$ d. $0.37f_{ci}$

7.7 As per IS: 1343, for M60 grade concrete, maximum permissible compressive stress due to bending and direct forces at the time of transfer of pre–stress for pre–tensioned works shall be determined by.......

a. $0.51f_{ci}$ b. $0.54f_{ci}$ c. $0.44f_{ci}$ d. $0.37f_{ci}$

7.8 In the determination of maximum permissible compressive stress in concrete under the action of live load and effective pre–stressing force at service conditions, which of the following statement/s is/are wrong?

1. For zone I, the straight line relation of permissible stress shall be determined by the straight line joining a point given by a permissible stress of $0.41f_{ck}$ for concrete of grade M30 to another point given by the permissible stress of $0.35f_{ck}$ for the concrete of grade M60.

2. For zone I, the straight line relation of permissible stress shall be determined by the straight line joining a point given by a permissible stress of $0.44f_{ck}$ for concrete of grade M30 to another point given by the permissible stress of $0.35f_{ck}$ for the concrete of grade M60.
3. For zone II, the determining points of the graph shall be reduced to $0.34f_{ck}$ and $0.27f_{ck}$ respectively.
4. For zone II, the determining points of the graph shall be reduced to $0.36f_{ck}$ and $0.27f_{ck}$ respectively.

a. 2, 4 b. 2 c. 4 d. 1, 3

7.9 For type 1 and type 2 beams, at transfer, the tensile stress at the top fibre must be less than or equal to the allowable tensile stress of the concrete at transfer. Algebraically, this requirement can be stated as ....

a. $\frac{F_o}{A}\left(1-\frac{ey_1}{K^2}\right)+\frac{M_G y_1}{I}=-f_{u1}$ b. $\frac{F_o}{A}\left(1-\frac{ey_1}{K^2}\right)+\frac{M_G y_1}{I}=f_{u1}$

c. $\frac{F_o}{K}\left(1-\frac{ey_1}{A^2}\right)+\frac{M_G y_1}{I}=-f_{u1}$ d. $\frac{F_o}{K}\left(1-\frac{ey_1}{A^2}\right)+\frac{M_G y_1}{I}=f_{u1}$

7.10 For type 1 and type 2 beams, at transfer, the compressive stress at the bottom concrete fibre must be less than or equal to the allowable compressive stress of concrete at transfer. This requirement can be stated as ....

a. $\frac{F_o}{A}\left(1-\frac{ey_2}{K^2}\right)-\frac{M_G y_2}{I}=f_{u2}$ b. $\frac{F_o}{A}\left(1+\frac{ey_2}{K^2}\right)+\frac{M_G y_2}{I}=f_{u2}$

c. $\frac{F_o}{A}\left(1+\frac{ey_2}{K^2}\right)-\frac{M_G y_2}{I}=f_{u2}$ d. $\frac{F_o}{A}\left(1-\frac{ey_2}{K^2}\right)+\frac{M_G y_2}{I}=f_{u2}$

7.11 For type 1 and type 2 beams, the compressive stress at the top fibre must be less than or equal to the allowable final compressive stress of concrete. This requirement can be stated as ....

a. $\frac{\eta F_o}{A}\left(1+\frac{ey_1}{K^2}\right)+\frac{M_G y_1}{I}+\frac{M_L y_1}{I}=fl_1$ b. $\frac{\eta F_o}{A}\left(1-\frac{ey_1}{K^2}\right)+\frac{M_G y_1}{I}+\frac{M_L y_1}{I}=fl_1$

c. $\frac{\eta F_o}{A}\left(1-\frac{ey_1}{K^2}\right)-\frac{M_G y_1}{I}+\frac{M_L y_1}{I}=fl_1$ d. $\frac{\eta F_o}{A}\left(1-\frac{ey_1}{K^2}\right)+\frac{M_G y_1}{I}-\frac{M_L y_1}{I}=fl_1$

7.12 For type 1 and type 2 beams, the tensile stress at the bottom fibre must be less than or equal to the allowable final tensile stress of concrete. This requirement can be stated as ....

a. $\frac{\eta F_o}{A}\left(1+\frac{ey_2}{K^2}\right)+\frac{M_G y_2}{I}-\frac{M_L y_2}{I}=-fl_2$ b. $\frac{\eta F_o}{A}\left(1+\frac{ey_2}{K^2}\right)-\frac{M_G y_2}{I}+\frac{M_L y_2}{I}=-fl_2$

c. $\frac{\eta F_o}{A}\left(1+\frac{ey_2}{K^2}\right)+\frac{M_G y_2}{I}-\frac{M_L y_2}{I}=-fl_2$ d. $\frac{\eta F_o}{A}\left(1+\frac{ey_2}{K^2}\right)-\frac{M_G y_2}{I}-\frac{M_L y_2}{I}=-fl_2$

7.13 Trial rectangular section shall be obtained using the equation of modulus of section as ....

a. $Z = \frac{BH^2}{6} = \frac{M_L}{fl_2 + (1.5\eta - 0.5)f_{u2}}$ b. $Z = \frac{BH^2}{6} = \frac{M_L}{fl_2 - (1.5\eta + 0.5)f_{u2}}$

c. $Z = \frac{BH^2}{6} = \frac{M_L}{fl_2 - (1.5\eta - 0.5)f_{u2}}$ d. $Z = \frac{BH^2}{6} = \frac{M_L}{fl_2 + (1.5\eta + 0.5)f_{u2}}$

7.14 In the design of PSC members for flexure, the thickness of the top flange may be taken to vary between ...., where D is the overall depth of member.

a. 0.05D to 0.10D b. 0.10D to 0.15D

c. 0.12D to 0.15D d. 0.15D to 0.18D

7.15 In the design of PSC members for flexure, thickness of web is computed by....... where D is the overall depth of member.

a. D/36+50mm+exteranal diameter of sheath or duct or about 0.10D to 0.15D

b. D/36+40mm+exteranal diameter of sheath or duct or about 0.12D to 0.15D

c. D/36+45mm+exteranal diameter of sheath or duct or about 0.125D to 0.15D

d. D/36+55mm+exteranal diameter of sheath or duct or about 0.125D to 0.15D

7.16 As per IS 1343, the minimum longitudinal reinforcement of .... of the total concrete area shall be provided in all cases except in the case of pre–tensioned units of small sections.

a. 0.20% b. 0.18% c. 0.15% d. 0.10%

7.17 As per IS: 1343, the minimum longitudinal reinforcement, when the depth of the web exceeds 500mm, is not less than .... of the area of the web shall be provided in each face.

a. 0.20% b. 0.15% c. 0.10% d. 0.05%

7.18 In post–tensioned work, when cables and large sized bars are used, the minimum clear cover from sheathing/duct is ....

a. At least 30mm or the size of the cable or bar whichever is greater

b. At least 30mm or the size of the cable or bar whichever is least

c. At least 40mm or the size of the cable or bar whichever is greater

d. At least 40mm or the size of the cable or bar whichever is least

7.19 In the design of PSC sections for flexure, in case of cables or large bars, the clear spacing (measured between sheathing/ducts, whichever used) shall not be less than the greater of

a) 40mm

b) maximum size of the cable or bar and

c) 5mm+maximum size of aggregate. Among three which is correct?

a. a, b b. b, c c. a, b, c d. a, c

7.20 The four fundamental conditions for stresses at transfer and service loads are as follows,

At transfer, (a) Top fibres $\left(f_{sup} + \frac{M_G}{Z_t}\right) \geq f_{tt}$, (b) Bottom fibres $\left(f_{inf} - \frac{M_G}{Z_b}\right) \leq f_{ct}$

At working load, (c) Top fibres $\left(\eta f_{sup} + \frac{M_G}{Z_t} + \frac{M_L}{Z_t}\right) \leq f_{cw}$,

(d) Bottom fibres $\left(\eta f_{inf} - \frac{M_G}{Z_b} - \frac{M_L}{Z_b}\right) \leq f_{tw}$, which of the following condition is wrong?

a. (a) b. (b) c. (c) d. (d)

7.21 Design formulae for the required section moduli are expressed as ....Where, f_{tr} and f_{br} are the ranges of stress at top and bottom fibres respectively and η is loss ratio.

a. $Z_t \geq \left[\frac{M_L + (1-\eta)M_G}{f_{tr}}\right]$ and $Z_b \geq \left[\frac{M_L + (1-\eta)M_G}{f_{br}}\right]$

b. $Z_t \geq \left[\frac{M_L + (1-\eta)M_G}{f_{tr}}\right]$ and $Z_b \leq \left[\frac{M_L + (1-\eta)M_G}{f_{br}}\right]$

c. $Z_t \leq \left[\frac{M_L + (1-\eta)M_G}{f_{tr}}\right]$ and $Z_b \geq \left[\frac{M_L + (1-\eta)M_G}{f_{br}}\right]$

d. $Z_t \leq \left[\frac{M_L + (1-\eta)M_G}{f_{tr}}\right]$ and $Z_b \leq \left[\frac{M_L + (1-\eta)M_G}{f_{br}}\right]$

7.22 For pre–tensioned members, the loss ratio, generally lies in the range of ....

a. 0.65 to 0.70 b. 0.70 to 0.75
c. 0.75 to 0.80 d. 0.80 to 0.85

7.23 For post–tensioned members, the loss ratio, generally lies in the range of ....

a. 0.65 to 0.70 b. 0.70 to 0.75
c. 0.75 to 0.80 d. 0.80 to 0.85

7.24 Expression for pre–stressing force is given by.........

a. $F_o = \frac{A(f_{inf}Z_b + f_{sup}Z_t)}{(Z_t + Z_b)}$ b. $F_o = \frac{A(f_{inf}Z_b - f_{sup}Z_t)}{(Z_t - Z_b)}$

c. $F_o = \frac{A(f_{inf}Z_b + f_{sup}Z_t)}{(Z_t - Z_b)}$ d. $F_o = \frac{A(f_{inf}Z_b - f_{sup}Z_t)}{(Z_t + Z_b)}$

7.25 Expression for maximum eccentricity is given by ....

a. $e = \dfrac{Z_t Z_b (f_{inf} + f_{sup})}{A(f_{inf} Z_b + f_{sup} Z_t)}$ b. $e = \dfrac{Z_t Z_b (f_{inf} - f_{sup})}{A(f_{inf} Z_b + f_{sup} Z_t)}$

c. $e = \dfrac{Z_t Z_b (f_{inf} + f_{sup})}{A(f_{inf} Z_b - f_{sup} Z_t)}$ d. $e = \dfrac{Z_t Z_b (f_{inf} - f_{sup})}{A(f_{inf} Z_b - f_{sup} Z_t)}$

7.26 The minimum stress developed in long span at transfer, f_{mint} is computed as ....

a. $f_{mint} = \left[\dfrac{M_{min}}{Z_t} + \dfrac{Z_b(Z_t - Ae)}{\eta Z_t (Z_t + Ae)} \left(\dfrac{M_T}{Z_b} - f_{tw} \right) \right]$

b. $f_{mint} = \left[\dfrac{M_{min}}{Z_t} - \dfrac{Z_b(Z_t - Ae)}{\eta Z_t (Z_t + Ae)} \left(\dfrac{M_T}{Z_b} + f_{tw} \right) \right]$

c. $f_{mint} = \left[\dfrac{M_{min}}{Z_t} + \dfrac{Z_b(Z_t - Ae)}{\eta Z_t (Z_t + Ae)} \left(\dfrac{M_T}{Z_b} + f_{tw} \right) \right]$

d. $f_{mint} = \left[\dfrac{M_{min}}{Z_t} - \dfrac{Z_b(Z_t - Ae)}{\eta Z_t (Z_t + Ae)} \left(\dfrac{M_T}{Z_b} - f_{tw} \right) \right]$

7.27 Pre–stressing force F_o acting at known eccentricity e, which can develop required pre–stress f_{inf} at the bottom fibre, is computed directly by the expression ...., where, A and Z_b are the cross sectional area and the section modulus of the cross–section actually provided.

a. $F_o = \left(\dfrac{AZ_b}{Z_b - Aef_{inf}} \right)$ b. $F_o = \left(\dfrac{AZ_b}{Z_b + Aef_{inf}} \right)$

c. $F_o = \left(\dfrac{Af_{inf} Z_b}{Z_b - Ae} \right)$ d. $F_o = \left(\dfrac{Af_{inf} Z_b}{Z_b + Ae} \right)$

7.28 The pre–stress equations for the top and bottom fibres given by............

a. $f_{sup} = \left(\dfrac{F_o}{A} + \dfrac{F_o e}{Z_t} \right)$ and $f_{inf} = \left(\dfrac{F_o}{A} + \dfrac{F_o e}{Z_b} \right)$

b. $f_{sup} = \left(\dfrac{F_o}{A} - \dfrac{F_o e}{Z_t} \right)$ and $f_{inf} = \left(\dfrac{F_o}{A} + \dfrac{F_o e}{Z_b} \right)$

c. $f_{sup} = \left(\dfrac{F_o}{A} - \dfrac{F_o e}{Z_t} \right)$ and $f_{inf} = \left(\dfrac{F_o}{A} - \dfrac{F_o e}{Z_b} \right)$

d. $f_{sup} = \left(\dfrac{F_o}{A} + \dfrac{F_o e}{Z_t} \right)$ and $f_{inf} = \left(\dfrac{F_o}{A} - \dfrac{F_o e}{Z_b} \right)$

7.29 Limiting zone for pre–stressing force determined by following four eccentricity equations, mention which one is wrong?

A. $e \geq \frac{-Z_t f_{tt}}{F_o} + \frac{Z_t}{A} + \frac{M_{min}}{F_o}$ B. $e \leq \frac{Z_b f_{ct}}{F_o} - \frac{Z_b}{A} + \frac{M_{min}}{F_o}$

C. $e \geq \frac{-Z_t f_{cw}}{\eta F_o} + \frac{Z_t}{A} + \frac{M_T}{\eta F_o}$ D. $e \geq \frac{Z_b f_{tw}}{\eta F_o} - \frac{Z_b}{A} + \frac{M_T}{\eta F_o}$

a. A   b. B   c. C   d. D

7.30 According to IS: 1343–2012, the maximum ultimate moment of resistance of a rectangular section is given by ....

a. $M_{ud} = 0.51 f_{ck} bd^2$ b. $M_{ud} = 0.41 f_{ck} bd^2$
c. $M_{ud} = 0.31 f_{ck} bd^2$ d. $M_{ud} = 0.21 f_{ck} bd^2$

7.31 According to IS: 1343–2012, the maximum ultimate moment of resistance of a flanged section is varies from ...., when a reasonable proportion of $h_f/d = 0.2$ to 0.25 and $b_w/d = 0.2$ to 0.3 are used.

a. $0.10 f_{ck} bd^2$ to $0.12 f_{ck} bd^2$ b. $0.06 f_{ck} bd^2$ to $0.10 f_{ck} bd^2$
c. $0.08 f_{ck} bd^2$ to $0.12 f_{ck} bd^2$ d. $0.12 f_{ck} bd^2$ to $0.15 f_{ck} bd^2$

7.32 As per IS: 1343–2012, the design rules applied to

a. Beams of solid rectangular cross–section (D > b),
b. Hollow rectangular beams with D > b and with a wall thickness, $t \geq b/4$, and
c. T–beam and I–beams. In all the three cases, the average intensity of pre–stress in the concrete shall be less than $0.3 f_{ck}$. Which of the statements are correct?

a. a, b   b. b, c   c. a, c   d. a, b, c

7.33 In case of design of PSC sections for torsion, the longitudinal reinforcement is calculated by equivalent ultimate bending moments,

(a) $M_{e1} = M + M_t$,   (b) $M_{e2} = M_t - M$,

(c) $M_{e3} = M_t\left(1 + \frac{x_1}{2e}\right)^2 \left(\frac{1 + \frac{2b}{D}}{1 + \frac{2D}{b}}\right)$

where, M = Applied ultimate bending moment at the cross–section acting in combination with ultimate torque T, x_1, is the smaller dimension of a closed hoop used as torsional shear reinforcement and $e = \frac{V}{T}$, where V is the ultimate shear force. Out of three which statement/s is/are correct?

a. (a)   b. (a), (b)   c. (a), (c)   d. (a), (b), (c)

7.34 In design of PSC sections for torsion for longitudinal reinforcement, torsional moment is given by ...., where, T = Ultimate torque, D = Overall depth of beam, b = Breadth of the member which for T and I beams shall be taken as the breadth of the web b_w.

a. $M_t = T\sqrt{1-\left(\frac{2D}{b}\right)}$ b. $M_t = T\sqrt{1+\left(\frac{2D}{b}\right)}$

c. $M_t = T\sqrt{1-\left(\frac{D}{2b}\right)}$ d. $M_t = T\sqrt{1+\left(\frac{D}{2b}\right)}$

7.35 In design of PSC sections for torsion for transverse reinforcement, torsional moment carried by concrete is given by ...., where, b = Breadth of the member, which for T and I beams shall be taken as the breadth of the web b_w, D = Overall depth of beam, f_{ck} = Characteristic compressive strength of concrete.

a. $T_c = 0.10b^2D\left(1-\frac{b}{3D}\right)\lambda_p\sqrt{f_{ck}}$ b. $T_c = 0.12b^2D\left(1-\frac{b}{3D}\right)\lambda_p\sqrt{f_{ck}}$

c. $T_c = 0.15b^2D\left(1-\frac{b}{3D}\right)\lambda_p\sqrt{f_{ck}}$ d. $T_c = 0.18b^2D\left(1-\frac{b}{3D}\right)\lambda_p\sqrt{f_{ck}}$

7.36 In design of PSC sections for torsion for transverse reinforcement, reduced torsional moment carried by concrete is given by ...., where, T_c = Torsional moment carried by concrete, b = Breadth of the member, which for T and I beams shall be taken as the breadth of the web b_w, D = Overall depth of beam, f_{ck} = Characteristic compressive strength of concrete, T = Torsional moment applied to a cross section under ultimate load conditions, V = Shearing force at a cross section calculated for the specified ultimate load, V_c = Theoretical shear strength at a cross section and $e = \frac{V}{T}$, $e_c = \frac{T_c}{V_c}$, $T_c = 0.15b^2D\left(1-\frac{b}{3D}\right)\lambda_p\sqrt{f_{ck}}$

a. $T_{c1} = T_c\frac{e}{e+e_c}$ b. $T_{c1} = T_c\frac{e+e_c}{e}$

c. $T_{c1} = T_c\frac{e}{e-e_c}$ d. $T_{c1} = T_c\frac{e-e_c}{e}$

7.37 In the torsional moment expression, $T_c = 0.15b^2D\left(1-\frac{b}{3D}\right)\lambda_p\sqrt{f_{ck}}$, λ_p = ...

a. $\sqrt{\left(1+\frac{10f_{cp}}{f_{ck}}\right)}$ b. $\sqrt{\left(1+\frac{12f_{cp}}{f_{ck}}\right)}$ c. $\sqrt{\left(1+\frac{15f_{cp}}{f_{ck}}\right)}$ d. $\sqrt{\left(1+\frac{18f_{cp}}{f_{ck}}\right)}$

7.38 In design of PSC sections for torsion for transverse reinforcement, shear force carried by the concrete is given by ..., where, V_c = Theoretical shear strength at a cross section. i.e., smaller of V_{co} and V_{cr} as per 22.4.1 and 22.4.2 of IS: 1343–2012 and $e = \frac{V}{T}$, $e_c = \frac{T_c}{V_c}$.

a. $V_{c1} = V_c \dfrac{e}{e + e_c}$ b. $V_{c1} = V_c \dfrac{e + e_c}{e}$

c. $V_{c1} = V_c \dfrac{e}{e - e_c}$ d. $V_{c1} = V_c \dfrac{e - e_c}{e}$

7.39 The area of cross section A_{sv} of the closed stirrup enclosing the corner longitudinal bars shall be taken as the larger of the following two values .... and..........., where, S_v = Spacing of stirrup reinforcement, b_1 = Centre to centre distance between corner bars in the direction of width, d_1 = Centre to centre distance between corner bars in the direction of depth and $A_v = \dfrac{(V - V_{c1})S_v}{0.87 f_y d_1}$, $A_t = \dfrac{(T - T_{c1})S_v}{0.87 f_y b_1 d_1}$

a. $A_{sv} = \left(\dfrac{M_t S_v}{0.5 b_1 d_1 f_y}\right), A_{sv} = A_v + 2A_t$ b. $A_{sv} = \left(\dfrac{M_t S_v}{0.5 b_1 d_1 f_y}\right), A_{sv} = A_v - 2A_t$

c. $A_{sv} = \left(\dfrac{M_t S_v}{1.5 b_1 d_1 f_y}\right), A_{sv} = A_v + 2A_t$ d. $A_{sv} = \left(\dfrac{M_t S_v}{1.5 b_1 d_1 f_y}\right), A_{sv} = A_v - 2A_t$

7.40 The area of cross section A_{sv} of the closed stirrup enclosing the corner longitudinal bars is $A_{sv} = A_v + 2A_t$. In this expression A_v is ....

a. $\dfrac{(V - V_{c1})S_v}{0.87 f_y d_1}$ b. $\dfrac{(V - V_{c1})S_v}{0.9 f_y d_1}$

c. $\dfrac{(V + V_{c1})S_v}{0.87 f_y d_1}$ d. $\dfrac{(V + V_{c1})S_v}{0.9 f_y d_1}$

7.41 The area of cross section A_{sv} of the closed stirrup enclosing the corner longitudinal bars is $A_{sv} = A_v + 2A_t$. In this expression A_t is ....

a. $\dfrac{(T + T_{c1})S_v}{0.87 f_y b_1 d_1}$ b. $\dfrac{(T + T_{c1})S_v}{0.9 f_y b_1 d_1}$

c. $\dfrac{(T - T_{c1})S_v}{0.87 f_y b_1 d_1}$ d. $\dfrac{(T - T_{c1})S_v}{0.9 f_y b_1 d_1}$

7.42 As per IS: 1343, maximum reinforcement should not be less than the given by ....

a. $\dfrac{A_{sv}}{bS_v} = \dfrac{0.4}{0.9 f_y}$ b. $\dfrac{A_{sv}}{bS_v} = \dfrac{0.4}{0.87 f_y}$

c. $\dfrac{A_{sv}}{bS_v} = \dfrac{0.5}{0.9 f_y}$ d. $\dfrac{A_{sv}}{bS_v} = \dfrac{0.5}{0.87 f_y}$

7.43 There shall be at least one longitudinal bar not less than .... in dia. in each corner of the stirrups.

a. 8mm b. 10mm c. 12mm d. 20mm

7.44 When a member is designed for torsion, torsion reinforcement shall be provided as below: a) All transverse reinforcement provided for torsion shall be in the form of closed stirrups perpendicular to the axis of the members. b) The spacing, S_v, of the stirrups shall not exceed $(x_1 - y_1)/4$ or 200 mm whichever is smaller, where x_1 and y_1 are respectively short and long dimensions of the stirrup. c) Each end of the bar forming the stirrup shall be anchored in accordance with IS: 456–2000. d) Torsional reinforcement shall be continued to a distance not less than $(D + b_w)$ beyond the point at which it is no longer than theoretically required, where D is the overall depth and b_w is the effective width of the web of a flanged member. As per IS: 1343, which statement is wrong?

a. a) b. b) c. c) d. d)

Answers

7.1 a	7.2 d	7.3 a	7.4 b	7.5 d
7.6 a	7.7 c	7.8 a	7.9 a	7.10 c
7.11 b	7.12 d	7.13 a	7.14 b	7.15 d
7.16 a	7.17 d	7.18 a	7.19 c	7.20 d
7.21 a	7.22 c	7.23 d	7.24 a	7.25 b
7.26 c	7.27 d	7.28 b	7.29 a	7.30 d
7.31 c	7.32 d	7.33 d	7.34 b	7.35 c
7.36 a	7.37 b	7.38 a	7.39 c	7.40 a
7.41 c	7.42 b	7.43 c	7.44 b	

❑

CHAPTER

8 End Blocks

8.1 INTRODUCTION

The large pre–stressing force concentrated over relatively small area cause transverse and shear stresses close to the ends of the member. The zone between the end of the beam and the section where only longitudinal stress varying linearly across the depth is set up is called the anchorage zone or end block or transmission zone.

In the anchorage zone or end block of a post–tensioned pre–stressed concrete element, the state of stress distribution is complex and three dimensional in nature. In most post–tensioned members, the pre–stressing wires are introduced in a cable holes or ducts, pre–formed in the members, and then stressed and anchored at the end faces. As result, large forces, concentrated over relatively small areas, area applied on the end blocks. These highly discontinuous forces which are applied at the end, while changing progressively to continuous linear distribution, develop transverse and shear stresses.

According to St. Venant's principle, the stress distribution at a distance far away from the loaded face (normally at a distance equal to or greater than the depth of the beam) can be computed from the simple bending theory. The zone between the end of the beam and the section where only longitudinal stresses exists is known as end block or anchorage zone.

The transverse stresses developed at the anchorage zone are tensile in nature over a large length and since concrete is weak in tension, adequate reinforcement should be provided to resist this tension. Hence from the point of view of the designer, it is essential to have a good knowledge of the distribution of stresses in the anchorage zone, so as that he can provide an adequate amount of steel, properly distributed to sustain the transverse tensile stresses.

8.2 STRESS DISTRIBUTION IN END BLOCK

Theoretical and experimental investigation show that, the length of the end block is not more than the depth of the beam. For practical design therefore it is essential to assume the length of the end block equal to the beam's overall depth. The stress distribution in the end block zone due to concentrated pre–stressing force in the anchorages is influenced by many factors such as,

a. Type of the anchorages and their deposition.
b. The area over which the force is applied.
c. The size and shape of the cross section of the member.

These lines of forces to be acting as individual struts, it is seen that they are first convex and then concave towards the axis of the member.

In addition to the tensile and compressive stresses, the corners are found to be subjected to high spalling tensile stresses. The spalling zone is, however, very shallow and the resultant force small. High lateral strains are caused due to high beating stresses and these strains are probably the main reasons for spalling of concrete.

The force on the end block of a post–tensioned pre–stressed concrete member is as shown in Fig. 8.1 (a). A physical concept of the state of stress in the transverse direction, that is, normal to planes parallel with the top and bottom surfaces of the beam, may be obtained by considering these lines of forces as individual fibres acting as curved struts inserted between end forces 2P and the main body of the beam. The curvature of the struts, being convex to the centre of the block, induces compressive stresses in zone A. in zone B, the curvature is reversed in direction and the struts tends to deflect outwards, separating from each other and consequently developing transverse tensile stresses. In zone C, the struts are straight and parallel so that no transverse stresses are induced, only longitudinal stresses developed in this zone.

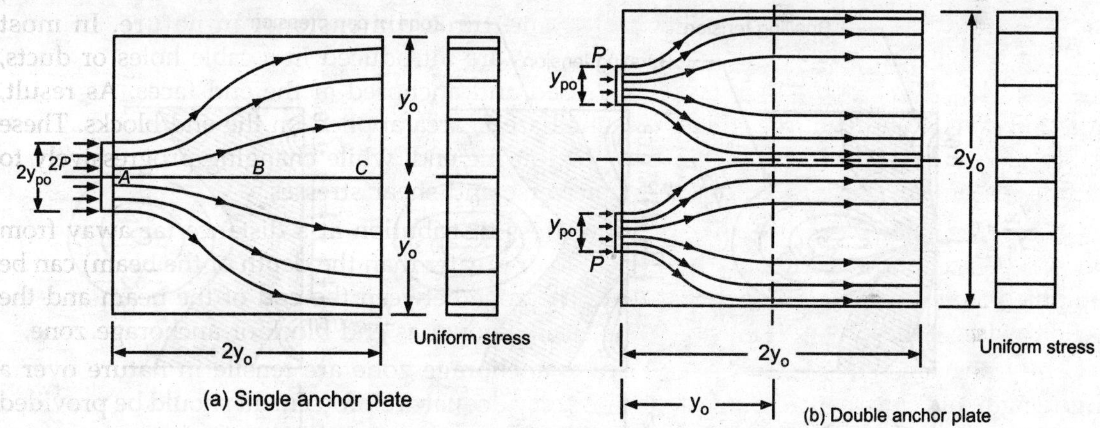

Fig. 8.1 Transmission of Forces in End Block

In Fig. 8.1 (b), the same end block is subjected to the same total load applied through two zones symmetrically disposed in the upper and lower halves of the beam. Since the lines of force follow the same pattern with half the radius of curvature, the length of the anchorage zone is halved. The transverse tension developed is also proportionately reduced. In a similar way, the greater the number of points of application of the pre–stressing force on the end block, the more uniform is the stress distribution.

The idealized stress distribution in the end lock with the compressive and tensile stress paths is shown in Fig. 8.2 (a). The effect of transverse tensile stress is to develop a zone of bursting tension in a direction perpendicular to the anchorage force, resulting in horizontal cracking as shown in Fig. 8.2 (b). Since concrete is weak in tension, suitable reinforcements are generally provided in the transverse direction to resist the bursting tension.

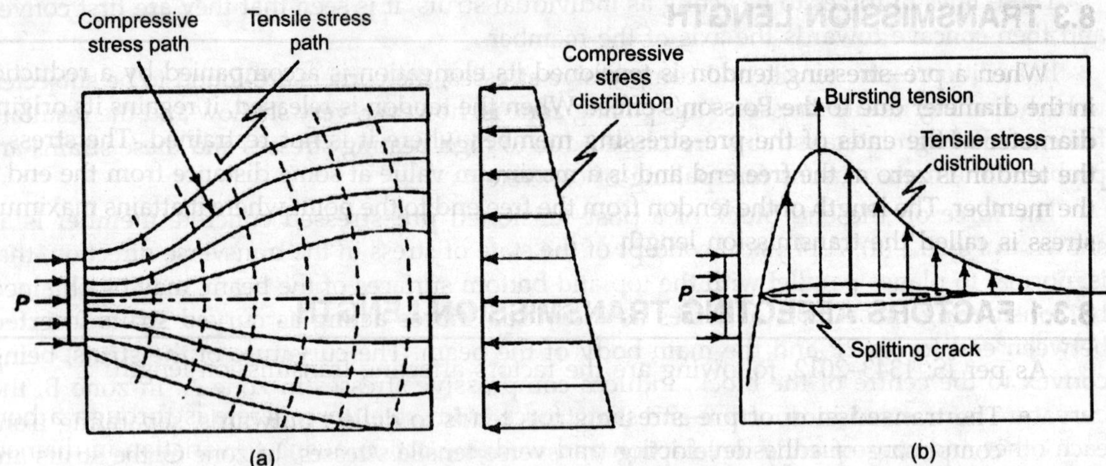

Fig. 8.2 End Blocks of Post–tensioned Beams (a) Idealized Stress Paths (b) Bursting Tension and Splitting Cracks

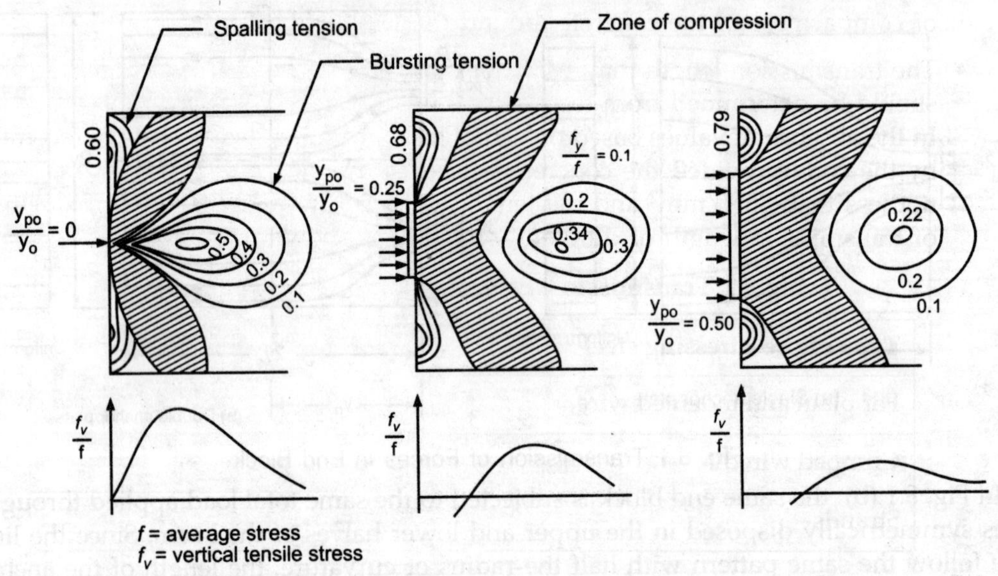

Fig. 8.3 Isobars of Transverse Tensile Stress

The distribution of transverse stresses in the anchorage zone subjected to a symmetrically placed pre–stressing force, which is distributed over a small area for increasing ratio of y_{po}/y_o varying from 0 to 0.50, has been investigated by Guyon. The results are shown in Fig. 8.3. The lines of equal transverse tensile stress are termed as isobars. The figure shows the influence of the height of the anchor plate on the distribution of compressive and tensile stresses in the transverse direction. The ratio of transverse tensile stress to the average compressive stress gradually decreases with the increase in the ratio of the depth of the anchor plate to that of the end block.

8.3 TRANSMISSION LENGTH

When a pre–stressing tendon is tensioned its elongation is accompanied by a reduction in the diameter due to the Poisson's effect. When the tendon is released, it regains its original diameter at the ends of the pre–stressing member where it is not restrained. The stress in the tendon is zero at the free end and is a maximum value at some distance from the end of the member. The length of the tendon from the free end to the point where it attains maximum stress is called the transmission length.

8.3.1 FACTORS AFFECTING TRANSMISSION LENGTH

As per IS: 1343–2012, following are the factors affecting transmission length.

- The transmission of pre–stressing force from steel to concrete is through a bond comprising of adhesion, friction and wedging action owing to variation in diameter of wire near the ends of the member. The transmission length has been found to depend on a number of factors, the most important being the strength of concrete at transfer, the size and type of tendon, the surface condition of the tendon and degree of compactness of the concrete around the tendon.
- The transmission length may vary depending upon the site conditions and therefore should be determined from tests carried out under the most unfavorable conditions. In the absence of values based on actual tests, the following values are recommended by the code provided the concrete is well compacted and its strength at transfer is not less than $35N/mm^2$ and the tendon released gradually. The approximate values of transmission length of different wires are as shown in Table 8.1.

Table 8.1 Transmission Length of Different Type of Wires

Type of pre–stressing steel	Transmission length
For plain and indented wire	100ϕ
Crimped wired	65ϕ
Strands	30ϕ

Where, ϕ = Diameter of wire

- The development of stress in the tendon may be assumed to vary parabolically along the length of the member.
- For general guidance, it is recommended that one–half of the transmission length shall overhang the support in a simply supported beam. Where there is end–fixing, the whole of the transmission length shall overhang.

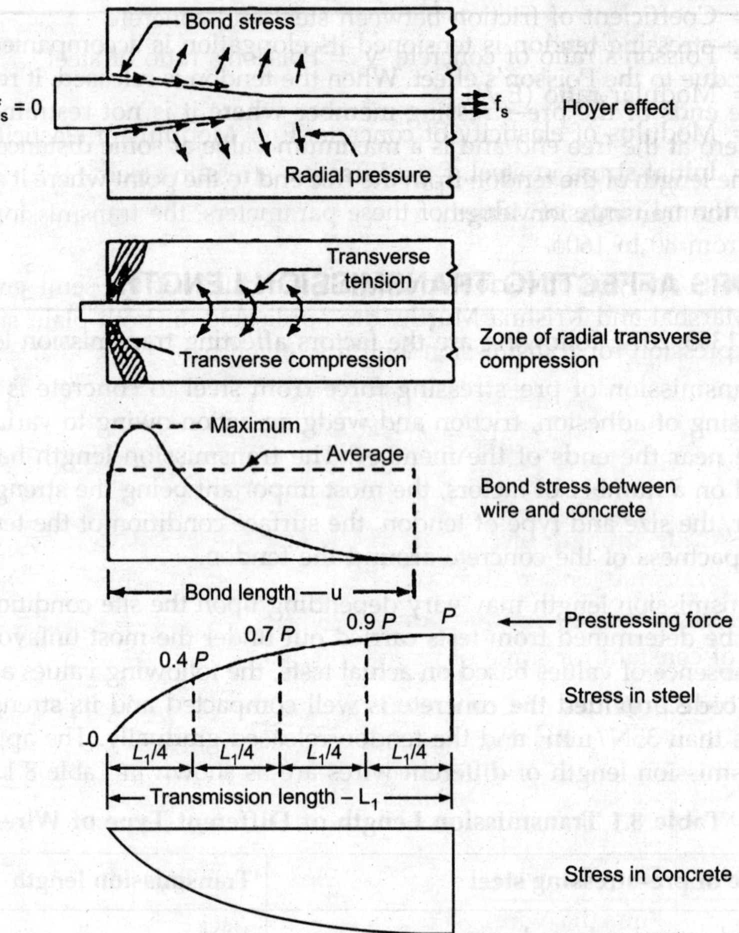

Fig. 8.4 Distribution of Bond Stresses

As per Hoyer, the length required at the ends of a pre–tensioned member for the build–up of stress in concrete is of great importance, particularly in short pre–tensioned units, since it controls the working bending moment and shear force allowable on the section. The transmission length mainly depends on the diameter and surface characteristics of the wire, the elastic properties of steel and concrete, and the coefficient of friction between steel and concrete. The distribution of bond stresses is as shown in Fig. 8.4.

Based on the wedge action, Hoyer has developed an expression for computing the transmission length, and is given by

$$L_t = \frac{\phi}{2\mu}(1+\nu_c)\left(\frac{\alpha_e}{\nu_s} - \frac{f_{pi}}{E_c}\right)\left(\frac{f_{pe}}{2f_{pi} - f_{pe}}\right) \quad ...(8.1)$$

Where, L_t = Transmission length

ϕ = Wire diameter

μ = Coefficient of friction between steel and concrete

ν_c = Poisson's ratio of concrete, ν_s = Poisson's ratio of steel

α_e = Modular ratio (E_s/E_c)

E_c = Modulus of elasticity of concrete, E_s = Modulus of elasticity of steel

f_{pi} = Initial stress in steel, f_{pe} = Effective stress in steel

Under the normal range of values of these parameters, the transmission length is likely to be varying from 80 to 160ϕ.

A comparative analysis of various proposals indicate that the semi–empirical relations suggested by Marshal and Krishna Murthy are applicable for both plain smooth wires and strands. The expression for transmission length is given by,

$$L_t = \sqrt{\frac{\sqrt{f_{cu}} \times 10^3}{\beta}} \quad \text{...(8.2)}$$

Where,

f_{cu} = Cube strength of concrete at transfer, expressed in MPa

L_t = Transmission length in mm

β = Constant depending upon the details of strand and wire

The values of constant for some of the typical wires and strands are shown in Table 8.2.

Table 8.2 Values of Constant β for Marshal and Krishna Murthy Relation for Transmission Length

Sl. No.	Details of wire or strand	β
1	2mm dia. wire	0.144
2	5mm dia. wire	0.0235
3	7mm dia. wire	0.0174
4	10mm dia., 7 wire strand	0.144
5	12.5mm dia., 7 wire strand	0.058
6	18mm dia., 19 wire strand	0.0235
7	19mm dia., 7 wire strand	0.0235
8	Twin twisted wires of 6.25mm dia., 7 wire strand	0.077

8.3.2 Solved Examples

Problem 8.1

Calculate the transmission length at the end of a pre–tensioned beam as per Hoyer's method using the following data:

Span of the beam = 50m, Diameter of the wires = 7mm

Coefficient of friction between steel and concrete = 0.1

Poisson's ratio of concrete = 0.15, Poisson's ratio of steel = 0.30

$E_c = 30kN/mm^2$, $E_s = 210kN/mm^2$

Ultimate tensile strength of steel wire = $f_{pu} = 1500N/mm^2$

f_{pi} = Initial stress in steel = $0.7f_{pu} = 1050N/mm^2$,

f_{pe} = Effective stress in steel = $0.6f_{pu} = 900N/mm^2$

Solution:

α_e = Modular ratio $(E_s/E_c) = 210/30 = 7$, Using Hoyer's expression,

$$L_t = \frac{\phi}{2\mu}(1+\nu_c)\left(\frac{\alpha_e}{\nu_s} - \frac{f_{pi}}{E_c}\right)\left(\frac{f_{pe}}{2f_{pi} - f_{pe}}\right)$$

$$L_t = \frac{7}{2\times 0.1}(1+0.15)\left(\frac{7}{0.30} - \frac{0.7\times 1500}{30\times 10^3}\right)\left(\frac{900}{2\times 1050 - 900}\right) = 700mm$$

If the beam is simply supported over a span of 50m, there should be at least 700mm of beam projection beyond the centre of the supports at each end. Therefore, the total overall length = (50 + 2 × 0.7) = 51.4m

Problem 8.2

Calculate the transmission length at the ends of a pre–tensioned beam pre–stressed by 7mm dia. wires. Assume the cube strength of concrete at transfer as $42N/mm^2$. (Adopt Krishna Murthy empirical relation)

Solution:

For 7mm diameter smooth wires, $\beta = 0.0174$ (from Table 8.1) and $f_{cu} = 42N/mm^2$, thus transmission length is given by

$$L_t = \sqrt{\frac{\sqrt{f_{cu}}\times 10^3}{\beta}} = \sqrt{\frac{\sqrt{42}\times 10^3}{0.0174}} = 610mm = 87\phi$$

8.4 INVESTIGATION ON THE ANCHORAGE ZONE STRESSES

A number of investigations have been done of stress distribution in the anchorage zone using empirical equations or theoretical solutions based on two or three dimensional elasticity and experimental techniques. The important investigations were those done by Guyon, Magnel. Iyengar, Yettram and Robins and Chandrashekhara etc. The main aim of the stress analysis in the anchorage zone is to obtain the transverse tensile stress distribution in the end block from which the total transverse bursting tension could be computed.

8.4.1 Magnel's Method

In this method the stresses due anchorage bearing are taken to be dispersed over a length of beam equal to the depth of beam. Consider any unit elemental unit situated within this end zone.

It is also stated that, the end block is considered as a deep beam subjected to concentrated loads due to anchorages on one side and to normal and tangential distributed loads from the linear direct stress and shear stress distribution from the other side. The forces acting on the end block and the stresses acting on any point on the horizontal axis parallel to the beam are shown in Fig. 8.5.

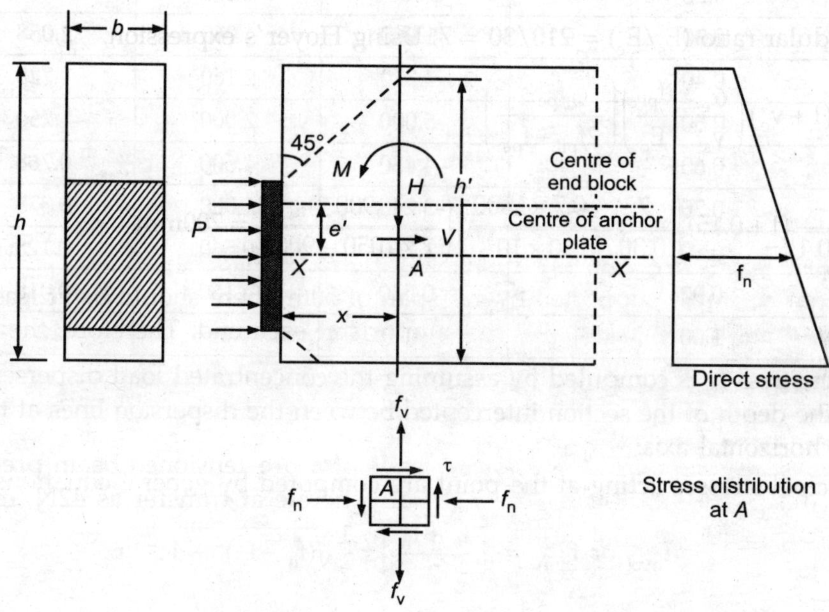

Fig. 8.5 Forces Acting on the End Block

Such an elemental unit is subjected to three stresses f_h, f_v and τ. Where, f_h and f_v are horizontal and vertical stresses and τ is shear stress. M = Bending moment, H = Direct force (vertical), V = Shear force (horizontal).

The stress distribution across the section can be approximated using following equations

$$f_v = K_1\left(\frac{M}{bh^2}\right) + K_2\left(\frac{H}{bh}\right) \quad ...(8.3)$$

$$\tau = K_3\left(\frac{V}{bh}\right) \quad ...(8.4)$$

$$f_h = \frac{P}{bh}\left(1 + 12\frac{e'^2}{h'^2}\right) \quad ...(8.5)$$

Constants K_1, K_2 and K_3 are shown in Table 8.3 for varying distance from the end face of the beam.

Table 8.3 Magnel's Coefficient of Stresses in End Blocks[6]

Distance from far end x/h	K_1	K_2	K_3
0	20.00	–2.000	0.000
0.10	9.720	0.000	1.458
0.20	2.560	1.280	2.048
0.30	–1.960	1.960	2.058
0.40	–4.320	2.160	1.728
0.50	–5.000	2.000	1.250
0.60	–4.480	1.600	0.768
0.70	–3.240	1.080	0.378
0.80	–1.760	0.560	0.128
0.90	–0.520	0.160	0.018
1.00	0	0	0

The direct stress f_h is computed by assuming the concentrated load disperses at 45° and considering the depth of the section intercepted between the dispersion lines at the required point on the horizontal axis.

The principal stresses acting at the point are computed by general equations;

$$f_{max} \text{ or } f_{min} = \left(\frac{f_v + f_h}{2}\right) \pm \frac{1}{2}\sqrt{(f_h - f_v)^2 + 4\tau^2} \quad \text{...(8.6)}$$

$$\tan 2\theta = \left(\frac{2\tau}{f_v - f_h}\right) \quad \text{...(8.7)}$$

The bursting tension is computed from the distribution of principal stresses on the required axis and suitable reinforcements designed to take this tension.

8.4.1.1 Solved Examples

Problem 8.3

The end block of a pre–stressed concrete beam, rectangular in section is 100mm wide and 200mm deep. The pre–stressing force of 100kN is transmitted to concrete by a distribution plate, 100mm wide and 50mm deep, concentrically placed at the ends. Calculate the position and magnitude of the maximum tensile stress on the horizontal section through the centre and edge of the anchor plate. Compute the bursting tension on these horizontal planes.

Solution:

P = 100kN, h = 200mm, b = 100mm

Direct stress, $f_h = \left(\frac{100 \times 10^3}{200 \times 100}\right) = 5\text{N/mm}^2$

Normally the vertical stress, f_v and the principal tensile stress are critical at x = 0.5h.

Referring the Fig. 8.6,

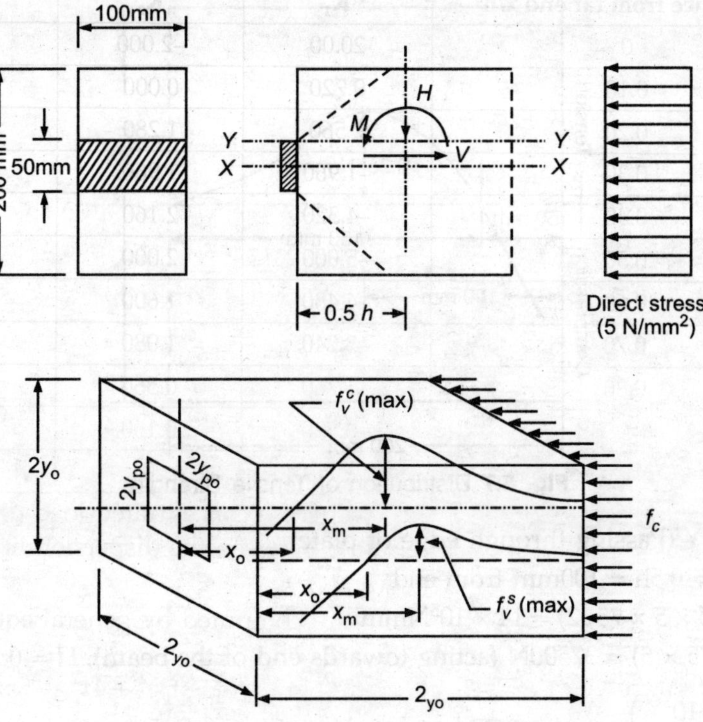

Fig. 8.6 Forces Acting on the End Block

For section XX

At $\frac{x}{h} = 0.5$, from Table 8.2, $K_1 = -5.00$, $K_2 = 2.00$, $K_3 = 1.25$

$$M = \left[(5\times100\times100)\left(\frac{100}{2}\right)-\left(\frac{100\times10^3}{2}\right)\left(\frac{50}{4}\right)\right] = 1875 \times 10^3 \text{Nmm}$$

$V = 0, H = 0$

$$\therefore\ f_v = -5\left(\frac{1875\times10^3}{100\times200^2}\right) = -2.35\text{N/mm}^2,\ f_h = +5\text{N/mm}^2$$

The principal stress (acting at 0.5h = 100mm from the end) is given by,

$$f_{min} = \left(\frac{5-2.35}{2}\right)-\frac{1}{2}\sqrt{(5+2.35)^2+0} = -2.35\text{N/mm}^2$$

Therefore, the total splitting tension, assuming parabolic distribution of stress as shown in Fig. 8.7, is given by

$$F_{bst} = [(2/3 \times 150 \times 2.35)100] = 23500\text{N}$$

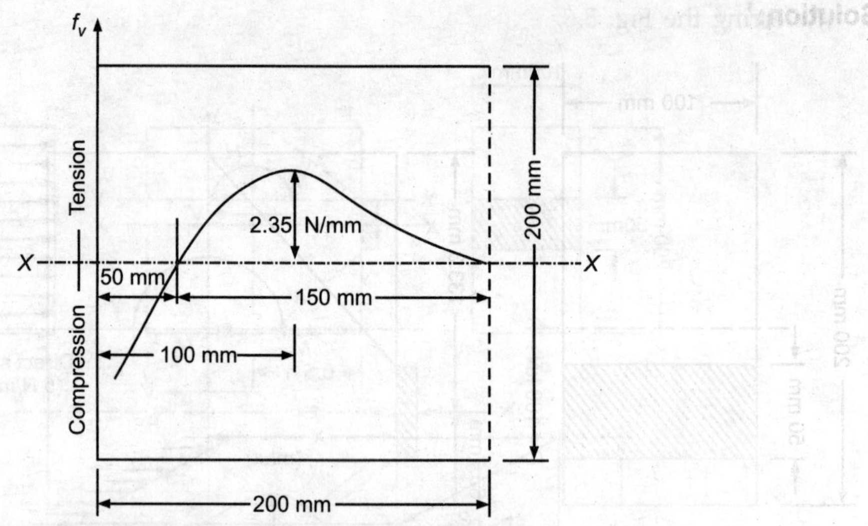

Fig. 8.7 Distribution of Tensile Strength

For section YY (Passing through edge of plate)

Stresses at x = 0.5h = 100mm from end

M = (100 × 75 × 5 × 75/2) = 14 × 10^5Nmm

V = –(100 × 75 × 5) = 37500N (acting towards end of the beam), H = 0

$$f_v = -5\left(\frac{14\times10^5}{100\times200^2}\right) + 0 = -1.75\text{N/mm}^2,\ f_h = +5\text{N/mm}^2$$

$$\tau = 1.25\left(\frac{-37500}{100\times200}\right) = -2.35\text{N/mm}^2$$

$$\text{Principal tensile stress} = \left(\frac{5-1.75}{2}\right) - \frac{1}{2}\sqrt{(5+1.75)^2+4(-2.35)^2} = -2.475\text{N/mm}^2$$

Angle of inclination of the plane of principal stress with respect to the vertical plane is,

$$\tan 2\theta = \left(\frac{-2\times2.35}{-1.75-5.0}\right) = 0.7,\ 2\theta = \tan^{-1}(0.7) = 35^\circ,\ \theta = 17.5^\circ$$

Tensile stress component in the vertical direction = 2.475 × sec 17.5° = 2.6N/mm²

Bursting tension, F_{bst} = [(2/3 × 150 × 2.6)100] = 26000N (on axis YY)

Problem 8.4

The end block of a pre–stressed concrete beam, 100mm wide and 200mm deep, supports an eccentric pre–stressing force of 100kN, the line of action which coincides with the bottom kern of the section. The depth of the anchor plate 50mm. Find the magnitude and position of the principal tensile stress on the horizontal plane passing through the centre of the anchor plate.

Solution:

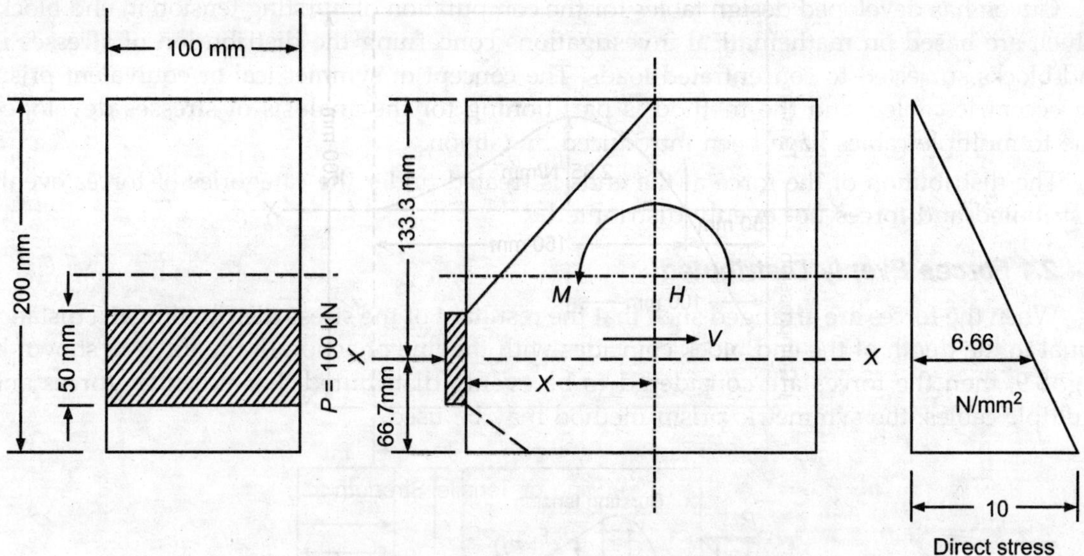

Fig. 8.8 Forces Acting on the End Block

Stresses on XX

$$M = \left[\left(\frac{1}{2}\times 6.66\times 133.3\times 100\right)1/3\times 133.3\right] - [(50\times 10^3)\ 12.5] = 1345\times 10^3\text{Nmm}$$

$$V = \left[\left(\frac{1}{2}\times 6.66\times 133.3\times 100\right) - (50\times 10^3)\right] = -5612\text{N}$$

$$H = 0$$

At $x = 0.5h = 100\text{mm from end}$

$$f_h = +6.666\text{N/mm}^2$$

$$f_v = -5\left(\frac{1345\times 10^3}{100\times 200^2}\right) = -1.66\text{N/mm}^2$$

$$\tau = 1.25\left(\frac{-5612}{100\times 200}\right) = -0.35\text{N/mm}^2$$

Principal tensile stress,

$$f_{min} = \left(\frac{6.66-1.66}{2}\right) - \frac{1}{2}\sqrt{(6.66+1.66)^2 + 4(-0.35)^2} = -1.7\text{N/mm}^2$$

Assuming the magnitude of the tensile stress in the vertical direction also as 1.7N/mm^2, bursting tension, $F_{bst} = [(2/3\times 150\times 1.7)100] = 17000\text{N}$.

8.4.2 Guyon's Method

Guyon has developed design tables for the computation of bursting tension in end blocks which are based on mathematical investigations concerning the distribution of stresses in end blocks subjected to concentrated loads. The concept of symmetrical or equivalent prism for eccentric cables, and the method is partitioning for the analysis of stresses developed due to multiple cables have been introduced by Guyon.

The distribution of the force at the ends is treated under the categories of forces evenly distributed and forces not evenly distributed.

8.4.2.1 Forces Evenly Distributed

When the forces are arranged such that the resultant of the stress distribution at a distance equal to the depth of the end block coincides with the line of action of the force as shown in Fig. 8.9, then the forces are considered to be evenly distributed. For eccentric forces and multiple cables, the symmetric prism method may be used.

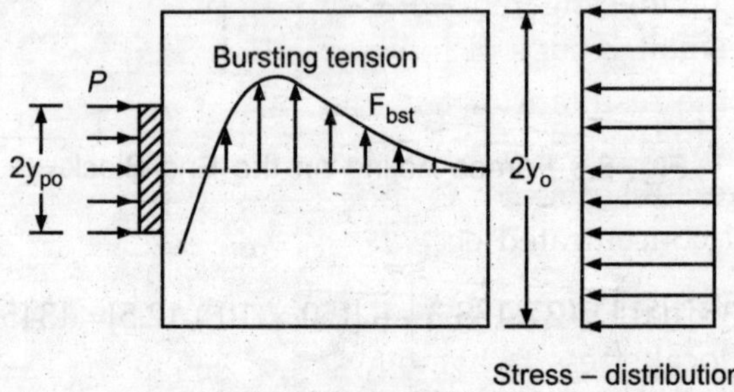

Fig. 8.9 Evenly Distributed Force Systems (Guyon)

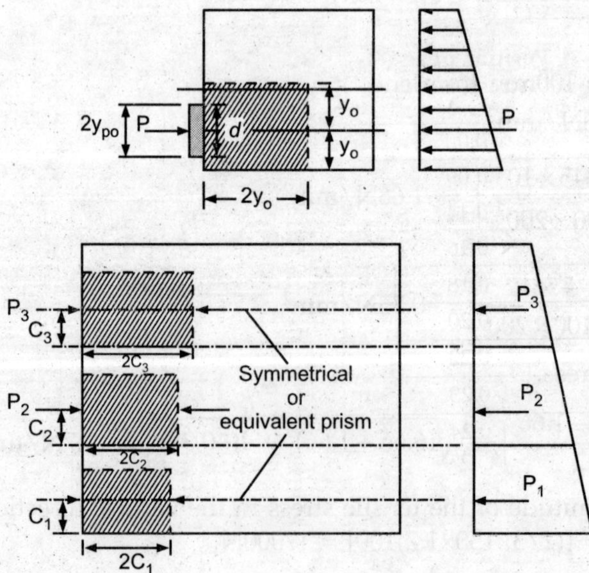

Fig. 8.10 Evenly Distributed Force Systems with Equivalent Prisms

This consist of a prism of concrete of side equal to twice the distance of pre-stressing force from the nearest free edge as shown in Fig. 8.10.

The position of zero stresses, maximum transverse stress and its magnitude for the forces which are evenly distributed are computed by using the coefficients given in the Table 8.3 under the category of distributed axial forces. According to Guyon, the bursting tension is expressed as,

$$F_{bst} = 0.3P[1-(y_{po}/y_o)^{0.58}] \quad ...(8.8)$$

Where, P = Anchorage force

y_{po}/y_o = Distribution ratio ...(8.9)

Where, $2y_{po}$ = Depth of the anchorage plate, $2y_o$ = Depth of the equivalent prism.

8.4.2.2 Forces Not Evenly Distributed

When it is not possible to arrange the end forces evenly, Guyon recommended that transverse tensile stresses be investigated along successive resultants, such as (i) resultant of all forces, (ii) resultant of small groups of forces and (iii) lines of action of individual forces.

The line of action of the resultant force is taken as the axis of an equivalent prism of length and depth equal to twice the distance of axis from the free edge or adjoining equivalent prism. The transverses stress distribution is computed by using the coefficients given Table 8.4, under the category of concentrated eccentric force and eccentric shear force.

Table 8.4 Vertical Stresses along Axis at Ends of Pre-stressed Beams (Guyon)

(Ratios of local stress to average stress across whole section)

+ = compression; – = tension

(a) Distributed Axial Force

Distribution ratio y_{po}/y_o	Position of zero stress, $x/2y_o$	Position of maximum stress, $x/2y_o$	Ratio of maximum tensile stress to average stress
0.00	0.00	0.17	0.50
0.10	0.09	0.24	0.43
0.20	0.14	0.30	0.36
0.30	0.16	0.36	0.33
0.40	0.18	0.39	0.27
0.50	0.20	0.43	0.23
0.60	0.22	0.44	0.18
0.70	0.23	0.45	0.13
0.80	0.24	0.46	0.09

(b) Concentrated Eccentric Force

Eccentricity of force, $e/2y_o$	Distance of stress from end of beam, $x/2y_o$						
	0	1/12	1/6	1/4	1/3	1/2	3/4
+1/2	−2.187	−0.913	−0.428	−0.014	+0.307	+0.399	+0.192
3/8	−1.222	−0.601	+0.125	+0.192	+0.250	+0.242	+0.122
+1/4	−0.758	−0.025	+0.238	+0.154	+0.062	−0.024	−0.016
+1/8	−0.566	+1.004	+0.074	−0.144	−0.266	−0.262	−0.128
0	0.000	−0.448	−0.500	0.462	−0.423	−0.314	−0.161
−1/8	−0.566	+1.004	+0.074	−0.144	−0.266	−0.262	−0.128
−1/4	−0.758	−0.025	+0.238	+0.154	+0.062	−0.024	−0.016
−3/8	−1.222	−0.601	+0.125	+0.192	+0.250	+0.242	+0.122
−1/2	−2.187	−0.913	−0.428	−0.014	+0.307	+0.399	+0.192

(c) Eccentric Shear Force

Eccentricity of force $e/2y_o$	Distance of stress from end of beam, $x/2y_o$						
	0	1/12	1/6	1/4	1/3	1/2	3/4
+1/2	—	0	0	0	0	0	0
3/8	+5.66	+2.96	+0.87	+0.19	−0.05	−0.14	−0.07
+1/4	+4.00	+3.10	+1.52	+0.44	−0.22	−0.20	−0.10
+1/8	+5.66	+2.96	+0.87	+0.19	−0.05	−0.14	−0.07
0	—	0	0	0	0	0	0
−1/8	−5.66	−2.96	−0.87	−0.18	+0.05	+0.14	+0.07
−1/4	−4.00	−3.10	−1.52	−0.44	+0.22	+0.20	+0.10
−3/8	−5.66	−2.96	−0.87	−0.18	+0.05	+0.14	+0.07
−1/2	—	0	0	0	0	0	0

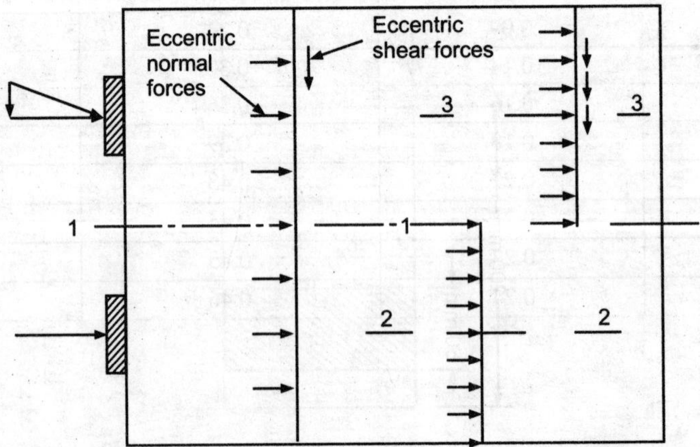

Fig. 8.11 Distribution of Normal and Shear Forces

Since the coefficients are applicable for forces at intervals of one–eighth of the prism depth, the end forces have to be replaced by a statically, equivalent system of normal and shear forces acting at these regular intervals as shown in Fig. 8.11.

8.4.2.3 Solved Examples

Problem 8.5

Using Guyon's method, compute the position and magnitude of maximum tensile stress and bursting tension for the end block with concentric anchor force of 100kN, The end block of a pre–stressed concrete beam, rectangular in section is 100mm wide and 200mm deep. The pre–stressing force transmitted to concrete by a distribution plate, 100mm wide and 50mm deep, concentrically placed at the ends.

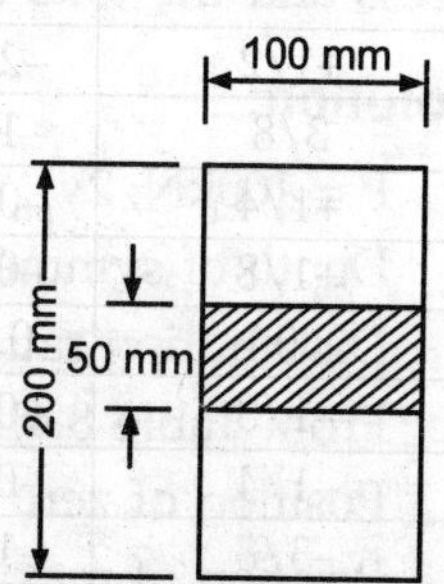

Fig. 8.12 Rectangular End Block with Concentric Distribution Plate

Solution:

$P = 100\text{kN}, 2y_{po} = 50\text{mm}, 2y_o = 200\text{mm}$

Distribution ratio = $y_{po}/y_o = 0.25$, from Table 8.4,

Position of zero stress from the end face = $0.15(2y_o) = 0.15(200) = 30\text{mm}$

Position of maximum stress = $0.33(2y_o) = 0.33(200) = 66\text{mm}$

$$\text{Maximum tensile stress} = 0.345(P/A) = 0.345\left(\frac{100\times10^3}{200\times100}\right) = 1.725\text{N/mm}^2$$

Bursting tension, $F_{bst} = 0.3P[1-(y_{po}/y_o)^{0.58}] = 0.3 \times 100 \times 10^3[1-(0.25)^{0.58}] = 16575\text{N}$

If the yield stress in the mild steel = 260N/mm², then

$$\text{Area of steel required} = \left[\frac{16575}{0.87\times260}\right] = 73\text{mm}^2$$

Problem 8.6

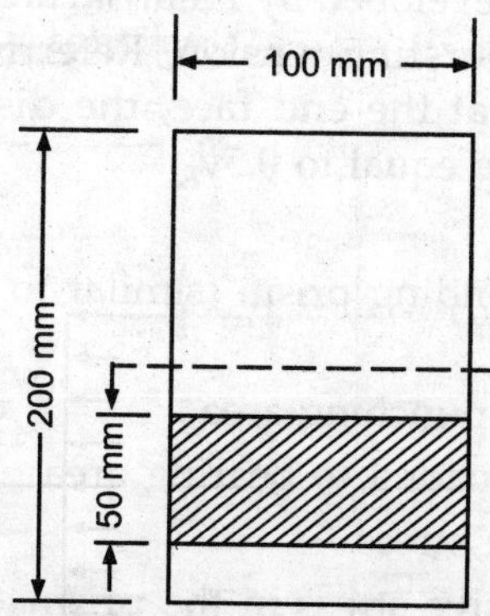

Fig. 8.13 Rectangular End Block With Eccentric Distribution Plate

The end block of a pre–stressed concrete beam, 100mm wide and 200mm deep, supports an eccentric pre–stressing force of 100kN, the line of action which coincides with the bottom kern of the section. The depth of the anchor plate 50mm. Compute the maximum tensile stress and the total splitting tension using Guyon's method.

Solution:

$P = 100kN$, $2y_{po} = 50mm$,

Depth of symmetric prism, $2y_o = 133mm$

Distribution ratio = $y_{po}/y_o = (50/133) = 0.375$,

From Table 8.4,

Position of zero stress = $0.175(2y_o) = 0.175(133) = 23.5mm$

Position of maximum stress = $0.382(2y_o) = 0.382(133) = 51mm$

$$\text{Maximum tensile stress} = 0.285(P/A) = 0.285\left(\frac{100\times10^3}{100\times133}\right) = 2.13\text{N/mm}^2$$

∴ Total splitting tension,

$$F_{bst} = 0.3P[1-(y_{po}/y_o)^{0.58}] = 0.3 \times 100 \times 10^3[1 - (0.375)^{0.58}] = 13015N$$

8.4.3 Zielinski and Rowe's Method

Experimental investigations on concrete prismatic specimens were conducted by Zielinski and Rowe using the technique of surface strain measurements. The concrete prism simulated the end blocks and the parameters investigated include the ratio of loaded to cross sectional area, the cable duct or hole, type of anchorages and the, cracking and ultimate load. The studies revealed that the distribution of transverse stress and ultimate load of the end block is not significantly affected by,

- The anchorage being either embedded or external
- The material of the anchorage, and
- The method of anchoring of wires.

Empirical relations have been developed by Zielinski and Rowe to compute the maximum transverse tensile stress and the bursting tension. Referring figure, when an end block is subjected to a concentrated load at the end face, the distribution of transverse stress is found to be maximum at a distance equal to $0.5y_o$.

Using the following notations,

$2y_o$ = Side of the surrounding prism (similar to the equivalent prism of Guyon's method)

$2y_{po}$ = Side of loaded or punching area

y_{po}/y_o = Ratio of sides of loaded to bearing area

f_v = Transverse tensile stress

f_c = Average compressive stress in the prism

P_k = Applied compressive force on the end block (tendon jacking force)

F_{bst} = Bursting tension

$f_{v(max)}$ = Maximum transverse tensile stress

The recommended equations are,

$$\text{Tensile stress, } f_{v(max)} = f_c\left[0.98 - 0.825\left(\frac{y_{po}}{y_o}\right)\right] \quad ...(8.10)$$

valid for ratio of $(y_{po}/y_o) = 0.3$ to 0.7

$$\text{Bursting tension, } F_{bst} = P_k\left[0.48 - 0.4\left(\frac{y_{po}}{y_o}\right)\right] \quad ...(8.11)$$

If allowance is made for tension taken by concrete, the corrected value of the bursting tension is given by

$$F_{bst}(\text{corrected}) = F_{bst}\left[1 - \left(\frac{f_t}{f_{v(max)}}\right)^2\right] \quad ...(8.12)$$

Where, f_t = Permissible tensile stress of concrete

The reinforcement required to resist the bursting tension is to be arranges between $0.2y_o$ to $2y_o$ where the intensity of stress is maximum.

8.4.3.1 Solved Examples

Problem 8.7

Estimate the position and magnitude of the maximum transverse tensile stress and bursting tension for the end block with a concentric anchor force of 100kN, using Rowe's method. The end block of a pre–stressed concrete beam, rectangular in section is 100mm wide and 200mm deep. The pre–stressing force transmitted to concrete by a distribution plate, 100mm wide and 50mm deep, concentrically placed at the ends.

Solution:

$P = 100\text{kN}$, $2y_{po} = 50\text{mm}$, $2y_o = 100\text{mm}$, $\therefore$ Distribution ratio $= y_{po}/y_o = 0.5$,

$$\text{Average compressive stress, } f_c = \left(\frac{100\times10^3}{100\times100}\right) = 10\text{N/mm}^2$$

$$F_{v(max)} = f_c\left[0.98 - 0.825\left(\frac{y_{po}}{y_o}\right)\right] = 10\,[0.98 - 0.825(0.5)] = 5.68\text{N/mm}^2$$

Acting at a distance equal to $(0.5 \times 50) = 25\text{mm}$ from the end face. The bursting tension is given by, $F_{bst} = P_k\left[0.48 - 0.4\left(\frac{y_{po}}{y_o}\right)\right] = F_{bst} = (100 \times 10^3)\,[0.48 - 0.4(0.5)] = 28000\text{N}$

If the permissible tensile stress in concrete is assumed to be 2N/mm^2, the corrected value of the bursting tension is given by

$$F_{bst}(\text{corrected}) = F_{bst}\left[1-\left(\frac{f_t}{f_{v(max)}}\right)^2\right] = 28000\left[1-\left(\frac{2}{5.68}\right)^2\right] = 24500N$$

Problem 8.8

The end block of a pre–stressed beam, 200mm wide and 300mm deep, has two Freyssinet anchorages (100mm diameter) with their centres at 75mm from the top and bottom of the beam. The force transmitted by each anchorage being 200kN, estimate the maximum tensile stress and the bursting tension developed.

Solution:

Anchorage diameter = 100mm

Equivalent side of square, $2y_{po} = \sqrt{\frac{\pi}{4}\times 100^2} = 89mm$

Side of the surrounding prism $2y_o = 150mm$

Therefore, $y_{po}/y_o = 89/150 = 0.593$,

Average compressive stress, $f_c = \left(\frac{200\times 10^3}{150\times 150}\right) = 8.9N/mm^2$

Tensile stress,

$$F_{v(max)} = f_c\left[0.98 - 0.825\left(\frac{y_{po}}{y_o}\right)\right] = 8.9[0.98 - 0.825(0.593)] = 4.45N/mm^2$$

Acting at a distance equal to (0.5 × 50) = 25mm from the end face.

The bursting tension,

$$F_{bst} = P_k\left[0.48 - 0.4\left(\frac{y_{po}}{y_o}\right)\right] = (200\times 10^3)\,[0.48 - 0.4(0.593)] = 50000N = 50kN$$

8.4.1 Indian Code Provisions (IS: 1343–2012)

8.4.4.1 Bearing Stress

(a) On the areas immediately behind external anchorages, the permissible unit bearing stress on the concrete, after accounting for all lossesdue to relaxation of steel, elastic shortening, creep of concrete, slip and/or seating of anchorages, etc, shall not exceed

$0.48f_{ci}\sqrt{\frac{A_{br}}{A_{pun}}}$ or $0.8f_{ci}$ whichever is smaller, where f_{ci} is the sun cube strength at transfer, A_{br} is the bearing area and A_{pun} is the punching area.

(b) During tensioning, the allowable bearing stress specified in a) may be increased by 25 percent, provided that this temporary value does not exceed f_{ci}.

(c) The bearing stress specified in (a) and (b) for permanent and temporary bearing stress may be increased suitably if adequate hoop reinforcement complying with IS: 456–2000 is provided at the anchorages.

(d) When the anchorages are embedded in concrete, the bearing stress shall be investigated after accounting for the surface friction between the anchorage and the concrete.

(e) The effective punching area shall generally be the contact area of the anchorage devices which, if circular in shape, shall be replaced by a square of equivalent area. The bearing area shall be the maximum area of that portion of the member which is geometrically similar and concentric to the effective punching area.

(f) Where a number of anchorages are used, the bearing area A_{br} shall not overlap. Where there is already a compressive stress prevailing over the bearing area, as in the case of anchorage placed in the body of a structure the total stress shall not exceed the limiting values specified in (a), (b) and (c). For stage stressing of cables, the adjacent unstressed anchorages shall be neglected when determining the bearing area.

8.4.4.2 Bursting Tensile Forces

a. The bursting tensile forces in the end blocks, or regions of bonded post–tensioned members, should be assessed on the basis of the tendon jacking load. For un–bonded members, the bursting tensile forces should be assessed on the basis of the tendon jacking load or the load in the tendon at the limit state of collapse, whichever is greater.

The bursting tensile force, F_{bst} existing in an individual square end block loaded by a symmetrically placed square anchorage or bearing plate, may be derived from the equation below:

$$\frac{F_{bst}}{P_k} = 0.32 - 0.3\frac{y_{po}}{y_o} \qquad ...(8.13)$$

Where,

F_{bst} = Bursting tensile force,

P_k = Load in the tendon assessed as above,

y_{po} = Side of loaded area, and

y_o = Side of end block.

b. The force F_{bst} will be distributed in a region extending from 0.1 y_o to y_o from the loaded face of the end block. Reinforcement provided to sustain the bursting tensile force may be assumed to be acting at its design strength (0.87 times characteristic strength of reinforcement) except that the stress should be limited to a value corresponding to a strain of 0.001 when the concrete cover to the reinforcement is less than 50 mm.

c. In rectangular end blocks, the bursting tensile forces in the two principal directions should be assessed on the basis of 18.6.2.2 of IS: 1343–2012. When circular anchorage or bearing plates are used, the side of the equivalent square area should be used.

Where groups of anchorages or bearing plates occur, the end blocks should be divided into a series of symmetrically loaded prisms and each prism treated in the above manner. For designing end blocks having a cross–section different in shape from that of the general cross–section of the beam, reference should he made to specialist literature.

d. Compliance with the requirements of (a), (b) and (c) will generally ensure that bursting tensile forces along the load axis are provided for. Alternative methods of design which make allowance for the tensile strength of the concrete may be used, in which case reference should be made to specialist literature.

e. Consideration should also be given to the spalling tensile stresses that occur in end blocks where the anchorage or bearing plate.

8.5 ANCHORAGE ZONE REINFORCEMENT

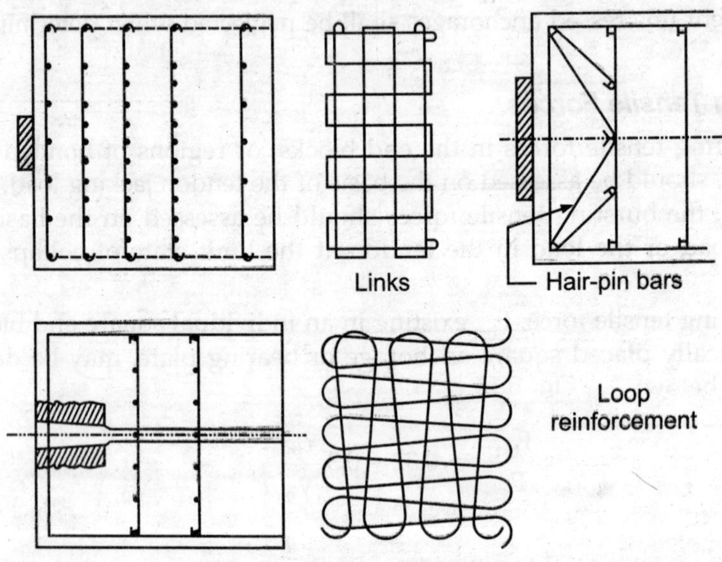

Fig. 8.14 Arrangements of Reinforcements in End Blocks

The main reinforcement in the anchorage zone should be designed to with stand the bursting tension, which is determined by the transverse stress distribution on the critical axis, usually coinciding with the line of action of the largest individual force. For plate and embedded (Freyssinet) type of anchorages, the typical arrangement of reinforcement in the end blocks is shown in Fig. 8.14. Mats, helics, loops or links are generally provided in perpendicular directions. Based on several tests, helical reinforcement is more efficient than mat reinforcement. In view of short available bond lengths, loops, hooks or right angle bents are necessary, even with deformed bars.

In case where spalling or secondary tension develops at the corners, suitable steel in the form of hair–pin bars should be provided to prevent the failure of corner zones. Suitable pockets are generally provided behind the anchorages so that the secondary reinforcements can be bent as shown in Fig. 8.15, and the pocket filled with mortar after pre–stressing operations. There must always be enough space for the fixing and handling of the hydraulic jack, especially at the soffits of beams when using cap cables, and this should be considered while designing the formwork. (Fig. 8.16)

In case of end blocks, where bearing plates are positioned closed to the edges of block as shown in Fig. 8.17, the steel cage should be arranged so that the bearing plates do not overlap with it. This precaution is necessary to prevent the spalling of concrete at the corners during stressing due to the different elastic modulus of the plane containing the reinforcement.

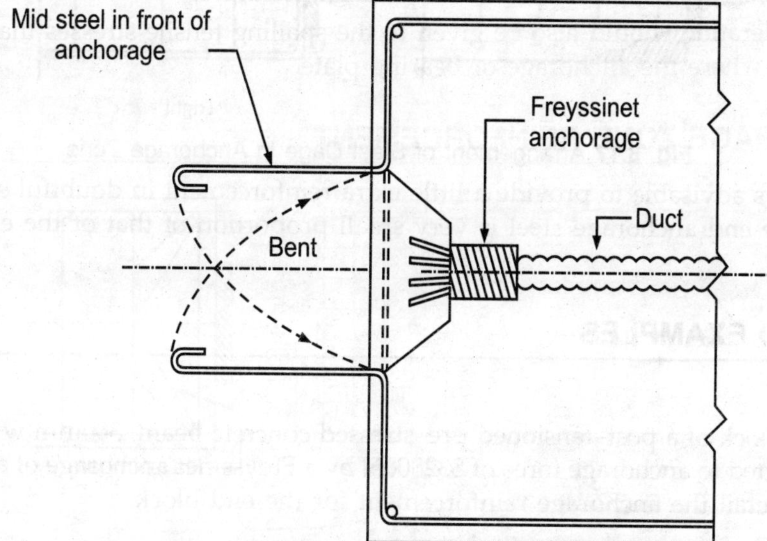

Fig. 8.15 Pockets behind Anchorages

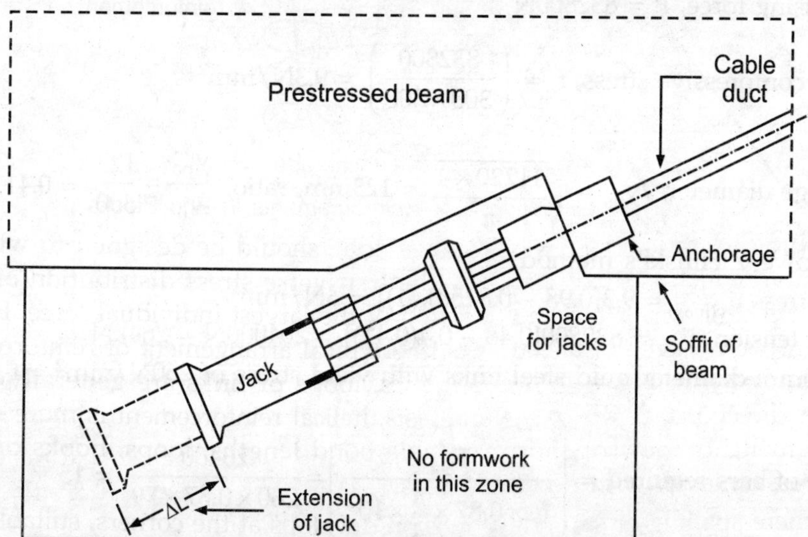

Fig. 8.16 Provision for Jack at Soffit or Beam

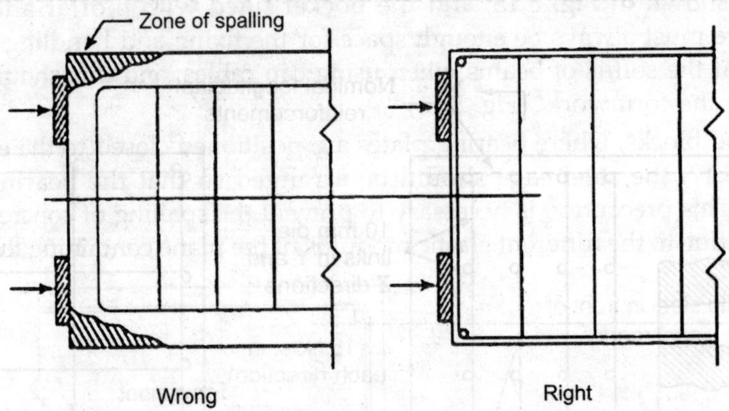

Fig. 8.17 Arrangement of Steel Cage in Anchorage Zone

It is always advisable to provide a little extra reinforcement in doubtful situations, since the cost of the end anchorage steel is very small proportion of that of the entire structural member.

8.6 SOLVED EXAMPLES

Problem 8.9

The end block of a post–tensioned pre–stressed concrete beam, 300mm wide and 300mm deep, is subjected to anchorage force of 832800N by a Freyssinet anchorage of area 11720 mm^2. Design and detail the anchorage reinforcement for the end block.

Solution:

Pre–stressing force, P = 832800N

Average compressive stress, $f_c = \left(\dfrac{832800}{300 \times 300}\right) = 9.3\text{N/mm}^2$

Anchorage diameter, $2y_{po} = \sqrt{\dfrac{11720 \times 4}{\pi}} = 123\text{mm}$, ratio, $\dfrac{y_{po}}{y_o} = \dfrac{123}{300} = 0.41$

Using Rowe–Zielinski's method,

Tensile stress, $f_{v(max)} = 9.3[0.98 - 0.825(0.41)] = 6\text{N/mm}^2$

Bursting tension, $F_{bst} = 832800[0.48 - 0.4(0.41)] = 264000\text{N} = 264\text{kN}$

Using 10mm diameter mild steel links with yield stress of $260\text{N/mm}^2 = f_y$

$$\text{Number of bars required} = \left(\frac{F_{bst}}{f_y \times 0.87 \times \frac{\pi}{4} \times 10^2}\right) = \frac{26400}{260 \times 0.87 \times 79} = 15$$

The reinforcements is to be arranged in the zone between $0.2y_o = (0.2 \times 150) = 30\text{mm}$ and

y_o = 150mm. The arrangement of reinforcement in the two perpendicular directions is shown in Fig. 8.18.

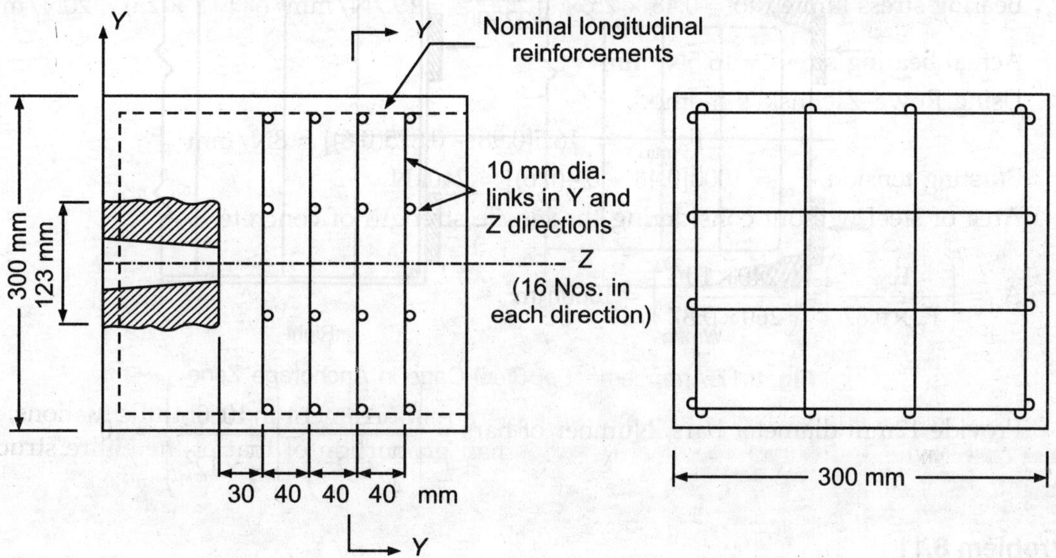

Fig. 8.18 Arrangement of Anchorage Zone Reinforcement

Problem 8.10

The end block of a post–tensioned pre–stressed member is 550mm wide and 550mm deep. Four cables, each made up of 7 wires of 12mm diameter strands carrying a force of 1000kN, are anchored by plate anchorages, 150mm by 150mm, located with their corners, at 125mm from the edges of the end block. The cable duct is of 50mm diameter. The 28 days cube strength of concrete = f_{cu} = 45N/mm^2. The cube strength of concrete at transfer = f_{ci} = 25N/mm^2. Permissible bearing stresses behind anchorages should confirm IS: 1343–2012. The characteristic yield stress in mild steel anchorage reinforcement is 260N/mm^2. Design suitable anchorages for the end block.

Solution:

P_k = 1000kN, $2y_{po}$ = 150mm, $2y_o$ = 250mm, y_{po}/y_o = 0.6

$$\text{Area of the cable duct} = \left(\frac{\pi \times 50^2}{4}\right) = 2000\text{mm}^2$$

$$\text{Net area of the surrounding prism, } A_{br} = [(250)^2 - (2000)] = 60500\text{mm}^2$$

$$\text{Average compressive stress, } f_c = \left(\frac{1000 \times 10^3}{60500}\right) = 16.5\text{N/mm}^2$$

According to IS: 1343, the bearing stress shall not exceed $0.48f_{ci}\sqrt{\frac{A_{br}}{A_{pun}}}$ or $0.8f_{ci}$, whichever is smaller, where,

A_{br} = Bearing area = 60500mm^2, A_{pun} = Punching area = 22500mm^2

Bearing stress limited to $= 0.48 \times 25 \times \sqrt{\dfrac{60500}{22500}} = 19.7\text{N/mm}^2$ or $(0.8 \times 25) = 20\text{N/mm}^2$

Actual bearing stress = 16.5N/mm^2

Using Rowe–Zielinski's method,

$$F_{v(max)} = 16.5[0.98 - 0.825(0.6)] = 8\text{N/mm}^2$$

Bursting tension, $F_{bst} = 1000[0.48 - 0.4(0.6)] = 240\text{kN}$

Area of steel without considering the tensile strength of concrete,

$$A_{st} = \left(\frac{F_{bst}}{F_y \times 0.87}\right) = \left(\frac{240 \times 10^3}{260 \times 0.87}\right) = 1060\text{mm}^2,$$

Provide 12mm diameter bars, Number of bars $= \left(\dfrac{A_{st}}{\dfrac{\pi}{4} \times 12^2}\right) = \left(\dfrac{1060}{\dfrac{\pi}{4} \times 12^2}\right) = 10.$

Problem 8.11

The end block of a post–tensioned beam is 80mm wide and 160mm deep. A pre–stressing wire, 7mm in diameter, stressed to 1200N/mm^2 has to be anchored against the end block at the centre. The anchorage plate is 50mm by 50mm. The wire bears on the plate through a female cone of 20mm diameter. Given the cube strength of concrete at transfer = f_{ci} = 25N/mm^2 and the permissible shear in steel as 94.5 N/mm^2, determine the thickness of the plate.

Solution:

Force in wire $= \left(\dfrac{38.5 \times 1200}{1000}\right) = 46.2$ kN,

$2y_{po} = 50\text{mm}$, $2y_o = 80\text{mm}$, $y_{po}/y_o = 0.625$

Average compressive stress, $f_c = \left(\dfrac{46.2 \times 10^3}{80 \times 80}\right) = 7.25\text{N/mm}^2$

Permissible bearing stress $= 0.48 \times 25 \times \sqrt{\dfrac{80 \times 80}{50 \times 50}} = 15.36\text{N/mm}^2$

or $(0.8 \times 20) = 16\text{N/mm}^2$, whichever is smaller.

The actual bearing stress = 7.25N/mm^2

Female cone diameter = 20mm

Punching circumference $= \pi \times 20 = 62.86$mm

If t = thickness of anchorage plate,

$(62.86 \times 94.5 \times t) = 46.2 \times 10^3$

$\therefore$ t = 7.8mm. Use an anchorage plate of 8mm thickness.

Problem 8.12

The solid end block of a post–tensioned pre–stressed beam of 25m span, with three cables, each of 7–15 mm strands, tensioned to 1200kN shown in Fig. 8.19. The anchorage plates are square with a side length of 180mm. Design the end block for bursting forces and sketch the details of reinforcements according to the provisions of the Indian standard code IS: 1343–2012.

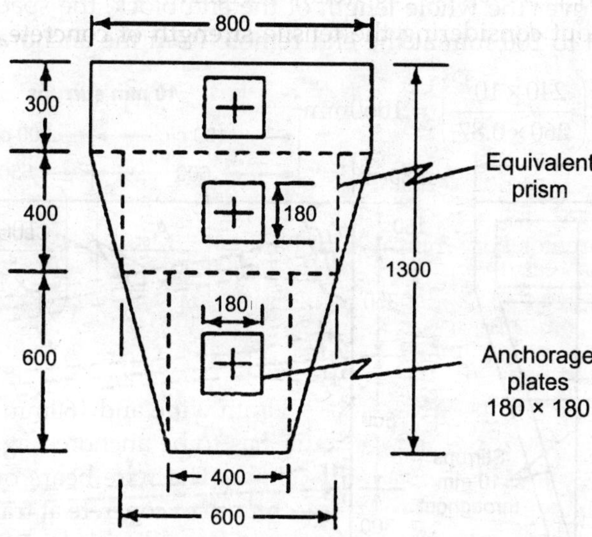

Fig. 8.19 End Block with Anchorages

Solution:

The end block has been divided into three equal parts, with one anchorage located approximately at the centroid of each area. The sections of the equivalent prisms corresponding to each other anchorage force are shown in Fig. 8.19. The bursting tension may be calculated for each prism as follows:

- Bottom Anchorage

For vertical bursting force $= \dfrac{y_{po}}{y_o} = \dfrac{180}{600} = 0.3$, as per IS: 1343–2012.

$$\frac{F_{bst}}{P_k} = 0.32 - 0.3\frac{y_{po}}{y_o},\ \frac{F_{bst}}{1200} = 0.32 - 0.3(0.3),\ F_{bst} = 276\text{kN}$$

Resistance of 5 – 10mm stirrups (100mm centres)

$$= \left(5 \times 2 \times \frac{\pi}{4} \times 10^2 \times 0.87 \times 415\right) = 285\text{kN}$$

For vertical bursting force $= \dfrac{y_{po}}{y_o} = \dfrac{180}{400} = 0.45$, $F_{bst} = 1200[0.32 - 0.3(0.45)] = 222\text{kN}$

Resistance of 4 – 10 mm stirrups (100mm centres)

$$= \left(4 \times 2 \times \frac{\pi}{4} \times 10^2 \times 0.87 \times 415\right) = 228\text{kN}$$

In the same way, the bursting force are evaluated for the centre and top anchorages and it is found that the required resistance can be provided by 10 mm stirrups at 100 mm centres. A possible arrangement of stirrups in the end block is shown in Fig. 8.20. Since the bursting forces do not extended over the whole length of the end block, the spacing of the stirrups is increased from 100 mm to 200 mm at the end remote from the anchorages.

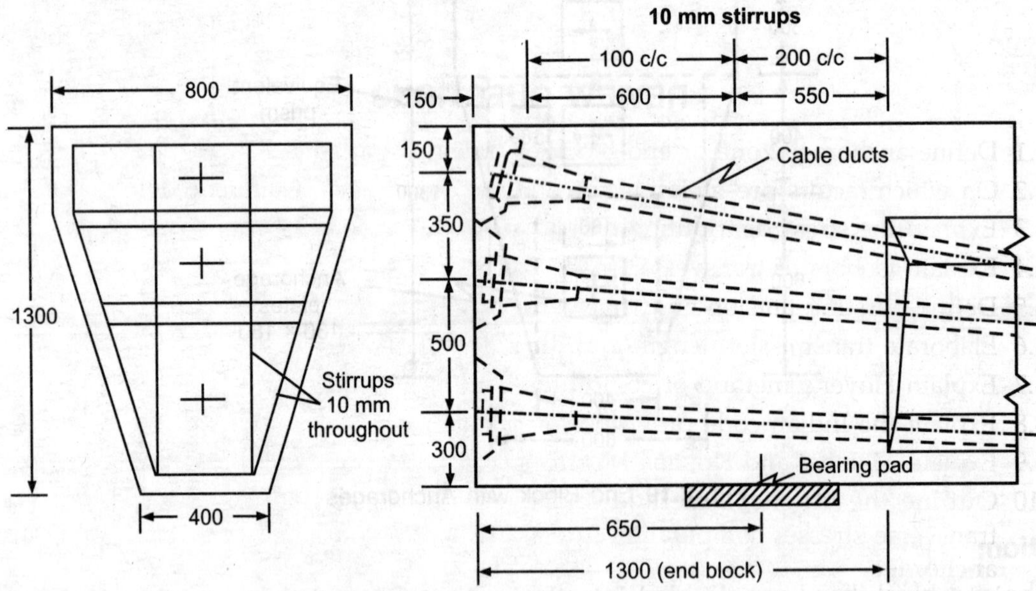

Fig. 8.20 Details of End Block Reinforcements

Problem 8.13

The end block of a post–tensioned bridge girder is 600mm wide by 1200 mm deep. Two cables, each comprising 97 high–tensile wires of 7mm diameter, are anchored using square anchor plates of side length of 410 mm with their centres located at 600 mm from the top and bottom edges of the beam. The jacking force in each cable is 4500kN. Design a suitable anchorage zone reinforcement for Fe 415 HYSD bars conforming to IS: 1343.

Solution:

Depth of the anchor plate = $2y_{po}$ = 410mm, Depth of the equivalent prism = $2y_o$ = 600mm

Jacking force in each cable = P_k = 4500kN

Ratio, $\dfrac{2y_{po}}{2y_o} = \dfrac{410}{600}$ = 0.683, as per IS: 1343–2012, bursting tension

$$F_{bst} = P_k\left[0.32 - 0.3\left(\frac{y_{po}}{y_o}\right)\right] = 4500\,[0.32 - 0.3(0.68)] = 517.5\text{kN}$$

Area of steel, $A_{st} = \left(\frac{F_{bst}}{F_y \times 0.87}\right) = \left(\frac{517.5}{415 \times 0.87}\right) = 1434\text{mm}^2$

Provide 10mm diameter bars, number of bars = $\left(\frac{A_{st}}{\frac{\pi}{4} \times 10^2}\right) = \left(\frac{1434}{79}\right) = 19$

Adopt 10 mm diameter bars at 150mm centres in the horizontal and vertical directions over a length of 600mm from the end face of the beam.

REVIEW QUESTIONS

8.1 Define anchorage zone or end block or transmission zone.

8.2 On which factors pre–stressing force in the anchorages is influenced?

8.3 Explain the stress distribution on end block.

8.4 Explain isobars of transverse tensile stress.

8.5 Define bursting tension and transverse tensile stress.

8.6 Elaborate transmission length and the factors affecting on it.

8.7 Explain Hoyer's method of computing transmission length.

8.8 Explain the Indian code provision for calculation of transmission length.

8.9 Explain Marshal and Krishna Murthy's computation method of transmission length.

8.10 Outline the Magnel's method and the coefficients used for the horizontal and transverse stresses computation in end blocks subjected to concentrated forces from anchorage.

8.11 Explain Guyon's method for evaluation of bursting force on end blocks when forces are evenly and not evenly distributed with multiple anchorages.

8.12 Explain Zielinski and Rowe's Method for computing the bursting force.

8.13 Outline the IS: 1343–2012 code provisions for evaluating bearing stress and bursting tensile force in case of end blocks.

8.14 Explain anchorage zone reinforcements with neat sketches.

8.15 Sketch the typical arrangement of reinforcements in end blocks of post–tensioned pre–stressed concrete beams with single and multiple anchorages.

EXERCISE PROBLEMS

8.1 A pre–tensioned beam of 160mm wide and 320mm deep, is pre–stressed by 4 plain wires of 7mm diameter at an eccentricity of 100mm. If the cube strength of concrete at transfer is 40MPa, estimate the transmission length at the ends of the pre–tensioned units using IS: 1343–2012 code provisions.

[Ans: Transmission length = 700mm]

8.2 A pre–tensioned beam of 200mm wide and 500mm deep, is pre–stressed by 7 wires of 15mm diameter strands at an eccentricity of 200mm. If the cube strength of concrete at transfer is 35MPa, estimate the transmission length at the ends of the pre–tensioned units using Indian code recommendations.

[Ans: Transmission length = 450mm]

8.3 Using Magnel's method, evaluate (a) vertical, (b) shear and (c) principal tensile stress stresses at a distance of 400mm from the end of the on the horizontal section (i) through the centre and (ii) at the level of edge of the anchor plate for the following data:

Pre–stressed concrete beam of 400mm wide by 800mm deep.

Anchor plate of 300mm wide and 200mm deep.

Pre–stressing force = 1000kN.

[Ans: (i) (a) = –1.10MPa, (b) = 1.46MPa, (c) = –1.55MPa
(ii) (a) = –1.86MPa, (b) = 0.00MPa, (c) = –1.86MPa]

8.4 The end block of a pre–stressed concrete beam, rectangular in section is of 120mm wide and 300mm deep. The pre–stressing force of 250kN is transmitted to concrete by a distributed plate, 120mm wide and 75mm deep, concentrically placed at the ends. Calculate the position and magnitude of the maximum tensile stress on the horizontal section through the centre of the end block using (i) Magnel, (ii) Guyon, and (iii) Rowe. Design the reinforcement for the end block for the maximum transverse tension. The yield stress in steel = 260MPa.

[Ans: (i) 3.3MPa (150mm), (ii) 2.415MPa (99mm),
(iii) 8MPa (30mm), A_s = 252mm^2]

8.5 A pre–stressing force of 250kN is transmitted through a distribution plate of 120mm by 120mm wide, the centre of which is located at 100mm from the bottom of the end block having a section 120mm wide and 300mm deep. Estimate the position and magnitude of the maximum tensile stress on a horizontal section passing through the centre of the distribution plate using the methods (i) Guyon, (ii) Magnel and (iii) Rowe. Calculate the area of steel necessary to resist the largest tensile force resulting from any of these methods. Yield stress in steel is 260N/mm^2.

[Ans: (i) 1.875MPa (88mm), (ii) 1.72MPa (150mm),
(iii) 5.04MPa (50mm), A_s = 265mm^2]

8.6 The end block of a pre–stressed concrete beam, 200mm wide and 400mm deep, has two anchor plates, 200mm by 50mm deep, at 80mm from the top and 200mm by 80mm deep located at 100mm from the bottom of the beam, transmitting forces of 250 and 300kN respectively.

(a) Evaluate the magnitude of the maximum tensile stress on a horizontal section using Guyon's method which passes through centre of the beam.

(b) Evaluate the maximum tensile stress on sections passing through the larger and smaller pre–stressing forces using Guyon's and Rowe's method

[Ans: (a) 5.515MPa, (b) 2.09MPa section through larger force 4.9MPa]

8.7 A Freyssinet anchorage (125 mm dia.) carrying 12 wires of 7mm diameter stressed to 950mm^2, is embedded concentrically in the web of an I–sections beam at the ends.

The thickness of 225mm. Evaluate (a) the maximum tensile stress, (b) the bursting force in the end block using Rowe's method. (c) Design the reinforcements for the end block.

[Ans: (a) 5.0MPa, (b) 125kN, (c) 550mm²]

8.8 The end block of a pre–stressed beam 500mm wide and 1050mm deep contains 6 Freyssinet cables, each carrying a force of 266kN anchored through 100mm diameter anchorages, which are spaced 150mm apart at the end of the beam. Calculate (i) the maximum tensile stress, (b) the bursting force in the end block using Rowe's method. (c) Design the reinforcements for the end block. Adopt yield stress of reinforcements as 260MPa.

[Ans: (a) 11.8MPa, (b) 65kN, (c) 287mm²]

8.9 A high tensile cable comprising 12 strands of 15mm diameter with an effective force of 2500kN is anchored concentrically in an end block of a post–tensioned beam. The end block is 400mm wide by 800mm deep and the anchor plate is 200mm by 260mm deep. Design suitable anchorage zone reinforcements using Fe 415 grade bars using Indian standard code provisions.

[Ans: Bursting tension = 425kN, A_{st} = 1177mm², 10mm diameter bars spaced at 100mm centres both ways over a length of 400mm from the end face]

8.10 The end block of a post–tensioned beam is 500mm wide by 1000mm deep. Two cables, each of 55 numbers of 7mm diameter high tensile wires carrying a force of 2800kN, are anchored using the anchor plate of side 305mm. The anchor plate centres are located symmetrically at 250mm from the top and bottom edges of the beam. Using Fe 415 grade high yield bars, design the suitable reinforcements in the end block using IS: 1343 recommendations.

[Ans: Bursting tension = 384kN, A_{st} = 1063mm², 10mm diameter bars spaced at 150mm centres both ways over a length of 500mm]

OBJECTIVE QUESTIONS

8.1 The length of the tendon from the free end to the point where it attains maximum stress is called .... length

a. elongation b. transmission
c. extraction d. contraction

8.2 Transmission length for plain and indented wire ...., where, ϕ = Diameter of wire

a. 100ϕ b. 85ϕ c. 65ϕ d. 30ϕ

8.3 Transmission length for crimped wired ...., where, ϕ = Diameter of wire

a. 100ϕ b. 85ϕ c. 65ϕ d. 30ϕ

8.4 Transmission length for plain and indented wire ...., where, ϕ = Diameter of wire

a. 100ϕ b. 85ϕ c. 65ϕ d. 30ϕ

8.5 Based on the wedge action, Hoyer has developed an expression for computing the transmission length, and is given by ...., Where, L_t = Transmission length, ϕ = Wire diameter, μ = Coefficient of friction between steel and concrete, ν_s = Poisson's ratio of concrete, ν_s = Poisson's ratio of steel, α_e = Modular ratio (E_s/E_c), E_c = Modulus of elasticity of concrete, E_s = Modulus of elasticity of steel, f_{pi} = Initial stress in steel, f_{pe} = Effective stress in steel.

a. $L_t = \frac{\phi}{2\mu}(1+\nu_c)\left(\frac{\alpha_e}{\nu_s}+\frac{f_{pi}}{E_c}\right)\left(\frac{f_{pe}}{2f_{pi}-f_{pe}}\right)$ b. $L_t = \frac{\phi}{2\mu}(1-\nu_c)\left(\frac{\alpha_e}{\nu_s}+\frac{f_{pi}}{E_c}\right)\left(\frac{f_{pe}}{2f_{pi}-f_{pe}}\right)$

c. $L_t = \frac{\phi}{2\mu}(1+\nu_c)\left(\frac{\alpha_e}{\nu_s}-\frac{f_{pi}}{E_c}\right)\left(\frac{f_{pe}}{2f_{pi}-f_{pe}}\right)$ d. $L_t = \frac{\phi}{2\mu}(1-\nu_c)\left(\frac{\alpha_e}{\nu_s}-\frac{f_{pi}}{E_c}\right)\left(\frac{f_{pe}}{2f_{pi}-f_{pe}}\right)$

8.6 Semi–empirical relations by Marshal and Krishna Murthy are applicable for both plain smooth wires and strands. The expression for transmission length is given by ...., Where, f_{cu} = Cube strength of concrete at transfer, expressed in MPa, L_t = Transmission length in mm, β = Constant depending upon the details of strand and wire

a. $L_t = \sqrt{\frac{\sqrt{f_{cu}}\times 10^2}{\beta}}$ b. $L_t = \sqrt{\frac{\sqrt{f_{cu}}\times 10^3}{\beta}}$

c. $L_t = \sqrt{\frac{\sqrt{f_{cu}}\times 10^6}{\beta}}$ d. $L_t = \sqrt{\frac{\sqrt{f_{cu}}\times 10^9}{\beta}}$

8.7 Value of constant β for Marshal and Krishna Murthy Relation for transmission length for 10mm dia., 7 wire strand is ....

a. 0.144 b. 0.0235 c. 0.0174 d. 0.058

8.8 What is the dimension of wire, which is having β = 0.058 for Marshal and Krishna Murthy Relation for transmission length?

a. 2mm dia. wire b. 10mm dia., 7 wire strand

c. 12.5mm dia., 7 wire strand d. 19mm dia., 7 wire strand

8.9 In Magnel's method for finding anchorage zone stresses, the horizontal stress distribution across the section can be approximated using .... equation. Where f_h is horizontal stress.

a. $f_h = \frac{P}{bh}\left(1+15\frac{e'^2}{h'^2}\right)$ b. $f_h = \frac{P}{bh}\left(1-15\frac{e'^2}{h'^2}\right)$

c. $f_h = \frac{P}{bh}\left(1-12\frac{e'^2}{h'^2}\right)$ d. $f_h = \frac{P}{bh}\left(1+12\frac{e'^2}{h'^2}\right)$

8.10 In Magnel's method for finding anchorage zone stresses, the vertical stress distribution across the section can be approximated using .... equation. Where f_v is vertical stress, M = Bending moment, H = Direct force (vertical), K_1, K_2 are Magnel's coefficients.

a. $f_v = K_1\left(\frac{M}{bh^2}\right) - K_2\left(\frac{H}{bh}\right)$ b. $f_v = K_1\left(\frac{M}{bh^2}\right) + K_2\left(\frac{H}{bh}\right)$

c. $f_v = K_1\left(\frac{bM}{h^2}\right) - K_2\left(\frac{H}{bh}\right)$ d. $K_1\left(\frac{bM}{h^2}\right) + K_2\left(\frac{H}{bh}\right)$

8.11 In Magnel's method for finding anchorage zone stresses, the shear stress distribution across the section can be approximated using .... equation. Where is vertical stress, V = Shear force (horizontal), K_3 is Magnel's coefficient.

a. $\tau = K_3\left(\frac{V}{bh}\right)$ b. $\tau = K_3\left(\frac{bh}{V}\right)$

c. $\tau = K_3\left(\frac{bV}{h}\right)$ d. $\tau = K_3\left(\frac{h}{bV}\right)$

8.12 In Magnel's method for finding anchorage zone stresses, Magnel's Coefficient K_2 = ... for distance from far end x/h = 0.30

a. –1.960 b. 1.960 c. 2.058 d. 2.160

8.13 In Magnel's method for finding anchorage zone stresses, Magnel's Coefficient K_1= .... for distance from far end x/h = 0.30

a. –1.960 b. 1.960 c. 2.058 d. 2.160

8.14 In Magnel's method for finding anchorage zone stresses, Magnel's Coefficient K_3= .... for distance from far end x/h = 0.30

a. –1.960 b. 1.960 c. 2.058 d. 2.160

8.15 In Magnel's method for finding anchorage zone stresses, Magnel's Coefficient K_3 = 0.768, find distance from far end x/h?

a. 0.40 b. 0.50 c. 0.60 d. 0.70

8.16 In Magnel's method for finding anchorage zone stresses, Magnel's Coefficient K_2 = 2.000, find distance from far end x/h?

a. 0.40 b. 0.50 c. 0.60 d. 0.70

8.17 In Magnel's method for finding anchorage zone stresses, Magnel's Coefficient K_1 = –3.240, find distance from far end x/h?

a. 0.40 b. 0.50 c. 0.60 d. 0.70

8.18 The principal stresses acting at the point are computed by general equations ...., where f_h and f_v are horizontal and vertical stresses and is shear stress.

a. f_{max} or $f_{min} = \left(\frac{f_v + f_h}{2}\right) \pm \frac{1}{2}\sqrt{(f_h + f_v)^2 + 4\tau^2}$

b. f_{max} or $f_{min} = \left(\frac{f_v + f_h}{2}\right) \pm \frac{1}{2}\sqrt{(f_h - f_v)^2 - 4\tau^2}$

c. f_{max} or f_{min} = $\left(\frac{f_v + f_h}{2}\right) \pm \frac{1}{2}\sqrt{(f_h - f_v)^2 + 4\tau^2}$

d. f_{max} or f_{min} = $\left(\frac{f_v + f_h}{2}\right) \pm \frac{1}{2}\sqrt{(f_h + f_v)^2 - 4\tau^2}$

8.19 The direction of principal stress is computed by ...., where f_h and f_v are horizontal and vertical stresses and is shear stress.

a. $\tan 2\theta = \left(\frac{2\tau}{f_v - f_h}\right)$ b. $\tan 2\theta = \left(\frac{2\tau}{f_v + f_h}\right)$

c. $\tan 2\theta = \left(\frac{f_v - f_h}{2\tau}\right)$ d. $\tan 2\theta = \left(\frac{f_v + f_h}{2\tau}\right)$

8.20 In an evenly distributed force system, according to Guyon's method of computation of anchorage zone stresses, the bursting tension is expressed as ...., Where, P = Anchorage force, y_{po}/y_o = Distribution ratio, $2y_{po}$= Depth of the anchorage plate, $2y_o$ = Depth of the equivalent prism.

a. $F_{bst} = 0.3P[1-(y_{po}/y_o)^{0.48}]$ b. $F_{bst} = 0.3P[1-(y_{po}/y_o)^{0.58}]$

c. $F_{bst} = 0.3P[1-(y_{po}/y_o)^{0.68}]$ d. $F_{bst} = 0.3P[1-(y_{po}/y_o)^{0.78}]$

8.21 When it is not possible to arrange the end forces evenly, Guyon recommended that transverse tensile stresses be investigated along successive resultants, such as (i) resultant of all forces, (ii) resultant of small groups of forces and (iii) lines of action of individual forces. Which of the above statement/s are correct?

a. (i) b. (i) (ii) c. (i) (iii) d. (i) (ii) (iii)

8.22 Experimental investigations on concrete prismatic specimens were conducted by Zielinski and Rowe using the technique of ....

a. surface strain measurements b. surface stress measurements

c. section strain measurements d. section stress measurements

8.23 In Zielinski and Rowe's Method of computation of anchorage zone stresses,distribution of transverse stress and ultimate load of the end block is not significantly affected by,

I. The anchorage being either embedded or external

II. The material of the anchorage, and

III. The method of anchoring of wires.

Which of the above statements are correct?

a. I, II b. I, III c. I, II, III d. II, III

8.24 Empirical relations have been developed by Zielinski and Rowe to compute the maximum transverse tensile stress ($f_{v(max)}$) is given by ...., where $2y_o$= Side of the surrounding prism, $2y_{po}$ = Side of loaded or punching area, y_{po}/y_o = Ratio of sides of loaded to bearing area, f_c = Average compressive stress in the prism.

a. $f_c\left[0.98-0.825\left(\frac{y_{po}}{y_o}\right)\right]$ b. $f_c\left[0.88-0.825\left(\frac{y_{po}}{y_o}\right)\right]$

c. $f_c\left[0.78-0.825\left(\frac{y_{po}}{y_o}\right)\right]$ d. $f_c\left[0.68-0.825\left(\frac{y_{po}}{y_o}\right)\right]$

8.25. Zielinski and Rowe empirical relation to compute the maximum transverse tensile stress is given by, $(f_{v(max)}) = f_c\left[0.98-0.825\left(\frac{y_{po}}{y_o}\right)\right]$, here y_{po}/y_o = Ratio of sides of loaded to bearing area ranges from ....

a. 0.3 to 0.4 b. 0.3 to 0.5 c. 0.3 to 0.6 d. 0.3 to 0.7

8.26 Zielinski and Rowe empirical relation to compute the bursting tension is given by ...., where, P_k = Applied compressive force on the end block (tendon jacking force), F_{bst} = Bursting tension, y_{po}/y_o = Ratio of sides of loaded to bearing area.

a. $F_{bst} = P_k\left[0.38-0.4\left(\frac{y_{po}}{y_o}\right)\right]$ b. $F_{bst} = P_k\left[0.48-0.4\left(\frac{y_{po}}{y_o}\right)\right]$

c. $F_{bst} = P_k\left[0.58-0.4\left(\frac{y_{po}}{y_o}\right)\right]$ d. $F_{bst} = P_k\left[0.68-0.4\left(\frac{y_{po}}{y_o}\right)\right]$

8.27 If allowance is made for tension taken by concrete, the corrected value of the bursting tension by Zielinski and Rowe's method is given by....., f_t = Permissible tensile stress of concrete and $f_{v(max)}$= Maximum transverse tensile stress.

a. $F_{bst}(\text{corrected}) = F_{bst}\left[1-\left(\frac{f_t}{f_{v(max)}}\right)^2\right]$ b. $F_{bst}(\text{corrected}) = F_{bst}\left[1-\left(\frac{f_t}{f_{v(max)}}\right)^3\right]$

c. $F_{bst}(\text{corrected}) = F_{bst}\left[1+\left(\frac{f_t}{f_{v(max)}}\right)^2\right]$ d. $F_{bst}(\text{corrected}) = F_{bst}\left[1+\left(\frac{f_t}{f_{v(max)}}\right)^3\right]$

8.28 According to Indian code provisions (IS: 1343–2012), bearing stress on the concrete, after accounting for all losses due to relaxation of steel, elastic shortening, creep of concrete, slip and/or seating of anchorages, etc, shall not exceed .... or .... whichever is smaller, where f_{ci} is the sun cube strength at transfer, A_{br} is the bearing area and A_{pun} is the punching area.

a. $0.8f_{ci}\sqrt{\frac{A_{br}}{A_{pun}}}, 0.48f_{ci}$ b. $0.58f_{ci}\sqrt{\frac{A_{br}}{A_{pun}}}, 0.48f_{ci}$

c. $0.48f_{ci}\sqrt{\dfrac{A_{br}}{A_{pun}}}, 0.8f_{ci}$ d. $0.48f_{ci}\sqrt{\dfrac{A_{br}}{A_{pun}}}, 0.58f_{ci}$

8.29 According to Indian code provisions (IS: 1343–2012), during tensioning, the allowable bearing stress $0.48f_{ci}\sqrt{\dfrac{A_{br}}{A_{pun}}}$ or $0.8f_{ci}$ whichever is less,may be increased by ....

a. 10% b. 15% c. 20% d. 25%

8.30 As per IS: 1343–2012, The bursting tensile force, F_{bst} existing in an individual square end block loaded by a symmetrically placed square anchorage or bearing plate, may be derived from the equation ...., where, F_{bst} = Bursting tensile force, P_k = Load in the tendon, y_{po} = Side of loaded area, and y_o = Side of end block.

a. $\dfrac{F_{bst}}{P_k} = 0.3 - 0.32\dfrac{y_{po}}{y_o}$ b. $\dfrac{F_{bst}}{P_k} = 0.32 - 0.3\dfrac{y_{po}}{y_o}$

c. $\dfrac{F_{bst}}{P_k} = 0.32 + 0.3\dfrac{y_{po}}{y_o}$ d. $\dfrac{F_{bst}}{P_k} = 0.3 + 0.32\dfrac{y_{po}}{y_o}$

Answers

8.1 b	8.2 a	8.3 c	8.4 d	8.5 c
8.6 b	8.7 a	8.8 c	8.9 d	8.10 b
8.11 a	8.12 b	8.13 a	8.14 c	8.15 c
8.16 b	8.17 d	8.18 c	8.19 a	8.20 b
8.21 d	8.22 a	8.23 c	8.24 a	8.25 d
8.26 b	8.27 a	8.28 c	8.29 d	8.30 b

□□

CHAPTER 9

Analysis and Design of PSC Continuous Beams

9.1 ANALYSIS OF PSC CONTINUOUS BEAMS

9.1.1 Methods of Achieving Continuity

Continuity in pre–stressed concrete construction is achieved by using a straight or curved cable which is continuous over several spans as shown in Fig. 9.1(a) and (b). It is also possible to develop continuity between two precast beams by using cap cables as shown in Fig. 9.1(c). Alternatively, short, straight tendons may be used over the supports to develop continuity between two precast pre–stressed beams as shown in Fig. 9.1(d).

Based on the method of construction, continuous beams may be classified as "fully continuous beam" in which tendons are generally continuous from one end to other, and "partially continuous beams" where each span is first precast as a simple beam and the elements are assembled to form a continuous member by using cap cables or short tendons over the supports.

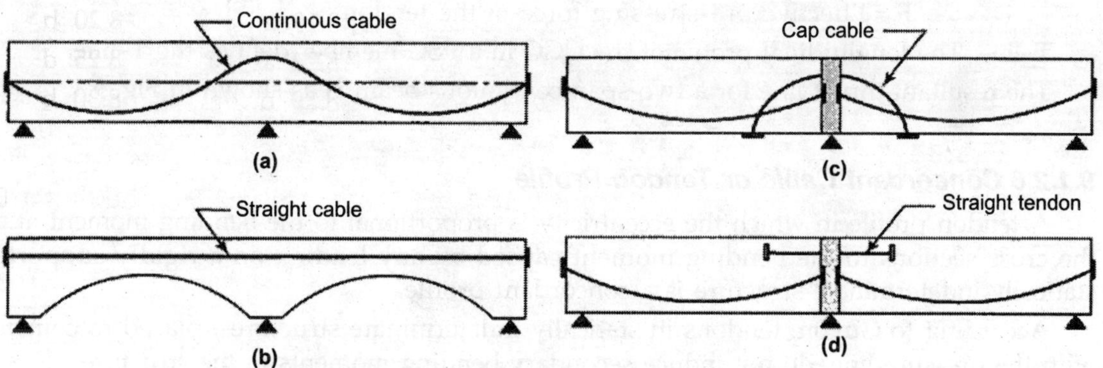

Fig. 9.1 Cable Layouts for Continuous Beams

9.1.2 Definition of Important Terms

9.1.2.1 Primary Moment (PM)

The primary moment at any section is defined by the product of F and e of the tendon at the given section. Referring to Fig. 9.2, PM = Fe, the variation of primary moment is identical to the variation of the eccentricity of the tendon force F. It is considered the effect of pre–stress alone.

9.1.2.2 Secondary Moment (Parasitic Bending Moment) (SM)

Secondary moments are additional moments induced at a section of a statically indeterminate structure due to the redundant reactions developed as a consequence of pre–stressing the structure. The variation of secondary moment in two span continuous beam is pre–stressed by a straight eccentric tendon is as shown in Fig. 9.2.

9.1.2.3 Resultant Moment (RM)

The resultant moment at any section is given by the sum of primary and secondary moments. i.e., RM = PM + SM.

9.1.2.4 Line of Pre–stress (CGS Line)

The locus of the centroid of the pre–stressing force along the structure is the line of pre–stress or centre of gravity of the steel line.

9.1.2.5 C–Line or Pressure Line or Thrust Line

The locus of the point at which the resultant thrust in concrete acts at all the section in a given PSC member is known as C–line. In a continuous beam, the distance of C–line from CGC at any section is given by $e = \frac{M}{F}$

Where, M = Regular bending moment across the given section

F = Effective pre–stressing force in the tendon

T–line: The longitudinal profile of the CGC in a PSC member defines the T–line.

The resultant thrust line for a two span continuous beam is as shown in Fig. 9.3.

9.1.2.6 Concordant Cable or Tendon Profile

A tendon profile in which the eccentricity is proportional to the bending moment at all the cross sections to the bending moment caused by any loading on a rigidly supported statically indeterminate structure is a concordant profile.

According to Guyon, tendons in statically indeterminate structures, placed to coincide with the pressure line, do not induce secondary bending moments in the structure.

The resultant trust line in a two span continuous beam, pre–stressed by a parabolic cable with zero eccentricity at all the supports is shown in Fig. 9.3 If a tendon profile is made to coincide with the resultant thrust line, the redundant reactions are completely eliminated and the cable profile may be considered as concordant.

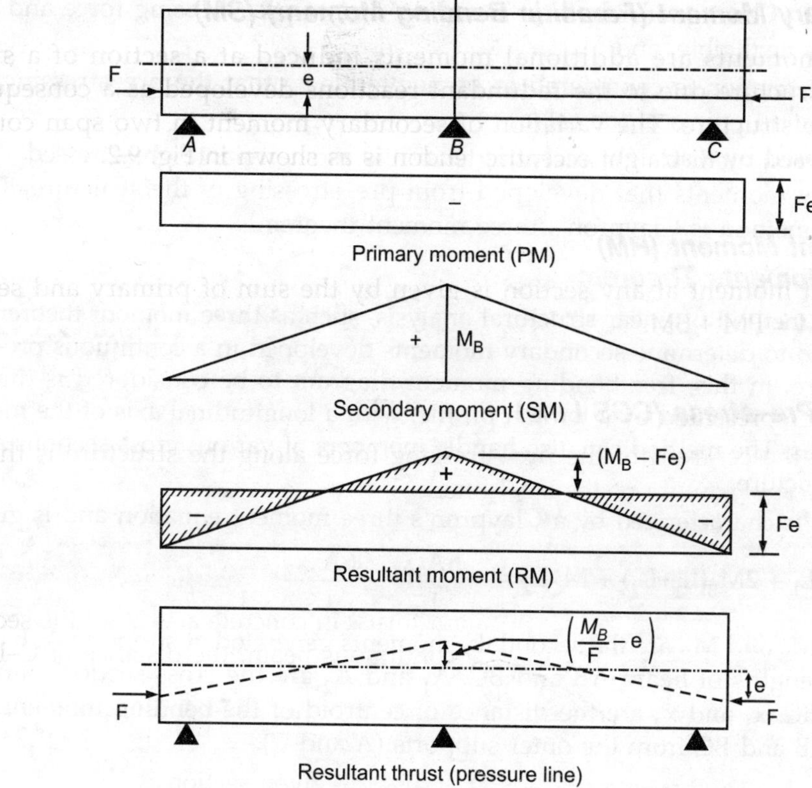

Fig. 9.2 Pressure Line in Pre–stressed Continuous Beam

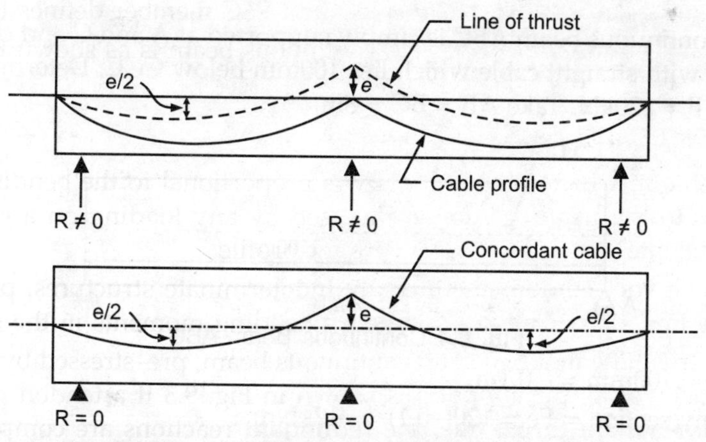

Fig. 9.3 Line of Thrust and Concordant Cable Profile

9.1.3 Methods of Analysis of Secondary Moments

In addition to the basic assumptions, such as elastic behavior of materials and linear strain distribution across the section, the following assumptions are made for the analysis of secondary moments in continuous beams or members.

1. The effect of change in length of members due to pre–stressing force and external loading is negligible, and
2. The cable friction is considered to be negligible so that the pre–stressing force is constant at all points of the cable.

There are several methods for analyzing statically indeterminate pre–stressed structures to find secondary moments that developed from pre–stressing of the structure. The most commonly used method is Claypron's three moment theorem.

9.1.3.1 Three Moments Theorem

The classical method of linear structural analysis, such as three moment theorem can be conveniently used to determine secondary moments developed in a continuous pre–stressed concrete structure. In this, free bending moment diagram to be considered is that due to primary moment represented by a tendon profile, with a longitudinal axis of the member as the horizontal axis. The method can also handle members of various cross sections along the length of the structure.

This method is characterized by a Claypron's three moment equation and is given by

$$M_A L_1 + 2M_B(L_1+L_2) + M_C L_2 = -\frac{6A_1x_1}{L_1} - \frac{6A_2x_2}{L_2} \qquad ...(9.1)$$

Where, M_A, M_B and M_C are the secondary moments generated at supports A, B and C. L_1 and L_2 ate the lengths of beam AB and BC. A_1 and A_2 are the cross sectional area of the beams AB and BC. x_1 and x_2 are the distance of centroid of the bending moment diagram area of beams AB and BC from the outer supports (A and C).

9.1.4 Solved Examples

Problem 9.1

A two span continuous beam ABC is simply supported at A and C and continuous over B. It is pre–stressed with straight cable which lies 100mm below CGC. Determine the secondary moment at 'B' if F = 120kN. Take AB = BC = 8m.

Solution:

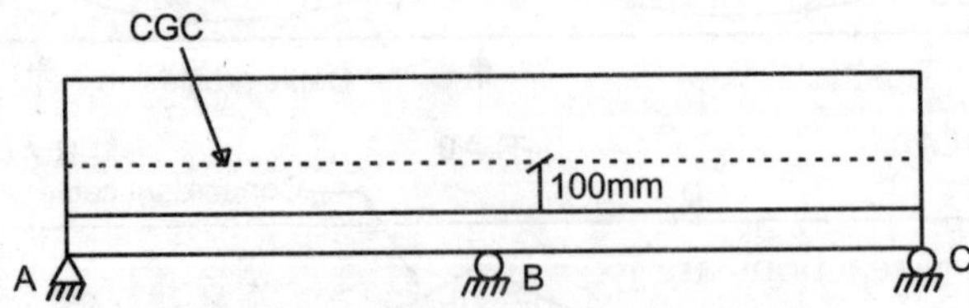

Fig. 9.4 Continuous Beam ABC

F = 120kN, e = 100mm = – 0.1m

The PM at any section = Fe = 120(–0.1) = –12kNm

Since the eccentricity is constant, the PM diagram is rectangle

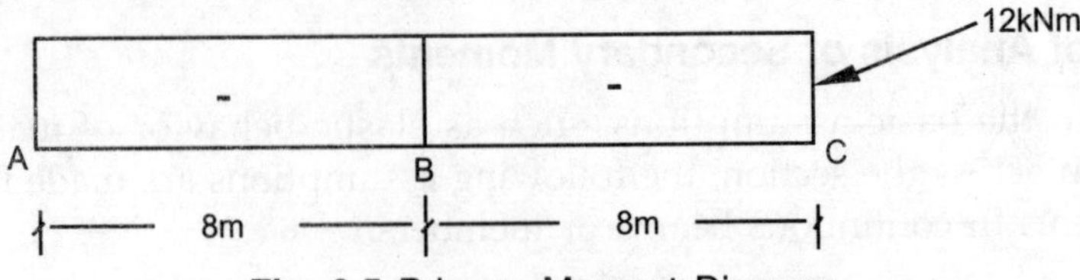

Fig. 9.5 Primary Moment Diagram

M_A, M_B and M_C be the secondary moments at A, B and C respectively. Here $M_A = M_C = 0$

∵ The ends at A and C are simply supported.

Using three moment equation or Claypron's three moment equation

$$M_A L_1 + 2M_B(L_1 + L_2) + M_C L_2 = -\frac{6A_1x_1}{L_1} - \frac{6A_2x_2}{L_2}$$

$$0 + 2M_B\,(8 + 8) + 0 = -2\left(\frac{6A_1x_1}{L_1}\right)$$

$$2M_B\,(8+8) = -2\left(\frac{6 \times (8 \times 12) \times 4}{8}\right)$$

$\therefore\; M_B = 18\text{kNm}$

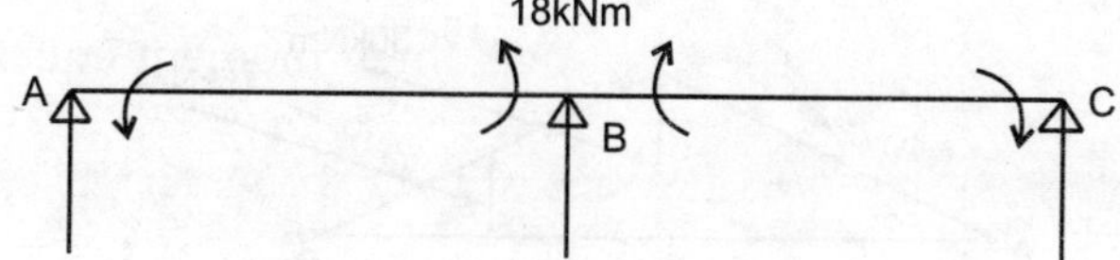

Fig. 9.6 Direction of Moments

The secondary moment at B is of sagging in nature, which causes compression at top fibres. The arrow side for the moments always indicates compression side.

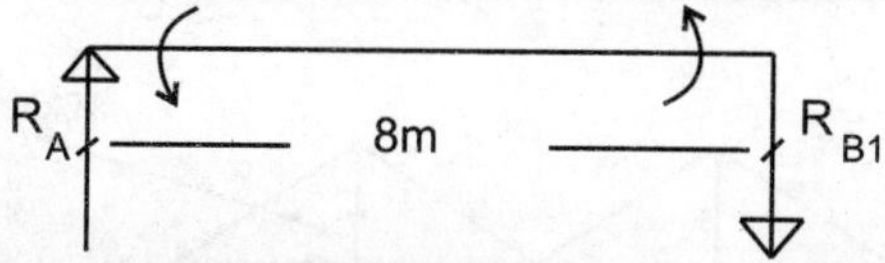

Fig. 9.7 Reactions in Part of Continuous Beam AB

Since the portion AB is under action of pure couple, the reactions R_A and R_{B1} are bound to be equal in magnitude and opposite in direction, so that they form an opposing couple.

$$R_A = R_{B1} = \frac{18}{8} = 2.25\text{kN}$$

Similarly, $\quad R_C = R_{B2} = 2.25\text{kN}$

Thus the secondary reactions are $R_A = R_C = 2.25\text{kN}\ (\uparrow)$ and $R_B = R_{B1} + R_{B2} = 4.5\text{kN}\ (\downarrow)$

Problem 9.2

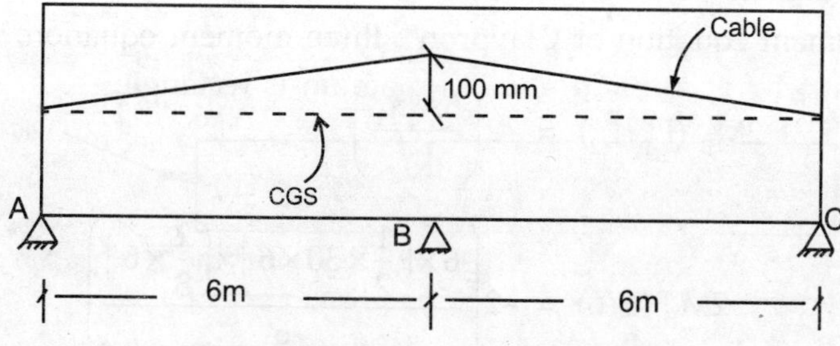

Fig. 9.8 Continuous Beam ABC

A two span continuous ABC with AB = BC = 6m is pre–stressed with a bent cable which has zero eccentricity at the support A and C and eccentricity of 100mm towards the top fibre at B. The beam is 150mm × 300mm in section. Effective pre–stressing force in the cable is 300kN. Determine 1) secondary moment at 'B', 2) extreme fibre stresses in the cross–section at B when the beam supports two point loads of 15kN each at mid–span section of each span, also locate the position of thrust line in concrete.

Solution:

1. Determination of secondary moment at 'B'

$$F = 30\text{kN at B, } e = 0.1\text{m}$$

$$\text{PM at B} = Fe = 3000.1 = 30\text{kNm}$$

$$\text{PM at A and C} = 0$$

So, M_A, M_B and M_C be the secondary moments at A, B and C respectively. Here $M_A = M_c = 0$

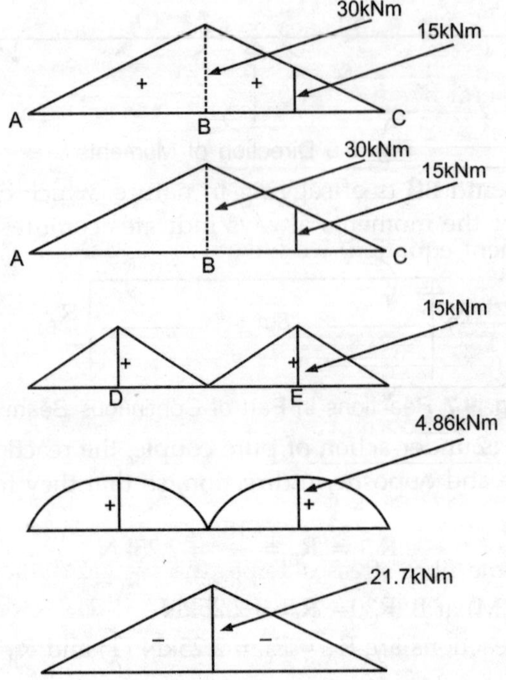

Fig. 9.9 Bending Moment Diagrams of Continuous Beam

Using three moment equation or Claypron's three moment equation, we have

$$2M_B\ (L_1+L_2) = -2\left(\frac{6A_1x_1}{L_1}\right)$$

$$2M_B\ (6+6) = -2\left(\frac{6\times\left(\frac{1}{2}\times 30\times 6\right)\times\left(\frac{2}{3}\times 6\right)}{6}\right)$$

$\therefore\ M_B = -30kNm$

2. Determination of extreme fibre stresses in the cross–section at B when the beam carries two point load of 15kN each at mid–span section of each span

$A = 0.15 \times 0.3 = 0.045m^2$

Self–weight = $0.045 \times 24 = 1.08kN/m$

$$I = \frac{0.15 \times 0.3^3}{12} = 3.375 \times 10^{-4}m^4 = 337.5 \times 10^6 mm^4$$

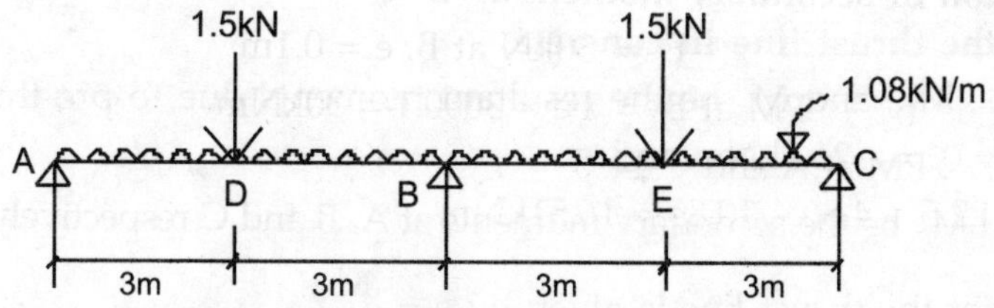

Fig. 9.10 Loads on Continuous Beam

$$\text{Max. free BM for AB and BC due to point loads} = \frac{W \times L}{4} = \frac{15 \times 6}{4} = 22.5kNm$$

$$\text{Max. free BM for AB and BC due to UDL} = \frac{W \times L^2}{8} = \frac{1.08 \times 6^2}{8} = 4.86kNm$$

Applying three moment equation, we have

$$2M_B\ (L_1 + L_2) = -2\left(\frac{6A_1x_1}{L_1}\right)$$

$$2M_B\ (6 + 6) = -2\left(\frac{6 \times \left(\frac{1}{2} \times 22.5 \times 6\right) \times 3}{6} + \frac{6 \times \left(\frac{2}{3} \times 4.86 \times 6\right) \times 3}{6}\right),\ \therefore\ M_B = -21.7kNm$$

To find out the extreme fibre stresses across the section (which lies at the support B)

Resultant moment (RM) at B = 30 – 30 +0+0 – 21.7 = –21.7 kNm

$$y_1 = +150mm \text{ and } y_2 = 150mm$$

Let f_1 and f_2 represents the stresses in the concrete at the topmost and bottom most fibres respectively.

Here, $f_1 = \dfrac{F}{A} + \dfrac{My_1}{I}$

Note: In case of continuous beam, the resultant moment M includes the effect of primary moment i.e., Fe and therefore the term $\dfrac{Fe}{I}$ is missing while computing extreme fibre stresses.

$$f_1 = \frac{300\times10^3}{150\times300} + \frac{(-21.7\times10^6\times150)}{337.5\times10^6} = -2.97\text{N/mm}^2 \text{ (Tension)}$$

and $f_2 = \frac{F}{A} + \frac{My_2}{I}$

$$f_2 = \frac{300\times10^3}{150\times300} + \frac{(21.7\times10^6\times150)}{337.5\times10^6} = 16.31\text{N/mm}^2 \text{ (Compression)}$$

3. Location of the thrust line in concrete

Let M_A, M_B, M_C, M_D, and M_E be the resultant moments due to pre thrust loads.
Then $M_A = M_C = 0, = -21.7$kNm and
$M_D = 15 - 15 + 22.5 + 4.86 - 10.85 = 16.51$kNm

Let eccentricity of the thrust line is given by $e = \frac{M}{F}$

$$e_A = e_c = 0,\ e_B = \frac{-21.7\times10^6}{300\times10^3} = -72 \text{ mm and}$$

$$e_D = e_E = \frac{16.51\times10^6}{300\times10^3} = 55\text{mm}$$

When once the beam carries the UDL over entire length, resultant moment M is bound to vary parabolically. Since the e of C–line depends upon the resultant moment RM. The longitudinal profile of the C–line should also be parabolic, when the variation of RM is parabolic.

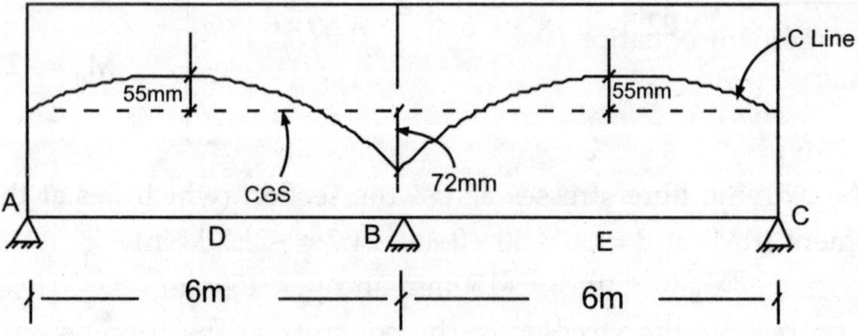

Fig. 9.11 Position of C–Line

Problem 9.3

A two span continuous beam ABC with AB = BC = 6m is pre–stressed with parabolic cable has zero e, at A, B and C and maximum eccentricity of 80mm towards the soffit at mid–section of each span. Determine

1. Secondary and resultant moment at B.
2. Position of the thrust line at B

Effective pre–stressing force in the cable 600kN

Solution:

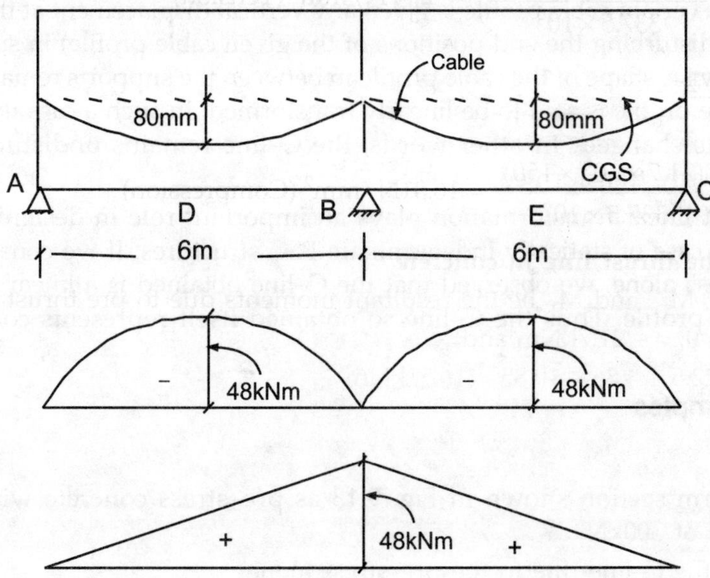

Fig. 9.12 Continuous Beam ABC with its BMD

Since the e of the cable varying parabolically, the primary moment also varying parabolically, F = 600kN

PM at A, B and C = 0

PM at D and E = Fe = 600(–0.08) = –48kNm

Let M_A, M_B and M_C be the secondary moments at A, B and C respectively.

Here $M_A = M_C = 0$.

Applying three moment equation, we have

$$2M_B\,(L_1 + L_2) = -2\left(\frac{6A_1x_1}{L_1}\right)$$

$$2M_B\,(6 + 6) = -2\left(\frac{-6\times\left(\frac{2}{3}\times 48\times 6\right)\times 3}{6}\right)$$

$\therefore\ M_B = 48\text{kNm}$

Thus, SM at B i.e., $M_B = 48\text{kNm}$

Further resultant moment at B = 0 + 48 = 48kNm

Lastly e of the thrust line at B is given by

$$e_B = \frac{48\times 10^6}{600\times 10^3} = 80\text{mm}$$

9.1.5 Guyon's Theorem of Linear Transformation

It states that if a certain cable profile is given any vertical displacement at the intermediate supports (without disturbing the end positions of the given cable profile) in such a way as to ensure that the intrinsic shape of the cable profile in between the supports remains unchanged, then the given cable profile is said to be linearly transformed. In such a case the thrust line in concrete remains unchanged. In other words, the C–line remains undisturbed when the cable profile is linearly transformed.

The theorem of linear transformation plays an important role in designing concordant cable profile in the case of statically indeterminate PSC structures. If we consider the C–line due to the pre–stress alone, we observed that the C–line obtained is a linear transformation of the given cable profile. Thus the C–line so obtained itself represents concordant cable profile.

9.1.6 Solved Examples

Problem 9.4

Beam of uniform section shown in Fig. 9.13 as pre–stress concrete with an effective pre–stressing force of 300kN.

a. Sketch the C–line due to the pre–stress alone
b. Suggest a concordant cable profile
c. Determine SM and secondary reactions
d. Also sketch the C–line when the span AB and BC carry imposed load of 15kN/m and 5kN/m respectively. Also determine the stress distribution across the section which lies at 'B'.

Solution:

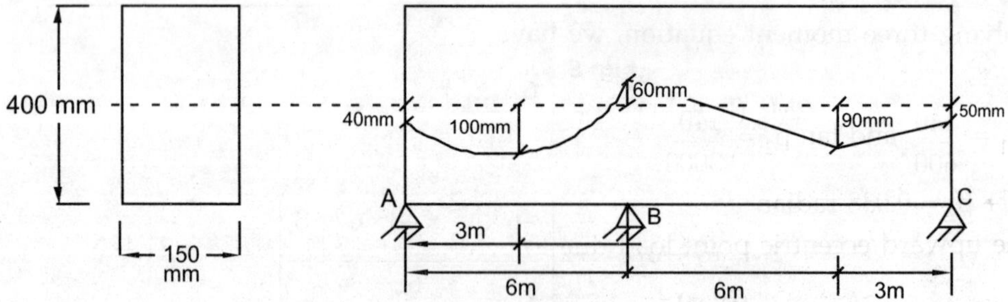

Fig. 9.13 Continuous Beam ABC with its Cross Section

Note: Whenever the given cable profile is unsymmetrical, the analysis by the use of BMD becomes inconvenient. In such a case we consider the equivalent loads due to the given cable profile and the resultant moments are determined. Since, the beam is freely supported at A and C, the resultant moments at A and C are due to PM alone.

Here

F = 300kN, PM at A = 300 (–0.04) = –12kNm, PM at B = 300 (0.06) = 18kNm

PM at C = 300 (–0.05) = –15kNm

a. To sketch the C–line due to the pre–stress alone

• Span AB

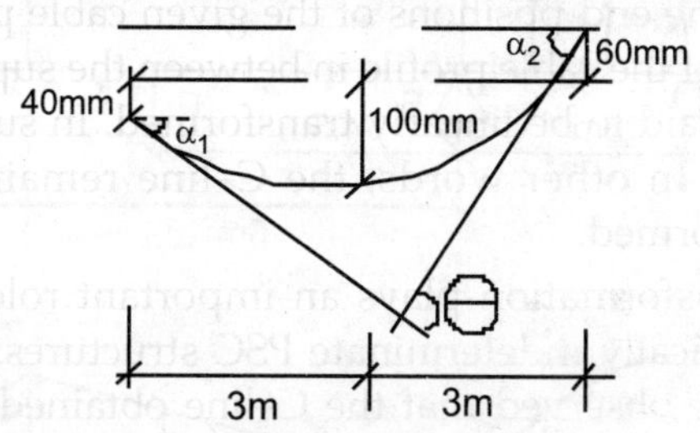

Fig. 9.14 Span AB

$$\alpha_1 = \frac{4x_1}{L_1} \text{ and } \alpha_2 = \frac{4x_2}{L_2}, \text{ where, } L_1 = L_2 = 6m$$

$$\theta = \alpha_1 + \alpha_2 = \frac{4\times60}{6000} + \frac{4\times160}{6000} = 0.147 \text{ Radian}$$

$$\text{Intensity of UDL due to parabolic cable} = \frac{F(\alpha_1+\alpha_2)}{L} = \frac{F\theta}{L} = \frac{300\times0.147}{6} = 7.35\text{kN/m}$$

• Span BC

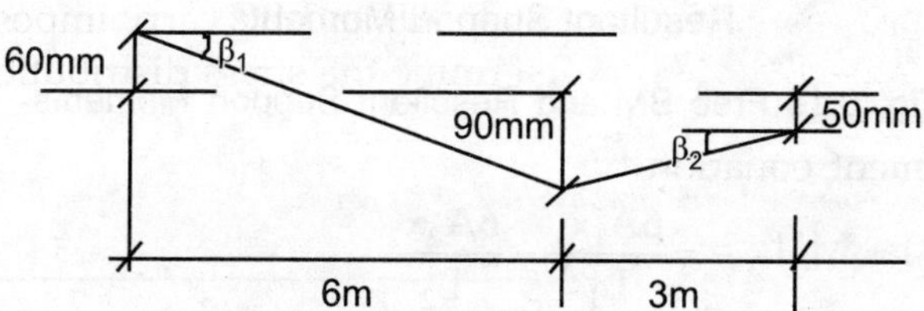

Fig. 9.15 Span BC

$$\tan\beta_1 = \frac{140}{600} \text{ and } \tan\beta_2 = \frac{40}{3000}$$

$\therefore\ \beta_1 + \beta_2 = 0.038$ radian

$\therefore$ The upward eccentric point load due to bent cable = 300 × 0.038 = 11.4kN

$$\text{Max. free BM for AB} = -\frac{W\times L^2}{8} = -\frac{7.35\times6^2}{8} = -33\text{kNm}$$

$$\text{Max. free BM for BC} = -\frac{Wab}{L} = -\frac{11.4\times6\times3}{9} = -27.8\text{kNm}$$

Note:

–ve sign in the BMD indicates hogging BM. Let RM be the resultant moment due to pre-stress alone.

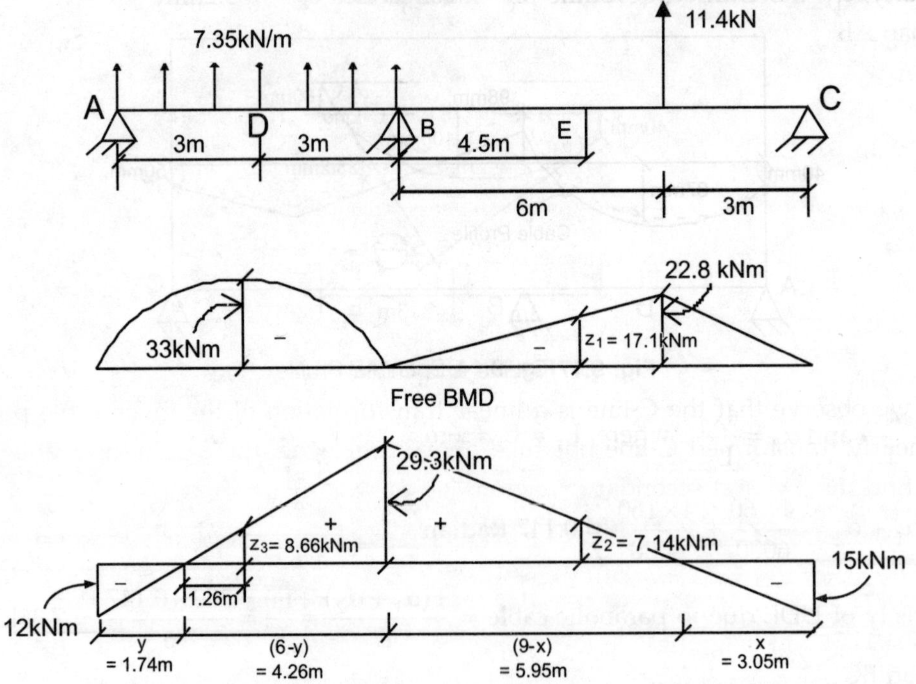

Fig. 9.16 Free BM and Resultant Support Moments

Applying three moment equation,

$$M_A L_1 + 2M_B(L_1 + L_2) + M_C L_2 = -\frac{6A_1x_1}{L_1} - \frac{6A_2x_2}{L_2}$$

$$-12 \times 6 + 2M_B\,(6 + 9) + (-15) \times 9$$

$$= -\left(\frac{6\times\left(-\frac{2}{3}\times 6\times 33\right)\times 3}{6}\right) - \left(\frac{6\times\left(\left(-\frac{1}{2}\times(3\times 22.8)\times\left(\frac{2}{3}\times 3\right)\right)-\left(\frac{1}{2}\times(6\times 27.8)\times 5\right)\right)}{9}\right)$$

$M_B = 29.3\text{kNm}$

Unless otherwise specified we can always locate the position of pre-stress line at mid-span points.

From similar triangle x = 3.05m and y = 1.74m

Thus the resultant moments are

$M_A = -12\text{kNm}$ $\quad M_B = 29.3\text{kNm}$ $\quad M_C = -15\text{kNm}$

$M_D = -33 + Z_3 = -33 + 8.66 = -24.4\text{kNm}$ $\quad M_E = Z_1 + Z_2 = -17.1 + 7.14 = -9.96\text{kNm}$

Thus the eccentricity of the thrust line w.r.t CGC is given by $e = \dfrac{M}{F}$

$$e_A = \frac{-12\times 10^6}{300\times 10^3} = -40\text{mm},$$

Similarly, e_D = –87mm, e_B = 98mm e_C = –50mm and e_E = –33.2mm

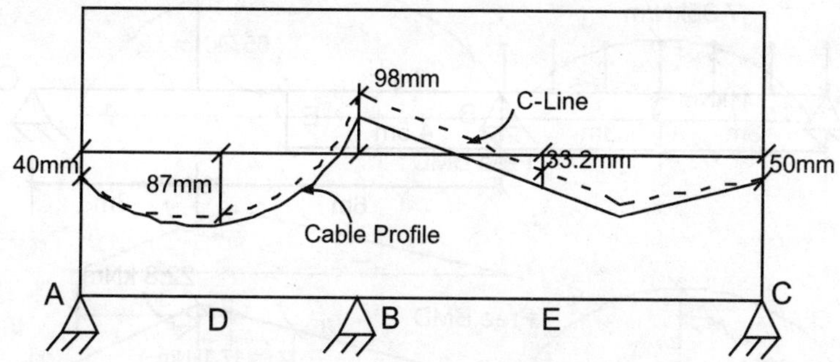

Fig. 9.17 C-Line and Cable Profile

Here we observe that the C-line is a linear transformation of the given cable profile.

b. Linearly transformed C-line obtained above represents the concordant cable profile.

c. To find the SM and secondary reactions

At 'B' $\quad$ RM = 29.3kNm

$\quad$ PM = 300 × 0.06 = 18kNm

But, $\quad$ RM = PM + SM, $\quad$ 29.3 = 18 + SM

∴ SM = 11.3kNm (Sagging)

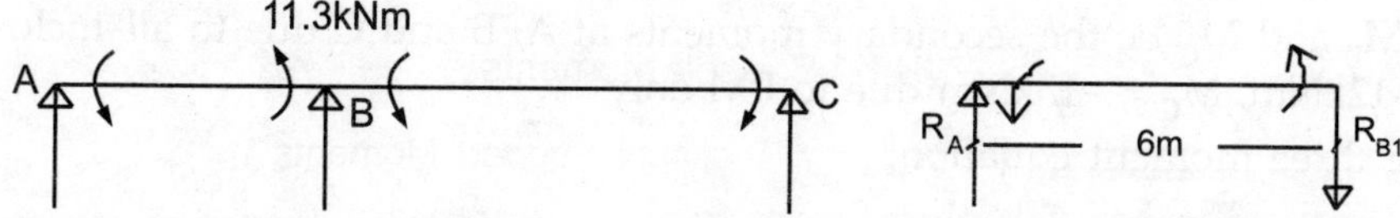

Fig. 9.18 Directions of Moments and Reactions in Part of Continuous Beam AB

$$R_A = R_{B1} = \frac{11.3}{6} = 1.88\text{kN}$$

Similarly,
$$R_C = R_{B2} = \frac{11.3}{6} = 2.25\text{kN}$$

Thus the secondary reactions are $R_A = R_C$ = 1.88kN (↑) and

$$R_B = R_{B1} + R_{B2} = 3.135\text{kN } (\downarrow)$$

$$A = 150 \times 400 = 0.06\text{m}^2$$

$$\text{Self-weight} = 0.06 \times 24 = 1.44\text{kN/m}$$

Net UDL on AB = 15 + 1.44 – 7.35 = 9.09kN/m

Net UDL on BC = 5 + 1.44 = 6.44kN/m

$$\text{Max. free BM for AB} = \frac{W \times L^2}{8} = \frac{9.09 \times 6^2}{8} = 41\text{kNm}$$

$$\text{Max. free BM for BC due to downward UDL} = \frac{W \times L^2}{8} = \frac{6.44 \times 9^2}{8} = 65.2\text{kNm}$$

$$\text{Maximum free BM for BC due to upward concentrated load} = \frac{Wab}{L} = -\frac{11.4 \times 6 \times 3}{9}$$

$$= -22.8\text{kNm}$$

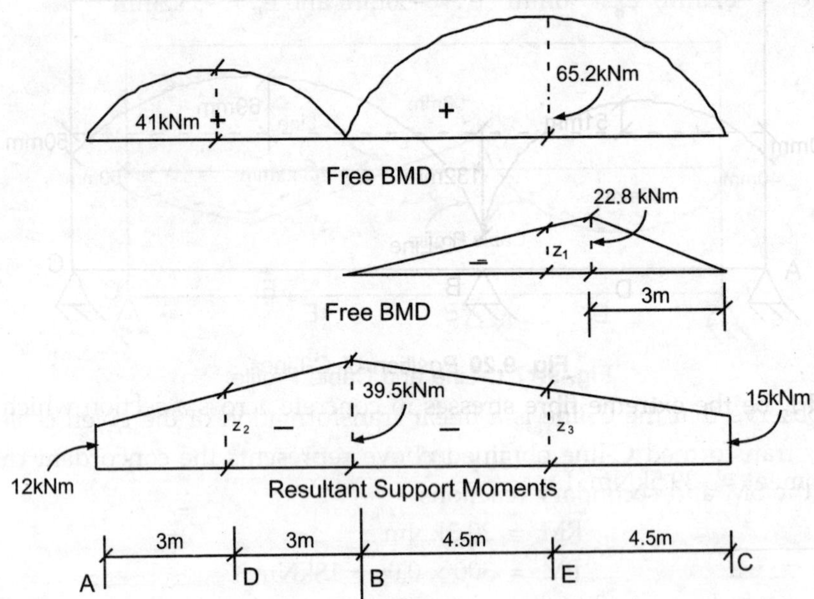

Fig. 9.19 Free BMD and Resultant Support Moments

Let M_A, M_B and M_C be the secondary moments at A, B and C due to all–inclusive loads
Here, $M_A = -12$kNm, $M_C = -15$kNm due to PM only.

Applying three moment equation,

$$M_A L_1 + 2M_B(L_1 + L_2) + M_C L_2 = -\frac{6A_1x_1}{L_1} - \frac{6A_2x_2}{L_2}, -12 \times 6 + 2M_B(6+9) + (-15) \times 9$$

$$= \left(\frac{6\times\left(\frac{2}{3}\times 6\times 41\right)\times 3}{6}\right) - \left(\frac{6\times\left(\left(\frac{2}{3}\times 65.2\times 9\right)\times 4.5\right) - \left(\frac{1}{2}\times(3\times 22.8)\times\left(\frac{2}{3}\times 3\right)\right) - \left(\frac{1}{2}\times(6\times 22.8)\times 5\right)}{9}\right)$$

$M_B = -39.5$kNm

The ordinates Z_1, Z_2 and Z_3 can be computed.

Thus the resultant moments at salient points are

$M_A = -12$kNm $M_D = 41 - Z_2 = 15.2$kNm

$M_B = -39.5$kNm $M_E = 65.2 - Z_1 - Z_3 = 20.8$kNm $M_C = -15$kNm

Eccentricity at salient points as follows

$e_A = -40$mm, $e_D = e_B = -132$mm, $e_C = -50$mm and $e_E = 69$mm

Since both the span carry UDL, the resultant moment varies parabolically and the profile of the C-line is also parabolic.

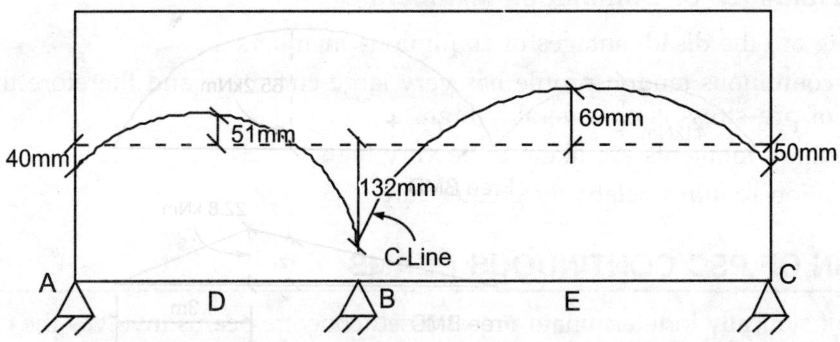

Fig. 9.20 Position of C-Line

Let f_1 and f_2 be the extreme fibre stresses in concrete across a section which lies at B.

$$\text{At 'B', moment} = -39.5\text{kNm},\ I = \frac{150 \times 400^3}{12} = 800 \times 10^6 \text{mm}^4$$

$$\text{Here, } f_1 = \frac{F}{A} + \frac{My_1}{I},\ f_1 = \frac{300 \times 10^3}{150 \times 400} + \frac{(-39.5 \times 10^6 \times 200)}{800 \times 10^6} = -4.875\text{N/mm}^2 \text{ (Tension)}$$

$$\text{and } f_2 = \frac{F}{A} + \frac{My_2}{I},\ f_2 = \frac{300 \times 10^3}{150 \times 400} + \frac{(-39.5 \times 10^6 \times ?200)}{800 \times 10^6} = 14.875\text{N/mm}^2 \text{ (Compression)}$$

9.1.7 Advantages of Continuous Members

Continuity in pre–stressed concrete construction is more advantageous in several aspects. Following are the advantages of continuous members.

1. The bending moments more evenly distributed between the centre of the span and the supports of members.
2. The magnitude of the greatest bending moments gets reduced and therefore the size of the structure also gets reduced.
3. Number of costly anchorage fixtures reduces to a minimum as compared with simply supported beams. Only one pair of posttensioning anchorages and a single stressing operation can be applied to several members.
4. The ultimate load carrying capacity is higher than in a statically determinate structure due to the phenomenon of redistribution of moments.
5. The continuous curved tendon profile can be suitably adjusted to ensure the effective neutralization of the superimposed loads, span moments and support moments.
6. Continuous girders are formed by segmental construction using precast units connected by pre–stressed cables.
7. In continuous members, deflection under working loads reduces to a minimum as compared to simply supported spans.
8. The ultimate strength gets increased.
9. There is greater stability in a framed structure.

9.1.8 Disadvantages of Continuous Members

Following are the disadvantages of continuous members

1. The continuous tendon profile has very large curvature and therefore the frictional loss of pre–stress is considerably high.
2. Secondary moments are likely to be very high.
3. Execution requires relatively greater care.

9.2 DESIGN OF PSC CONTINUOUS BEAMS

Design of statically indeterminate pre–stressed concrete beams involves the computation of maximum and minimum moments at various cross sections of the members so as to obtain the range of moments which generally determines the cross sectional dimensions of the member. The various steps involved in the design of PSC continuous beams are as follows.

1. Calculate the maximum positive and negative moments due to live load and dead load moments at various sections of the member and hence compute the range of moment, which is obtained as the difference of the difference of the maximum and minimum moments at a cross–section.

$$M_r = M_{max} - M_{min} \qquad ...(9.2)$$

2. The overall cross sectional dimensions are fixed using the permissible compressive stress in concrete, f_c, by the equation,

$$Z = \frac{BH^2}{6} = \frac{M_r}{f_c} \qquad ...(9.3)$$

Assume the suitable value of B = 0.4 – 0.5H, the overall depth H is obtained.

3. The maximum pre–stressing force required is estimated by the expression,

$$F = \frac{M_r}{\frac{H}{3}} \qquad ...(9.4)$$

4. The limiting zone of thrust is obtained by plotting (M_{max}/F) and (M_{min}/F) at each section measured from the upper and lower kern respectively.
5. The profile of a cable lying within the limiting zone and suitable for a concordant profile is determined.
6. The stress developed at working and transfer loads are checked at critical sections.
7. The cable profile, if necessary, may be linearly transformed to reduce the slopes at supports with due regard to cover reinforcements.

9.2.1 Determination of Concordant Tendon Profile

In the design of continuous PSC members, it is often necessary to determine a cable profile lying within the limiting zone of thrust and also satisfying the conditions of concordancy. A general method based on the principal of virtual work is outlined for determining the concordant cable profile.

The deflection at any point in a beam due to the loading is expressed as

$$a = \int \frac{Mmdx}{EI} \qquad ...(9.5)$$

Where, M = Primary moment (moment due to actual eccentricity of the tendon at any cross section)

m = Moment due to unit load applied at the point where deflection is required.

EI = Flexural rigidity of the beam

Using finite differences, the deflection a can be expressed as,

$$a = \sum \frac{KMm}{E} \qquad ...(9.6)$$

Where,

$$K = \frac{dx}{I} \qquad ...(9.7)$$

That is a summation constant.

In a pre–stressed beam, the primary moment is given by M = Fe, where, F is the pre–stressing force located at an eccentricity e.

Hence,

$$a = \sum \frac{FemK}{E} \qquad ...(9.8)$$

Since F and E are constant along the length of the beam,

$$a = \frac{F}{E}\sum Kme \qquad ...(9.9)$$

At a concordant cable profile a = 0.

Consequently, to obtain concordancy, the eccentricity of the cable along the length of the beam is arranged to satisfy the condition $\sum Kme = 0$.

9.2.2 Solved Examples

Problem 9.5

Design a continuous PSC beam of two spans AB and BC of 12m in length to support a UDL of 10kN/m. Tensile stresses are not permitted in concrete and the compressive strength on concrete is not to exceed 13MPa. Sketch the details of the cable profile and check the stressed developed at the support and span sections.

Solution:

1. Maximum Positive and Negative Moments under Different Combinations of Live Load

The various load combinations are considered to determine the maximum and minimum moments at various sections along the span are as shown in Fig. 9.21.

The maximum range of live load moment at support section B,

$$M_{qr} = M_r = \frac{q \times L^2}{8} = \frac{10 \times 12^2}{8} = 180\text{kNm}$$

Assuming the width of the rectangular section to be used B = 250mm, and the overall depth H is obtained as,

$$Z = \frac{250H^2}{6} = \frac{M_r}{f_c} = \frac{180 \times 10^6}{13}, H = \sqrt{\frac{6 \times 180 \times 10^6}{13 \times 250}} = 570\text{mm}$$

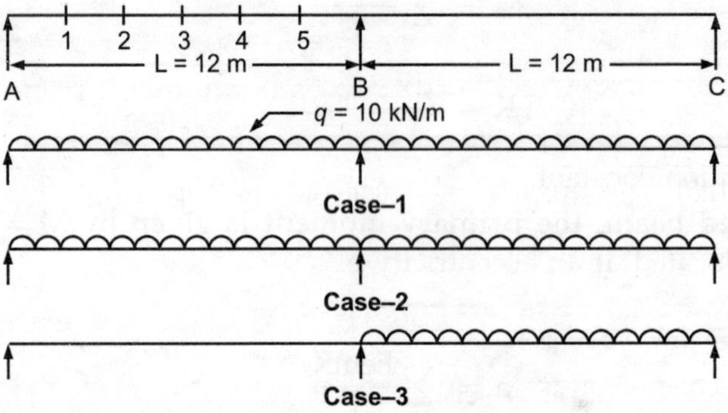

Fig. 9.21 Loading Patterns for Maximum Positive and Negative Moments

Adopt rectangular section of 250mm wide and 600mm deep.

Self–weight of the beam = g = (0.25 × 0.60 × 24) = 3.6kN/m

The maximum and minimum moments developed due to the loading patterns at various points from A to B are kept in Table 9.1.

The moments at each of the points are expressed as $M = kw_dL^2$, where, k is the moment coefficient whose values are shown in brackets in the Table 9.1 and can be obtained by Reynold's design handbook. Using the values of the maximum positive (M_{Lp}) and negative (M_{Ln}) live load and dead load moments, the minimum pre–stressing force required is given by,

$$F = \frac{M_r}{\frac{H}{3}} = 180 \times \frac{10^3}{\frac{600}{3}} = 900\text{kN}$$

Table 9.1 Computation of Maximum Positive and Negative Moments

g = 3.6kN/m L = 12m

q = 10kN/m $M = kw_dL^2$

Location	Loading pattern			Maximum moments due to live load (kNm)		Dead load moments (kNm)
	Case 1	Case 2	Case 3	M_{Lp}	M_{Ln}	
A	(0.000) 0	(0.000) 0	(0.000) 0	0	0	0
1	(0.048) 69.0	(0.059) 85.0	–(0.010) 14.4	85.0	–14.4	25.0
2	(0.070) 101.0	(0.091) 130.0	–(0.021) 30.2	130.0	–30.2	36.0
3	(0.062) 90.5	(0.093) 135.0	–(0.036) 45.0	135.0	–45.0	32.0
4	(0.033) 47.5	(0.070) 101.0	–(0.042) 60.5	101.0	–60.5	17.4
5	–(0.036) 52.0	(0.017) 24.5	–(0.052) 75.0	24.5	–75.0	–18.8
B	–(0.125) 180	–(0.063) 90.5	–(0.063) 90	0	–180.0	–65.0

2. Limiting Zone for Thrust

Using the values of maximum positive and negative live load and dead load moments at each points from A to B, the minimum (M_1) and maximum (M_2) moments, the range of moments and the position of the limits of thrust zone is determined as shown in Table 9.2.

3. Determination of Concordant Profile

Using the trial and error procedure explained in section 9.2.1, a suitable profile lying within the limiting zone and satisfying the requirements of concordancy is determined using the procedure outlined in Table 9.3. The concordant profile of trial 2 is shown in Fig. 9.22.

Table 9.2 Determination of Limiting Zone for Thrust Line

Location point	M_{Lp} (kNm)	M_{Ln} (kNm)	M_G (kNm)	Maximum moment = M_2 = $M_{Lp} + M_G$ (kNm)	Maximum moment = M_1 = $M_{Ln} + M_G$ (kNm)	Range of moments = $M_2 - M_1$ (kNm)	(M_2/F) From upper kern (mm)	(M_1/F) From lower kern (mm)	Eccentricity e_2 = (M_2/F)–100 (mm)	Eccentricity e_1 = (M_1/F)+100 (mm)
A	0	0	0	0	0	0	0	0	–100.0	+100
1	85.0	–14.4	25.0	110.0	10.6	99.4	122.0	11.8	22.0	111.8
2	130.0	–30.2	36.4	166.0	5.8	160.2	184.0	6.4	84.0	106.4
3	135.0	–45.0	32.0	167.0	–13.0	180.2	186.0	–14.0	86.0	86.0
4	101.0	–60.5	17.4	118.4	–43.1	161.5	132.0	–48.0	32.0	52.0
5	24.5	–75.0	–18.8	5.7	–93.8	99.5	6.30	–104.0	–93.7	–4.0
B	0	–180.0	–35.0	–65.0	–245.0	180.0	–72.0	–272.0	–172.0	–172.0

Table 9.3 Determination of Concordant Profile

Point	L (m)	Simpson's rule coefficient q	$\frac{qdx}{3L}$	$K = \frac{dx}{3}$	Unit moment diagram, m	Km	First trial e (cm)	Kme	Second trial e (cm)	Kme
A	1	1	2/3	0.667	0	0	0	0	0	0
1	1	4	2/3	2.667	0.167	0.440	+6.0	+2.64	+6.7	+2.97
2	1	2	2/3	1.333	0.333	0.440	+9.0	+3.96	+9.7	+4.30
3	1	4	2/3	2.667	0.500	1.333	+8.6	+10.66	+8.6	+11.50
4	1	2	2/3	1.333	0.667	0.880	+4.0	+3.52	+4.0	+3.52
5	1	4	2/3	2.667	0.833	2.230	–5.0	–11.15	–4.8	–10.70
B	1	1	2/3	0.067	1.000	0.667	–17.2	–11.50	–17.2	–11.50
								ΣKme = –1.87		Σ Kme = +0.09

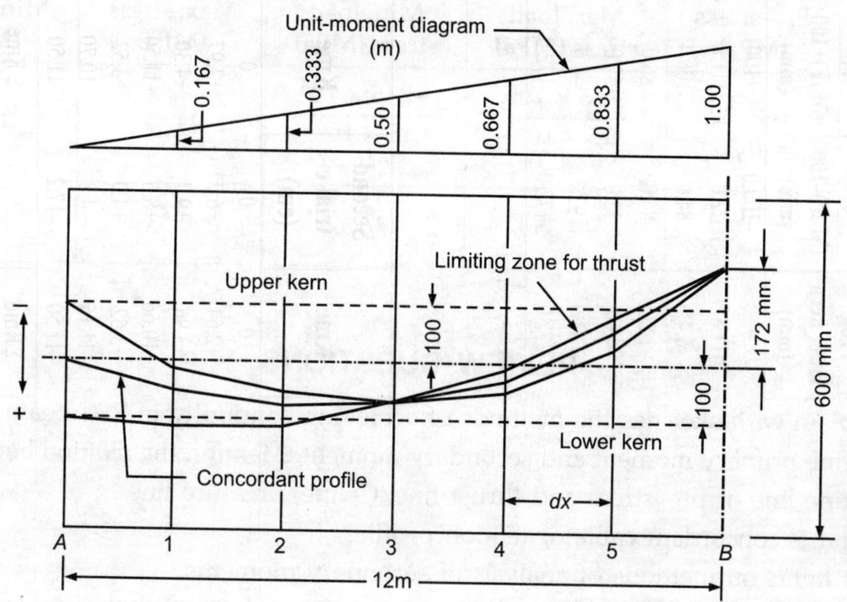

Fig. 9.22 Limiting Zone for Thrust and Concordant Profile

4. Check for Stresses at Section 3 and B

- Section 3

$P = 900\text{kN}, \quad \frac{F}{A} = 6\text{N/mm}^2, \qquad Z = 15 \times 10^6\text{mm}^3, \quad \frac{Fe}{Z} = 5.16\text{N/mm}^2$

$A = 15 \times 10^2\text{mm}^2, \quad e = 86\text{mm}$

$M_{max} = 167\text{kNm}$ and $M_{min} = -13\text{kNm}$

Fibre	Pre–stress (MPa)	Max. load stress (MPa)	Min. load stress (MPa)	Max. stress (MPa)	Min. stress (MPa)
Top	6–5.16 = 0.84	$\frac{167 \times 10^6}{15 \times 10^6}$ = 11.2	$\frac{-13 \times 10^6}{15 \times 10^6}$ = –0.87	12.04	–0.03
Bottom	6+5.16 = 11.16	–11.2	+0.87	–0.04	12.03

- Section B

$P = 900\text{kN}, \qquad \frac{F}{A} = 6\text{N/mm}^2, \quad Z = 15 \times 10^6\text{mm}^3, \qquad \frac{Fe}{Z} = 10.32\text{N/mm}^2$

$A = 15 \times 10^2\text{mm}^2, \qquad e = 172\text{mm}$

$M_{max} = -65\text{kNm}$ and $M_{min} = -245\text{kNm}$

Fibre	Pre–stress (MPa)	Max. load stress (MPa)	Min. load stress (MPa)	Max. stress (MPa)	Min. stress (MPa)
Top	6+10.32 = 16.32	$\frac{-65 \times 10^6}{15 \times 10^6}$ = –4.35	$\frac{-245 \times 10^6}{15 \times 10^6}$ = –16.32	11.97	0.00
Bottom	6–10.32 = –4.32	+4.35	+16.32	0.03	12.00

REVIEW QUESTIONS

9.1 Explain with sketches the methods of achieving continuity in PSC beams.

9.2 Define primary moment and secondary moment. Mention the relation between them.

9.3 Define line of pre–stress and thrust line/C–line/pressure line.

9.4 What is concordant cable or tendon profile?

9.5 Put lights on methods of analysis of secondary moments.

9.6 Explain Claypron's theorem of three moments and mention the terms used in the three moment equation.

9.7 Elaborate Guyon's theorem of linear transformation.

9.8 Mention the advantages and disadvantages of continuous members.

9.9 What is the method of finding concordant tendon profile?

9.10 Outline the steps used in the design of continuous beams.

EXERCISE PROBLEMS

9.1 A continuous concrete beam ABC (AB = BC = 10m) has a uniform rectangular cross section of 100mm wide and 300mm deep. A cable carrying an effective pre–stressing force of 360kN varies linearly with an eccentricity of 50mm towards the soffit at the end supports to 50mm towards top of the beam at mid support B.

(a) Determine the resultant moment at B due to pre–stressing only.

(b) If the eccentricity of the cable at B is +25mm, show that cable is concordant.

[Ans: (a) RM = +9kNm, (b) Cable is concordant]

9.2 A two span continuous beam ABC (AB = BC = 12m) has rectangular section, 300mm wide and 800mm deep. The beam is pre–stressed by a cable carrying an effective pre–stress of 700kN. The cable has a linear profile in the span Ab and parabolic in span BC. The eccentricities of the cable are +50mm at A, –100mm at a distance 7m from support A and +200mm at support B and –200mm at mid–span of span BC (– means below and + means above centroidal axis).

(a) Evaluate the resultant moment developed at B due to the pre–stressing force.

(b) Sketch the line of thrust in the beam if it supports, an UDL of 5kN/m which includes the self–weight of the beam.

(c) Find the resultant stress distribution at the mid support section for condition (b).

[Ans: (a) RM at B = +170kNm, (b) Pressure line at A = +50mm, at 7m from support A = –24.4mm, at B = +115.4mm, at centre of BC = –113.7mm (c) Stress at top = 5.44N/mm^2 and stress at bottom = 0.39N/mm^2]

9.3 A two span continuous beam ABC (AB = BC = 10m) is of rectangular section, 200mm wide and 500mm deep. The beam is pre–stressed by a parabolic cable, concentric at ends supports and having an eccentricity of 100mm towards the soffit of the beam at centre spans and having an eccentricity of 100mm towards the mid support B. The effective force in the cable is 500kN.

(a) Show that the cable is concordant

(b) Locate the pressure line in the beam when, in addition to its self–weight, it supports an imposed load of 5.6kN/m.

[Ans: (a) SM at B = 0, (b) Pressure line coincides with the centroidal axis]

9.4 A three span continuous prismatic pre–stressed concrete beam of depth 2m is pre–stressed by a design pre–stressing force of 3300kN. The spans are AB = CD = 30m and BC = 40m. The eccentricities of the pre–stressing force at different locations along the beam are shown in following table.

X measured from A	0	15	30	50	70	85	100
Eccentricity e, mm	0	–260	+450	–410	+450	–260	0

Note: – = below the centroid, + = above the centroid.

(a) If the cable profile is parabolic between the supports, evaluate the equivalent load and hence compute the moments produced at supports of mid–span points, due to pre–stressing.

(b) Also locate the position of the pressure line.

(c) Suggest a suitable concordant profile.

[Ans: (a) Equivalent load = 14.2N/mm^2, moments at B and D = 1800kNm, moment at mid–point of AB = –700kNm, moment at mid–point of BC = –1050kNm (b) Position of pressure line at B and C = +545mm, at centre of mid–span = –317mm, at centre of end spans = –212mm, (c) Same as the pressure line]

9.5 A two span continuous pre–stressed concrete beam ABC (AB = BC = 15m) has a uniform cross–section with a width of 250mm and depth of 600mm. A cable carrying a uniform pre–stressing force of 500kN is parallel to the axis of the beam and located at an eccentricity of 200mm.

(a) Determine the secondary resultant moment developed at mid support section B

(b) If the beam supports an imposed load of 2.4kN/m, calculate the resultant stresses developed at the top and bottom of the beam at B. Also locate the thrust line through the beam AB.

[Ans: (a) SM at B = +150kNm, RM at B = +150kNm (b) Stress at top = –4.58N/mm^2 (Tension) Stress at soffit = +11.22N/mm^2 (Compression)]

9.6 Design a pre–stressed concrete continuous beam over two equal spans of 9m to support a live load of 30kN each at the centre of span. The loads may be applied independently or jointly. Permissible stress being zero in tension and $15N/mm^2$ in compression. Loss ratio = 0.85. Determine the concordant cable profile and show it on the elevation of the beam. Allowing for minimum cover of 100mm, sketch a suitable transformed profile to reduce the slope of the tendons at the central support to a minimum. Check for a limit states of serviceability and collapse.

[Ans: Rectangular section of 200mm wide and 400mm deep, Pre–stressing force = 510kN]

OBJECTIVE QUESTIONS

9.1 Among below which is not the cable layout for continuous beam?

a. Continuous
b. Cap
c. Conquered
d. Straight

9.2 Additional moments induced at a section of a statically indeterminate structure due to the redundant reactions developed as a consequence of pre–stressing are called .....

a. primary moments
b. secondary moments
c. resultant moments
d. continuous moments

9.3 The relation between primary (PM), secondary (SM) and resultant moment (RM) is ...

a. RM= PM × SM
b. RM = PM – SM
c. RM = PM/SM
d. RM = PM + SM

9.4 In a continuous beam, the distance of C–line or thrust line or pressure line from line of pre–stress is given by .... Where, M = Regular Bending Moment across the given section, F = Effective pre–stressing force in the tendon

a. $e = \frac{M}{F}$
b. $e = \frac{F}{M}$
c. $e = \frac{M-F}{F}$
d. $e = \frac{M}{F-M}$

9.5 Claypron's three moment equation and is given by ....

a. $M_AL_1 + 2M_B(L_1 + L_2) - M_CL_2 = -\frac{6A_1x_1}{L_1} - \frac{6A_2x_2}{L_2}$

b. $M_AL_1 - 2M_B(L_1 + L_2) + M_CL_2 = -\frac{6A_1x_1}{L_1} - \frac{6A_2x_2}{L_2}$

c. $M_AL_1 + 2M_B(L_1 + L_2) + M_CL_2 = -\frac{6A_1x_1}{L_1} - \frac{6A_2x_2}{L_2}$

d. $M_AL_1 - 2M_B(L_1 + L_2) - M_CL_2 = -\frac{6A_1x_1}{L_1} - \frac{6A_2x_2}{L_2}$

9.6 If a certain cable profile is given any vertical displacement at the intermediate supports (without disturbing the end positions of the given cable profile) in such a way as to ensure that the intrinsic shape of the cable profile in between the supports remains unchanged, then the given cable profile is said to be linearly transformed. It is known as ....

a. Lin's theorem
b. Guyon's theorem
c. Havel's theorem
d. Gabriel theorem

9.7 In design of continuous beam, the range of moment M_r, which is obtained as ....

a. $M_r = M_{max} \times M_{min}$
b. $M_r = M_{max} - M_{min}$
c. $M_r = M_{max} + M_{min}$
d. $M_r = M_{max}/M_{min}$

9.8 In continuous beam design, overall cross sectional dimensions are fixed using the permissible compressive stress in concrete, f_c, by the equation ....

a. $Z = \frac{BH^2}{6} = \frac{M_r}{f_c}$
b. $Z = \frac{BH^2}{8} = \frac{M_r}{f_c}$
c. $Z = \frac{BH^2}{6} = \frac{f_c}{M_r}$
d. $Z = \frac{BH^2}{8} = \frac{f_c}{M_r}$

9.9 In continuous beam design, the maximum pre–stressing force required is estimated by the expression ....

a. $F = \frac{M_r}{\frac{H}{12}}$
b. $F = \frac{M_r}{\frac{H}{9}}$
c. $F = \frac{M_r}{\frac{H}{6}}$
d. $F = \frac{M_r}{\frac{H}{3}}$

9.10 A general method based on the principal of virtual work is outlined for determining the concordant cable profile. The deflection at any point in a beam due to the loading is expressed as

a. $a = \int \frac{Mmdx}{3EI}$
b. $a = \int \frac{Mmdx}{EI}$
c. $a = \int \frac{Mmdx}{2EI}$
d. $a = \int \frac{Mmdx}{6EI}$

9.11 To obtain concordancy, the eccentricity of the cable along the length of the beam is arranged to satisfy the condition ....

a. $\sum Kme = 0$
b. $\sum Kme < 0$
c. $\sum Kme > 0$
d. $\sum Kme \neq 0$

Answers

9.1 c	9.2 b	9.3 d	9.4 a	9.5 c
9.6 b	9.7 b	9.8 a	9.9 d	9.10 b
9.11 a				

❑

CHAPTER 10

Analysis and Design of Composite Sections

10.1 INTRODUCTION

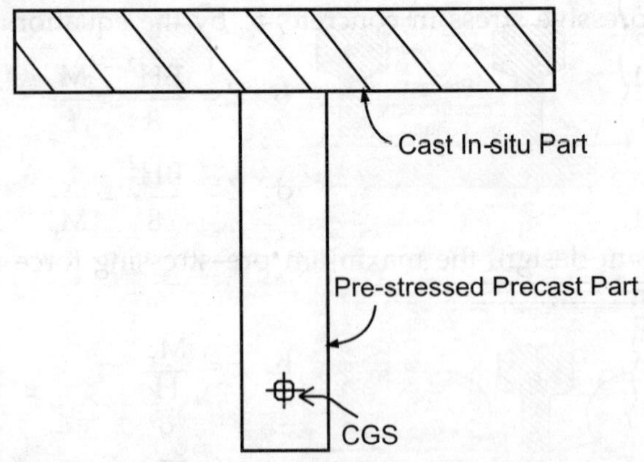

Fig. 10.1 Composite Section

In the case of composite construction, a precast pre–stressed part is used in combination with a cast in–situ part. So that two together behave like a monolithic member under working loads. It is usual practice to employ low strength concrete in the cast in–situ part. Generally, the high strength pre–stressed units are used in tension zone while the concrete, which is the cast in–situ of relatively lower compressive strength, is used in the compression zone of the composite sections.

The composite action between the two components is achieved by roughening the surface of the pre–stressed part on to which the concrete is cast in–situ, thus gives the better frictional resistance. Fig. 10.1 shows the typical composite section.

10.1.1 Types of Composite Construction

The most common type of composite construction consists of a number of precast pre–stressed inverted T–beams, placed side by side and connected by a continuous top slab of in–situ concrete. This type of construction is widely used in bridge decks construction. Transverse pre–stressing is also used to develop monolithic action in the lateral direction. The dead weight of the deck is reduced by using voids or lightweight longitudinal cores in the space between the precast pre–stressed units. For large span composite bridge decks of span more than 30m, the most commonly used precast pre–stressed units are I, unsymmetrical

T or box sections. Typical cross–section of the bridge decks with different types of precast units is shown in Fig. 10.2.

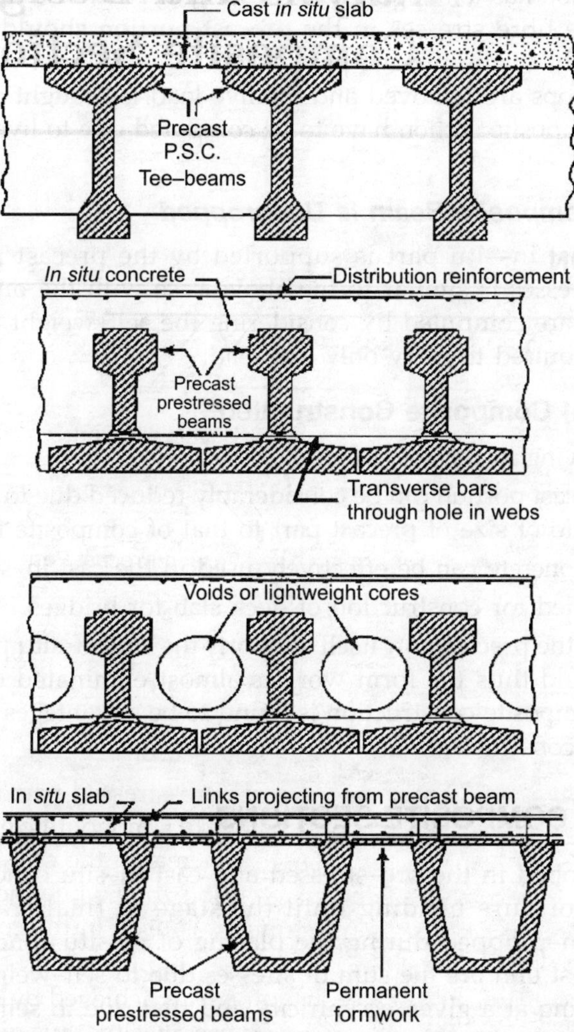

Fig. 10.2 Composite Bridge Decks with Precast Pre–stressed Elements

There are mainly following two types of composite construction:

1. When the cast in–situ part is supported by means of additional supports during the process of hardening. This type of construction is referred to as the propped construction.
2. When the cast in–situ part is supported by the precast part during the process of hardening. In this case it is said to be un–propped construction.

Note: While solving problem on a composite section, we must first decide whether the beam is propped or un–propped.

10.1.1.1 When the Composite Beam is Propped

In this case the cast in–situ slab is supported by the additional supports called the props. Hence the precast portion has to support its own weight. When the cast in–situ part is being hardened, the extreme fibre stresses in the precast portion should be compared due to pre–stress + self–weight of the precast part. When once the cast in–situ slab attains the desired strength the props are removed and the live load is brought on. Hence the extreme fibre stresses in the composite section have to be computed due to live load + self–weight of the cast in–situ slab.

10.1.1.2 When the Composite Beam is Un–propped

In this case the cast in–situ part is supported by the precast part itself. Hence the computation of the stresses is similar to the above case with the only exception that, the stresses in the precast are computed by considering the self–weight of both the parts. The composite section is required to carry only live load.

10.1.2 Advantages of Composite Construction

1. Appreciable saving in the cost of steel.
2. Sizes of the precast portion can be considerably reduced due to the effect of composite action. Low ratio of size of precast part to that of composite member.
3. Low strength concrete can be effectively used in the cast in–situ part.
4. It is ideally suited for construction of deck slab for bridges.
5. In many cases, the precast port itself supports the cast in–situ part during the process of hardening and thus the form work is almost eliminated considering the above said factors, composite construction is found to be advantages and economical in the case of bridge construction.

10.2 ANALYSIS OF COMPOSITE SECTIONS

The stresses developed in the pre–stressed and cast in–situ concrete are evaluated by using simple theory of pure bending until the stage of final cracking. If the precast pre–stressed unit is un–propped during the placing of in–situ concrete, the stresses that developed in the precast unit are the sum of stresses due to self–weight of the member, the pre–stressing force acting at a given eccentricity and that due to self–weight of the in–situ cast concrete. After the in–situ concrete is hardened, the whole section is assumed to be monolithic and the stressed that developed due to subsequent live loads are computed using the properties of composite section.

However, if the precast unit is propped during the placing of in–situ concrete, the stresses developed due to self–weight of the in–situ concrete are computed using section modulus of the composite section. In all cases live load stresses are based on composite sections. The typical stress distribution at various stages of loading for propped and un–propped constructions is shown in Fig. 10.3.

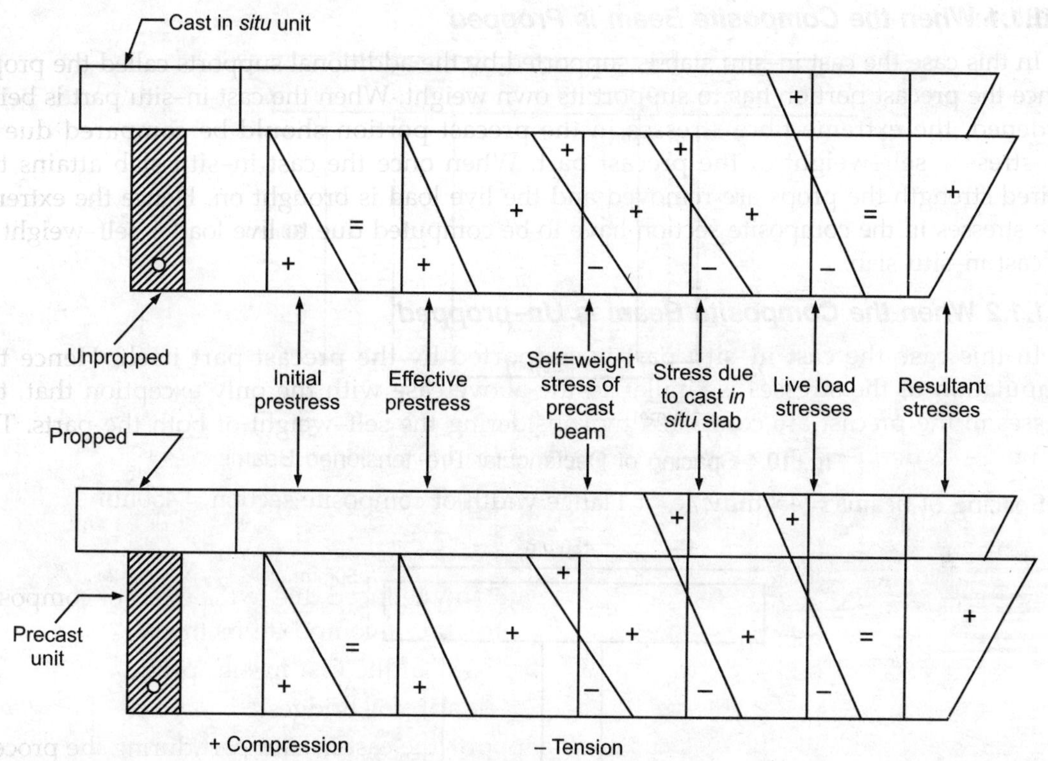

Fig. 10.3 Stress Distribution in Un–propped and Propped Composite Construction

In most composite constructions which involve pre–stressed precast units and in–situ cast concrete the latter is invariably of low or medium strength concrete while the former are generally made of high strength concrete of grade more than M35. Consequently the modulus of elasticity of the two parts will be different. For computing live load stresses, the effect of different moduli between the cast in–situ and precast pre–stressed concrete units is considered by using modular ratio of precast to in–situ concrete for calculating the area, centroid, second moment of area and second modulus of the equivalent composite section.

In most practical instances, the modulus of elasticity of in–situ concrete of grade M20 will be about 25kN/mm^2, while the modulus of concrete in precast pre–stressed units could vary from 28 to 36kN/mm^2 for concrete grades of M30 to M60. Hence the modular ratio varies in the range of 1.1 to 1.5.

10.2.1 Solved Examples

Problem 10.1

A rectangular pre–tensioned beam 100mm × 250mm in section are spaced at 450mm c/c. A 50mm thick slab is cast in–situ over the pre–tensioned beam. The effective stress or pre–stress in the pre–tensioned beam varies linearly from zero at top to 12.6N/mm^2 compressing at bottom. The LL is 1.8kN/m. Assuming the beam is propped. Determine the resultant stress distribution in composite section. Span of the beam is 7.5m.

Solution:

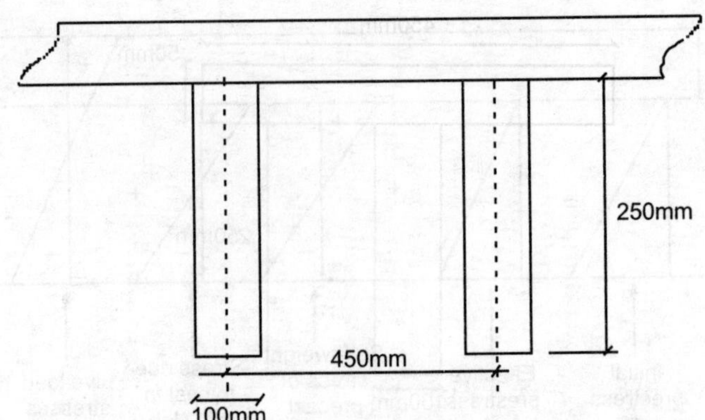

Fig. 10.4 Spacing of Rectangular Pre–tensioned Beams

Spacing of beams = 450mmc/c, ∴ Flange width of composite section = 450mm

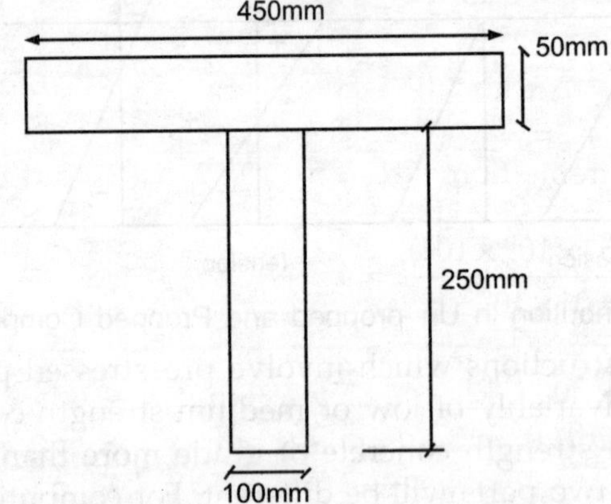

Fig. 10.5 Pre–tensioned Section

It is given that effective pre–stress in the pre–tensioned beam varies linearly from zero at top to 12.6N/mm² compressing at bottom (Due to pre–stress alone). Since the given composite section is propped one. The DL due to cast in–situ slab is not carried by pre–tensioned beam.

A = 100 × 250 = 25000mm², Self–weight = 0.025 × 24 = 0.6kN/m

$$M_G = \frac{W \times L^2}{8} = \frac{0.6 \times 7.5^2}{8} = 4.22\text{kNm}$$

Let f_1 and f_2 be the extreme fibre stresses in the pre–tensioned beam due to M_G alone.

$$I = \frac{100 \times 250^3}{12} = 1.3 \times 10^8\text{mm}^4,\ f_1 = \frac{M_G y_1}{I} = \frac{4.22 \times 10^6 \times 125}{1.3 \times 10^8} = 4.05\text{N/mm}^2 \text{ (Compression)}$$

and f_2 = –4.05N/mm² (Tension)

∴ Resultant stress at top = 0 + 4.05 = 4.05N/mm²

Resultant stress at bottom = 12.6 – 4.05 = 8.55N/mm²

- **Composite Section**

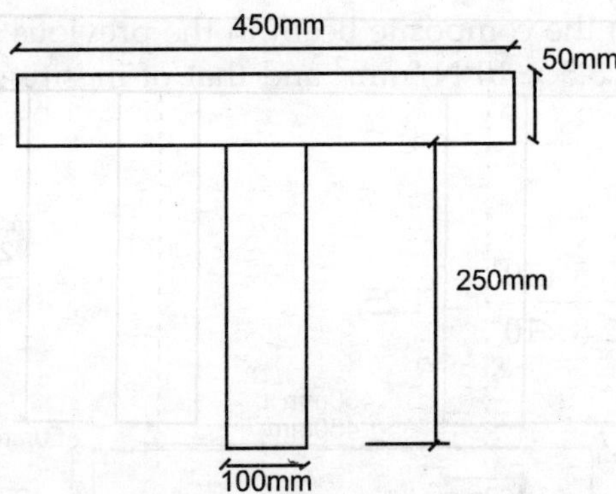

Fig. 10.6 Composite Section

$Y_1 = 104\text{mm}$, $Y_2 = -196\text{mm}$, $I = 4.01 \times 10^8 \text{mm}^4$

Self–weight of slab = $0.05 \times 0.45 \times 24 = 0.54\text{kN/m}^2$, LL = 1.8kN/m^2,

Total load = $0.54 + 1.8 = 2.34\text{kN/m}^2$, Max. BM = $\dfrac{W \times L^2}{8} = \dfrac{2.34 \times 7.5^2}{8} = 16.453\text{kNm}$

Let f_1' and f_2' be the extreme fibre stresses due to M in the composite section.

Here, $f_1' = \dfrac{My_1}{I} = \dfrac{(16.453 \times 10^6 \times 104)}{4.01 \times 10^8} = 4.27\text{N/mm}^2$ and

$$f_2' = \frac{My_2}{I} = \frac{(16.453 \times 10^6 \times (-196)}{4.01 \times 10^8} = -8.04\text{N/mm}^2$$

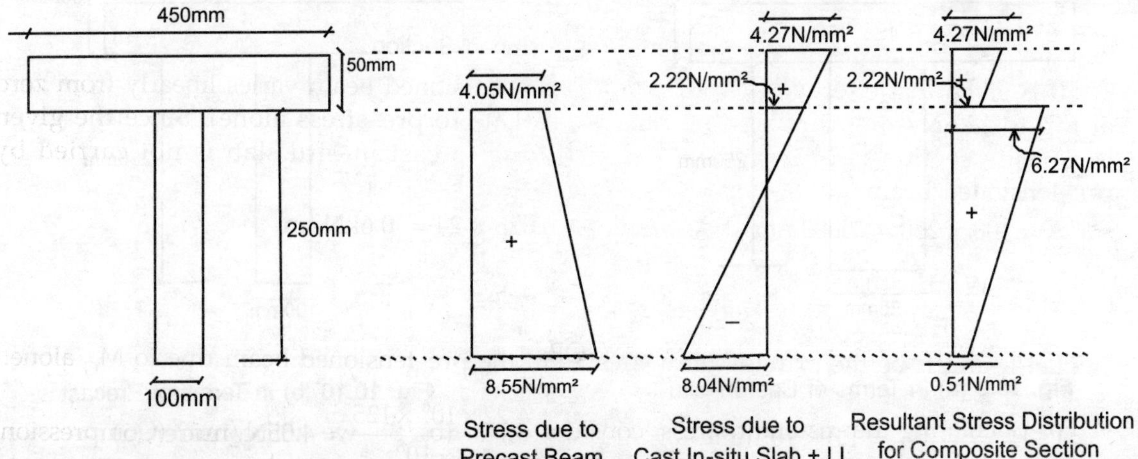

Fig. 10.7 Resultant Stress Distribution for Composite Beam

Problem 10.2

Calculate the stresses in the composite beam in the previous example, if the precast unit has a modulus of elasticity $3.5 \times 10^4 N/mm^2$ and that of in–situ slab a modulus of elasticity $2.8 \times 10^4 N/mm^2$.

Solution:

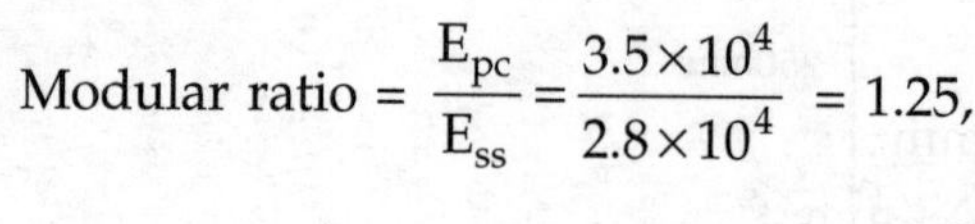

$$\text{Modular ratio} = \frac{E_{pc}}{E_{ss}} = \frac{3.5 \times 10^4}{2.8 \times 10^4} = 1.25,$$

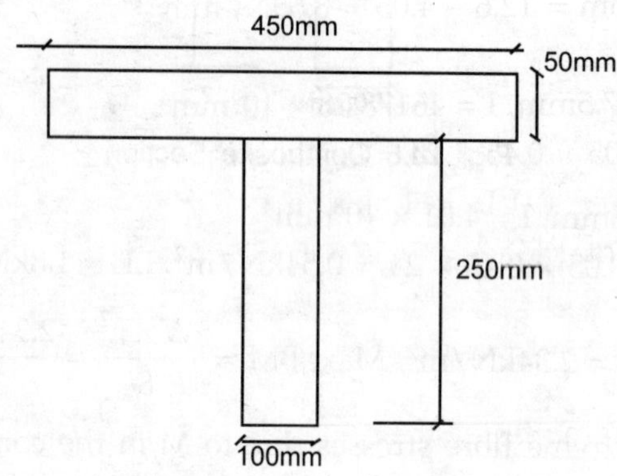

Fig. 10.8 Given Composite Section

(a) In Terms of Cast In–situ

Width of Beam = $100 \times m$

= $100 \times 1.25 = 125$mm

OR

(b) In Terms of Precast

Width of Slab = $450 \times 1/m$

= $450 \times 1/1.25 = 360$mm

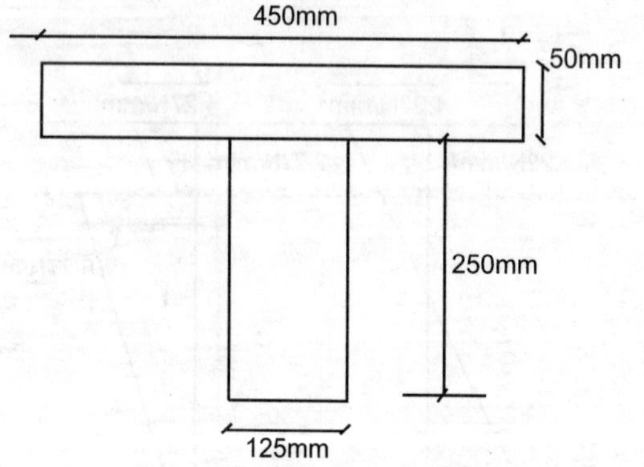

OR

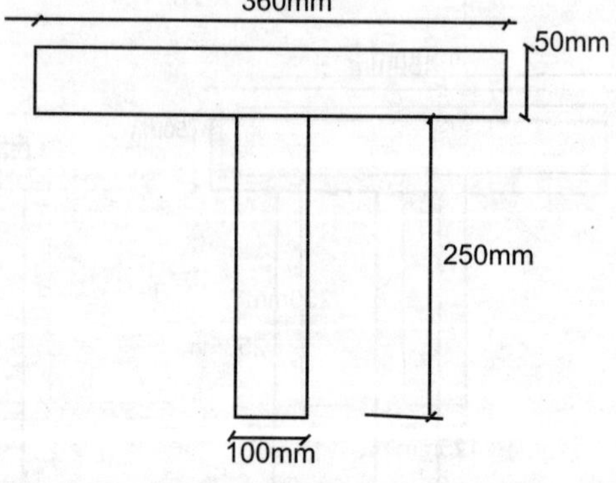

Fig. 10.9 (a) In Terms of Cast In–situ

Fig. 10.10 (b) In Terms of Precast

For calculating the resultant stress, consider (a) or (b). So, we will consider (a)

$$A = 0.125 \times 0.25 = 0.03125\text{mm}^2$$

$$\text{Self–weight} = 0.03125 \times 24 = 0.75\text{kN/m}$$

$$M_G = \frac{W \times L^2}{8} = \frac{0.75 \times 7.5^2}{8} = 5.273\text{kNm}$$

Let f_1 and f_2 be the extreme fibre stresses in the pre–tensioned beam due to M_G alone.

$$I = \frac{125 \times 250^3}{12} = 16.276 \times 10^7 mm^4$$

Here, $f_1 = \frac{M_G y_1}{I} = \frac{5.273 \times 10^6 \times 125}{16.276 \times 10^7} = 4.05N/mm^2$ (Compression)

and $f_2 = -4.05N/mm^2$ (Tension)

∴ Resultant stress at top = 0 + 4.05 = 4.05N/mm²

Resultant stress at bottom = 12.6 – 4.05 = 8.55N/mm²

- **Composite Section**

$Y_1 = 112.5mm$, $Y_2 = -187.5mm$, $I = 46178.28 \times 10^4 mm^4$

Self–weight of slab = 0.05 × 0.45 × 24 = 0.54kN/m²

LL = 1.8kN/m²

Total load = 0.54 + 1.8 = 2.34kN/m²

$$\text{Max. BM} = \frac{W \times L^2}{8} = \frac{2.34 \times 7.5^2}{8} = 16.453kNm$$

Let f_1' and f_2' be the extreme fibre stresses due to moment in the composite section.

Here, $f_1' = \frac{My}{I} = \frac{(16.453 \times 10^6 \times 112.5)}{46178.28 \times 10^4} = 4N/mm^2$ and

$$f_2' = \frac{My_2}{I} = \frac{(16.453 \times 10^6 \times (-187.5))}{46178.28 \times 10^4} = -6.679N/mm^2$$

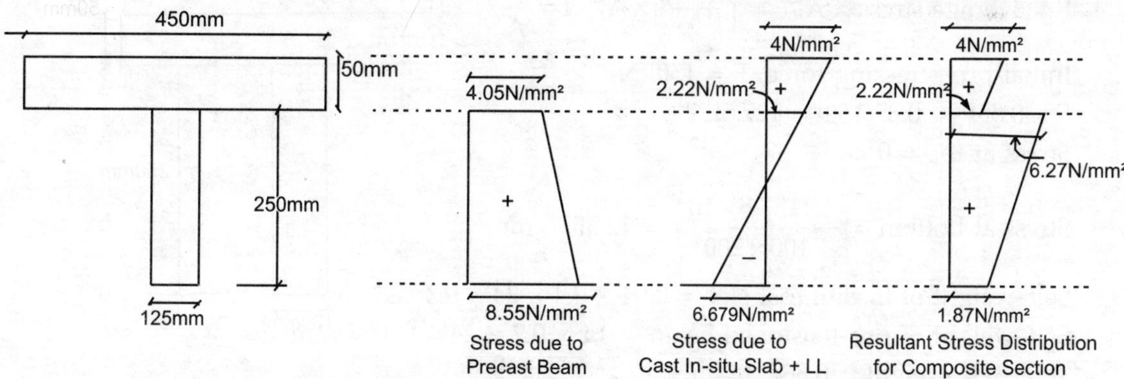

Fig. 10.11 Resultant Stress Distribution for Composite Beam

Problem 10.3

A precast pre–tensioned beam of rectangular section has a breadth of 100mm and depth of 200mm. The beam with an effective span of 5m is pre–stressed by tendons with their centroids coinciding with the bottom kern. The initial force in the tendon is 150kN. The loss of pre–stress may be assumed to be 15%. The beam is incorporated in composite T–beam by casting a top flange of breadth 400mm and thickness of 40mm. if the composite beam supports a live load of 8kN/m², calculate resultant stress developed in the precast and in–situ cast concrete assuming the pre–tensioned beam as un–propped during the casting of the slab. Assume the same modulus of elasticity for concrete in precast beam and in–situ cast slab.

Solution:

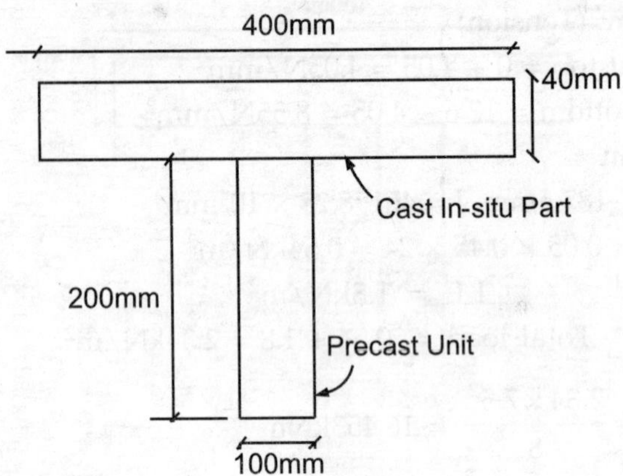

Fig. 10.12 Rectangular Pre–tensioned Beam with Cast In–situ Slab

Note: As the beam is un–propped, the stresses in the precast are computed by considering the self–weight if both the parts and here the pre–stressing force passes through the bottom kern points. For this condition, final stress at the top edge should reach zero and bottom edge = $\frac{2F_o}{A}$.

$$F = \text{average stress} \times A,\ F = \left(\frac{0+f}{2}\right) \times A, \therefore f = \frac{2F}{A}$$

Initial pre–stressing force, F_o= 150kN

$F = 0.85F_o = 0.85(150) = 127.5\text{kN}$

Stress at top = 0

$$\text{Stress at bottom} = \frac{2 \times 127.5 \times 10^3}{100 \times 200} = 12.75\text{N/mm}^2$$

Self–weight of in–situ cast slab = 0.04 × 0.4 × 24 = 0.384kN/m

Self–weight of pre–tensioned beam = 0.1 × 0.2 × 24 = 0.48kN/m

Total self–weight = 0.864kN/m

Let f_1 and f_2 be the extreme fibre stresses in the pre–tensioned beam due to M_G alone.

$$I = \frac{100 \times 200^3}{12} = 6666.66 \times 10^4\text{mm}^4$$

$$f_1 = \frac{M_G y_1}{I} = \frac{2.7\times10^6\times100}{6666.66\times10^4} = 4.05\text{N/mm}^2 \text{ (Compression) and } f_2 = -4.05\text{N/mm}^2 \text{ (Tension)}$$

- **Composite Section**

$Y_1 = 87\text{mm}$, $Y_2 = -153\text{mm}$, $I = 19480 \times 10^4\text{mm}^4$

$LL = 8 \times 0.4 = 3.2\text{kN/m}$

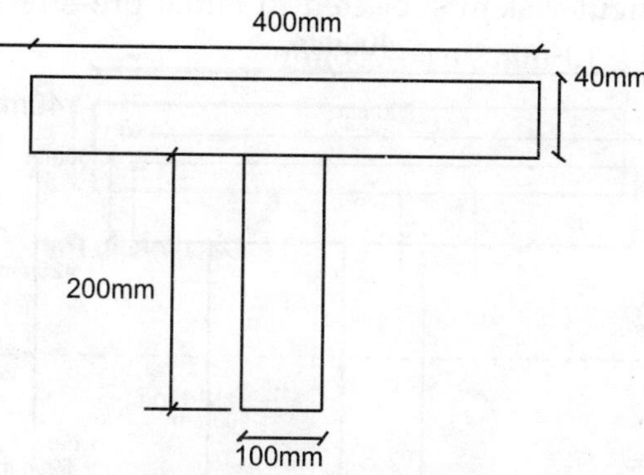

Fig. 10.13 Composite Section

$$\text{Max. BM due to LL} = \frac{W\times L^2}{8} = \frac{3.2\times5^2}{8} = 10\text{kNm}$$

Let f_1' and f_2' be the extreme fibre stresses due to moment in the composite section.

$$f_1' = \frac{My_1}{I} = \frac{(10\times10^6\times87)}{19480\times10^4} = 4.466\text{N/mm}^2, f'_2 = \frac{My_2}{I} = \frac{(10\times10^6\times(-153))}{19480\times10^4} = -7.854\text{ N/mm}^2$$

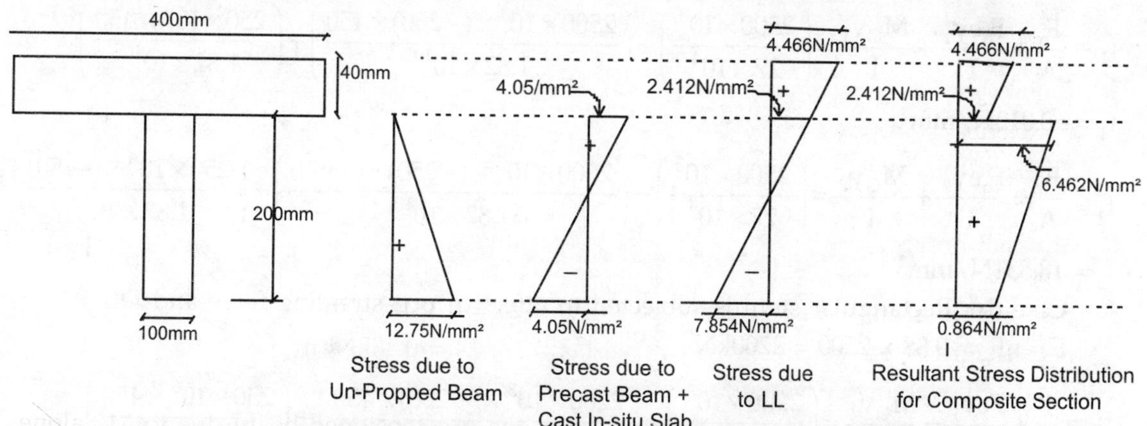

Fig. 10.14 Resultant Stress Distribution for Composite Beam

Problem 10.4

Fig. 10.14 shows the cross section of a composite beam. The beam is pre–tensioned with an initial pre–stressing force of 2500kN. The loss of pre–stress may be assumed to be 12%.

The DL moment at mid–span due to the weight of the precast section is 250kNm. The DL moment due to the weight of flange is 125kNm. After the hardening of the flange concrete, the composite section has to carry a live load which produces a bending moment of 700kNm. Determine the stresses in the concrete at the various stages of loading.

Solution:

Case (a): When rectangular stem subjected to initial pre–stressing force and DL

$F_o = 2500\text{kN}$, $A = 2700 \times 10^2\text{mm}^2$, $e = -250\text{mm}$

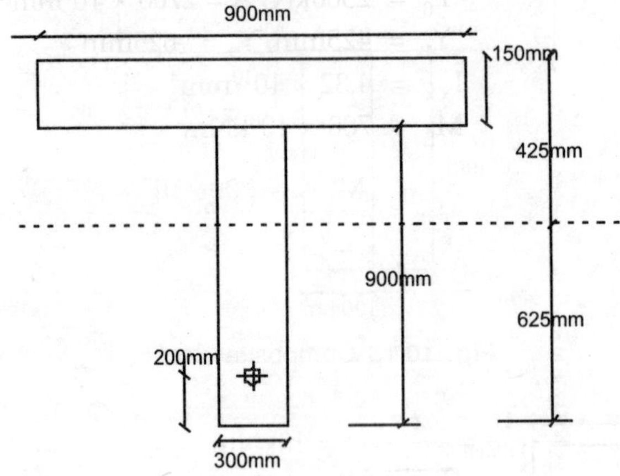

Fig. 10.15 Cross–section of Composite Section

$Y_1 = 450\text{mm}$, $Y_2 = -450\text{mm}$

$I = 1.82 \times 10^{10}\text{mm}^4$

$M_G = 250 \times 10^6\text{kNm}$

$$f_1 = \frac{F_o}{A} + \frac{F_o e y_1}{I} + \frac{M_G y_1}{I} = \left(\frac{2500 \times 10^3}{27 \times 10^4}\right) + \left(\frac{2500 \times 10^3 \times (-250) \times 450}{1.82 \times 10^{10}}\right) + \left(\frac{250 \times 10^6 \times 450}{1.82 \times 10^{10}}\right)$$

$f_1 = -0.013\text{N/mm}^2$

$$f_2 = \frac{F_o}{A} + \frac{F_o e y_2}{I} + \frac{M_G y_2}{I} = \left(\frac{2500 \times 10^3}{27 \times 10^4}\right) + \left(\frac{2500 \times 10^3 \times (-250) \times (-450)}{1.82 \times 10^{10}}\right) + \left(\frac{250 \times 10^6 \times (-450)}{1.82 \times 10^{10}}\right)$$

$f_2 = 18.531\text{N/mm}^2$

Case (b): Rectangular stem is subjected to effective pre–stressing force and DL

$F = \eta F_o = 0.88 \times 2500 = 2200\text{kN}$

$$f_1 = \frac{F}{A} + \frac{F e y_1}{I} + \frac{M_G y_1}{I} = \left(\frac{2200 \times 10^3}{27 \times 10^4}\right) + \left(\frac{2200 \times 10^3 \times (-250) \times 450}{1.82 \times 10^{10}}\right) + \left(\frac{250 \times 10^6 \times 450}{1.82 \times 10^{10}}\right)$$

$f_1 = 0.730\text{N/mm}^2$

$$f_2 = \frac{F}{A} + \frac{F e y_2}{I} + \frac{M_G y_2}{I} = \left(\frac{2200 \times 10^3}{27 \times 10^4}\right) + \left(\frac{2200 \times 10^3 \times (-250) \times (-450)}{1.82 \times 10^{10}}\right) + \left(\frac{250 \times 10^6 \times (-450)}{1.82 \times 10^{10}}\right)$$

$f_2 = 15.565 N/mm^2$

Case (c): Rectangular stem is subjected to effective pre-stressing force, DL of stem and DL of flange

Total DL moment = DL moment due to stem + DL moment due to flange

$= 250 \times 10^6 + 125 \times 10^6 = 375 \times 10^6 Nmm$

Therefore, $f_1 = 3.821 N/mm^2$ and $f_2 = 12.471 N/mm^2$

Case (d): When composite section is subjected to live load

$$F_0 = 2500kN, A = 2700 \times 10^2 mm^2, e = -250mm$$

$$Y_1 = 425mm, Y_2 = -625mm$$

$$I_{XX} = 4.32 \times 10^{10} mm^4$$

$$M_L = 700 \times 10^6 kNm$$

$$f_1 = \frac{M_L y_1}{I} = \frac{700 \times 10^6 \times 425}{4.32 \times 10^{10}} = 6.873 N/mm^2$$

$$f_2 = \frac{M_L y_2}{I} = \frac{700 \times 10^6 \times (-625)}{4.32 \times 10^{10}} = -10.127 N/mm^2$$

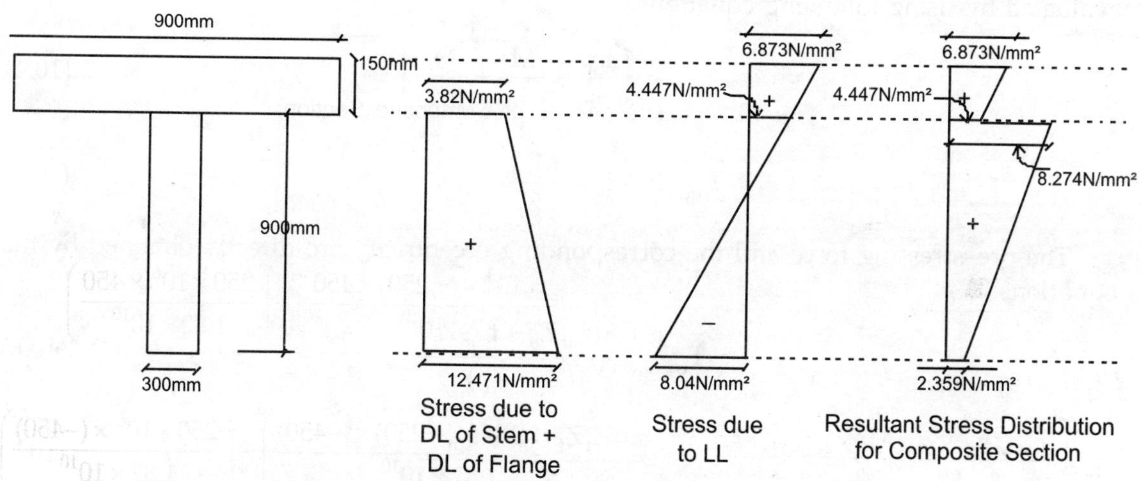

Fig. 10.16 Resultant Stress Distribution for Composite Beam

10.3 DESIGN OF COMPOSITE SECTIONS

The dimensioning of the composite sections involves determining the required size of the composite section using a standard precast pre-stressed beam of known section properties in order to support the required design service loads. Alternatively, it is necessary to determine the section modulus of the precast pre-stressed section for a composite slab of given depth.

The critical stress condition generally occurs at the soffit of the precast pre-stressed element under minimum and maximum moments. Hence, at the stage of transfer, when the minimum moment (self-weight of precast beam) is acting on the precast pre-stressed beam, the stress condition is

$$\left(f_{inf} - \frac{M_{min}}{Z_b} \right) \leq f_{ct} \qquad ...(10.1)$$

If Z_b' = Section modulus of the bottom fibre of the composite section.

M = Moment acting on the precast part of a composite section during construction

M′ = Moment acting on the composite section which is generally due to imposed loads

η = Loss ratio

We then have the stress condition at the soffit of the composite section,

$$\left(\eta f_{inf} - \frac{M}{Z_b} - \frac{M'}{Z_b'} \right) \geq f_{tw} \qquad ...(10.2)$$

By eliminating the pre–stress, f_{inf}, from Eq. 10.1 and Eq. 10.2, the required section modulus of the composite section is given by,

$$Z_b' \geq \left[\frac{Z_b M'}{Z_b(\eta f_{ct} - f_{tw}) - (M - \eta M_{min})} \right] \qquad ...(10.3)$$

The pre–stress required at the bottom and top fibres of the precast pre–stressed beam is evaluated by using following equations.

$$f_{inf} \geq \left(\frac{f_{tw}}{\eta} + \frac{M}{\eta Z_b} + \frac{M'}{\eta Z_b'} \right) \qquad ...(10.4)$$

$$f_{sup} \geq \left(f_{tt} - \frac{M_{min}}{Z_t} \right) \qquad ...(10.5)$$

The pre–stressing force and the corresponding eccentricity are directly obtained by the equations.

$$F_o = \frac{A(f_{inf} Z_b + f_{sup} Z_t)}{(Z_t + Z_b)} \qquad ...(10.6)$$

$$e = \frac{Z_t Z_b (f_{inf} - f_{sup})}{A(f_{inf} Z_b + f_{sup} Z_t)} \qquad ...(10.7)$$

If it is required to determine the section modulus of the precast pre–stressed section in a composite slab of given depth, Eq. 10.2 is rearranged in another form, as

$$Z_b \geq \left[\frac{Z_b'(M - \eta M_{min})}{Z_b'(\eta f_{ct} - f_{tw}) - M'} \right] \qquad ...(10.8)$$

The required pre–stress is calculated using Eq. 10.3 and Eq. 10.4 and the stress developed in the in–situ and pre–stressed components are checked under working loads.

10.3.1 Solved Examples

Problem 10.5

Design a composite slab for the bridge deck using a standard inverted T–section. The top flange is 250mm wide and 100mm thick. The bottom flange is 500mm wide and 250mm deep. The web thickness is 100mm and overall depth of inverted T–section = 655mm. The bridge deck has to support a characteristic imposed load of 50kN/m^2 over an effective span of 12m. M40 grade concrete is specified for precast pre–tensioned T with a compressive strength at transfer of 36MPa. Concrete M30 grade is used for in–situ part. Determine the minimum pre–stress and check for safety under serviceability limits.

Solution:

Section Properties:

Area of pre–tensioned T = A = 180500mm^2

Position of centroid = 220mm from the soffit

Second moment of area = I = 81 × 10^{8}mm^4

Section modulus, Z_t = 18.7 × 10^{6}mm^3

Section modulus, Z_b = 37 × 10^{6}mm^3

Using straight, parallel pre–tensioned tendons with a loss ratio of $\eta = 0.8$, the minimum moment for propped construction $M_{min} = 0$.

The overall depth of the composite slab is estimated as (655 + 95) = 750mm

Load due to self–weight of the precast beam and the in–situ concrete is given by,

$$g = 0.75 \times 0.5 \times 24 = 9\text{kN/m}$$

$$\text{Live load} = q = 50 \times 0.5 = 25\text{kN/m}$$

Corresponding moment $M = gL^2/8 = 9 \times 12^2/8 = 162\text{kNm}$

Live load moment, $M' = qL^2/8 = 25 \times 12^2/8 = 450\text{kNm}$

Permissible compressive stress in concrete at transfer, $f_{ct} = 0.5f_{ci} = 0.5 \times 36 = 18\text{N/mm}^2$

Permissible tensile stress in concrete under service loads = $f_{tw} = -1\text{N/mm}^2$

The minimum section modulus required for the composite section is given by,

$$Z'_b \geq \left[\frac{Z_b M'}{Z_b(\eta f_{ct} - f_{tw}) - (M - \eta M_{min})}\right] \geq \left[\frac{37 \times 10^6 \times 450 \times 10^6}{37 \times 10^6(0.8 \times 18 - (-1)) - (162 \times 10^6 - 0.8 \times 0)}\right]$$

$Z'_b \geq 41 \times 10^6\text{mm}^3$

If H be the total depth of composite slab

$$\left(\frac{BH^2}{6}\right) = 41 \times 10^6, \text{ where } B = 500\text{mm}, \therefore\ H = 705\text{mm}$$

The total depth of 750mm is adequate.

$$Z'_b = \frac{500 \times 750^2}{6} = 46.87 \times 10^6\text{mm}^3$$

The minimum pre–stress required is obtained as,

$$f_{inf} = \left(\frac{f_{tw}}{\eta} + \frac{M}{\eta Z_b} + \frac{M'}{\eta Z'_b}\right)$$

$$= \left(\frac{-1}{0.8} + \frac{162\times10^6}{0.8\times37\times10^6} + \frac{450\times10^6}{0.8\times46.87\times10^6}\right) = 16.2\text{N/mm}^2$$

$$f_{sup} = \left(f_{tt} - \frac{M_{min}}{Z_t}\right) = (-1-0) = -1\text{N/mm}^2$$

Minimum pre–stressing force and eccentricity are computed as follows

$$F_o = \frac{A(f_{inf}Z_b + f_{sup}Z_t)}{(Z_t + Z_b)} = \frac{180500(16.32\times37\times10^6 - 1\times18.7\times10^6)}{(37\times10^6 + 18.7\times10^6)\times1000} = 1881\text{kN}$$

$$e = \frac{Z_tZ_b(f_{inf} - f_{sup})}{A(f_{inf}Z_b + f_{sup}Z_t)} = \frac{18.7\times10^6\times37\times10^6(16.32-(-1))}{180500(16.32\times37\times10^6 - 1\times18.7\times10^6)} = 113\text{mm}$$

A suitable tendon is arranged with straight, parallel wires distributed in the bottom and top flanges as to achieve the required eccentricity.

The stresses due to pre–stress and the loads developed in the precast and in–situ cast concrete are shown in Fig. 10.17.

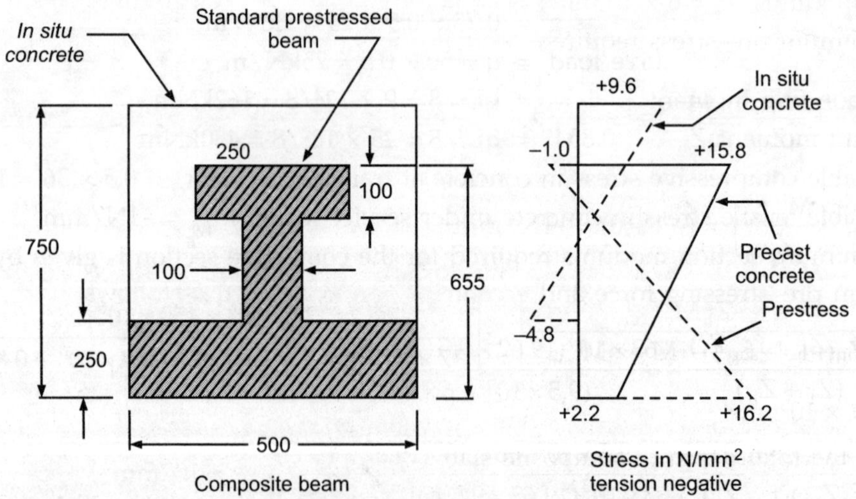

Fig. 10.17 Design of Composite Deck Slab using a Standard Pre–stressed Beam

Problem 10.6

Design a suitable precast pre–stressed inverted T–section used in a composite slab of total depth 600mm and with 300mm. The composite slab is required to support an imposed load of 16kN/m² over a span of 14m. The compressive stress in concrete at transfer and tensile stress under working loads may be assumed to be 20MPa and 1MPa respectively. The loss ratio is 0.85. Determine the pre–stressing force required for the section.

Solution:

Load due to self–weight of the precast beam and the in–situ concrete is given by,
$g = 0.3 \times 0.6 \times 24 = 4.32$kN/m, Live load = $q = 16 \times 0.3 = 4.8$kN/m
Corresponding moment, $M = gL^2/8 = 4.32 \times 14^2/8 = 105$kNm
Live load moment, $M' = qL^2/8 = 4.8 \times 12^2/8 = 118$kNm

$$\text{Section modulus, } Z'_b = \frac{300 \times 600^2}{6} = 18 \times 10^6 \text{mm}^3$$

The minimum section modulus required for the composite section is given by,

$$Z_b \geq \left[\frac{Z'_b(M - \eta M_{min})}{Z'_b(\eta f_{ct} - f_{tw}) - M'}\right] = \left[\frac{18 \times 10^6 \times (105 \times 10^6 - 0)}{18 \times 10^6 (0.85 \times 20 - (-1)) - 118 \times 10^6}\right] \geq 9.2 \times 10^6 \text{ mm}^3$$

An inverted T with the following dimensions and section properties will provide the required section modulus.

Thickness and width of top flange = 100mm and 150mm respectively
Thickness and width of bottom flange = 100mm and 300mm respectively
Thickness of the web = 60mm, Overall depth of the T–section = 500mm
Area = $A = 63 \times 10^3$mm²
Section modulus, $Z_t = 6 \times 10^6$mm³, Section modulus, $Z_b = 9.5 \times 10^6$mm³
The minimum pre–stress required is obtained as,

$$f_{inf} = \left(\frac{f_{tw}}{\eta} + \frac{M}{\eta Z_b} + \frac{M'}{\eta Z'_b}\right) = \left(\frac{1}{0.85} + \frac{105 \times 10^6}{0.85 \times 9.5 \times 10^6} + \frac{118 \times 10^6}{0.85 \times 18 \times 10^6}\right) = 19.62 \text{N/mm}^2$$

$$f_{sup} = \left(f_{tt} - \frac{M_{min}}{Z_t}\right) = (-1 - 0) = -1 \text{N/mm}^2$$

Minimum pre–stressing force and eccentricity are computed as follows

$$F_o = \frac{A(f_{inf} Z_b + f_{sup} Z_t)}{(Z_t + Z_b)} = \frac{63 \times 10^3 (19.62 \times 9.5 \times 10^6 - 1 \times 6 \times 10^6)}{(9.5 \times 10^6 + 6 \times 10^6) \times 1000} = 730 \text{kN}$$

$$e = \frac{Z_t Z_b (f_{inf} - f_{sup})}{A(f_{inf} Z_b + f_{sup} Z_t)} = \frac{9.5 \times 10^6 \times 6 \times 10^6 (19.62 - (-1))}{63 \times 10^3 (19.62 \times 9.5 \times 10^6 - 1 \times 6 \times 10^6)} = 102 \text{mm}$$

A suitable tendon is arranged with straight, parallel wires distributed in the bottom and top flanges as to achieve the required eccentricity.

REVIEW QUESTIONS

10.1 Give introduction about composite construction.
10.2 Explain types of composite construction.
10.3 Elaborate composite construction when the composite beam is propped.
10.4 Explain composite construction when the composite beam is un–propped.
10.5 List the advantages of composite construction.
10.6 Explain the steps used in the analysis of composite section.
10.7 Explain the steps used in the design of composite section.

EXERCISE PROBLEMS

10.1 A composite bridge deck of span 12m is made up of precast pre–stressed symmetrical I–section and an in–situ cast slab. The precast I–beams are placed at 750mm c/c and the top slab of the in–situ concrete is 120mm thick. The cross sectional details if the precast I–section is given as follows.

Width of top and bottom flanges = 200mm, Thickness of web = 75mm

Thickness of the top and bottom flanges = 110mm

Depth of precast I–section = 500mm

Self–weight of precast concrete = 24kN/m^3

Self–weight of cast in–situ concrete = 23.5kN/m^3

The pre–stressed beam is un–propped during the placing of in–situ concrete.

The formwork load is estimated to be 0.2kN/m of span.

If the compressive pre–stress in the beam is 15kN/mm^2 at the bottom and zero at the top, calculate the maximum stresses developed in the precast and in–situ cast concrete under the imposed load of 5kN/mm^2, assuming,

(a) Modular ratio of cast in–situ concrete to precast concrete to be 1.0

(b) Modular ratio of cast in–situ concrete to precast concrete to be 0.8

[Ans: In–situ cast concrete, (a) +2.3N/mm^2 (at top) (b) +2.1N/mm^2 (at top)
Precast Concrete, (a) +10.2N/mm^2 (junction) (b) +10.5N/mm^2 (junction)]

10.2 The deck of a pre–stressed concrete bridge girder with an overall depth of 300mm is made up of inverted T–section, with in–situ concrete laid over it. The precast T–section has the following dimensions and properties.

Width and depth of stem = 70 and 160mm respectively

Width and depth of slab = 300 and 80mm respectively

Height of centroid from soffit = 80mm

Second moment of area = 1472 × 10^{5}mm^4

Pre–stress at bottom = 11MPa (Compression), Pre–stress at top = –1MPa (Tension)

Modulus of elasticity of concrete = 35kN/mm^2

The bridge has a span of 6m and the precast beams are required to support the weight of web concrete infill without any propping. When the infill, which may be assumed to have a modulus of elasticity of $28kN/m^2$, has hardened, a uniformly distributed live load of $13kN/m^2$ is applied.

Calculate the resultant final stress at

(a) the top and bottom of the precast beams

(b) the highest and lowest points in the concrete infill.

[Ans: (a) Top: $12.4N/mm^2$ (Compression), Bottom: $1.3N/mm^2$ (Compression)

(b) Top: $3.65N/mm^2$ (Compression), Bottom: $-1.54N/mm^2$ (Tension)]

10.3 The mid–span section of a composite T–beam comprises a pre–tensioned beam, 300mm wide and 900mm deep and an in–situ concrete slab of 900mm wide and 150mm deep. The effective pre–stressing force located 200mm above the soffit of the beam is 2180kN. The moment due to dead weight of the precast section is 273kNm at mid–span. After this erected in place, the top slab is cast producing a moment of 136.5kNm at mid–span.

After the slab is hardened, the composite section is to carry maximum live load of 750kNm. Compute the resultant final stresses at

(a) the top of slab, (b) the top and bottom of the precast section.

[Ans: (a) $7.35N/mm^2$ (b) $9.75N/mm^2$ at top and $0.63N/mm^2$ at bottom]

10.4 Design the required depth of composite deck slab of a bridge using the standard inverted T–beam to support an imposed load of $15kN/m^2$ on an effective span of 15m. Determine the minimum pre–stressing force required and its corresponding eccentricity. Assume M40 grade concrete in the precast beam with a compressive strength at transfer of 35MPa. The compressive strength of concrete in the in–situ cast slab is 30MPa, f_{ci} = 17.5MPa, f_{tw} = –2.9MPa.

[Ans: Depth of slab = 740mm, Pre–stressing force = 2750kN and Eccentricity = 112mm]

OBJECTIVE QUESTIONS

10.1 When the cast in–situ part is supported by means of additional supports during the process of hardening. This type of construction is referred to as the …….. construction.

a. propped b. un–propped

c. composite d. under propped

10.2 When the cast in–situ part is supported by the precast part during the process of hardening. In this case it is said to be ………….. construction.

a. propped b. un–propped

c. composite d. under propped

10.3 The pre–stress required at the bottom and top fibres of the precast pre–stressed beam is evaluated by using following equations.

a. $f_{inf} \leq \left(\frac{f_{tw}}{\eta} + \frac{M}{\eta Z_b} + \frac{M'}{NZ'_b}\right)$ and $f_{sup} \geq \left(f_{tt} - \frac{M_{min}}{Z_t}\right)$

b. $f_{inf} \geq \left(\frac{f_{tw}}{\eta} + \frac{M}{\eta Z_b} + \frac{M'}{NZ'_b}\right)$ and $f_{sup} \leq \left(f_{tt} - \frac{M_{min}}{Z_t}\right)$

c. $f_{inf} \leq \left(\frac{f_{tw}}{\eta} + \frac{M}{\eta Z_b} + \frac{M'}{NZ'_b}\right)$ and $f_{sup} \leq \left(f_{tt} - \frac{M_{min}}{Z_t}\right)$

d. $f_{inf} \geq \left(\frac{f_{tw}}{\eta} + \frac{M}{\eta Z_b} + \frac{M'}{\eta Z'_b}\right)$ and $f_{sup} \geq \left(f_{tt} - \frac{M_{min}}{Z_t}\right)$

10.4 The required section modulus of the composite section is given by ....

a. $Z'_b \geq \left[\frac{Z_b M'}{Z_b(\eta f_{ct} + f_{tw}) - (M + \eta M_{min})}\right]$ b. $Z'_b \geq \left[\frac{Z_b M'}{Z_b(\eta f_{ct} - f_{tw}) - (M + \eta M_{min})}\right]$

c. $Z'_b \geq \left[\frac{Z_b M'}{Z_b(\eta f_{ct} - f_{tw}) - (M - \eta M_{min})}\right]$ d. $Z'_b \geq \left[\frac{Z_b M'}{Z_b(\eta f_{ct} + f_{tw}) - (M - \eta M_{min})}\right]$

10.5 The section modulus of the precast pre–stressed section in a composite slab of given depth is given by ....

a. $Z_b \geq \left[\frac{Z'_b(M - \eta M_{mim})}{Z'_b(\eta f_{ct} - f_{tw}) - M'}\right]$ b. $Z_b \geq \left[\frac{Z'_b(M - \eta M_{mim})}{Z'_b(\eta f_{ct} + f_{tw}) + M'}\right]$

c. $Z_b \geq \left[\frac{Z'_b(M + \eta M_{mim})}{Z'_b(\eta f_{ct} - f_{tw}) + M'}\right]$ d. $Z_b \geq \left[\frac{Z'_b(M + \eta M_{mim})}{Z'_b(\eta f_{ct} + f_{tw}) - M'}\right]$

Answers

10.1 a	10.2 b	10.3 d	10.4 c	10.5 a

❑

CHAPTER

11 Design of PSC One Way, Two Way & Flat Slabs and Grid Floors

11.1 DESIGN OF PSC ONE WAY SLABS

The design of pre–stress concrete one way slab spanning between parallel supports is based on the principle of designing members for flexure. One way slabs may be supported across the entire width of the slab by beams, piers or abutments or bearing walls, which is positioned perpendicular to the longitudinal axis of the span, or the supports may be at an angle to the span directions. One way slabs may be continuous over one or several supports. Single or continuous slabs are analyzed for designed moments by considering a unit width of slab.

The pre–stressing force and the eccentricity of the cables required at prominent sections to resist the dead and live load moments are determined and the spacing of the cables or wire fixed based on the availability of the type of tendons. In case of slabs subjected to heavy concentrated loads, transverse reinforcement in the form of mild steel or pre–stressed cables will be required to resist transverse moments. An approximate methods for these moments may be by using data of Westergaard or Kawai or Kist or Bouma. Concentric tendons are generally preferred for transverse pre–stressing of one way slabs to prevent objectionable deflections in the transverse direction. The slabs designed should confirm to the requirement of serviceability and strength. The serviceability requirements include limiting deflections and cracking under working loads, as prescribed in the codes.

11.2 SOLVED EXAMPLES

Problem 11.1

A deck slab of a road bridge of road bridge of span 10m is to be designed as a one way pre–stressed concrete slab, with parallel post–tensioned cables in each of which the force at transfer is 500kN. If the deck slab is required to support a uniformly distributed live load of 25kN/m^2, with the compressive and tensile stress in concrete at any stage not exceeding 15 and zero N/mm^2 respectively. Calculate the maximum horizontal spacing of the cables and their position at the mid–span section. Assume the loss ratio as 0.80.

Solution:

Bending moment due to live load = $M_L = \dfrac{qL^2}{8} = \dfrac{25\times 10^2}{8}$ = 312.5kNm

At transfer, $f_{u1} = 0, fl_1 = 15N/mm^2,$

$f_{u2} = 15N/mm^2, fl_2 = 0$

Section modulus, $Z = \dfrac{BH^2}{6} = \dfrac{M_L}{fl_2 + (1.5\eta - 0.5)f_{u2}} = \dfrac{312.5\times10^6}{0(1.5\times0.8-0.5)15}$

$$\frac{BH^2}{6} = \frac{312.5\times10^6}{10.5}$$

$$H^2 = \frac{312.5\times10^6\times6}{1000\times10.5}$$

$$H = 422.577\text{mm Say } 425\text{mm}$$

$$A = 1000\times425 = 425000\text{mm}^2$$

$$I = \frac{1000\times425^3}{12} = 6.397\times10^9\text{mm}^4$$

$$y_1 = y_2 = H/2 = 212.5\text{mm}$$

$$K^2 = I/A = 15051.76\text{ mm}^2$$

$$DL = g = 0.425\times25 = 10.625\text{kN/ m}^2$$

$$M_G = \frac{gL^2}{8} = \frac{10.625\times10^2}{8} = 132.8125\text{kNm}$$

$$M_T = M_G + M_L = 132.8125 + 312.5 = 445.3125\text{kNm}$$

- **Calculation of F_o and e**

At transfer

$$f_{u1} = \frac{F_o}{A} - \frac{F_oey_1}{I} + \frac{M_Gy_1}{I} \quad \text{...(1)}$$

$$f_{u2} = \frac{F_o}{A} + \frac{F_oey_2}{I} - \frac{M_Gy_2}{I} \quad \text{...(2)}$$

At working

$$fl_1 = \frac{F}{A} - \frac{Fey_1}{I} + \frac{M_Ty_1}{I} \quad \text{...(3)}$$

$$fl_2 = \frac{F}{A} + \frac{Fey_2}{I} - \frac{M_Ty_2}{I} \quad \text{...(4)}$$

Considering (1) and (4)

$$f_{u1} = \frac{F_o}{A} - \frac{F_oey_1}{I} + \frac{M_Gy_1}{I} \ldots \times\eta$$

$$fl_2 = \frac{F}{A} + \frac{Fey_2}{I} - \frac{M_Ty_2}{I}$$

$$\frac{\eta F_o}{A} - \eta\frac{F_o e y_1}{I} + \eta\frac{M_G y_1}{I} = 0 \quad ...(5)$$

$$\frac{\eta F_o}{A} + \eta\frac{F_o e y_2}{I} - \frac{M_T y_2}{I} = 0 \quad ...(6)$$

since $y_1 = y_2$, add (5) and (6)

$$2\frac{\eta F_o}{A} + \eta\frac{M_G y_1}{I} - \frac{M_T y_2}{I} = 0$$

$$2\times\frac{0.8\times F_o}{425000} + 0.8\times\frac{132.8425\times10^6\times212.5}{6.397\times10^9} - \frac{445.3125\times10^6\times212.5}{6.397\times10^9} = 0$$

$F_o = 2994155N$

$F_o = 2994kN$

From (1) we get

$$\frac{2994\times10^3}{425000} - \frac{2994\times10^3\times e\times212.5}{6.397\times10^9} + \frac{138.8125\times10^6\times212.5}{6.397\times10^9} = 0$$

e = 115.19 mm

Assume that the wires are initially stretched to $1100N/mm^2$

This should not be greater than 80% of ultimate strength.

$\therefore$ Required area of high tensile steel wires $= \frac{2994\times10^3}{1100} = 2721mm^2$

Number of 7mmϕ tensile wires $= \frac{2721}{\frac{\pi}{4}\times7^2} = 73$ No.s

Provide 6 Freyssinet cables each having 12 No.s of 7mmϕ.

11.3 DESIGN OF PSC TWO WAY SLABS

The design of two–way slab supported in all four sides involves the computation of bending moments in the principal direction of the slab. The slabs may be supported on masonry wall or beam and may or may not be continuous over the supports. Transverse loads are resisted by the development of two way slab action, resulting in moments in the longer and shorter span directions. The magnitude and nature of moments developed in the two way slabs depend mainly upon the type of load, ratio of the sides of the slab and the degree of restraint at the supports. Bending moment coefficients complied in IS: 456–2000[4] one useful for the design of restrained slabs supported on four sides with provision for torsion at corners.

In case of two–way slabs which are fully pre–stressed (Class I), the serviceability criterion to be satisfied is generally the constraint imposed on upward deflections under pre–stress and self–weight, so that tensile stresses are completely eliminated under service loads. The question of limiting cracks within permissible limits arises only in slabs which are designed as partially pre–stressed systems.

The moments developed in two way slab may be determined by using ultimate load methods such as Johanson Yield line method or Hillerborg's strip method.

11.4 SOLVED EXAMPLES

Problem 11.2

Design a post–tensioned concrete two way slab, 6m × 9m, with discontinuous edges, to support an imposed load of 3kN/m². Cables of four wires of 5mm diameter carrying an effective force of 100kN are available for use. Design the spacing of cables in the two directions. Assume = 40N/mm², f_p = 1600N/mm², E_c = 38kN/mm².

Solution:

$L_x = 6m, L_y = 9m$

$$\frac{L_x}{L_y} = \frac{9}{6} = 1.5; \text{ Thickness of slab} = \frac{\text{Span}}{50} = \frac{6000}{50} = 120\text{mm}$$

Self–weight of slab = 0.12 × 24 × 1 = DL = 2.44kN/m², LL = 3kN/m², FF = 0.12kN/m²

∴ Total load = 6kN/m²

Ultimate load = 1.4DL + 1.6LL = 1.4 × 3 + 1.6 × 3 = 9kN/m²

Referring IS: 456 – 2000, working moments in the middle strips

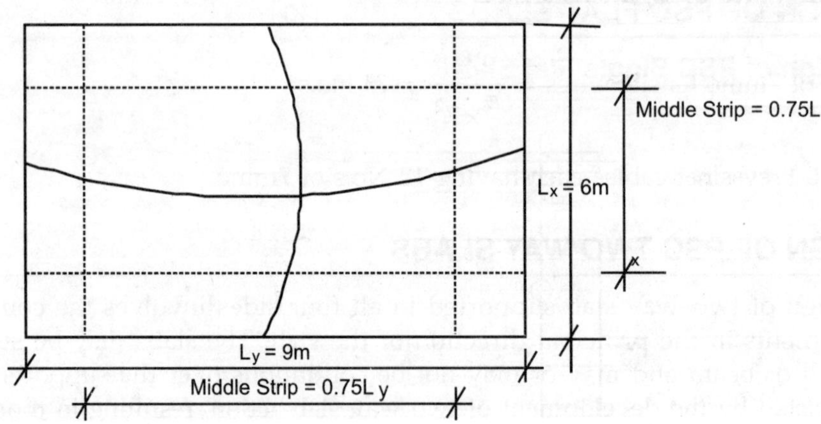

Fig. 11.1 Plan View of Two Way Slab

Referring Table 26 of IS: 456 – 2000,

Working moments in the middle strip

$M_x = 0.089 \times 6 \times 6^2 = 19.3\text{kN/m}, M_y = 0.056 \times 6 \times 6^2 = 12.1\text{kN/m}$

Total moment in the middle strip (X–direction) = 19.3 × 0.75 × 9 = 130.275kNm

Say 130kNm

Using a minimum cover of 30mm for the tendons at the center of slab, the distance between the top kern and the centroid of cable = 120 – 30 – 40 = 50mm

If F_o is the total pre–stressing force required in the X–direction

$F_o \times 50 = 130 \times 10^6$, $F_o = 26 \times 10^5 N = 2600kN$

∴ Force in each cable = 100kN

∴ Number of cables in X–direction (middle strip) = 26 No.s

$$\text{Spacing of cables} = \frac{0.75 \times 9 \times 1000}{26} = 260mm$$

Adopt a spacing of 250mm (four wires per cable)

Total moment in the middle strip (Y–direction) = 12.1 × 0.75 × 6 = 55kNm

Provide a cover of 40mm

Distance between cable and top kern = 120 – 40 – 40 = 40mm

$$\therefore \text{ Pre–stressing force required} = \frac{55 \times 10^6}{40 \times 10^3} = 1380kN$$

$$\therefore \text{ Number of cables in Y–direction (middle strip)} = \frac{1380}{100} = 14 \text{ No.s}$$

$$\text{Spacing of cables} = \frac{0.75 \times 6 \times 1000}{14} = 320mm$$

The cable profile is parabolic with maximum eccentricity at the centre and concentric to the supports.

11.5 DESIGN OF PSC FLAT SLABS

11.5.1 Design of PSC Single Flat Slab

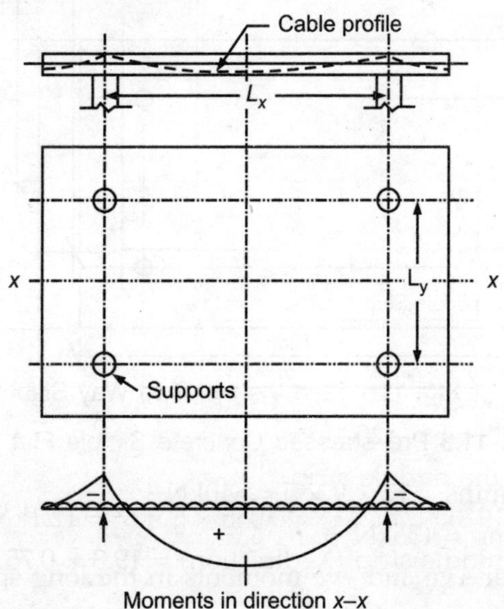

Fig. 11.2 Pre–stressed Concrete Simple Flat Slab

A simple pre–stress flat slab is generally supported by a network of columns without beam and pre–stressed in two perpendicular directions. The design of a typical single flat slab shown in Fig. 11.2 involves the analysis of moments in the two principal directions, so that cables may be arranged to resist these moments. The slab is analyzed as a one way slab and the total no. of cables required to resist the moments in each of the two principal directions are determined. The column strip being stiffer than the middle strip, a greater percentage of tendons are housed in the column strips. Column strip share a higher proportion of the total moment. Since it is not generally possible to vary the spacing of cables for positive and negative moments in either the column or the middle strip middle strips, the total no. of cables required in any direction is proportioned in the ratio 65% and 35% between column and middle strips.

11.5.2 Solved Examples

Problem 11.3

A single flat slab 12m × 9m is supported by four columns so placed to form a symmetrical rectangular grid 7m × 6m. The cantilevers formed are 2.5m and 1.5m in the long and short directions of the slab. The live load on the slab is 1kN/m². Pre–stressing cables consist of four wires of 5mm carrying an effective force of 100kN are available for use. Design the number of cables required and arranges them suitably in the two principal directions.

Solution:

$$\text{Thickness of slab} = \frac{\text{Span}}{40} = \frac{6000}{40} = 150\text{mm}$$

Self–weight of slab = 0.15 × 11 × 24 = DL = 3.6kN/m², LL = 1kN/m², FF = 0.4kN/m²

$$\therefore \quad \text{Total load} = 5\text{kN/m}^2$$

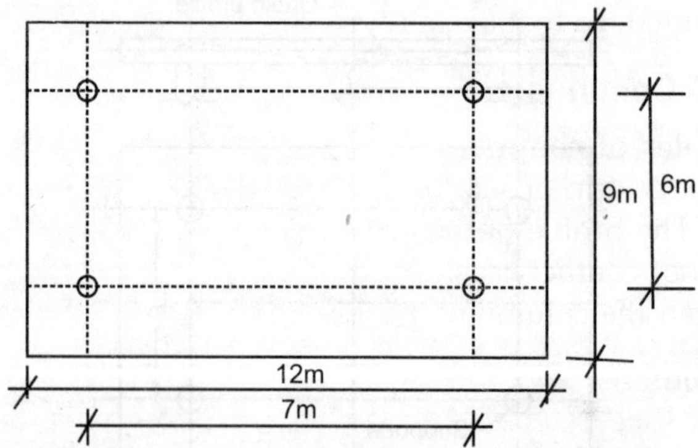

Fig. 11.3 Pre–stressed Concrete Simple Flat Slab

Total load on four columns = 12 × 9 × 5 = 540kN

Reaction on each column = 135kN

The slab is analyzed for +ve and –ve moments in the long span (X–direction) and short span (Y–direction).

Moments in the direction of long span

+ve BM at center of slab = (135 + 135) 3.5 – 5 × 6 × 9 × 3 = 135kNm

–ve BM at center of slab = 9 × 2.5 × 5 × 2.5/2 = 141kNm

The pre–stressing force required is designed to resist the maximum moment of 141kNm in the X–direction. The cable profile is parabolic along the span, so that the eccentricity is proportional to the moment at the section.

Assume distance between the top kern and centroid of cable = 70mm

The total pre–stressing force required in X–direction

$$= F_o = \frac{141 \times 10^3}{70} = 2020\text{kN}$$

$$\therefore \text{ Number of cables in X–direction} = \frac{2020}{100} = 21 \text{ No.s}$$

Similarly, in the short span (Y–direction),

The maximum moment occurs at the Centre of the slab

= 270 × 3 – 5 × 12 × 4.5 × 4.5/2 = 203kNm

The total pre–stressing force required in Y–direction

$$F_o = \frac{203 \times 10^3}{70} = 2900\text{kN, Say } 3000\text{kN}$$

$$\therefore \text{ Number of cables in X–direction} = \frac{3000}{100} = 30 \text{ No.s}$$

(The column and middle strip are 7m and 5m width in the Y direction). The cables are suitably arranged in the column and middle strips in the X and Y–directions. Along the X and Y–direction the width of middle strip and column strip is 3m. The total no. of cables in each direction is distributed, so that 65% are in column strip and 35% are in the middle strip.

11.5.3 Design of PSC Continuous Flat Slab

The design principles of continuous flat slab floors are similar to those of two way reinforced concrete slab. A strip of slab of unit width, continuous over supports, is analyzed as a continuous beam. The profile of the cables in the two principal directions should be so arranged that they do not conflict each other. The no. of cables required to resist moment is suitably divided between the column and middle strips as in case of flat slabs.

11.5.4 Solved Examples

Problem 11.4

Design a continuous pre–stressed flat slab of overall size 16m × 16m. The columns are spread at 7.5m intervals in the perpendicular direction. The floor slab has to support a superimposed load of 3kN/m^2. Freyssinate cables, consisting of 12 wires of 5mm diameter stressed to 1000N/mm^2, are available for use. Determine the no. of cables required and their spacing in each direction.

Solution:

Thickness of slab = $\frac{\text{Span}}{40} = \frac{7.5 \times 1000}{40}$ = 187mm Say 200mm

Self–weight of slab = 0.2 × 1 × 1 × 24 = 4.8kN/m², LL = 3kN/m²

∴ Total load = 7.8kN/m²

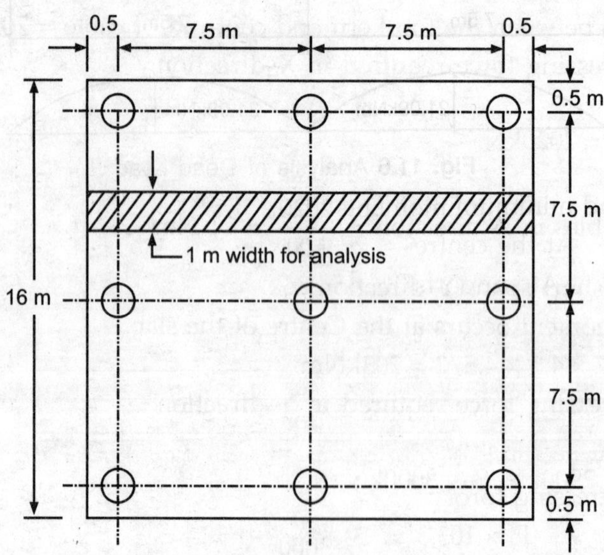

Fig. 11.4 Pre–stressed Concrete Continuous Flat Slab

- **Analysis for Dead load:**

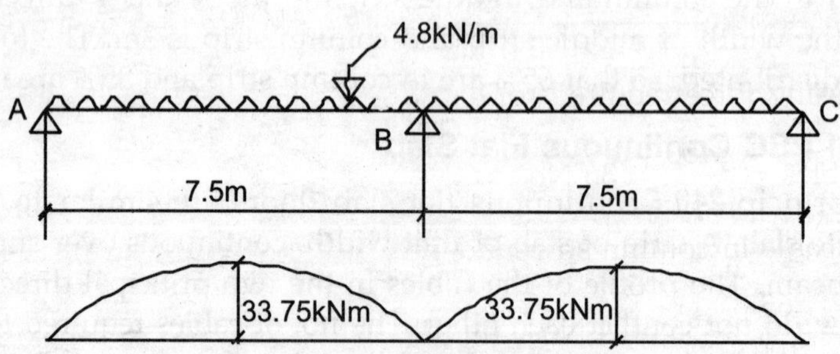

Fig. 11.5 Analysis of Dead Load

Using three moment equation or Claypron's three moment equation

$$M_A L_1 + 2M_B(L_1+L_2) + M_C L_2 = -\frac{6A_1 x_1}{L_1} - \frac{6A_2 x_2}{L_2}$$

$$0 + 2M_B\,(7.5 + 7.5) + 0 = -2\left(\frac{6A_1 x_1}{L_1}\right),\ 2M_B\,(7.5 + 7.5) = -2\left(\frac{6\times\left(\frac{2}{3}\times 7.5\times 33.65\right)\times\frac{7.5}{2}}{7.5}\right)$$

$$\therefore\ M_B = 33.75\text{kNm}$$

- **Analysis for Live load**

Again applying three moment equation

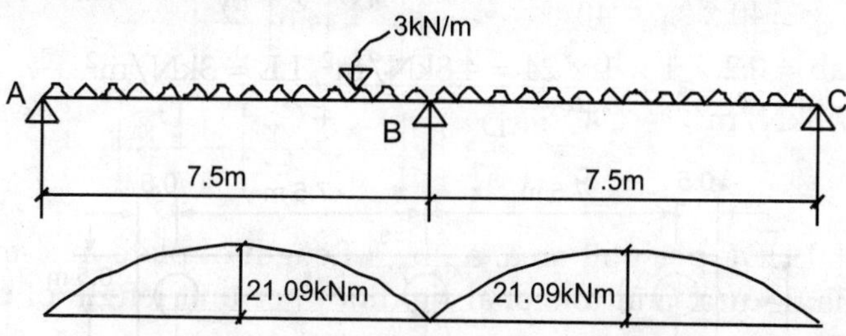

Fig. 11.6 Analysis of Dead Load

The maximum and minimum moments

At the centre = 27.42kNm,

At supports = – 54.84kNm

- **Pre–stressing Force**

The absolute moment occurs at the mid–support section. Using minimum cover of 30mm to the cable, if no tensile stresses are permitted in the section,

The distance between the cable and the bottom kern = 70 + 33.3 = 103.3mm

If F_o is the pre–stressing force

$$F_o \times 103.3 = 54.84 \times 10^6$$

$$F_o = 530\text{kN}$$

Using 12 wires of 5mmϕ stressed to 1000N/mm^2,

The force in each cable = $\frac{12 \times 20 \times 1000}{1000}$ = 240kN

Total pre–stressing force required for a width of 16m = 16 × 530 = 8480kN

Number of cables = $\frac{8480}{240}$ = 36 Nos.

Using 65% of cable in column strip and 35% cable in middle strip,

Column strip = $\frac{65}{100}$ × 36 = 24 Nos., middle strip = $\frac{35}{100}$ × 36 = 14 Nos.

The spacing of cables in column strip = (7.5 × 1000)/24 = 312.5mm

and spacing of cables in middle strip = (7.5 × 1000)/12 = 625mm

11.6 DESIGN OF PSC GRID FLOORS

Pre–stressed concrete grid floor consists of ribs at close intervals in two mutually perpendicular directions and connected by slab in between the ribs can be considered as orthotropic plate freely supported on four sides. Timoshenko's analysis may be used to evaluate the moments and shears in the grid which depend upon the deflection surface.

The vertical deflection 'a' at any point of the grid as shown in Fig. 11.7, is expressed as

$$a = \frac{16w_d}{\pi^6} \frac{\sin\left(\frac{\pi x}{a_x}\right)\sin\left(\frac{\pi y}{a_y}\right)}{\left(\frac{D_x}{a_x^4}\right)+\left(\frac{C_x + C_y}{a_x^2\, b_y^2}\right)+\left(\frac{D_y}{b_y^4}\right)} \quad ...(11.1)$$

Where, w_d = Total UDL per unit area, a_x, b_y = Length of plate in x and y directions, D_x, D_y, C_x and C_y are the flexural and torsional rigidities per unit length of plate along x and y directions respectively.

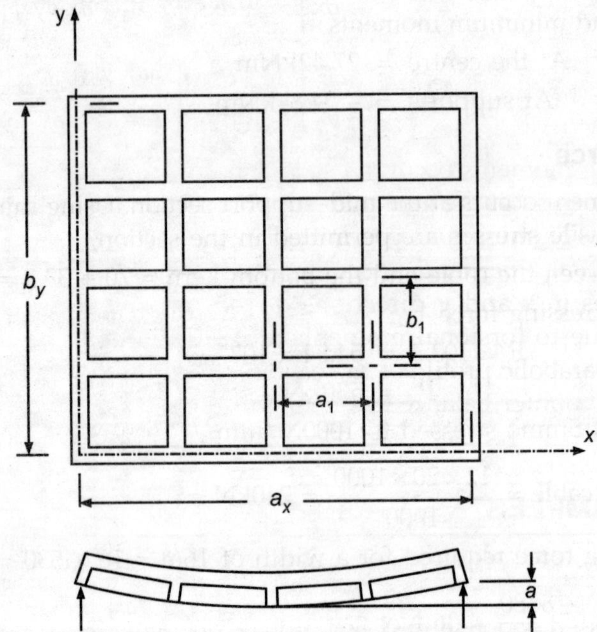

Fig. 11.7 Deflection characteristics of Grid Floors

If a_1 and b_1 are the spacing of the ribs in x and y direction then,

$$D_x = \frac{EI_1}{b_1} \quad ...(11.2)$$

$$D_y = \frac{EI_2}{a_1} \quad ...(11.3)$$

$$C_x = \frac{C_1}{b_1} \quad ...(11.4)$$

$$C_y = \frac{C_2}{a_1} \quad ...(11.5)$$

Where, EI_1, EI_2, $C_1$and C_2 are the flexural and torsional rigidities of the effective section in x and y directions.

The moments and shears are computed using the following formulas

$$M_x = -D_x\left(\frac{\partial^2 a}{\partial x^2}\right), M_y = -D_y\left(\frac{\partial^2 a}{\partial y^2}\right) \quad ...(11.6)$$

$$T_{xy} = -\frac{c_1}{b_1}\left(\frac{\partial^2 a}{\partial x\,\partial y}\right) \qquad T_{yx} = -\frac{C_2}{a_1}\left(\frac{\partial^2 a}{\partial x\,\partial y}\right) \quad ...(11.7)$$

$$Q_x = -\frac{\partial}{\partial x}\left[D_x\left(\frac{\partial^2 a}{\partial x^2}\right) + \frac{C_2}{a_1}\left(\frac{\partial^2 a}{\partial x\,\partial y}\right)\right] \quad ...(11.8)$$

$$Q_y = -\frac{\partial}{\partial x}\left[D_y\left(\frac{\partial^2 a}{\partial y^2}\right) + \frac{C_1}{b_1}\left(\frac{\partial^2 a}{\partial x\,\partial y}\right)\right] \quad ...(11.9)$$

Maximum bending moments occur at centre of span while maximum torsional moments are generated at the corners of the grid and maximum shear forces develop at mid points of longer side supports. The minimum pre–stressing force and eccentricity required at centre of span along the mutually perpendicular directions are determined care being taken to see that the cable profiles in x and y directions do not interfere with each other. The shear stresses developed due to torsional moments and shear should be checked at the critical sections. Due to the parabolic profile of the cables, which are concentric to the supports, the shear due to loads is counter balanced to a considerable extent by a vertical component of the pre–stressing force.

11.7 SOLVED EXAMPLES

Problem 11.5

Design a pre–stressed concrete grid floor to cover a rectangular panel of an office floor, of 12m × 7m. The LL = 3kN/mm^2, M40 grade concrete and Freyssinate cables of 18–5mm and 10–7mm are available for use. The arrangement of ribs in principal directions and the cross sectional details of the ribs are as shown in Fig. 11.8. Design the grid and sketch the details of profiles of the cables in ribs.

Solution:

The ribs are spaced at interval of 1.5m in the y direction and 1.75m in the x direction so that a_1 = 1.75m and b_1 = 1.5m. The second moment of inertia of T–section in x and y directions may be computed by using Reynolds [9] tables.

Area of the section, A = 157200mm^2, $I = Kb_wh^3$

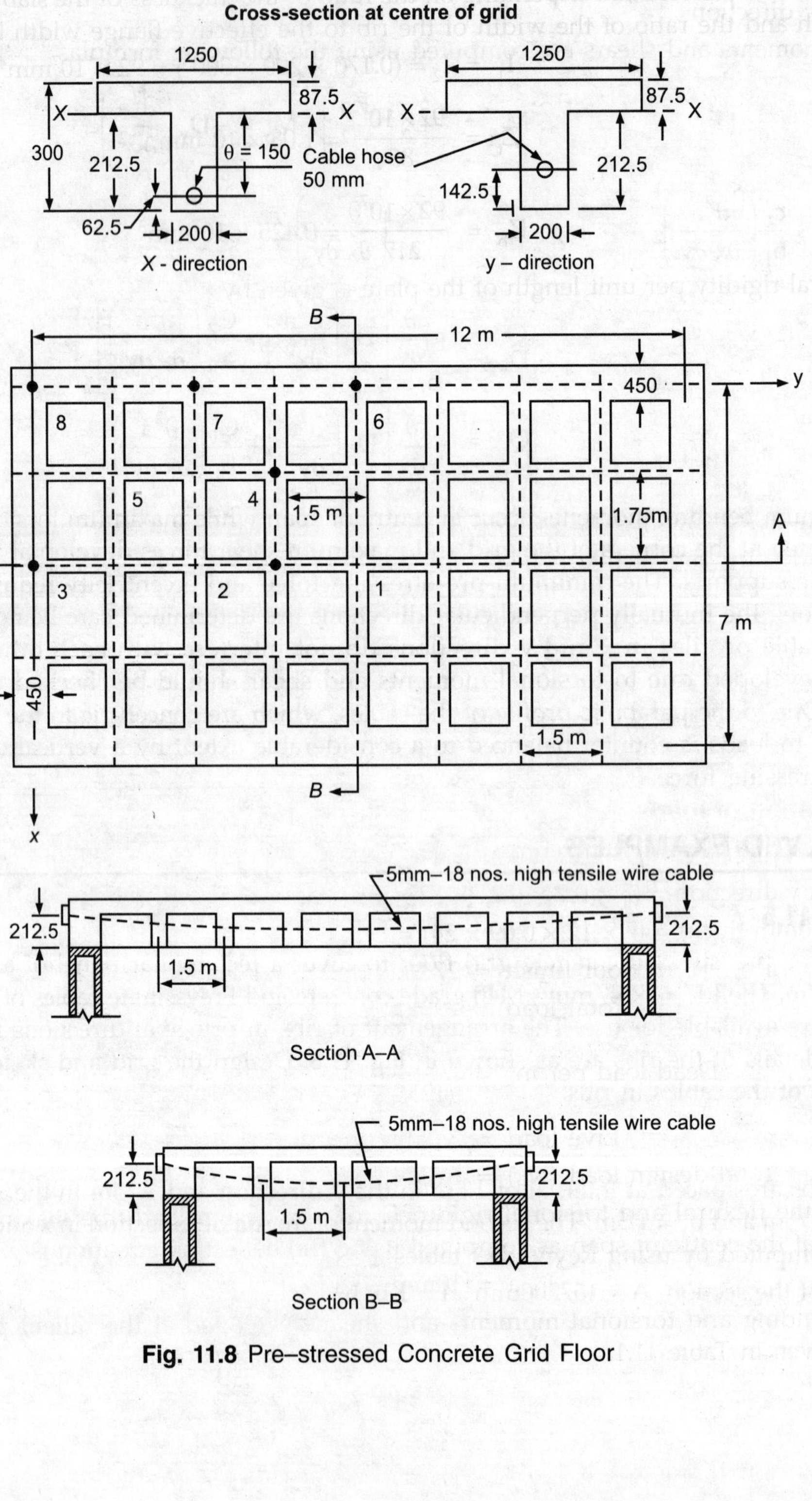

Cross-section at centre of grid

1250
87.5
X
X
300
212.5
θ = 150
Cable hose
50 mm
62.5
200
X - direction

1250
87.5
X
X
212.5
142.5
200
y – direction

B
12 m
450
y
8
7
6
5
4
1.5 m
1.75m
A
A
3
2
1
7 m
450
1.5 m
B
x

5mm–18 nos. high tensile wire cable
212.5
212.5
1.5 m
Section A–A

5mm–18 nos. high tensile wire cable
212.5
212.5
1.5 m
Section B–B

Fig. 11.8 Pre–stressed Concrete Grid Floor

Where, K is the constant depending on the ratio of the thickness of the slab to the overall depth h_f/h and the ratio of the width of the rib to the effective flange width b_w/b. Hence,

$$I_1 = I_2 = (0.170 \times 200 \times 300^3) = 92 \times 10^7 mm^4$$

$$Z_t = \frac{92\times10^7}{85} = 1.08 \times 10^7 mm^3,$$

$$Z_b = \frac{92\times10^7}{217} = 0.425 \times 10^7 mm^3$$

Flexural rigidity per unit length of the plate is given by

$$D_x = \frac{EI_1}{b_1} = \frac{92\times10^7\times E}{1.5\times10^3} = 61 \times 10^4 E,$$

$$D_y = \frac{EI_2}{a_1} = \frac{92\times10^7\times E}{1.75\times10^3} = 53 \times 10^4 E$$

Torsional rigidity of the effective section is computed by using the coefficients recommended by Timoshenko[10]

$$C_1 = C_2 = K_1 G(b_w)^3 h = 0.196 \frac{E}{2(1+0.15)}(200)^3(300) = 2.04 \times 10^8 E$$

$$C_x + C_y = \left[\frac{C_1}{b_1}+\frac{C_2}{a_1}\right] = 2.04\times10^8 E\left[\frac{1}{1.5\times10^3}+\frac{1}{1.75\times10^3}\right] = 2.55 \times 10^5 E$$

- **Computation of loads**

Ribs in x direction = (7 × 0.2 × 0.21 × 7 × 24) = 49.5kN

Ribs in y direction = (3 × 0.2 × 0.21 × 12 × 24) = 36.5kN

Slab (90mm thick) = (7 × 12 × 0.09 × 24) = 182.0kN

Floor finish = 27.0kN

Total load = 295kN

$$\text{Dead load per } m^2 = \frac{295}{(12\times7)} = 3.5 kN/m^2$$

Live load = $3.0 kN/m^2$

Total design load (w_d) = $6.5 kN/m^2$

Using the flexural and torsional rigidities and the design load, the maximum vertical deflection at the centre of span as computed using the deflection equation is given by

a = 10mm

The bending and torsional moments and shear developed at the salient points of the grid are given in Table 11.1.

Table 11.1 Moments and Shear per Metre Width of Grid Floor

Point	x (m)	y (m)	M_x (kNm)	M_y (kNm)	T_{xy} (kNm)	T_{yx} (kNm)	V_x (kN)	V_y (kN)
1	3.5	6.0	39.0	15.1	0	0	0	0
2	3.5	3.0	28.0	10.5	0	0	0	4.30
3	3.5	0	0	0	0	0	0	6.10
4	1.75	4.5	25.5	9.0	1.86	1.25	13.00	1.65
5	1.75	1.5	12.2	4.1	4.52	3.00	5.35	3.95
6	0	6.0	0	0	0	0	19.60	0
7	0	3.0	0	0	–4.90	3.25	14.00	0
8	0	0	0	0	–6.90	4.60	0	0

- **Central rib in x direction**

Total bending moment to be resisted by the effective section at the centre of the grid in x direction is given by

$$M_x = 39 \times 1.5 = 58.5\text{kNm}$$

Dead load moment, $M_G = 0.54 \times 58.5 = 31.5\text{kNm}$

Live load moment, $M_L = 0.46 \times 58.5 = 27.0\text{kNm}$

The minimum section modulus Z_b required to safely resist the dead and live load moments is expressed as,

$Z_b \geq$. Assuming a loss ratio, $\eta = 0.85$, $f_{ct} = 15\text{N/mm}^2$ and $f_{tw} = 0$,

$$f_{br} = (\eta f_{ct} - f_{tw}) = (0.85 \times 15 - 0) = 12.75\text{N/mm}^2$$

$$Z_b \geq \left[\frac{27 + (1 - 0.85)5}{12.75}\right] \geq 2.5 \times 10^6\text{mm}^3$$

Which is less than the value of $Z_b = 4.25 \times 10^6\text{mm}^3$ provided by the section.

The minimum pre–stressing force required is to be obtained from,

$$F_o = \frac{A(f_{inf} Z_b + f_{sup} Z_t)}{(Z_t + Z_b)}$$

$$\text{Where, } f_{sup} = \left(-\frac{M_G}{Z_t}\right) = \left(-\frac{31.5 \times 10^6}{10.8 \times 10^6}\right) = -2.9\text{N/mm}^2$$

$$f_{inf} = \left[\frac{(M_L + M_G)}{\eta Z_b}\right] = \left[\frac{-58.5 \times 10^6}{0.85 \times 4.25 \times 10^6}\right] = 16.3\text{N/mm}^2$$

$$F_o = \frac{A(f_{inf} Z_b + f_{sup} Z_t)}{(Z_t + Z_b)} = \frac{157200(16.3 \times 4.25 - 2.9 \times 10.8)\, 10^6}{15.05 \times 10^6} = 0.395 \times 10^6\text{N} = 395\text{kN}$$

Using Freyssinet cables of 18–5mm diameter stressed to 1100N/mm^2,

$$\text{Force in each cable} = \frac{18 \times 20 \times 1100}{1000} = 396\text{kN}$$

One cable is provided in central rib at an eccentricity of

$$e = \frac{Z_t Z_b (f_{inf} - f_{sup})}{A(f_{inf} Z_b + f_{sup} Z_t)} = \frac{10.8 \times 10^6 \times 4.28 \times 10^6 (16.3 + 2.9)}{157200(16.3 \times 4.25 - 2.9 \times 10.8)\, 10^6} = 150\text{mm}$$

- **Central rib in y direction**

Total bending moment to be resisted by the effective section at the centre of the grid in y direction is given by

$$M_y = 15.1 \times 1.75 = 26.3\text{kNm}$$

Dead load moment, $M_G = 0.54 \times 26.3 = 14.2$kNm

Live load moment, $M_L = 0.46 \times 58.5 = 12.1$kNm

$$f_{sup} = \left(-\frac{M_G}{Z_t}\right) = \left(-\frac{14.1 \times 10^6}{10.8 \times 10^6}\right) = -1.32\text{N/mm}^2$$

$$f_{inf} = \left[\frac{(M_L + M_G)}{\eta Z_b}\right] = \left[\frac{26.3 \times 10^6}{0.85 \times 4.25 \times 10^6}\right] = 7.3\text{N/mm}^2$$

$$F_o = \frac{A(f_{inf} Z_b + f_{sup} Z_t)}{(Z_t + Z_b)} = \frac{157200(7.3 \times 4.25 - 1.32 \times 10.8)\, 10^6}{15.05 \times 10^6} = 0.175 \times 10^6\text{N} = 175\text{kN}$$

The corresponding eccentricity of

$$e = \frac{Z_t Z_b (f_{inf} - f_{sup})}{A(f_{inf} Z_b + f_{sup} Z_t)} = \frac{10.8 \times 10^6 \times 4.28 \times 10^6 (7.3 + 1.32)}{157200\,(7.3 \times 4.25 - 1.32 \times 10.8)\, 10^6} = 150\text{mm}$$

This eccentricity is not practicable as it obstructs the cables in x direction. Hence the maximum possible eccentricity is

$$e = 150 - 20 - 40 - 20 = 70\text{mm}$$

The modified pre–stressing force at the the reduced eccentricity is given by,

$$F_o = \left(\frac{A f_{inf}\, Z_b}{Z_b + Ae}\right) = \left(\frac{157200 \times 7.3 \times 4.25 \times 10^6}{4.25 \times 10^6 + 157200 \times 70}\right) = 320 \times 10^3\text{N} = 320\text{kN}$$

REVIEW QUESTIONS

11.1 Explain the salient design features of PSC one way and two way slabs.

11.2 Explain the design steps in pre–stressed concrete simple flat slabs.

11.3 Mention the design steps in pre–stressed concrete continuous flat slabs.

11.4 Briefly elaborate the design procedure of pre–stressed concrete grid floors.

EXERCISE PROBLEMS

11.1 The floor slab of an industrial structure spanning over 8m is to be designed as a one way pre–stressed concrete slab with parallel post–tensioned cables. The slab is supporting a LL of 10kN/m^2 with the compressive and tensile stresses in concrete at any stage is not exceeding 14MPa and 0MPa respectively. Design a suitable thickness of slab and find the maximum horizontal spacing of the Freyssinate cables of 12–5mm diameter initially stressed to 1200MPa and their position at mid–span section. Take loss ratio as 0.8.

[Ans: Thickness of slab = 220mm, spacing of cables = 187mm]

11.2 The design a two way slab of post–tensioned pre–restressed concrete of 6m × 8m, to support a LL of 3kN/m^2. If cables are 4–5mm diameter stressed to 1000N/mm^2, determine the number of cables in the two principal directions. The stress in concrete is not to exceed 15MPa in compression and tensile stress is not permitted under service loads. Take loss ratio as 0.80.

**[Ans: Thickness of slab = 160mm,
number of cables = 33 (short span) and 14 (long span]**

11.3 A simple flat slab of 10m × 8m, is supported by four columns which are spaced to form a symmetrical grid of 8m × 6m with cantilevers of 1m on all sides. The load of slab is 1.5kN/m^2. Pre–stressing cables consists of 4–4mm diameter stressed to 1000MPa. Design the number of cables and indicate their arrangement in two principal directions.

**[Ans: Thickness of slab = 150mm,
number of cables = 70 (short span) and 103 (long span]**

11.4 A PSC continuous flat slab of 16m by 13m is supported by nine columns arranged in three rows. The columns are spaced 7.5m and 6m in directions of long and short spans respectively with a cantilever of 0.5m all around. The 20cm thick slab continuous over two bays in transverse directions, supports a LL of 3kN/m^2. Assume the tensile stresses in slab is zero under full live load and maximum cover to the centre of the cable is to be 30mm. Determine the magnitude of pre–stressing force in the direction of long span. Estimate the number of cables required for the long span and arrange them suitably in the column and middle strips using 12–5mm diameter cables stressed to 1000MPa.

[Ans: Pre–stressing force = 530kN and number of cables = 29]

11.5 A pre–stressed concrete grid floor for a panel of 15m by 12m has the ribs at 1.5m centres in both the directions. Design the grid floor by assuming it as an orthotropic plate freely supported over four sides. LL over the floor = 3kN/m^2, thickness of slab = 40mm, overall depth of floor = 400mm, width of ribs = 200mm, loss ratio = 0.85. Also sketch the arrangement of cables for the central ribs in transverse directions. Permissible compressive stress in concrete is 15N/mm^2 and tensile stress is zero at transfer and working loads respectively.

**[Ans: Short span, F_o = 710kN, e = 170mm, 18–7mm cables,
Long span, F_o = 645kN, e = 100mm, 18–7mm cables]**

OBJECTIVE QUESTIONS

11.1 The vertical deflection 'a' at any point of the grid is expressed as

a. $a=\dfrac{16w_d}{\pi^6}\dfrac{\sin\left(\dfrac{\pi x}{a_x}\right)\sin\left(\dfrac{\pi y}{a_y}\right)}{\left(\dfrac{D_x}{a_x^4}\right)+\left(\dfrac{C_x+C_y}{a_x^2b_y^2}\right)+\left(\dfrac{D_y}{b_y^4}\right)}$

b. $a=\dfrac{18w_d}{\pi^6}\dfrac{\sin\left(\dfrac{\pi x}{a_x}\right)\sin\left(\dfrac{\pi y}{a_y}\right)}{\left(\dfrac{D_x}{a_x^4}\right)+\left(\dfrac{C_x+C_y}{a_x^2b_y^2}\right)+\left(\dfrac{D_y}{b_y^4}\right)}$

c. $a=\dfrac{20w_d}{\pi^6}\dfrac{\sin\left(\dfrac{\pi x}{a_x}\right)\sin\left(\dfrac{\pi y}{a_y}\right)}{\left(\dfrac{D_x}{a_x^4}\right)+\left(\dfrac{C_x+C_y}{a_x^2b_y^2}\right)+\left(\dfrac{D_y}{b_y^4}\right)}$

d. $a=\dfrac{32w_d}{\pi^6}\dfrac{\sin\left(\dfrac{\pi x}{a_x}\right)\sin\left(\dfrac{\pi y}{a_y}\right)}{\left(\dfrac{D_x}{a_x^4}\right)+\left(\dfrac{C_x+C_y}{a_x^2b_y^2}\right)+\left(\dfrac{D_y}{b_y^4}\right)}$

11.2 If a_1 and b_1 are the spacing of the ribs of grid floor in x and y direction then, flexural rigidities per unit length of plate along x and y directions is given by

a. $D_x=\dfrac{EI_2}{b_1}$ and $D_y=\dfrac{EI_1}{a_1}$

b. $D_x=\dfrac{EI_1}{b_1}$ and $D_y=\dfrac{EI_2}{a_1}$

c. $D_x=\dfrac{EI_1}{a_1b_1}$ and $D_y=\dfrac{EI_2}{a_1b_1}$

d. $D_x=\dfrac{EI_2}{a_1b_1}$ and $D_y=\dfrac{EI_1}{a_1b_1}$

11.3 If a_1 and b_1 are the spacing of the ribs of grid floor in x and y direction then, flexural rigidities per unit length of plate along x and y directions is given by

a. $C_x=\dfrac{C_1}{a_1b_1}$ and $C_y=\dfrac{C_2}{a_1b_1}$

b. $C_x=\dfrac{C_2}{a_1b_1}$ and $C_y=\dfrac{C_1}{a_1b_1}$

c. $C_x=\dfrac{C_1}{b_1}$ and $C_y=\dfrac{C_2}{a_1}$

d. $C_x=\dfrac{C_2}{b_1}$ and $C_y=\dfrac{C_1}{a_1}$

11.4 The bending moments in grid floors is computed by using .... formulae

a. $M_x=D_x\left(\dfrac{\partial^2 a}{\partial x^2}\right)$ and $M_y=-D_y\left(\dfrac{\partial^2 a}{\partial y^2}\right)$

b. $M_x=-D_x\left(\dfrac{\partial^2 a}{\partial x^2}\right)$ and $M_y=D_y\left(\dfrac{\partial^2 a}{\partial y^2}\right)$

c. $M_x=D_x\left(\dfrac{\partial^2 a}{\partial x^2}\right)$ and $M_y=D_y\left(\dfrac{\partial^2 a}{\partial y^2}\right)$

d. $M_x=-D_x\left(\dfrac{\partial^2 a}{\partial x^2}\right)$ and $M_y=-D_y\left(\dfrac{\partial^2 a}{\partial y^2}\right)$

11.5 The torsional moments in grid floors is computed by using .... formulae

a. $T_{xy} = -\frac{c_1}{b_1}\left(\frac{\partial^2 a}{\partial x\,\partial y}\right)$ and $T_{yx} = -\frac{c_2}{a_1}\left(\frac{\partial^2 a}{\partial x\,\partial y}\right)$

b. $T_{xy} = \frac{c_1}{b_1}\left(\frac{\partial^2 a}{\partial x\,\partial y}\right)$ and $T_{yx} = -\frac{c_2}{a_1}\left(\frac{\partial^2 a}{\partial x\,\partial y}\right)$

c. $T_{xy} = ?\frac{c_1}{b_1}\left(\frac{\partial^2 a}{\partial x\,\partial y}\right)$ and $T_{yx} = \frac{c_2}{a_1}\left(\frac{\partial^2 a}{\partial x\,\partial y}\right)$

d. $T_{xy} = \frac{c_1}{b_1}\left(\frac{\partial^2 a}{\partial x\,\partial y}\right)$ and $T_{yx} = \frac{c_2}{a_1}\left(\frac{\partial^2 a}{\partial x\,\partial y}\right)$

11.6 The shear force in grid floors is computed by using .... formulae

a. $Q_x = \frac{\partial}{\partial x}\left[D_x\left(\frac{\partial^2 a}{\partial x^2}\right) + \frac{c_2}{a_1}\left(\frac{\partial^2 a}{\partial x\,\partial y}\right)\right]$ and $Q_y = \frac{\partial}{\partial x}\left[D_y\left(\frac{\partial^2 a}{\partial y^2}\right) + \frac{c_1}{b_1}\left(\frac{\partial^2 a}{\partial x\,\partial y}\right)\right]$

b. $Q_x = -\frac{\partial}{\partial x}\left[D_x\left(\frac{\partial^2 a}{\partial x^2}\right) + \frac{c_2}{a_1}\left(\frac{\partial^2 a}{\partial x\,\partial y}\right)\right]$ and $Q_y = -\frac{\partial}{\partial x}\left[D_y\left(\frac{\partial^2 a}{\partial y^2}\right) + \frac{c_1}{b_1}\left(\frac{\partial^2 a}{\partial x\,\partial y}\right)\right]$

c. $Q_x = -\frac{\partial}{\partial x}\left[D_x\left(\frac{\partial^2 a}{\partial x^2}\right) + \frac{c_2}{a_1}\left(\frac{\partial^2 a}{\partial x\,\partial y}\right)\right]$ and $Q_y = \frac{\partial}{\partial x}\left[D_y\left(\frac{\partial^2 a}{\partial y^2}\right) + \frac{c_1}{b_1}\left(\frac{\partial^2 a}{\partial x\,\partial y}\right)\right]$

d. $Q_x = \frac{\partial}{\partial x}\left[D_x\left(\frac{\partial^2 a}{\partial x^2}\right) + \frac{c_2}{a_1}\left(\frac{\partial^2 a}{\partial x\,\partial y}\right)\right]$ and $Q_y = -\frac{\partial}{\partial x}\left[D_y\left(\frac{\partial^2 a}{\partial y^2}\right) + \frac{c_1}{b_1}\left(\frac{\partial^2 a}{\partial x\,\partial y}\right)\right]$

Answers

11.1 a	11.2 b	11.3 c	11.4 d	11.5 a
11.6 b				

❑

CHAPTER

12 Compression Members and Piles

12.1 COMPRESSION MEMBERS

Columns, bridge piers, chord members of trusses etc., are subjected to longitudinal compression. For axially loaded compression members, introducing a pre–stress may appear to be unnecessary. But in most cases, in addition to axial compression, such members are also subjected to considerable bending moments. Gravity loads are floor systems produce moments in columns framing to them. Wind and other lateral loads which are acting on the building produce bending moments in columns. A member which is supposed to be axially loaded member, is liable to subjected to bending moment due to slight eccentricity of the loads. IS code has also suggested that, a compression member should be designed for longitudinal loading allowing a certain eccentricity.

Any tension that can be caused due to bending can be overcome by pre–stressing the member. If a member is in pure axial compression, no doubt, pre–stressing the member decreases the load carrying capacity. But in most cases of pre–stressed concrete columns, the extent of pre–stress is in the range of $1.5 N/mm^2$ to $5 N/mm^2$. The pre–stress is so low, that it does not cause any significant change in strength. In case of load of large eccentricities, there will be considerable bending and hence pre–stressing enhances the moment capacity.

Consider a column of gross area A_g. Let A_c and A_{st} are the area of concrete and the reinforcements respectively then,

$$A_g = A_c + A_{st} \qquad \text{...(12.1)}$$

Let F_o be the pre–stressing force with eccentricity e.

Let M be the bending moment caused by lateral loading.

The extreme fibre stresses are given by

$$f = \frac{F_o}{A_c} \pm \frac{F_o e}{z} + \frac{W}{A_e} \mp \frac{M}{Z} \qquad \text{...(12.2)}$$

Where, A_e = Equivalent concrete area and is given by

$$A_e = A_g + (m - 1) A_{st} \qquad \text{...(12.3)}$$

and

$$Z = \frac{I_e}{y} \qquad \text{...(12.4)}$$

Where, I_e = Moment of inertia of equivalent concrete and m = modular ratio

12.2 SOLVED EXAMPLES

Problem 12.1

A PSC column is 400mm × 400 mm and 6m high and is provided with 8 high tension wires of 10mmϕ, the effective pre–stress in wires being 700N/mm^2. The column carries an axial compressive load of 360kN and is also subjected to a transverse load of 36kN acting at the mid–height of the column. Determine the extreme stresses on the column section. m = 6, assume the column to be hinged at both the ends.

Solution:

Axial force, W = 360kN, transverse load = q = 36kN, height of column = L = 6m

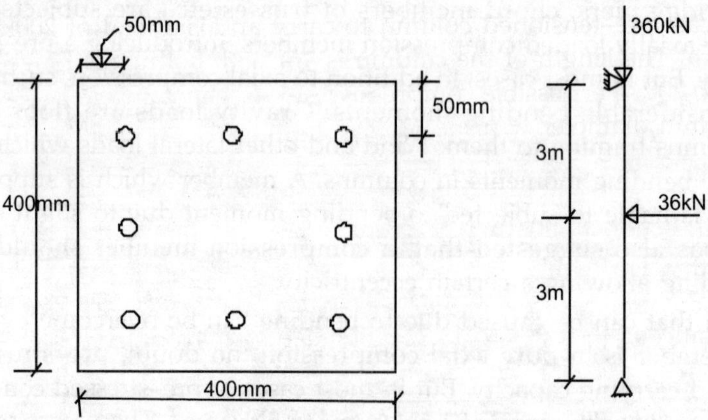

Fig. 12.1 Cross–section and Elevation of Column

$$\text{Pre–stressing force} = F_o = A_{st} \times f_{se} = 8 \times \frac{\pi \times 10^2}{4} \times 700 = 439999.9\text{N}$$

$$\text{Area of concrete} = A_c = 400 \times 400 - 8 \times \frac{\pi \times 10^2}{4} = 159371.42\text{mm}^2$$

$$\text{Stress due to pre–stressing force} = \frac{F_o}{A_c} = \frac{439999.9}{159371.42} = 2.76\text{N/mm}^2$$

Equivalent concrete area, $A_e = A_g + (m - 1) A_{st}$

$$A_e = 400 \times 400 + (6 - 1) \times 8 \times \frac{\pi \times 10^2}{4} = 163142.85\text{mm}^2$$

$$\text{Stress due to external axial force} = \frac{W}{A_e} = \frac{360 \times 10^3}{163142.85} = 2.76\text{N/mm}^2$$

$$\text{Max. BM} = M = \frac{qL}{4} = \frac{36 \times 6}{4} = 54\text{kNm}$$

Moment of inertia of the equivalent concrete section,

$$I_e = \frac{400 \times 400^3}{12} + \frac{\pi \times 10^2}{4} \times (6 - 1) \times 150^2 = 2.186 \times 10^9 mm^4$$

$$\text{Extreme fibre stress due to BM} = \frac{My}{I_e} = \frac{54 \times 10^6 \times 200}{2.186 \times 10^9} = 4.94 N/mm^2$$

Hence,

Max. stress = 2.78 + 2.20 + 4.94 = 9.92N/mm² (C)

Min. stress = 2.78 + 2.20 – 4.94 = 0.04N/mm² (C)

Problem 12.2

Design a precast pre–tensioned column to carry an axial load of 200kN and a bending moment of 15kNm. The length of the column is 4m with both ends hinged. Assume loss of pre–stress equal to 20%. Permissible compressive strength in concrete is 12N/mm² and hence tensile stress of 3mm diameter wires is 1200N/mm².

Solution:

Assume a trail section = 300mm × 300mm

Effective length = l_{eff} = 4m

$$\text{Slenderness Ratio} = \frac{l_{eff}}{b} = \frac{4}{0.3} = 13.22 > 12. \text{ Hence it is long column.}$$

$$\text{Reduction coefficient} = C_r = 1.25 - \frac{l_{eff}}{48b} = 1.25 - \frac{4}{48 \times 0.3} = 0.972$$

$$\text{Equivalent short column load} = W = \frac{200}{0.972} = 205.76 kN$$

$$\text{Equivalent short column moment} = \frac{15}{0.972} = 15.43 kNm$$

$$\text{Stress due to axial load} = \frac{W}{A} = \frac{205.76}{300 \times 300} = 2.286 N/mm^2$$

$$\text{Stress due to BM} = \frac{My}{I} = \frac{15.43 \times 10^6 \times 150}{\left(\frac{300 \times 300^3}{12}\right)} = 3.428 N/mm^2$$

If F is the equivalent pre–stressing force, then, maximum compressive stress is given by

$$\frac{F}{A} + \frac{W}{A} + \frac{My}{I} = 12$$

$$\frac{F}{300 \times 300} + 2.286 + 3.428 = 12$$

$\therefore$ $F = 565650N$

Since loss of pre–stress = 20%, $\eta = 0.8$, we know that, $F = \eta F_o$

Therefore the pre–stressing force required

$$F_o = \frac{F}{\eta} = \frac{565650}{0.8} = 707063N$$

$$\text{Number of 3mm}\phi\text{ wires} = \frac{707063}{1200 \times \left(\frac{\pi \times 3^2}{4}\right)} = 83.35$$

Provide 84 Nos. of 3mmϕ wires.

12.3 PSC PILES

12.3.1 Types of PSC Piles

Depending upon the functional behavior and locations used, pre–stressed concrete piles are categorized in 8 different parts for the purpose of design.

1. Bearing piles
2. Sheet piles
3. Combined bearing and sheet piles
4. Pier trestle and jelly bent piles
5. High tower and stack piles
6. Caisson piles
7. Anchor piles
8. Fender piles

Depending upon cross–section, piles are classified as follows,

1. Triangular piles
2. Square piles
3. Pentagon or octagon piles
4. Circular piles
5. Rectangular with or without semicircular end piles
6. I and star piles

12.3.2 Advantages of PSC Piles

According to Gerwick, the advantages of PSC piles over traditional reinforced concrete and steel piles are as follows.

1. Standardization in design for mass production.
2. Good resistance to hard driving loads and penetration into hard strata.
3. Ease of handling, transporting and driving.
4. Advantageous in deep foundations to carry heavy loads in weak soils.

5. Ease of connections with pile caps to form pier, trestle and jetty bents to support bridge.
6. Use of solid and hallow cross sectional configurations to suit design requirements.
7. Overall economy in production and installation.
8. Piles can be lengthened by splicing.
9. Good resistance to tensile loads due to uplift.
10. Combined load moment capacity.
11. Crack free characteristics under handling and driving.
12. Excellent durability under adverse environmental conditions.
13. High load and moment carrying capacity.

According to Shu–Tien Li, the full potential of pre–stressed concrete as a composite material in piling would be attained only through optimum analysis, optimum pre–stress and optimum design, in view of world demand for piles exceeding 57 million metres per year.

Pre–stressed concrete piles have been widely used in marine structures for fender piles, sheet piles and soldier piles mainly due to their excellent durability under both freeze thaw and salt spray attack.

12.3.3 Design Considerations

The primary considerations in the design of PSC piles are the stresses developed under different handling and driving, which are temporary in nature and stresses developed due to permanent loads (DL and LL), repetitive live loads, and transient loads such as wind, earthquake etc. The permissible stresses in concrete and steel after different conditions of loading as provided by PCI (Pre–stressed Concrete Institute) committee on pre–stressed concrete piling are kept in Table 12.1.

• *Handling Stresses and Impact Allowance*

The flexural stresses developed under different handling conditions, such as lifting from casting bed, storage and transportation, should be investigated. The stress analysis is generally is based on the weight of the piles + 50% allowance factor for impact with tensile stresses limited to $0.5(f_{cy})^{1/2}$.

• *Driving Stresses*

Dynamic stresses developed during pile driving are complex functions of pile and soil properties are influenced by driving resisting, hammer weight and stroke, cushioning materials and other parameters. The driving stresses are alternatively compression and tension, reaching the values of about 7–28N/mm^2 (compressive) and 10N/mm^2 (tensile) and even higher under certain conditions. Tensile stresses are likely to develop in piles, which are longer than 12m, under soft driving conditions at the tip. In case of shorter piles, tensile driving stresses are rarely develop and the minimum effective pre–stress values, as provided by various authorities throughout the world to resist driving stresses, ranges from 2.8–8.4N/mm^2.

Table 12.1 Permissible Stresses in Concrete and steel for PSC Piles

Concrete Stresses	
1. Uniform axial compression for foundation piles fully embedded in soils providing lateral support	$0.33f_{cy} - 0.27f_{cp}$ Where, f_{cp} = Effective compressive pre-stress
2. Uniform axial tension: Permanent and repetitive Transient	 0 $0.5(f_{cy})^{1/2}$
3. Compression due to bending: Normal Marine works and bridges	 $0.45f_{cy}$ $0.40f_{cy}$
4. Tension due to bending: Permanent and repetitive Transient	 $0.35(f_{cy})^{1/2}$ or 0 in corrosive environment $0.5(f_{cy})^{1/2}$
5. For combined loading, concrete stresses should be checked in interaction: $\left(\frac{f_a}{F_a}\right) = \left(\frac{f_b}{F_b}\right) \leq 1$	Where, F_a = Allowable direct stress F_b = Allowable compressive in stress bending
6. Effective pre-stress (recommended values): For piles 12-52m For piles <12m	 4.9–8.4N/mm² 2.8–4.9N/mm²
Steel Stresses	
1. Temporary stresses: Due to temporary jacking force, but not greater than the maximum value recommended by the steel manufacturer Pre-tensioning tendons immediately after transfer or post-tensioning tendons immediately after anchoring: Tension due to transient loads	 $0.8f_{pu}$ $0.7f_{pu}$ Concrete governs
2. Effective pre-stress	$0.6f_{pu}$ or $0.80f_{py}$ (whichever is smaller)
3. Unstressed pre-stressing steel	$0.5f_{py}$ (Maximum 210N/mm²)

• *Service Load Stresses*

The stresses that are developed in the pile due to primary service loads like axial compression or tension, either independently or in combination with moments, should conform to the safe permissible values shown Table 12.1. The safe permissible axial load on a pre–stressed pile is given by,

$$N = A_c\,(0.33f_{cy} - 0.27f_{cp}) \qquad \ldots(12.5)$$

Where, f_{cp} = Effective pre–stress after all losses

f_{cy} = Cylinder compressive strength

A_c = Area of concrete

The design of piles subjected to axial load and moment is simplified by the use of load moment interaction diagrams prepared by the Pre–stressed Concrete Institute. The interaction diagrams have been prepared for the ultimate load capacity of common pile sizes and cross sectional shapes (square, octagonal and round) for concrete strengths of 35, 42, 49 and 56N/mm^2 and for effective pre–stress levels of 4.9 – 8.4N/mm^2. A typical interaction diagram involves parameters such as ultimate axial load, moment and slenderness ratio of 350mm solid octagonal pile is shown in Fig. 12.2.

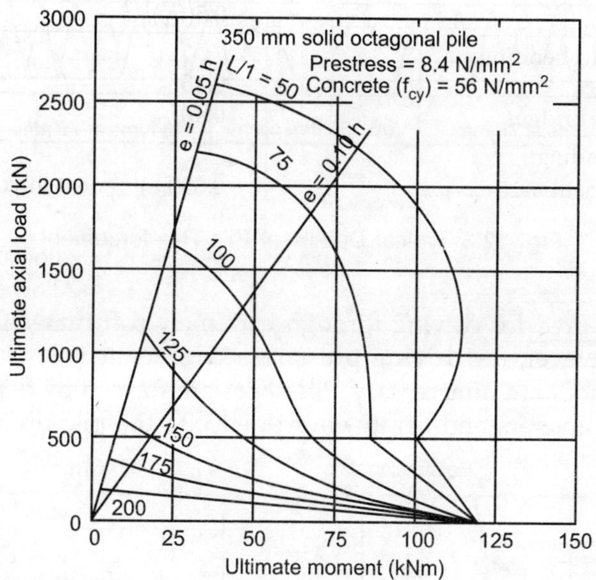

Fig. 12.2 Load Moment Characteristics of Pre–stressed Piles

The allowable service loads and section properties of various shapes and sizes of piles are compiles by Pre–stressed Concrete Institute. The data useful for the design of piles of specific shapes for concentric service loads.

12.3.4 Pile Reinforcements

The rebound tensile stresses are resisted by effective pre–stress together with the tensile strength of concrete. Hence the steel tendon area at yield should have an equal or greater force than the pre–stress + concrete tensile strength to prevent failure due to low–cycle fatigue. Based on theory and experience, Gerwick recommends a minimum tendon area not less than 0.5% of the concrete section.

The tendons are distributed in a circular pattern with circular spiral ties in case of round, hexagon and octagonal piles. Both circular and square patterns have been used square piles. Since there is no significance difference in the behavior of square piles with square or circular pattern of reinforcements, latter is predominant in current practice in the view of savings and increased efficiency of circular spiral and ease of manufacturing.

Spiral binding is provided throughout the length of the pile and at a closer pitch towards the head and tip of the pile to counteract the bursting forces developed during driving conditions. A minimum of 0.6% spiral over a length equal to three times the least dimension of the pile for the end zones. According to Gerwick, the amount of spiral reinforcements should be about 1% at the head and tip for a length of 300mm and 0.6% for next 600mm which reduces to 0.3% for solid pies and 0.4% for hollow core piles in the body of the pile. The typical reinforcements in the pile are as shown in Fig. 12.3.

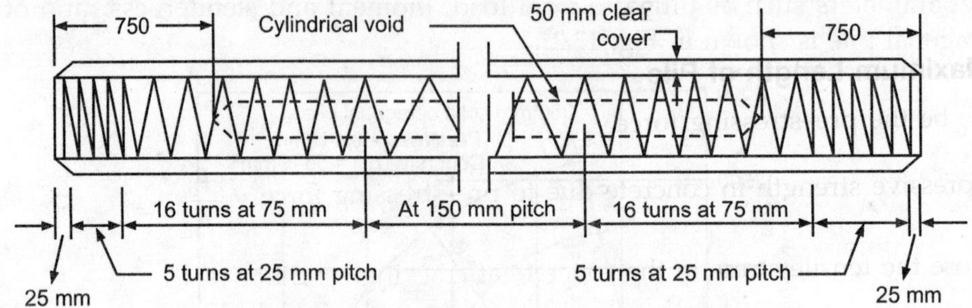

Fig. 12.3 Typical Details of Pile Reinforcement

12.3.5 Pile Shoes

Pile shoes are required for driving through extremely soft materials like buried timbers and rocky strata. However, for driving pre-stressed concrete piles into sands, silts, clays and soft shales, pile shoes are unnecessary. Pile shoes are formed by thick steel plates welded to the reinforcing bar anchors and firmly embedded into the pile tip as shown in Fig. 12.4.

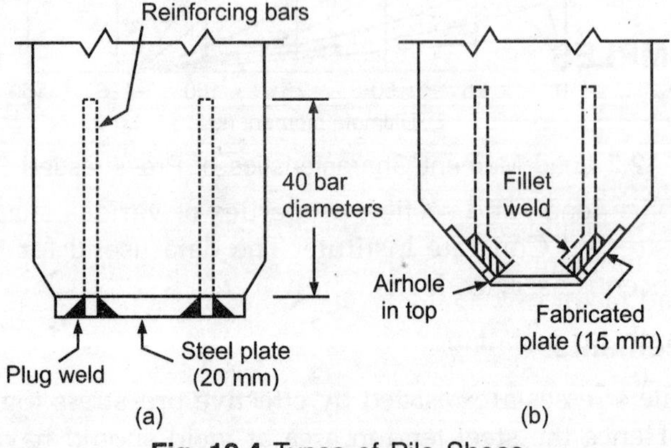

Fig. 12.4 Types of Pile Shoes

The shape of the tip may be varied to suit driving conditions. Field experience indicates that a square tip with chamfered corners is preferable to a pointed or wedge shaped tip since the latter causes the pile to deflect and develops high bending stresses which results in the failure of the shoe and pile.

12.3.6 Convenient Ways of Lifting a Pile

In order to reduce the maximum bending moment while handling the pile, a two point suspension, is generally adopted. Each lift points are at a distance of 0.207L from the appropriate ends. If the dead load of the pile is w per unit length, the maximum bending moment equal

to $\frac{wL^2}{47}$. ...(12.6)

While lifting a pile (hoisting it to the vertical position) the lower end of the pile rests on the ground and the lift force is applied at a distance of 0.293L from the upper end so that the maximum bending moment is minimized. When lifted in this manner, the maximum bending moment for the pile is equal to $\frac{wL^2}{23}$. ...(12.7)

12.3.7 Maximum Length of Pile

Let F_o be the pre–stressing force.

Compressive strength in concrete due to pre–stressing force = $\frac{F_o}{A_c}$ ...(12.8)

Suppose the tensile stress in the concrete at cracking condition = F_o'

This means cracking condition reached when the bending moment produces a stress

$$= F = \frac{F_o}{A_c} + F_o' \quad ...(12.9)$$

Corresponding bending moment = M = FZ ...(12.10)

After calculating the moment M the maximum length of the pile can be calculated corresponding to two point's suspension or for lifting the pile for hoisting.

12.4 SOLVED EXAMPLES

Problem 12.3

A multi storeyed building is to be supported on pre–stressed concrete pile foundations. The piles have an effective height of 5m and they have to support a total axial service load of 1100kN together with a moment of 37.5kNm. Design a suitable pile to support these loads, assuming a uniform load factor of 2 against collapse. The pile is to be designed to be lifted at any point along its length for installation.

Solution:

Service loads: N = 1100kN,

M = 37.5kNm

Ultimate loads: $N_u = 2 \times 1100 = 2200$kN,

$M_u = 2 \times 37.5 = 75$kNm

Referring the interaction diagram (Fig. 12.2),

Select the 350mm solid octagonal pile.

From the table given by the PCI, section properties of the pile are

$A = 1.02 \times 10^5 \text{mm}^2$ $w_d = 2.45\text{kN/m}$ $Z = 4.70 \times 10^6 \text{mm}^3$ $i = 90\text{mm}$

$\therefore$ Slenderness ratio $= \frac{L}{i} = \frac{5000}{90} = 55.5$

Eccentricity, $e = \frac{M_u}{N_u} = \frac{75 \times 1000}{2200} = 34mm$

$\therefore \quad \frac{e}{h} = \frac{34}{350} = 0.097$

The ultimate moment and load capacity of the 350mm solid octagonal pile for $M_u = 75kN$ and $\frac{L}{i} = 55.5$, from Fig. 12.2,

Axial load, $P_u = 2250kN$, effective pre–stress = 8.4MPa

$\therefore$ Initial pre–stressing force $= \frac{8.4 \times 1.04 \times 10^5}{0.85 \times 10^3} = 1030kN$

Use eight strands of 7–12.5mmϕ wires initially stressed to 1120N/mm²,

- ***Handling Stresses***

Self–weight of pile = 2.45kN/m, impact allowance = 1.23kN/m

Total load (w_d) = 3.68kN/m

Maximum moment in the pile when lifted at centre is given by

$$M = 3.68 \times \frac{2.5^2}{2} = 11.5kNm$$

Tensile bending stress $= \frac{11.5 \times 10^6}{4.70 \times 10^6} = 2.45N/mm^2$

Resultant stress = 8.4 – 2.45 = 5.95N/mm² (Compression)

Hence, the pile is safe against handling stresses.

Problem 12.4

The ground floor columns of an industrial shed are to be supported on pre–stressed piles of 10m length. Each pile is subjected to an axial load of 2500kN. The specific cylinder compressive strength of concrete = 35MPa. The permissible effective pre–stress is not to be exceeding 4.9N/mm². Design a suitable pile of (i) square section, and (ii) octagonal section. Also design the number of strands of 7–12.5mm required for the piles if the ultimate tensile strength of the strand is 165kN.

Solution:

$N = A_c (0.33f_{cy} - 0.27f_{cp})$

Assuming the effective pre–stress, f_{cp} = 4.9MPa,

$2500 \times 10^3 = A_c (0.33 \times 35 - 0.27 \times 4.9)$, $\therefore A_c = 245 \times 10^3 mm^2$

Referring to the table provided by PCI, adopt a square pile of 500mm size and a solid octagonal pile of 550mm.

Pre–stressing force = $\dfrac{4.9\times245\times10^6}{1000}$ = 1200kN

Effective force in each strand of 7–12.5mmϕ wires under service load

= 0.6 × 165 = 99kN. Number of strands = $\dfrac{1200}{99}$ = 12

Problem 12.5

A pre–stressed concrete pile 300mm × 300mm is pre–stressed with 18–5mm diameter to an initial stress of 1200MPa. Allowing 15% loss of pre–stress, calculate the maximum length of pile from handling considerations for the following cases. Take cracking stress of concrete = 3MPa

1. When the pile is lifted by two point suspension
2. When the pile has to be hoisted from the ground to vertical position

Solution:

A = 300 × 300 = 90000mm², $A_{st} = 18 \times \dfrac{\pi}{4}$ = 353mm², A_c = 90000 – 353 = 89647mm²

Final Pre–stressing force, F_o= 0.85 × 1200 × 353 = 360060N

Stress in concrete due to final pre–stressing force = $\dfrac{F_o}{A_c} = \dfrac{360060}{89647}$ = 4N/mm²

Cracking stress in tension, F_o' = 3N/mm²

The maximum stress that can be induced due to bending moment

= F = $\dfrac{F_o}{A_c} + F_o'$ = 4 + 3 = 7N/mm²

Section modulus of concrete section, Z = $\dfrac{300\times300^2}{6}$ = 45 × 10⁵mm³

Maximum bending moment to produce the above stress of 7N/mm²

M = FZ = 7 × 45 × 10^5 = 315 × 10^5Nmm = 31500Nm

DL of pile = 0.3 × 0.3 × 24000 = 2160Nm

Allowing 50% impact, effective DL = w = 2160 × 1.5 = 3240Nmm

Case 1: When two point suspension is adopted

Maximum bending moment, M = $\dfrac{wL^2}{47}$, 31500 = $\dfrac{3240\times L^2}{47}$, L = 21.37m

Case 2: When the pile is hoisted

Maximum bending moment, M = $\dfrac{wL^2}{23}$, 31500 = $\dfrac{3240\times L^2}{23}$, L = 14.95m

REVIEW QUESTIONS

12.1 What are compression members? Explain their salient features.

12.2 Explain the method of calculating equivalent concrete area of compression members.

12.3 Mention the different types of piles based on their functional behavior and locations used.

12.4 Give the different types of piles based on their cross–section.

12.5 What are the advantages of PSC piles?

12.6 Explain the design considerations used in piles.

12.7 Elaborate permissible stresses in concrete and steel for PSC piles.

12.8 What is load moment interaction diagram in case of PSC piles?

12.9 Write a short note on pile reinforcements.

12.10 Explain pile shoes.

EXERCISE PROBLEMS

12.1 Design a precast pre–tensioned column to carry an axial load of 200kN and a bending moment of 15kNm. The length of the column is 4m with both ends hinged. Assume loss of pre–stress equal to 20%. Permissible compressive strength in concrete is 12N/mm^2 and use of 5mm diameter wires which are initially stressed to 1200N/mm^2.

[Ans: Number of wires = 30]

12.2 The ground floor columns of an industrial shed are to be supported on pre–stressed piles of 10m length. Each pile is subjected to an axial load of 2500kN. The specific cylinder compressive strength of concrete = 35MPa. The permissible effective pre–stress is not to be exceeding 5.2N/mm^2. Design a suitable pile of (i) square section, and (ii) octagonal section. Also design the number of strands of 7–12.5mm required for the piles if the ultimate tensile strength of the strand is 165kN.

[Ans: Number of strands = 13]

12.3 A pre–stressed concrete pile 400mm × 400mm is pre–stressed with 20–5mm diameter to an initial stress of 1200MPa. Allowing 15% loss of pre–stress, calculate the maximum length of pile from handling considerations for the following cases. Take cracking stress of concrete = 3MPa

1. When the pile is lifted by two point suspension
2. When the pile has to be hoisted from the ground to vertical position

[Ans: Length of pile: Case 1: 21.88m, Case 2: 15.30m]

OBJECTIVE QUESTIONS

12.1 For pre–stressed concrete columns, the extent pre–stress is .....

a. 1.5N/mm^2 to 3N/mm^2
b. 1.5N/mm^2 to 4N/mm^2
c. 1.5N/mm^2 to 5N/mm^2
d. 1.5N/mm^2 to 6N/mm^2

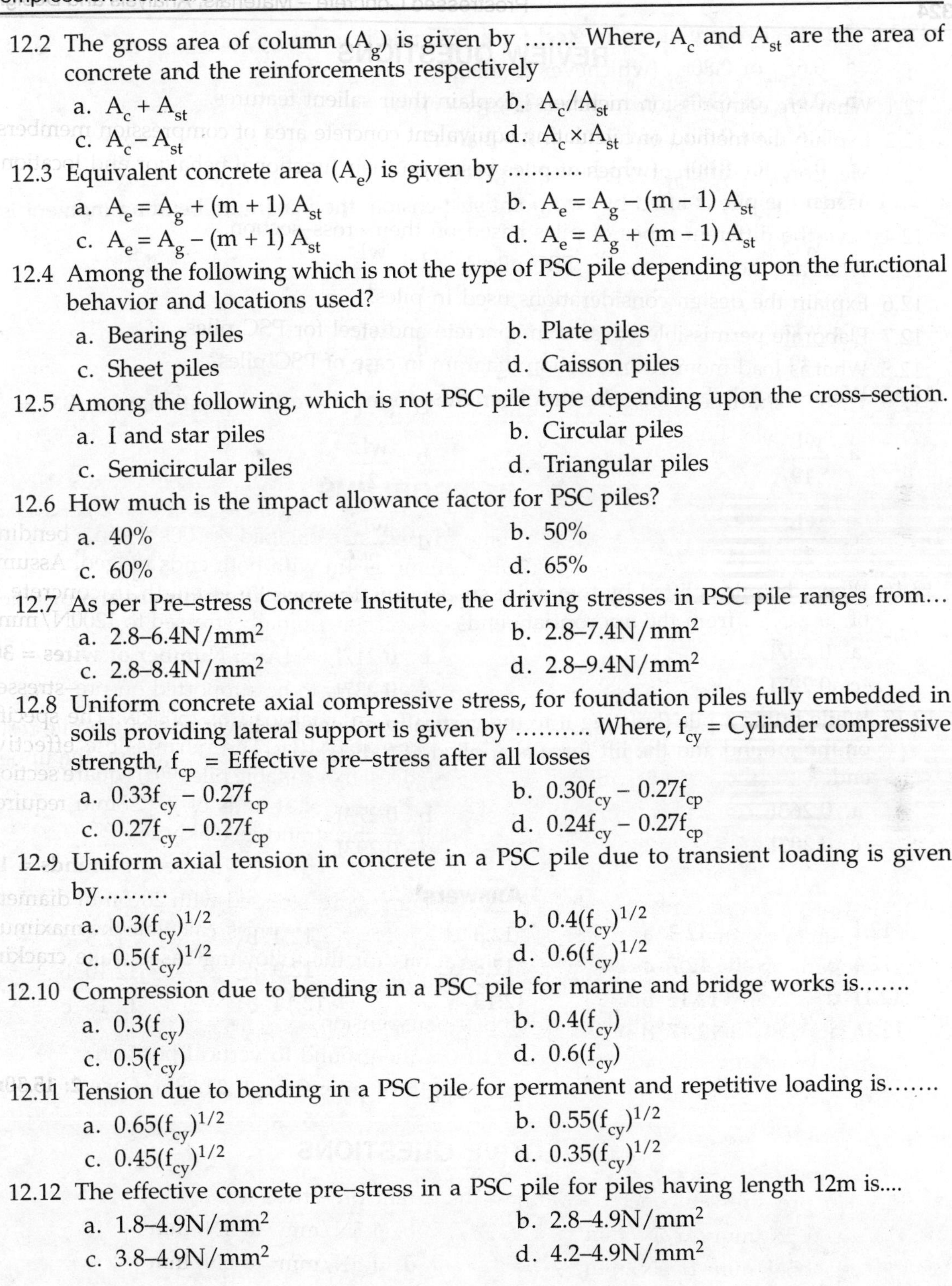

12.2 The gross area of column (A_g) is given by .... Where, A_c and A_{st} are the area of concrete and the reinforcements respectively

a. $A_c + A_{st}$
b. A_c/A_{st}
c. $A_c - A_{st}$
d. $A_c \times A_{st}$

12.3 Equivalent concrete area (A_e) is given by ....

a. $A_e = A_g + (m + 1) A_{st}$
b. $A_e = A_g - (m - 1) A_{st}$
c. $A_e = A_g - (m + 1) A_{st}$
d. $A_e = A_g + (m - 1) A_{st}$

12.4 Among the following which is not the type of PSC pile depending upon the functional behavior and locations used?

a. Bearing piles
b. Plate piles
c. Sheet piles
d. Caisson piles

12.5 Among the following, which is not PSC pile type depending upon the cross–section.

a. I and star piles
b. Circular piles
c. Semicircular piles
d. Triangular piles

12.6 How much is the impact allowance factor for PSC piles?

a. 40%
b. 50%
c. 60%
d. 65%

12.7 As per Pre–stress Concrete Institute, the driving stresses in PSC pile ranges from...

a. 2.8–6.4N/mm^2
b. 2.8–7.4N/mm^2
c. 2.8–8.4N/mm^2
d. 2.8–9.4N/mm^2

12.8 Uniform concrete axial compressive stress, for foundation piles fully embedded in soils providing lateral support is given by .... Where, f_{cy} = Cylinder compressive strength, f_{cp} = Effective pre–stress after all losses

a. $0.33f_{cy} - 0.27f_{cp}$
b. $0.30f_{cy} - 0.27f_{cp}$
c. $0.27f_{cy} - 0.27f_{cp}$
d. $0.24f_{cy} - 0.27f_{cp}$

12.9 Uniform axial tension in concrete in a PSC pile due to transient loading is given by........

a. $0.3(f_{cy})^{1/2}$
b. $0.4(f_{cy})^{1/2}$
c. $0.5(f_{cy})^{1/2}$
d. $0.6(f_{cy})^{1/2}$

12.10 Compression due to bending in a PSC pile for marine and bridge works is.......

a. $0.3(f_{cy})$
b. $0.4(f_{cy})$
c. $0.5(f_{cy})$
d. $0.6(f_{cy})$

12.11 Tension due to bending in a PSC pile for permanent and repetitive loading is.......

a. $0.65(f_{cy})^{1/2}$
b. $0.55(f_{cy})^{1/2}$
c. $0.45(f_{cy})^{1/2}$
d. $0.35(f_{cy})^{1/2}$

12.12 The effective concrete pre–stress in a PSC pile for piles having length 12m is....

a. 1.8–4.9N/mm^2
b. 2.8–4.9N/mm^2
c. 3.8–4.9N/mm^2
d. 4.2–4.9N/mm^2

12.13 The effective pre–stress in steel for a PSC pile is given by ....

a. $0.6f_{pu}$ or $0.80f_{py}$ (whichever is smaller)
b. $0.6f_{pu}$ or $0.80f_{py}$ (whichever is greater)
c. $0.8f_{pu}$ or $0.60f_{py}$ (whichever is smaller)
d. $0.8f_{pu}$ or $0.60f_{py}$ (whichever is greater)

12.14 When the pile is lifted by two point suspension, the maximum bending moment is

a. $\frac{wL^2}{45}$ b. $\frac{wL^2}{47}$

c. $\frac{wL^2}{49}$ d. $\frac{wL^2}{51}$

12.15 When the pile has to be hoisted from the ground to vertical position, max. BM =

a. $\frac{wL^2}{19}$ b. $\frac{wL^2}{21}$

c. $\frac{wL^2}{23}$ d. $\frac{wL^2}{25}$

12.16 When the pile is lifted by two point suspension, the each lift points are at a distance of .... from the appropriate ends.

a. 0.207L b. 0.217L
c. 0.227L d. 0.237L

12.17 While lifting a pile (hoisting it to the vertical position) the lower end of the pile rests on the ground and the lift force is applied at a distance of .... from the upper end.

a. 0.263L b. 0.273L
c. 0.283L d. 0.293L

Answers

12.1 c	12.2 a	12.3 d	12.4 b	12.5 c
12.6 b	12.7 c	12.8 a	12.9 c	12.10 b
12.11 d	12.12 b	12.13 a	12.14 b	12.15 c
12.16 a	12.17 d			

❑

CHAPTER 13 Tension Members, Pipes and Tanks

13.1 TENSION MEMBERS

PSC tension members are used in a variety of situations. They are used as long stay backs of retaining walls, tie rods of arches, anchors for footings, tie member of trusses, hangers, tank walls, stays in suspension bridges etc.

By pre–stressing the concrete, it is possible to provide strength as well as rigidity to a tension member. By preventing the formation of cracks the pre–stressing steel is protected corrosion. It is possible to control the deformation of a pre–stressed concrete tension member.

Fig. 13.1 shows a pre–stressed concrete member subjected to an external force.

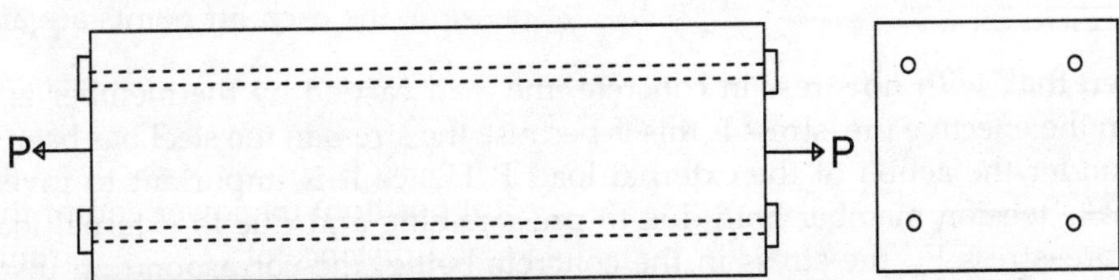

Fig. 13.1 Pre–stressed Concrete Member Subjected to an External Force

Initial pre–stress is F_o, effective pre–stress is F

Then, stress in concrete = $f_c = \dfrac{F_o}{A_c}$ .... for the initial pre–stress

$= f_c = \dfrac{F}{A_c}$ .... for the effective pre–stress

As P is applied externally, both steel and concrete undergo elongation of same amount. Hence, transform the cross–section into an equivalent area of concrete.

A_c is the actual area of concrete

A_{st} is the actual area of steel

$$A_e = A_c + mA_{st} \quad \text{...(13.1)}$$

If A_g is the gross area of the section

Then equivalent area of concrete = $A_e = mA_{st} + A_g - A_{st}$

$$A_e = A_g + (m - 1)\,A_{st} \quad \text{...(13.2)}$$

The above formula is valid only when the section is grouted otherwise the hole in the concrete will be greater than A_{st} and the formula (13.1) can be conveniently used, with A_c referring to the net area of concrete.

Then, the tensile stress produced by P will be,

For concrete, $f_c = -\dfrac{F}{A_e}$ (Tensile) and for steel, $f_s = -m\dfrac{P}{A_e}$, where, $m = \dfrac{f_s}{f_c}$

Then the resultant stresses due to effective pre–stress + the external load are

$$\text{For concrete} = \frac{F}{A_c} - \frac{P}{A_e} \qquad \text{...(13.3)}$$

$$\text{For steel} = -f_s - m\frac{P}{A_e} \qquad \text{...(13.4)}$$

If it is desired to determine the load P, which will produce zero stress in concrete, then equate (13.3) to zero. $\dfrac{F}{A_c} - \dfrac{P}{A_e} = 0, \dfrac{F}{A_c} = \dfrac{P}{A_e}$,

$$P = F \times \frac{A_e}{A_c} \qquad \text{...(13.5)}$$

It is seen that, with no stress in concrete, the load carried by the member is somewhat greater than the effective pre–stress F. this is because the stress in the steel has been somewhat increased under the action of the external load P. Hence it is important to investigate the strains in PSC tension member both due to pre–stressing and due to external loads. Under the initial pre–stress F_o, the stress in the concrete being, the corresponding instantaneous unit strain will be, $\dfrac{F_o}{EA_c}$ which will reduce to $\dfrac{F}{EA_c}$, after the losses have taken place.

Under the action of external load P, the instantaneous strain is given by $\dfrac{P}{EA_c}$

13.2 SOLVED EXAMPLES

Problem 13.1

A PSC member is 200mm × 200mm, the tendons being subjected to a stress of 900N/mm². The tendons provided an area of 500mm². Find the external tension the member can carry if the stress in concrete under the action of external loading is zero. Find also the tensile stress in tendon after the application of external tension on the member. Take m = 6.

Solution:

Equivalent area of concrete, $A_e = A_g + (m - 1) A_{st} = 200 \times 200 + (6 - 1) 500 = 42500\text{mm}^2$

Actual area of concrete = $A_c = 200 \times 200 - 500 = 39500\text{mm}^2$

Pre–stressing force, F = 900 × 500 = 450000N

External tension, $P = F \times \frac{A_e}{A_c} = 450000 \times \frac{42500}{39500} = 484177N$

Given stress in tendon, $f_s = 900N/mm^2$ and m = 6

Resultant stress in tendon $= -f_s - m\frac{P}{A_e} = -900 - 6 \times \frac{484177}{42500} = -968.4N/mm^2$

Problem 13.2

A straight concrete member is pre–stressed with high tension steel through the centroid of the section. Taking $A_c = 50000mm^2$, $A_{st} = 500mm^2$, m = 7.5, ultimate tensile strength of concrete = $2.5N/mm^2$, final pre–stress in steel = $890N/mm^2$. Calculate what working load it can carry using a factor of safety of 2 against the cracking of concrete. Find also the residual compressive stress in concrete under working load.

Solution:

Final pre–stressing force, F = 500 × 890 = 445000N

Equivalent concrete area = $A_e = A_c + mA_{st}$ = 50000 + 7.5 × 500 = $53750mm^2$

$$f_c = \frac{F}{A_c} - \frac{P}{A_e}, -2.5 = \frac{445000}{50000} - \frac{P}{53750}$$

∴ P = 612750N (Cracking load)

With a factor of safety of 2, Safe tension $= \frac{612750}{2} = 306375N$

Residual compressive stress under the above working load

$$= \frac{F}{A_c} - \frac{P}{A_e}$$

$$= \frac{445000}{50000} - \frac{306375}{53750}$$

$$= 3.20N/mm^2$$

Problem 13.3

Design a pre–tensioned member to carry a tension of 1000kN. Safe compressive stress in concrete = $10.5N/mm^2$, m = 5, initial pre–stress in steel = $1200N/mm^2$. Assume 20% loss of pre–stress.

Solution:

At transfer, F_o= Tension in steel = Compression in concrete

$$F_o = 1200A_{st} = 10.5A_c$$

$$\therefore \quad A_{st} = 0.00875A_c$$

Equivalent area of concrete, $A_e = A_c + mA_{st} = A_c + 8 \times 0.00875A_c = 1.07A_c$

At working condition, to produce zero stress in concrete

$$0 = \frac{F}{A_c} - \frac{P}{A_e} \text{ and } F = 0.8F_o = 0.8 \times 10.5A_c = 8.4A_c$$

$$0 = \frac{8.4A_c}{A_c} - \frac{1000 \times 10^3}{A_e}$$

$$\therefore\ A_e = \frac{1000 \times 10^3}{8.4} = 119048\text{mm}^2 \text{ and } \therefore\ A_c = \frac{119048}{1.07} = 111260\text{mm}^2$$

$A_{st} = 0.00875 \times 111260 = 974\text{mm}^2$

$\therefore\ A = A_c + A_{st} = 111260 + 974 = 112234\text{mm}^2$

Provide 335mm × 335mm in section

Problem 13.4

A straight tension member 250 mm × 250 mm is provided with 750mm^2 of high tension steel wires which are subjected to a stress of 900N/mm^2. Allow a factor of safety of 1.5 against cracking. Find the safe tension for the member assuming a cracking tensile stress of 3N/mm^2. Find also the residual compressive stress under working load. Take m = 6.

Solution:

Actual area of concrete, $A_c = 250 \times 250 - 750 = 61750\text{mm}^2$

Equivalent area of concrete, $A_e = A_g + (m - 1)\ A_{st} = 250 \times 250 + (6 - 1)\ 750 = 66250\text{mm}^2$

Pre–stressing force, $F = 900 \times 750 = 675000\text{N}$

Let P be the tension at which concrete will just develop cracking stress.

$$f_c = \frac{F}{A_c} - \frac{P}{A_e},\ -3 = \frac{675000}{61750} - \frac{P}{66250}$$

$\therefore\ P = 922940\text{N}$

With a factor of safety of 1.5

$$\text{Safe tension, } P = \frac{922940}{1.5} = 615290\text{N}$$

When the member is subjected to safe tension

$$\text{Residual compressive stress} = \frac{F}{A_c} - \frac{P}{A_e} = \frac{675000}{61750} - \frac{615290}{66250} = 1.64\ \text{N/mm}^2$$

Problem 13.5

Design a tie member to carry a load of 500kN. Use the following data,

Permissible tensile stress in concrete = 2N/mm^2

Permissible compressive stress initially = 14N/mm^2

Loss of pre–stress = 20%

Stress in the steel at transfer = 1100N/mm^2

m = 6

The design must ensure a residual compressive stress of $2N/mm^2$. Also calculate the load factor against cracking if the tensile strength of concrete = $5N/mm^2$.

Solution:

$f_{si} = 1100N/mm^2$, $f_c = 14N/mm^2$, P = 500kN, $\eta = 0.8$, $f_{sc} = 0.80 \times 1100 = 880N/mm^2$,

We know that, $f_{sc} = \dfrac{F}{A_{sc}}$, $\therefore$ The pre–stressing force, $F = 880 \times A_{st}$

At transfer, Tensile force in steel = Compressive force in concrete

$f_{si} \times A_{st} = f_c \times A_c$, $1100 \times A_{st} = 14 \times A_c$, $\therefore \dfrac{A_c}{A_{st}} = 78.57$

Under working load, there must be residual compressive stress of $2N/mm^2$

$$\frac{F}{A_c} - \frac{P}{A_e} = 2, \frac{880 \times A_{st}}{A_c} - \frac{500 \times 10^3}{A_c + mA_{st}} = 2, \frac{880}{78.57} - \frac{500 \times 10^3}{A_c + 6 \times \dfrac{A_c}{78.57}} = 2$$

On solving, $A_c = 50509mm^2$

$$A_{st} = \frac{A_c}{78.57} = \frac{50509}{78.57} = 642.85mm^2$$

Gross area, $A_g = A_c + A_{st} = 50509 + 642.85 = 51151.85mm^2$

Size = 226.2×226.2 mm^2

$$\text{Number of 5mm diameter wires} = \frac{642.85}{\pi \times \dfrac{5^2}{4}} = 33 \text{ No.s}$$

- **Checking load factor**

The cracking load factor can be determined as follows

$$\frac{F}{A_c} - \frac{P}{A_e} = -5, \frac{880 \times 642.85}{50509} - \frac{P}{50509 + 6 \times 642.85} = -5$$

On solving, P = 880.729kN

$$\text{Cracking load factor} = \frac{\text{Cracking load}}{\text{Working load}} = \frac{880.729}{500} = 1.76$$

Problem 13.6

Design a tie member subjected to 400kN axial tension with the following data

i. Ultimate cube strength of concrete = $45N/mm^2$

ii. $\dfrac{\text{Compressive stress in concrete at transfer}}{\text{Ultimate cube strength}} = 0.5$

iii. $\dfrac{\text{Compressive stress in concrete at service}}{\text{Ultimate cube strength}} = 0.1$

iv. Stress in steel at transfer = 900N/mm²

v. $\dfrac{\text{Final pre – stress}}{\text{Pre – stress at transfer}} = 0.8$

vi. Modular ratio = m = 6

vii. Ultimate tensile stress in concrete = 4N/mm²

Find also the factor of safety in the design.

Solution:

At transfer

$$\text{Tensile force in steel} = \text{Compressive force in concrete}$$

$$f_{si} \times A_{st} = f_c \times A_c,\ 900 \times A_{st} = 0.5 \times 45 \times A_c$$

$$\therefore \quad A_{st} = 0.025A_c$$

$$A_e = A_c + mA_{st} = A_c + 6 \times 0.025A_c = 1.15A_c$$

Working load condition

$$\frac{F}{A_c} - \frac{P}{A_e} = 0.1 \times 45 = 4.5$$

$$F = \eta F_o = 0.8F_o = 0.8f_{si}A_{st} = 0.8 \times 900 \times 0.025A_c$$

$$F = 18A_c$$

$$\therefore \quad \frac{18A_c}{A_c} - \frac{400\times10^3}{1.15\times A_c} = 4.5$$

$$\therefore \quad A_c = 25764.9\text{mm}^2$$

$$A_{st} = 0.025A_c = 0.025 \times 25764.9 = 644.1\text{mm}^2$$

Gross area, $A = A_c + A_{st} = 25764.9 + 644.1 = 26409.1\text{mm}^2$

Provide 162.5 × 162.5 mm²

$A_e = A_c + (m - 1)A_{st} = 25764.9 + (6 - 1)\ 644.1 = 29626.75\text{mm}^2$

Let P be the cracking load

$$\frac{F}{A_c} - \frac{P}{A_e} = -4$$

Here, $F = \eta F_o = 0.8F_o = 0.8f_{si}A_{st} = 0.8 \times 900 \times 644.1 = 463752\text{N}$

$$\frac{463752}{25764.9} - \frac{P}{29626.75} = -4,$$

On solving, P = 651827N

$$\therefore \text{Factor of safety} = \frac{651827}{400000} = 1.63$$

13.3 PRE–STRESSED CONCRETE PIPES

Pre–stressed concrete pipes are ideally suited for 0.4 to 2N/mm^2 pressure. For this range, cast iron or mild steel pipes are not economical and RCC pipes are not suitable. There are mainly two types of pre–stressed concrete pipes in use today, namely cylinder type and non–cylinder type. Cylinder type has a steel cylinder core over which concrete is shot–creted and pre–stressed. A non–cylinder type has a suitably compacted concrete core longitudinally pre–stressed with pre–tensioned high tensile steel wire embedded in the concrete, circumferentially pre–stressed and coated with cement mortar or concrete to protect the circumferential pre–stressing wire, to withstand internal pressure and external design loads.

13.3.1 Non–cylindrical Type Pre–stressed Concrete Pipes

Non–cylindrical pre–stressed concrete pipes are similar for all types of water supply schemes. Under normal conditions, this is economical for medium heads. The joints of this pipe are made with flexible rubber rings. Two types of rubber ring joints are commonly used. One is with circular chord either rolled or fixed gasket and other is with a V section gasket. Both the joints have been successfully used for pressure up to 2N/mm^2. Non cylinder pipes are manufactured using high quality concrete having the following properties. (a) impermeability even at high pressure, (b) high structural strength and (c) high resistance against attack by deleterious substances and therefore less friction.

13.3.1.1 Properties of Non–cylindrical Pre–stressed Pipes

1. Under hydraulic shock, excessive pressure developed brings about bursting the pipe material. Normally, such a burst pipe has to be replaced, but this does not applied to pre–stressed concrete pipe. When the pressure drops, cracks in the pipe close and the pipe becomes water tight as before. The pipe undergoes a self–healing process due subsequent calcification.
2. Under normal working pressure, the pipe wall is always in a state of compression, a condition most favorable for impermeability.
3. Under internal hydrostatic pressure, the mortar coat acts integrally with the core pipe. The ability of the cover coat to act integrally with the core is due to the fact that the coat is also pre–stressed to some extent due to creep of concrete and loss of stress in wire winding.

13.3.2 Cylindrical Type Pre–stressed Concrete Pipes

In cylindrical type there are two basic designs. The first design was introduced in 1942 in USA and termed as the pre–stressed concrete lined cylinder type pipe. They are suitable up to 1200mm in dia. It consists of a concrete lined steel cylinder with joint/rings welded to their ends and wrapped with a helix of highly stressed wire coated with a dense cement mortar to protect the wire. In 1950, a second type of pre–stressed cylinder concrete pipe was developed. In this, the steel cylinder with joint rings is embedded in the concrete core so that the stressed wire helix is in contact with concrete rather than with the steel cylinder. The protective coating is either cement mortar or concrete. These are found suitable for dia up to 3600mm.

13.3.3 Advantages of PSC Pipes

1. Installed cost is generally less than that for pipes of other materials such as CI or steel.
2. PSC pipes resist external load to a much higher degree than the steel pipes.
3. Laying of PSC pipes is faster.
4. In the design of PSC pipes, both concrete and steel are utilized to their maximum possible extent; and therefore, overall economy can be effected in the consumption of steel and cement.
5. PSC pipes possess excellent hydraulic properties and resistance corrosion similar to RCC pies.
6. From the experts estimate it is said that, life of PSC pipe is generally more than 100 years when buried in non–corrosive soil environment.
7. For large requirements of pipes it is more often economical to install a special manufacturing unit near the site of the pipeline than to transport the pipe to the site from the existing plant.
8. The service record of these pipes is very satisfactory, although a few failures of earlier pipelines were reported. These are mainly due to the use of oil tempered wire, use of calcium chloride as additive for achieving early strength and use of the different cement for the core and coat. If proper materials are used, service life of the pipes could be around 100 years with least maintenance.

13.3.4 Design of Pre–stressed Concrete Pipes

Design of PSC pipes is made as per the specifications of IS: 784–2001.

- The maximum grade of concrete should be M45 for core concrete, the minimum cover on all reinforcement in any finished pipes should be 12mm.
- The circumferential pre–stressing steel should be protected against corrosion due to ingress on water by the outer cover coating of cement mortar or concrete or any approved material with a minimum thickness of 18mm.
- Outer protective coating should be applied within 16 hours on circumferential wire winding.
- The tensile stress in cover coating due to any reason shall not cross the modulus of rupture of coating material.
- Inner protective coatings in the form of non–cementitious materials like epoxy; bitumen etc. should be used after checking basic material properties for its ingredients. There should not be any element which is injurious to health.
- Longitudinal pre–stressing may be done using high tensile wires anchored in permanent anchorages.

13.3.4.1 Design of Non–cylindrical Pipes

• Circumferential Wire Winding

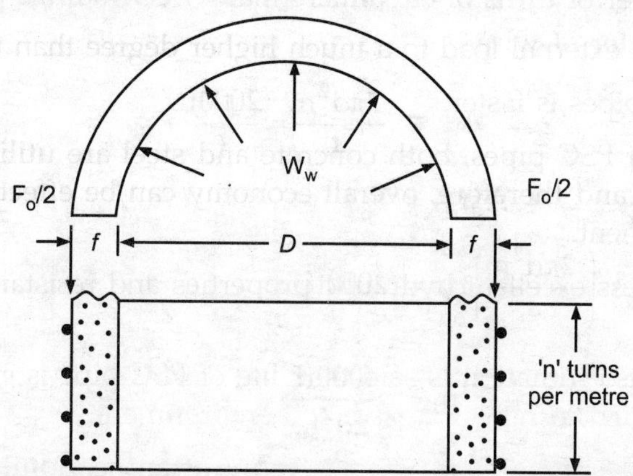

Fig. 13.2 Circumferential Winding in Pre–stressed Pipes

If N_d = Hoop tension developed under working pressure

t = Thickness of concrete pipe, D = Diameter of pipe

W_w = Hydrostatic pressure

f_{ct} = Permissible compressive stress in concrete at transfer

$f_{min, w}$ = Permissible stress in concrete under working pressure

Considering the equilibrium of pipe shown in Fig. 13.2.

$$\frac{W_w D}{2t} < \eta f_{ct} - f_{min, w}$$

$$\therefore \quad t > \left[\frac{W_w D/2}{\eta f_{ct} - f_{min, w}}\right] > \left[\frac{N_d}{\eta f_{ct} - f_{min, w}}\right] \qquad ...(13.6)$$

In case of liquid retaining structures, to ensure water tightness, the value of $f_{min, w}$ is either zero or minimum compressive strength of 20% of the ultimate compressive strength of concrete as per IS: 784–2001.

If the wan thickness provided is greater than the minimum value t, the actual stress in concrete can be reduced and hence the amount of reinforcement is also correspondingly reduced.

f_c = Actual compressive strength in concrete

$$f_c = \frac{N_d}{\eta t} + \frac{f_{min, w}}{\eta} \qquad ...(13.7)$$

At transfer, the pre–stressing force F_o per metre length of the pipe is given by

$$F_o = 1000 \times 2t \times f_c \qquad ...(13.8)$$

Where, t is in mm and f_c is in MPa

If A_{st} = Cross sectional area of wire per m length of pipe

f_{st} = Stress in wire at transfer

n = Number of turns of circumferential wire windings per m length of pipe

d = Diameter of wire

Then $$A_{st} = \frac{2\pi d^2 n}{4} = \frac{2000tf_c}{f_{st}}$$

Since $$A_{st}f_{st} = F_o$$

$$\left(\frac{2\pi d^2 n}{4}\right) f_{st} = 2000tf_c$$

$\therefore$ $$n = \frac{4000tf_c}{\pi d^2 f_{st}} \quad ...(13.9)$$

• *Losses of Pre–stress*

During circumferential wire winding, due to elastic deformation of concrete, there is a loss of pre–stress which depends upon modular ratio (α_e) and the reinforcement ratio (ρ).

If f_{si} = Initial pre–stress in steel

f_{se} = Effective stress in steel after winding for compatibility of strains

Then, $$\frac{f_{si} - f_{se}}{E_s} = \frac{f_c}{E_c} = \frac{1}{E_c} \times \frac{A_{st}f_{se}}{2000t}$$

If $$\frac{A_{st}}{2000t} = \rho = \frac{f_c}{f_{st}} \text{ and } \frac{E_s}{E_c} = \alpha_e$$

Then, $$f_{si} = (1 + \alpha_e\rho)\, f_{se} \quad ....(13.10)$$

For pre–stressed concrete pipes, the percentage of reinforcements varies from, 0.5 to 1.0 and the modular ratio varies between 5 and 6. Hence the loss due to elastic deformation is about 3–6% of initial stress.

• *Longitudinal Stresses due to Winding*

The longitudinal bending stress during winding are transient, reaching the maximum at $0.355f_v$ near the ends of the pipe and $0.283f_v$ at the middle region of the pipe. Where, f_v is the average hoop compressive stress due to circumferential wire winding and is given by,

$$f_v = \frac{pR}{t} = \frac{F_o R}{R_o x t} \quad ...(13.11)$$

Where, p = External pressure due to wire wrapping

R = Internal radius, R_o = External radius

t = Thickness of the pipe shell, x = Pitch of the wire winding

F_o = Pre–stressing force per wire

Wire wrapping generally starts at a distance from the very end of the spigot and then covers the whole pile length. Due to partial loading, a transient stress of $0.6f_v$ and a permanent stress $0.355f_v$ develops at the spigot end of the pipe.

In the design of longitudinal pre–stressing, the permissible tensile strength (without any additional reinforcements and with induced longitudinal compression), should not exceed, $0.8\sqrt{f_{ci}}$ for transient stress and $0.5\sqrt{f_{ci}}$ for permanent stress, where f_{ci} is the compressive strength of concrete at wire winding.

As per IS: 784–2001, the longitudinal steel should be designed such that, the pipe should acts as a hallow circular beam which supports, without cracking, three times its own weight together with the weight of full volume of water in the pipe when it is supported by knife edges at the external ends.

13.3.4.2 Solved Examples

Problem 13.7

A non–cylindrical pre–stressing pipe of 1.6m in diameter with a core thickness of 100mm, is required to withstand a working pressure of $1N/mm^2$. Determine the pitch of 5mm diameter wire winding which is initially stressed to 1000MPa. The permissible maximum and minimum stresses in concrete are 12MPa and zero. The loss ratio is 0.8. If the direct tensile strength of concrete is 2MPa, estimate the load factor against cracking.

Solution:

Minimum thickness of pipe,

$$t > \left| \frac{W_w D/2}{\eta f_{ct} - f_{min,\,w}} \right| > \frac{1\times1600/2}{0.8\times12-0} > 84mm \approx \text{Provide, } t = 100mm$$

$$\therefore\ f_c = \frac{N_d}{\eta t} + \frac{f_{min,\,w}}{\eta} = \frac{1\times1600/2}{0.8\times100} = 10N/mm^2$$

$$\text{No. of wires per m, } n = \frac{4000tf_c}{\pi d^2 f_{st}} = \frac{4000\times100\times10}{\pi\times5^2\times1000} = 51 \text{ turns per m}$$

$$\text{Pitch of winding} = \frac{1000}{51} = 19.6mm$$

$$\text{Hoop tension due to fluid pressure} = \frac{1\times1600}{2\times100} = 8N/mm^2$$

Hoop compression due to pre–stress = $10N/mm^2$

∴ Resultant compressive stress in concrete = 10 – 8 = $2N/mm^2$

Tensile strength of concrete = $2N/mm^2$

Additional fluid pressure required to develop tensile stress of $4N/mm^2$ in concrete is,

$$= \frac{2\times100\times4}{1600} = 0.5N/mm^2$$

Cracking fluid pressure = 1 + 0.5 = 1.5N/mm² and working pressure = 1N/mm²

Load factor = 1.5/1 = 1.5.

Problem 13.8

Design a non–cylindrical pre–stressed concrete pipe of 600mm internal diameter to hold a working hydrostatic pressure of 1.05MPa, using a 2.5mm high tensile strength wires initially stressed to 1000MPa at transfer. Permissible maximum and minimum stresses in concrete at transfer and service loads are 14 and 0.7MPa respectively. Take loss of pre–stress be 20%. Calculate the test pressure required to produce a tensile stress of 0.7MPa in concrete when applied immediately after tensioning and also the winding stress in steel. Take E_s = 210kN/mm² and E_c = 35kN/mm²

Solution:

$$t > \left| \frac{W_w D/2}{\eta f_{ct} - f_{min,\,w}} \right| > \frac{1.05 \times 600/2}{0.8 \times 14 - 0.7} > 30\text{mm}$$

For 30mm thick concrete pipe, the actual compressive strength in concrete, f_c = 14MPa. The number of turns of 2.5 mm wires stressed 1000MPa per metre length of pipe is,

$$n = \frac{4000 t f_c}{\pi d^2 f_{st}} = \frac{4000 \times 30 \times 14}{\pi \times 2.5^2 \times 1000} = 86 \text{ turns per m}$$

$$\text{Pitch of circumferential winding} = \frac{1000}{86} = 11.6\text{mm}$$

If W_w is the test pressure required immediately after winding (η = 1)

$$f_c = \frac{N_d}{\eta t} + \frac{f_{min,\,w}}{\eta} = \frac{W_w D/2}{\eta t} + \frac{f_{min,\,w}}{\eta}$$

$$\therefore W_w = \frac{2t}{D}(f_c - f_{min,\,w}) = \frac{2 \times 30}{600}(14 - (-0.7)) = 1.47\text{N/mm}^2$$

If f_{si} is winding stress in steel then

$$f_{si} = (1 + \alpha_e \rho) f_{se}$$

$$\alpha_e = \frac{E_s}{E_c} = \frac{210}{35} = 6, \quad \rho = \frac{f_c}{f_{st}} = \frac{14}{1000} = 0.014$$

$$\therefore f_{si} = (1 + 6 \times 0.014)1000 = 1084\text{MPa}$$

Problem 13.9

A 75mm thick and 1000mm internal diameter non–cylindrical PSC pipe is required to hold water at a working pressure of 1.5MPa. The length of each pipe is 6m. The maximum direct compressive stresses in concrete are 15 and 2MPa. Take loss of pre–stress as 20%.

(a) Design the circumferential wire winding using 5mm high ensile wires initially stressed to 1000MPa

(b) Design the longitudinal pre–stressing using 7mm wires tensioned to 10MPa. The maximum permissible tensile stress under the critical transient loading should not exceeding $0.8\sqrt{f_{ci}}$, where, f_{ci} = 40MPa = The compressive strength of concrete.

(c) Check safety against longitudinal stresses that developed, considering the pipe as hollow circular beam as per IS: 784–2001 specifications.

Solution:

D = 1000mm f_{ct} = 15MPa W_w = 1.5MPa

$f_{min, w}$ = 2MPa t = 75mm f_{st} = 1000MPa L = 6m

(a) Circumferential Wire Winding

Compressive strength in concrete,

$$f_c = \frac{N_d}{\eta t} + \frac{f_{min,w}}{\eta} = \frac{1.5\times 1000/2}{0.8\times 75} + \frac{2}{0.8} = 15\text{N/mm}^2$$

$$\text{No. of wires per m, } n = \frac{4000 t f_c}{\pi d^2 f_{st}} = \frac{4000\times 75\times 15}{\pi\times 5^2\times 1000} = 57 \text{ turns per m}$$

$$\text{Pitch of winding} = \frac{1000}{57} = 17.5\text{mm}$$

(b) Longitudinal Pre–stressing

Critical transient stress at spigot = $0.6f_v$ = 0.6 × hoop stress = 0.6 × 15 = 9N/mm²

Maximum permissible tensile stress = $0.8\sqrt{f_{ci}} = 0.8\sqrt{40}$ = 5N/mm²

Hence the tensile stress = 9 – 5 = 4N/mm² should be counterbalanced by longitudinal pre–stressing.

Cross sectional area of the pipe = (π × 1.075 × 0.075) m²

If F_o is the longitudinal pre–stressing force required

$$F_o = \frac{\pi\times 1.075\times 0.075\times 10^6\times 4}{10^3} = 1013\text{kN}$$

Using 7mmϕ wires stressed to 1000MPa,

Force in each wire = 38.5kN

$$\therefore \text{ Number of wires} = \frac{1013}{38.5} = 27$$

(c) Check for Flexural Stresses as per IS: 784–2001

Considering the pipe as hollow circular section over a span of 6m.

Three times the self–weight = 3 (π × 1.075 × 0.075) × 24 = 18.3kN/m

Weight of water = (π × 1² × 10)/4 = 7.9kN/m

Total UDL on pipe = 26.2kN/m

Maximum bending moment = $\dfrac{26.2\times 6^2}{8}$ = 118kNm

Second moment area, I = $\dfrac{\pi(1.15^2 - 1^2)}{64}$ = 0.0365m^4

Flexural tensile stress = $\dfrac{118\times 10^6 \times 575}{0.0365\times 10^{12}}$ = 1.88N/mm^2 (Tension)

Longitudinal pre–stress = 4N/mm^2

∴ Resultant stress on concrete = 4 – 1.88 = 2.12N/mm^2 (Compression)

The resultant stress being compressive,

The pipe is safe against cracking.

13.3.4.3 Design of Cylindrical Pipes

The design principles of cylinder pipes are similar to those of non–cylindrical ones, except that the required thickness of concrete is computed by the equivalent area of the light gauge steel pipe embedded in the concrete.

If t_s= thickness of the steel pipe and $\dfrac{E_s}{E_c} = \alpha_e$ = modular ratio

The thickness of concrete pipe is given by

$$t > \left[\frac{N_d}{\eta f_{ct} - f_{min,\,w}}\right] - \alpha_e t_s \qquad ...(13.12)$$

The pre–stress required in concrete at transfer,

$$f_c = \frac{N_d}{\eta(t+\alpha_e t_s)} + \frac{f_{min,\,w}}{\eta} \qquad ...(13.13)$$

The number of turns in circumferential wire winding per metre length of pipe is given by

$$n = \frac{4000(t+\alpha_e t_s)f_c}{\pi d^2 f_{st}} \qquad ...(13.14)$$

There are possibilities of pipe bursting due to the yielding of steel cylinder. The bursting fluid pressure is calculated by the equation

$$P_u = \frac{f_{pu}A_{ps} + f_y A_{cs}}{D} \qquad ...(13.15)$$

Since $A_{ps} = \dfrac{\pi d^2 n}{2}$ = 1.57d^{2}n in mm^2/m = 0.00157d^{2}n in mm^2/mm ...(13.16)

And $A_{cs} = 2t_s$ ....(13.17)

$$\therefore \quad P_u = \frac{0.00157d^2nf_{pu} + 2t_sf_y}{D} \quad ...(13.18)$$

Where, P_u = Bursting pressure in N/mm^2

d = Diameter of the wire winding in mm

n = Number of turns per m length of pipe

f_{pu} = Tensile strength of wire winding in N/mm^2

f_y = Yield stress of steel cylinder in N/mm^2

t_s = Thickness of the steel cylinder in mm

D = Diameter of the steel cylinder in mm

13.3.4.4 Solved Examples

Problem 13.10

A pre–stressed concrete cylindrical pipe is to be designed using a steel cylinder of 1000mm internal diameter and thickness of 1.6mm.

The circumferential wire winding consists of 4mm high tensile wires initially stressed to $1000N/mm^2$. Ultimate tensile strength of wire = 1600MPa. Yield stress of steel cylinder = $280N/mm^2$. The maximum permissible compressive stress in concrete at transfer is 14MPa and tensile stress = 0 under working pressure of $0.8N/mm^2$. Determine the thickness of the concrete lining. Also calculate the number of turns of circumferential wire winding and the factor of safety against bursting. Assume modular ratio = 6.

Solution:

$$t > \left| \frac{N_d}{\eta f_{ct} - f_{min,\,w}} \right| - \alpha_e t_s > \frac{0.8 \times 1000/2}{0.8 \times 14 - 0} - 6 \times 1.6 > 25.9mm$$

Using 26mm thick concrete lining,

$f_c = 14N/mm^2$

$$n = \frac{4000(t + \alpha_e\, t_s)f_c}{\pi d^2 f_{st}} = \frac{4000 \times (26 + 6 \times 1.6) \times 14}{\pi \times 4^2 \times 1000} = 40 \text{ turns per m}$$

$$\text{Bursting pressure, } P_u = \frac{0.00157d^2nf_{pu} + 2t_sf_y}{D}$$

$$P_u = \frac{0.00157 \times 4^2 \times 40 \times 1600 + 2 \times 1.6 \times 280}{1000} = 2.516N/mm^2$$

$$\text{Factor of safety against bursting} = P_u = \frac{\text{bursting pressure}}{\text{working pressure}} = \frac{2.516}{0.8} = 3.14$$

13.4 PRE–STRESSED CONCRETE TANKS

13.4.1 Shapes of PSC Tanks

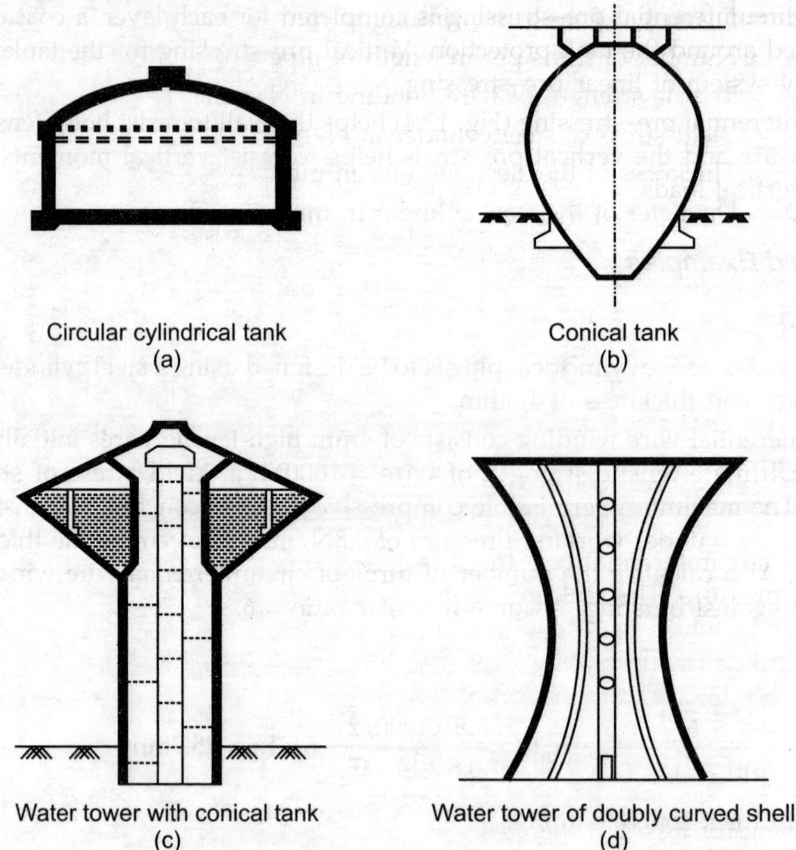

Fig. 13.3 Shapes of Pre–stressed Concrete Tanks

Cylindrical tanks are most commonly used types of PSC tanks. Some of the largest pre–stressed concrete tanks constructed are of circular in shape. A cylindrical shape is well suited for circumferential wire windings, which contributes more in PSC applications. Square or rectangular tanks spanning either vertically or horizontally, are required in industrial sectors. Some of the major types of shapes of PSC tanks are shown in Fig. 13.3.

13.4.2 PSC Circular Tanks

The term "circular pre–stressing" denotes the pre–stressing of circular structures such as tanks and pipes where the pre–stressing wires are wound in circles. In mast pre–stressed circular structures, pre–stress applied both circumferentially and longitudinally. The circumferential pre–stressing being circular and the longitudinal pre–stress is linear. Many pre–stressed concrete tanks have been constructed by the preload method. In this method the walls of the tanks are built of either concrete or pneumatic mortar, mortar being generally used if the walls are less than 150mm thick. Often, the walls are poured in at least vertical slices keyed together. After the concrete walls have attained sufficient strength they are pre–stressed circumferentially by a self–propelled machine, which winds the wire around the walls in a continuous operation, stressing it and spacing it at the same time.

After the circumferential pre–stressing is completed for each layer, a coat of pneumatic mortar is placed around the tank protection. Vertical pre–stressing for the tank walls can be applied by any system of linear pre–stressing.

The circumferential pre–stressing (Fig. 13.4) helps the wall to resist hoop tension imposed by liquid pressure and the vertical pre–stress helps to resist vertical moments imposed on the walls by vertical loads.

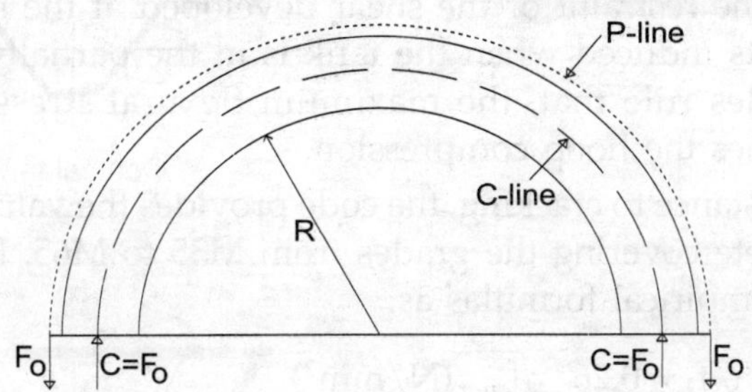

Fig. 13.4 Circumferential Pre–stressing

In a tank a circumferential pre–stress is provided to resist hoop tension produced by internal liquid pressure. The tank may be taken to consist of a number of circular rings, of unit height each of which is designed for hoop tension. Fig. 13.4 shows one half of such a ring. This ring is surrounded by pre–stressing wires carrying initial circumferential pre–stressing force F_o. The compressive stress in concrete due to initial pre–stressing force = $\frac{F_o}{A_c}$, where, A_c is the actual area of concrete

After allowing for losses, if the effective pre–stressing force is F, then the corresponding compressive stress in concrete = $\frac{F}{A_c}$

If the internal liquid pressure p is applied, Hoop tension due to this liquid pressure = pR

$\therefore$ Tensile stress in concrete due to internal liquid pressure = $\frac{pR}{A_e}$

Where, A_e = Equivalent area of concrete = $A_c + mA_{st}$

Resultant stress in concrete under the action of the final pre–stressing force and internal liquid pressure = $\frac{F}{A_c} - \frac{pR}{A_e}$ ...(13.19)

The practice followed is to provide a little residual compression in the concrete under working load conditions.

13.4.3 Design of PSC Circular Tanks

13.4.3.1 Design Criteria

Design of tank walls to resist the hoop tension and moments developed are based on the considerations of load factors against the cracking and collapse. According to IS: 3370 (Part III)–1967. It is desirable to have of a least load factor of 1.2 against cracking and 2 against ultimate collapse.

The ring pre–stressing is designed in all cases to counteract the maximum hoop tension developed, based on the assumption that the wall foot is free to slide without frictional resistance. Vertical pre–stressing is necessary to cater for the moments developed in the walls, depending upon the restraint or the shear developed at the base, as well as to resist the longitudinal moments induced when the tank is in the partially wound stage. For this case, the IS code provides rule that, the maximum flexural stress in the tank should be assumed equal to 0.3 times the hoop compression.

For estimation of resistance to cracking, the code provides the values of direct and bending tensile strength of concrete covering the grades from M35 to M65. However the values can be calculated using the empirical formulas as,

Direct tensile strength, $f_t = 0.267\sqrt{f_{cu}}$ (N/mm²) ...(13.20)

Flexural tensile strength, $f_{cr} = 2f_t$ ...(13.21)

13.4.3.2 Design Procedure of PSC Circular Tanks

Following is the procedure of PSC circular tank followed by design equations for the determinations of minimum wall thickness, circumferential pre–stress, spacing of wires and vertical pre–stress.

i. Estimation of maximum ring tension N_d and bending moment M_w, in the walls of the tank using tables of IS: 3370 (Part IV)–1967.

ii. Minimum wall thickness, $t > \left[\dfrac{N_d}{\eta f_{ct} - f_{min,\,w}}\right]$

The thickness of the wall provided should be such that a minimum cover of 35mm is available to the vertical pre–stressing cables.

iii. The circumferential pre–stress required is given by,

$$f_c = \frac{N_d}{\eta t} + \frac{f_{min\,w}}{\eta} \text{ in N/mm}^2$$

iv. Spacing of wires required in any section is obtained by considering the hoop tension due to fluid pressure and hoop compression due to circumferential wire winding is as follows,

If A_{st} = Cross sectional area of wire winding in mm²

w_t = Average radial pressure of wires at transfer at a given section in N/mm²

D = Diameter of the tank in mm

s = Spacing of the wires at the given section in mm

f_{st} = Stress in wires at transfer in N/mm^2

t = Thickness of the wall in mm

f_c = Compressive stress in concrete in N/mm^2

$\therefore$ Hoop compress due to pre–stressing = $\dfrac{W_t D}{2}$

$$\text{Equating } \frac{W_t D}{2} = \frac{f_{st} A_{st}}{s}, \quad \therefore\ w_t = \frac{2 f_{st} A_{st}}{sD} \qquad \text{...(13.22)}$$

If N_d = Hoop tension due to hydrostatic working pressure, W_w

N_t = Hoop compression due to radial pressure of wires, w_t

$$\text{Then} \qquad N_t = N_d \left(\frac{w_t}{W_w} \right) \qquad \text{...(13.23)}$$

Also $N_t = tf_c$ and from Eqs 13.22 and 13.23, the spacing of wire winding

$$s = \frac{2 N_d f_{st} A_{st}}{W_w f_c D t} \qquad \text{...(13.24)}$$

v. The vertical pre–stress required to resist the bending moments in the walls due to circumferential wire winding and hydrostatic pressure as a consequence of end restraint is,

If M_t = Vertical moment due to pre–stress at transfer

M_w = Vertical moment due to hydrostatic pressure.

$$\text{Then} \quad M_t = M_w \left(\frac{w_t}{W_w} \right)$$

The compressive pre–stress required in concrete is, $f_c = \dfrac{f_{min\,w}}{\eta} + \dfrac{M_w}{\eta Z}$ ...(13.25)

Where, Z = Section modulus of a unit length of wall about an axis in the tangential direction and passing through centroid.

When the tank is empty, the pre–stress required, $f_c = \dfrac{f_{min,\,w}}{\eta} + \dfrac{M_t}{Z}$ ...(13.26)

The vertical pre–stressing force required

$$F_o = f_c A_c \qquad \text{...(13.27)}$$

Where, A_c = Cross sectional area of concrete per unit length along the circumference. According to IS code standards, the vertical pre–stress force to be designed for 30% of the hoop compression.

13.4.4 Solved Examples

Problem 13.11

A pre–stressed circular cylindrical water tank of internal diameter 30m is to be hold the water at a depth of 7.5m. The permissible compressive stress of concrete at transfer is 13MPa and minimum compressive stress at working pressure is 1MPa. Loss of pre–stress is 25%. Wires of 5mm diameter with an initial stress of 1000MPa are used for circumferential winding and Freyssinet cables are made up of 12–8mm wires stressed to 1200MPa are used for vertical pre–stressing. Design the tank walls assuming the base is fixed. The cube strength of concrete is 40MPa.

Solution:

For the required depth of 7.5m and diameter of 30m, an average wall thickness of 150mm is assumed based on table of IS: 3370 (Part 4)–1967.

$D = 30m,\ H = 7.5m,\ t = 150mm,\ \eta = 0.75$

$$\frac{H^2}{Dt} = \frac{7.5^2}{30 \times 0.15} = 12.5 \text{ and } W_w = wH = (10 \times 7.5) = 75kN/m^2 = 0.075N/mm^2$$

Referring table of IS: 3370 (Part IV)–1967, the maximum ring tension and moment in tank walls for the fixed base conditions are

$$N_d = 0.64 \times 10 \times 7.5 \times 15 = 720kN/m = 720N/mm$$

$$M_w = 0.01 \times 10 \times 7.5^2 = 42.5kNm/m = 42500Nmm/mm$$

$$\text{Minimum wall thickness, } t = \left[\frac{N_d}{\eta f_{ct} - f_{min,w}}\right] = \left[\frac{720}{0.75 \times 13 - 1}\right] = 82.3mm$$

Net thickness available (Allowance for vertical cables of diameter 30mm)

$$= 150 - 30 = 120mm$$

$$\text{Required circumferential pre–stress, } f_c = \frac{N_d}{\eta t} + \frac{f_{min,w}}{\eta} = \frac{720}{0.75 \times 120} + \frac{1}{0.75} = 9.4N/mm^2$$

Spacing of circumferential wire winding at base,

$$s = \frac{2N_d f_{st} A_{st}}{W_w f_c Dt} = \frac{2 \times 720 \times 1000 \times 20}{0.075 \times 9.4 \times 30000 \times 120} = 11.4mm$$

$\therefore$ Number of wires per metre at base = 87

Ring tension N_d at 0.1H (0.75m) from top

$$N_d = 0.097 \times 10 \times 7.5 \times 75 = 109kN/m = 109N/mm$$

$$f_c = \frac{N_d}{\eta t} + \frac{f_{min,w}}{\eta} = \frac{109}{0.75 \times 120} + \frac{1}{0.75} = 2.5N/mm^2$$

$$s = \frac{2N_d f_{st} A_{st}}{W_w f_c Dt} = \frac{2 \times 109 \times 1000 \times 20}{0.075 \times 2.5 \times 30000 \times 120} = 64mm$$

Number of wires per metre at top = 16

Maximum radial pressure due to pre–stress,

$$w_t = \frac{2f_{st}A_{st}}{sD} = \frac{2\times1000\times20}{11.4\times30000} = 0.117\text{N/mm}^2$$

Maximum vertical moment,

$$M_t = M_w\left(\frac{W_t}{W_w}\right) = 42500\left(\frac{0.117}{0.075}\right) = 67000\text{Nmm/mm} = 67\times10^6\text{Nmm/m}$$

Considering 1 metre length of tank along the circumference, the section modulus,

$$Z = \frac{1000\times150^2}{6} = 375\times10^4\text{mm}^3$$

$$\therefore \text{ Vertical pre–stressing required, } f_c = \frac{f_{min,w}}{Z} + \frac{M_t}{Z} = \frac{1}{0.75} + \frac{67\times10^6}{375\times10^4} = 19.2\text{N/mm}^2$$

Since this exceeds the permissible value of f_{ct}= 13N/mm², the thickness of wall at base increased to 200mm. Thus, $Z = \frac{1000\times200^2}{6} = 666\times10^4\text{mm}^3$

$$\text{Vertical pre–stressing required, } f_c = \frac{f_{min,w}}{\eta} + \frac{M_t}{Z} = \frac{1}{0.75} + \frac{67\times10^6}{666\times10^4} = 12\text{N/mm}^2$$

$$\text{Vertical pre–stressing force, } F_o = f_cA_c = \frac{12\times1000\times200}{1000} = 2400\text{kN}$$

Using 12–8mm diameter Freyssinet cables

$$\text{Force per cable} = \frac{50\times12\times1200}{1000} = 720\text{kN}$$

$$\therefore \text{ Spacing of vertical cables} = \frac{1000\times720}{2400} = 300\text{mm}$$

The approximate vertical pre–stress required to counteract winding stresses as per IS code = $0.3f_c = 0.3\times9.4 = 2.82\text{N/mm}^2$

$$\therefore \text{ Vertical pre–stressing force required} = \frac{2.82\times1000\times200}{1000} = 564\text{kN}$$

$$\text{Ultimate tensile force in wires at base of the tank} = \frac{87\times20\times1500}{1000} = 2610\text{kN}$$

$$\therefore \text{ Load factor against collapse} = \frac{2610}{720} = 3.6$$

Direct tensile strength of concrete = $f_t = 0.267\sqrt{f_{cu}} = 0.267\sqrt{40} = 1.7\text{N/mm}^2$

$$\text{Cracking load} = 1000 \times 200 \times \frac{(0.75 \times 9.4 + 1.7)}{1000} = 1760\text{kN}$$

$$\therefore \text{ Factor of safety against cracking} = \frac{1760}{720} = 2.45$$

Nominal reinforcements of 0.2% of the cross sectional area are to be provided in the circumferential and longitudinal directions. Provide 8mm diameter mild steel bars at 300mm spacing c/c in both faces with nominal cover of 20mm.

REVIEW QUESTIONS

13.1 What are tension members and explain the salient features?

13.2 Distinguish the pre–stressed concrete pipes and elaborate them.

13.3 Put a light on non–cylindrical type pre–stressed concrete pipes.

13.4 Explain cylindrical type pre–stressed concrete pipes.

13.5 What are the advantages of PSC pipes?

13.6 Give the design specifications for PSC pipes as per IS: 784–2001.

13.7 Explain the design steps of non–cylindrical type pre–stressed concrete pipes.

13.8 Elaborate the design procedure of cylindrical type pre–stressed concrete pipes.

13.9 Explain the different shapes of PSC tanks with neat diagrams.

13.10 Write a points on pre–stressed circular tank

13.11 What is the design criteria in design of PSC circular tank?

13.12 Explain the design procedure for circular pre–stressed tank.

EXERCISE PROBLEMS

13.1 A pre–stressed concrete member is 250mm by 250mm, the tendons being subjected to a stress of 950MPa. The tendons provide an area of 550mm^2. Find the external tension that can the member carry if the stress in concrete under the action of external loading is 0MPa. Also find the tensile stress in the tendons after the application of the external tension on the members. Take modular ratio = 6

[Ans: External tension in member = 550.333kN and Tensile stress in tendons = 1003N/mm^2]

13.2 Design a pre–stressed tension member to carry a tension load of 1200kN. Safe compressive load stress in concrete = 10MPa, take modular ratio = 7.5, initial pre–stress in steel 1250MPa and loss of pre–stress = 18%.

[Ans: Section size: 373mm by 373mm]

13.3 A stressed tension member 275mm by 275mm is provided with 900mm^2 of high strength steel initially stressed to 950MPa. Allowing factor of safety of 1.5 against cracking of concrete. Find the safe tension for the member, assuming the cracking tensile stress of 3MPa. Find also the residual compressive stress under the working load. m = 6.5

[Ans: Safe tension in member = 775.77kN and Residual compressive stress = 1.81N/mm^2]

13.4 A pre–stressed concrete cylindrical type is formed by lining of steel cylinder of 7500mm internal diameter and thickness of 2.5mm with a layer of spun concrete of 38mm thick. If the pipe is required to with stand the hydraulic pressure of 0.85N/mm^2 without developing any tensile stress in concrete. Calculate (a). The required pitch of 4mm wires, wound round to the cylinder at a tensile stress of 980MPa, (b). The test pressure to produce a tensile stress of 1.4MPa in concrete immediately after winding and (c). The approximate bursting pressure. Take m = 6, tensile strength of wire 1680MPa, yield stress of cylinder = 280N/mm^2, loss of pre–stress is 15%

[Ans: (a). 32.5mm, (b). 1.2MPa, (c). 3.6MPa]

13.5 Design a pre–stressed non–cylindrical pipe with internal diameter 500mm to withstand a water working pressure of 1N/mm^2. The permissible maximum and minimum compressive stresses in concrete are 14N/mm^2 and 0.7N/mm^2, loss of pre–stress = 20%. High tensile wires of 3mm diameter initially stressed to 1000MPa. Modulus of elasticity of steel and concrete are 250kN/mm^2 and 35kN/mm^2 respectively. Calculate (a). The minimum thickness of the concrete pipe, (b). Number of turns of wire winding per metre length of pipe, (c). Test pressure required to produce a tensile stress of 0.7MPa in concrete, (d). Winding stress in steel wire.

[Ans: (a). 25mm, (b). 48 turns, (c). 1.47MPa, (d). 1084MPa]

13.6 A cylindrical pre–stressed concrete tank with internal diameter of 28m to hold water to a depth of 8m is to be designed using M40 grade concrete. The thickness of the concrete wall is 150mm and 5mm diameter high strength tensile wires are used. The permissible compressive stress in concrete at transfer is 12MPa. During service, a residual compressive stress of 1MPa. The initial stress in tensile wires is 1000MPa. The loss ratio = 0.8. Assuming the base slab and tank walls are monolithically cast, determine the number of turns of circumferential wire winding per metre height of the tank at base.

[Ans: Number of turns at base = 87 turns per m]

OBJECTIVE QUESTIONS

13.1 In a pre–stressed tension member, the equivalent area of concrete (A_e) is

a. $A_g - (m+1)\,A_{st}$ b. $A_g + (m-1)\,A_{st}$

c. $A_g + (m+1)\,A_{st}$ d. $A_g - (m-1)\,A_{st}$

13.2 Modular ratio is given by

a. $m = \dfrac{f_s}{f_c - f_s}$ b. $m = \dfrac{f_s - f_c}{f_c}$

c. $m = \dfrac{f_s}{f_c}$ d. $m = \dfrac{f_c}{f_s}$

13.3 In a tension, the resultant stresses due to effective pre–stress + the external load are for concrete ……..

a. $-\dfrac{F}{A_c} - \dfrac{P}{A_e}$ b. $\dfrac{F}{A_c} + \dfrac{P}{A_e}$

c. $\frac{F}{A_c} - \frac{P}{A_e}$ d. $-\frac{F}{A_c} + \frac{P}{A_e}$

13.4 In a tension, the resultant stresses due to effective pre–stress + the external load are for steel ....

a. $f_s + m\frac{P}{A_e}$ b. $-f_s - m\frac{P}{A_e}$

c. $f_s - m\frac{P}{A_e}$ d. $-f_s + m\frac{P}{A_e}$

13.5 Design of PSC pipes is made as per the specifications of ....

a. IS: 754–2001 b. IS: 774–2001

c. IS: 764–2001 d. IS: 784–2001

13.6 Thickness of concrete pipe in design of non–cylindrical PSC pipe is given by

a. $t > \left[\frac{N_d}{\eta f_{ct} + f_{min,w}}\right]$ b. $t > \left[\frac{N_d}{\eta f_{ct} - f_{min,w}}\right]$

c. $t < \left[\frac{N_d}{\eta f_{ct} - f_{min,w}}\right]$ d. $t < \left[\frac{N_d}{\eta f_{ct} + f_{min,w}}\right]$

13.7 Actual compressive strength in concrete, in design of non–cylindrical pipe is .....

a. $f_c = -\frac{N_d}{\eta t} + \frac{f_{min,w}}{2\eta}$ b. $f_c = \frac{N_d}{\eta t} - \frac{f_{min,w}}{\eta}$

c. $f_c = \frac{N_d}{\eta t} + \frac{f_{min,w}}{\eta}$ d. $f_c = -\frac{N_d}{\eta t} - \frac{f_{min,w}}{2\eta}$

13.8 At transfer, the pre–stressing force F_o per metre length of the non–cylindrical PSC pipe is given by

a. $F_o = 1000 \times 2t \times f_c$ b. $F_o = 1000 \times 4t \times f_c$

c. $F_o = 1000 \times 6t \times f_c$ d. $F_o = 1000 \times 8t \times f_c$

13.9 Number of turns of circumferential wire windings per m length of non–cylindrical pipe is given by ....

a. $n = \frac{3000tf_c}{\pi d^2 f_{st}}$ b. $n = \frac{2000tf_c}{\pi d^2 f_{st}}$

c. $n = \frac{4000tf_c}{\pi d^2 f_{st}}$ d. $n = \frac{1000tf_c}{\pi d^2 f_{st}}$

13.10 During circumferential wire winding, in case of PSC non–cylindrical pipes, due to elastic deformation of concrete, there is a loss of pre–stress which depends upon A. modular ratio (α_e) and B. reinforcement ratio (ρ).

Among A and B which is the right answer.

a. A b. B

c. Both A & B d. Neither A nor B

13.11 The reinforcement ratio (ρ) is given by......

a. $\frac{A_{st}}{1000t}$ b. $\frac{A_{st}}{2000t}$

c. $\frac{A_{st}}{3000t}$ d. $\frac{A_{st}}{4000t}$

13.12 In design of non–cylindrical pipes, the relation between initial pre–stress in steel (f_{si}) and effective stress in steel after winding for compatibility of strains (f_{se}) is ....

a. $f_{si} = (1 + \alpha_e\rho)\, f_{se}$ b. $f_{si} = (1 - \alpha_e\rho)\, f_{se}$

c. $f_{si} = (1 \times \alpha_e\rho)\, f_{se}$ d. $f_{si} = (1/\alpha_e\rho)\, f_{se}$

13.13 During design of non–cylindrical pipes, the longitudinal bending stress during winding are transient, reaching the maximum at ....near the ends of the pipe and ....at the middle region of the pipe. Where, f_v is the average hoop compressive stress due to circumferential wire winding. Fill the blanks with following options.

a. $0.325f_v$, $0.283f_v$ b. $0.335f_v$, $0.283f_v$

c. $0.345f_v$, $0.283f_v$ d. $0.355f_v$, $0.283f_v$

13.14 Average hoop compressive stress due to circumferential wire winding is given by ....

a. $f_v = \frac{F_o R}{R_o x t}$ b. $f_v = \frac{F_o R_o}{R x t}$

c. $f_v = \frac{F_o R x}{R_o t}$ d. $f_v = \frac{F_o R t}{R_o x}$

13.15 In the design of longitudinal pre–stressing for non–cylindrical pipes, the permissible tensile strength (without any additional reinforcements and with induced longitudinal compression), should not exceed, ... for transient stress and .... for permanent stress, where, f_{ci} is the compressive strength of concrete at wire winding. Fill the blanks.

a. $0.8\sqrt{f_{ci}}, 0.6\sqrt{f_{ci}}$ b. $0.6\sqrt{f_{ci}}, 0.5\sqrt{f_{ci}}$

c. $0.7\sqrt{f_{ci}}, 0.5\sqrt{f_{ci}}$ d. $0.8\sqrt{f_{ci}}, 0.5\sqrt{f_{ci}}$

13.16 The thickness of cylindrical PSC concrete pipe is given by ....

a. $t > \left[\frac{N_d}{\eta f_{ct} + f_{min,w}}\right] + \alpha_e t_s$ b. $t > \left[\frac{N_d}{\eta f_{ct} - f_{min,w}}\right] - \alpha_e t_s$

c. $t > \left[\frac{N_d}{\eta f_{ct} - f_{min,w}}\right] + \alpha_e t_s$ d. $t > \left[\frac{N_d}{\eta f_{ct} + f_{min,w}}\right] - \alpha_e t_s$

13.17 The pre–stress required in concrete at transfer, in cylindrical pipes is ....

a. $f_c = \dfrac{N_d}{\eta(t - \alpha_e t_s)} - \dfrac{f_{min,\,w}}{\eta}$ b. $f_c = \dfrac{N_d}{\eta(t + \alpha_e t_s)} - \dfrac{f_{min,\,w}}{\eta}$

c. $f_c = \dfrac{N_d}{\eta(t + \alpha_e t_s)} + \dfrac{f_{min,\,w}}{\eta}$ d. $f_c = \dfrac{N_d}{\eta(t - \alpha_e t_s)} + \dfrac{f_{min,\,w}}{\eta}$

13.18 The number of turns in circumferential wire winding per metre length of cylindrical PSC pipe is given by ....

a. $n = \dfrac{4000(t + \alpha_e t_s)f_c}{\pi d^2 f_{st}}$ b. $n = \dfrac{4000(t - \alpha_e t_s)f_c}{\pi d^2 f_{st}}$

c. $n = \dfrac{2000(t + \alpha_e t_s)f_c}{\pi d^2 f_{st}}$ d. $n = \dfrac{2000(t - \alpha_e t_s)f_c}{\pi d^2 f_{st}}$

13.19 The bursting fluid pressure in PSC cylindrical pipe is calculated by ....

a. $P_u = \dfrac{0.00147 d^2 n f_{pu} + 2 t_s f_y}{D}$ b. $P_u = \dfrac{0.00157 d^2 n f_{pu} + 2 t_s f_y}{D}$

c. $P_u = \dfrac{0.00167 d^2 n f_{pu} + 2 t_s f_y}{D}$ d. $P_u = \dfrac{0.00177 d^2 n f_{pu} + 2 t_s f_y}{D}$

13.20 Which of the following is not the shape of PSC tank?

a. Circular cylindrical b. Conical

c. Double curved shell d. Single curved shell

13.21 Design of PSC water tanks are done as per .... specifications.

a. IS: 3270 (Part III)–1967 b. IS: 3370 (Part III)–1967

c. IS: 3470 (Part III)–1967 d. IS: 3570 (Part III)–1967

13.22 In the design of pre–stress circular tank, direct tensile strength and flexural tensile strength is given by .... and .... respectively.

a. $f_t = 0.267\sqrt{f_{cu}}$, $f_{cr} = 2f_t$ b. $f_t = 0.267\sqrt{f_{cu}}$, $f_{cr} = 4f_t$

c. $f_t = 0.247\sqrt{f_{cu}}$, $f_{cr} = 2f_t$ d. $f_t = 0.247\sqrt{f_{cu}}$, $f_{cr} = 4f_t$

13.23 In design of PSC circular tanks, average radial pressure of wires at transfer (w_t) at a given section is given by ......

a. $w_t = \dfrac{f_{st} A_{st}}{sD}$ b. $w_t = \dfrac{4 f_{st} A_{st}}{sD}$

c. $w_t = \dfrac{3 f_{st} A_{st}}{sD}$ d. $w_t = \dfrac{2 f_{st} A_{st}}{sD}$

13.24 In design of PSC circular tanks, the spacing of wire winding is given by ....

a. $s = \frac{2N_d f_{st} A_{st}}{3W_w f_c Dt}$ b. $s = \frac{2N_d f_{st} A_{st}}{W_w f_c Dt}$

c. $s = \frac{N_d f_{st} A_{st}}{2W_w f_c Dt}$ d. $s = \frac{3N_d f_{st} A_{st}}{2W_w f_c Dt}$

13.25 In design of PSC circular tanks, according to IS code standards, the vertical pre–stress force to be designed for ...... of the hoop compression.

a. 10% b. 20% c. 30% d. 40%

Answers

13.1 b	13.2 c	13.3 c	13.4 b	13.5 d
13.6 b	13.7 c	13.8 a	13.9 c	13.10 c
13.11 b	13.12 a	13.13 d	13.14 a	13.15 d
13.16 b	13.17 c	13.18 a	13.19 b	13.20 d
13.21 b	13.22 a	13.23 d	13.24 b	13.25 c

❑

CHAPTER 14

Poles, Pavements & Sleepers

14.1 PRE–STRESSED CONCRETE POLES

PSC poles are used for railways power and signal lines, lighting poles, antenna masts, telephone transmission towers, electric transmission towers etc. The main advantages of PSC poles are,

1. Free from corrosion in humid and temperate climates and hence durable.
2. Freezing and thawing resistance in cold regions.
3. Lighter in weight due to reduced cross–section when compared with reinforced concrete poles.
4. Requires negligible maintenance for a number of years.

14.1.1 Shapes of PSC Poles

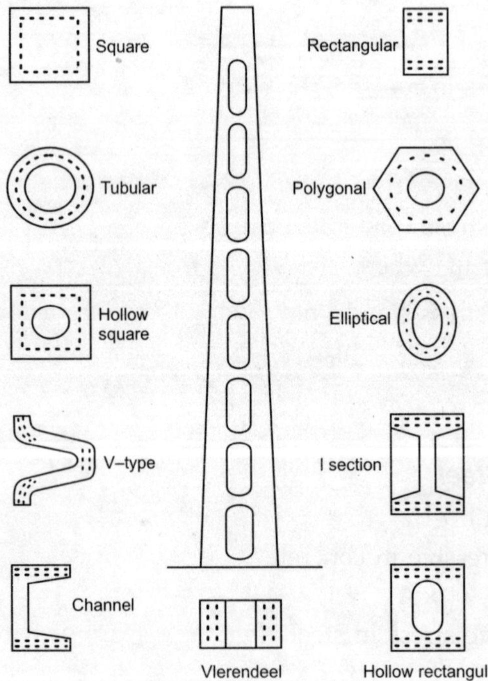

Fig. 14.1 Cross–sections of Pre–stressed Concrete Poles

As maximum bending moment occurs at the base of the pole, the cross sectional area required will also be maximum here. Generally poles are tapered with a hallow core to reduce the self–weight and to provide way for electric wires. For small lengths of up to 12m, the square or rectangular sections are preferable, since they can be easily manufactured and transported. Tubular sections are ideally suited for longer poles. The different types of pre–stressed concrete poles are shown in Fig. 14.1.

14.1.2 Design Considerations

PSC poles for power transmission lines are generally designed as members with uniform pre–stress since they are subjected to bending moments of equal magnitude in opposite directions. The poles are generally designed for the following ultimate and critical load conditions.

- Bending due to wind load on the cable and on the exposed face.
- Maximum torsion due to skew snapping of wires.
- Handling and erection stresses.
- Combined bending and torsion due to eccentric snapping of wires.
- Bending due to failure of all the wires on one side of the pole.

The load factor required for strength and serviceability criteria and in Indian code it is prescribed as 2.5 in IS: 1678–1998.

Note: Since the moment is fully reversible, the pre–stress has to be uniform over the whole section so that no tension occurs even after taking into account of full losses. So the bending stress must be equal to the residual stress.

14.1.3 Design Procedure of PSC Poles

Let f_c be the uniform stress in concrete due to initial pre–stressing force. Since the moment is fully reversible, the pre–stress has to be uniform over the whole section so that no tension occurs even after taking into account of full losses. So the bending stress must be equal to the residual stress in concrete.

Maximum compressive stress = f_c + Bending stress

Where, Bending stress = ηf_c, where, η is the loss ratio.

Maximum compressive stress should not exceed the permissible stress i.e., f_c'

So, $$f_c' = f_c + \eta f_c$$

$$\therefore \quad f_c' = f_c(1 + \eta) \qquad ...(14.1)$$

• *Calculation of Area of Steel*

For internal equilibrium

Tensile in steel = Compressive in concrete

$$f_{se} \times A_{st} = f_c \times A_c \qquad ...(14.2)$$

Where, f_{se} = The effective stress in steel after the release of wire

$$\therefore \quad A_{st} = \frac{f_c A_c}{f_{se}} \qquad ...(14.3)$$

• *Calculation of Depth of the Cross–section*

Using simple bending relation,

$$\frac{M}{I} = \frac{f_{cb}}{y},\ M = \frac{f_{cb} I}{y} = \frac{BD^3}{12}\frac{f_{cb}}{D/2}$$

$$M = \frac{BD^2 f_{cb}}{6}$$

$$D = \sqrt{\frac{6M}{Bf_{cb}}} \qquad ...(14.4)$$

Where, $f_{cb} = \eta f_c$

As the bending moment decreases towards the top the section of the pole can be reduced suitably.

14.1.4 Solved Examples

Problem 14.1

An electric pole is 12.5m high above ground level and supports a weight of 2000N due to transmission cables and has to withstand a reversible wind force of 4000N acting at its top. Losses in pre–stress may be taken as 12% due to shrinkage and creep. The pre–stressing wires are initially to be stressed to 1000N/mm^2. Allowable compressive stress in concrete is 13N/mm^2. Assume for soil, $\theta = 30^o$ and unit weight of soil = 17000N/m^3. m = 6. Design the pole.

Solution:

Let us provide a width, B = 300mm for the pole. Let the penetration of the pole below ground level be 'h' metres.

Coefficient of passive earth pressure, $K_p = \frac{1+\sin\phi}{1-\sin\phi} = 3$

Taking moment about C,

$$K_p \times \frac{Wh}{2} \times h \times \frac{h}{3} \times B = 4000(12.5 + h)$$

$$3 \times \frac{17000h^3}{6} \times 0.3 = 4000(12.5 + h)$$

$$2550h^3 = 50000 + 4000h$$

$$2550h^3 - 4000h - 50000 = 0$$

Solving for 'h', h = 2.89m

Let the maximum bending moment in the pole occurs at depth 'y' metres below the ground level.

At the section of maximum bending moment, SF = 0

i.e.,
$$\frac{Wy}{2} \times y \times B \times K_p = 4000$$

$$\frac{17000y^2}{6} \times 0.3 \times 3 = 4000$$

$$\therefore \quad y = 0.723\text{m}$$

Let f_c be the uniform stress in concrete due to initial pre–stressing force.

Then loss due to shrinkage and creep = $0.12f_c$

Pre–stress in concrete after shrinkage and creep = $0.88f_c$

So, Bending stress = $0.88f_c$

$$f_c + 0.88\,f_c = 13$$

$$\therefore \quad f_c = 6.915\text{N/mm}^2$$

Loss of stress in the tendon due elastic shortening of concrete = mf_c

$$= 6 \times 6.915 = 41.49\text{N/mm}^2$$

Stress in steel after release of wire = 1000 – 41.49 = 958.51N/mm²

For internal equilibrium

Tensile in steel = Compressive in concrete

$$f_{se} \times A_{st} = f_c A_c$$

$$958.51 \times A_{st} = 6.915A_c$$

$$\therefore \quad A_{st} = 0.00721A_c$$

$$M = \text{Maximum BM} = 4000(12.5 + 0.723) - (\text{Resistance offered by earth})$$

$$= 52892 - \left(K_p \times \frac{Wh}{2} \times h \times B\right)\frac{h}{3}$$

$$= 52892 - \left(3 \times \frac{17000 \times 0.723}{2} \times 0.723 \times 0.3\right)\frac{0.723}{3}$$

$$M = 51928.28\text{Nm}$$

$$f_{cb} = 0.88f_c = 0.88 \times 6.915 = 6.085\text{N/mm}^2,\ B = 0.3\text{m}$$

As the bending moment decreases towards the top the section of the pole can be reduced suitably.

$$\text{Depth of the section, } D = \sqrt{\frac{6M}{Bf_{cb}}} = \sqrt{\frac{6 \times 51928.28 \times 1000}{300 \times 6.085}}$$

$$\therefore \quad D = 413.33\text{mm}$$

$$\text{Provide } D = 415\text{mm}$$

$$\therefore A_{st} = 0.00721A_c = 0.00721BD = 0.00721 \times 300 \times 415 = 897.645\text{mm}^2$$

$$\text{No. of 5mm}\phi\text{ wires required} = \frac{897.645}{(\pi \times 5^2/4)} = 46\text{ No.s}$$

14.2 PRE–STRESSED CONCRETE PAVEMENTS

Roads and runways may be pre–stressed in the direction of the centre line by jacking, using abutments to take the reactions at the ends of a length. Road slabs can be individually pre–stressed between joints at great intervals. The interval may be up to 200m. Usually the main cables are provided in the longitudinal direction (i.e., parallel to the direction of the road). In addition, in a transverse direction secondary pre–stress is introduced using secondary reinforcements. In some designs, it is also a practice to provide the cables forming a diagonal grid.

Pre–stressed concrete provides smoother surface, a water tight covering for the subgrade, a longer life and significant reduction in the construction joints in taxiways and airfield runways.

14.2.1 Pavement Pre–stressing Methods

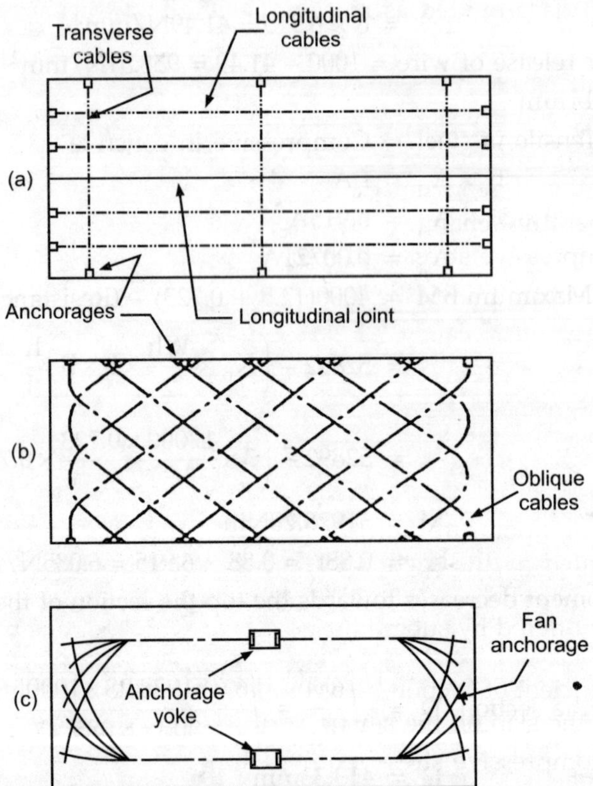

Fig. 14.2 Internal Pre–stressing Methods for Pavement

Longitudinal pre–stressing of slabs is achieved by external pre–stressing against rigid abutments or by internal pre–stressing by means of cables or tensioned bars. The external pre–stressing method is having many disadvantages, so that, method of internal pre–stressing become more popular.

The typical layouts of cables and anchorages used in internal pre–stressing is as shown in Fig. 14.2. The simplest and mostly used arrangement of longitudinal and transverse cables used for pre–stressing a two lane road slab as shown in Fig. 14.2 (a). Transverse pre–stressing cables eliminates need of using dowel bars across the longitudinal joint. The oblique cables are used as shown in Fig. 14.2 (b), the ratio of longitudinal to transverse pre–stress depends upon angle of inclination of cables and can be varied to suit the length/width ratio of pavement. Alternatively slabs can be pre–stressed by two anchorage yokes positioned at the centre of the slab as shown in Fig. 14.2 (c).

14.2.2 Design of Pre–stress in Pavements

It is preferable to provide a smooth layer over which the slab is placed in order to overcome the subgrade frictional resistance to compression of slabs (i.e., polythene sheet is spread over smooth finished layer of sand). The design is usually based on the requirement to overcome the subgrade resistance and then over and above this condition an additional pre–stress of $1.5 N/mm^2$ is provided as margin. Generally a thickness of 150mm is found sufficient.

The advantages of using pre–stressed concrete pavement are smooth surface, longer life, reduction in construction joint, water tight covering for the subgrade etc.

Tensile stresses are usually produced in a road slab due to frictional resistance offered by the subgrade preventing the movement of the slab, which is affected by as a consequence of shrinkage and temperature changes. The pre–stress applied to the slab should be such that, some residual compressive stress will remain finally.

Consider a slab of length 'L', width 'B' and thickness 't'

$$\therefore \quad \text{Weight of slab} = LBt\,(\text{Density})$$

$$W = LBtD_c \qquad \text{...(14.5)}$$

Consider one half of the slab on either side of the central axis Y–Y. So, weight of this part of slab = $\frac{W}{2}$

Suppose there is a temperature fall, so that there is a tendency for the slab to move inwards. This is prevented by the friction offered by the subgrade.

Frictional resistance offered by subgrade against movement of one half of the slab = $u\frac{W}{2}$

Where, u is the coefficient of friction between slab and subgrade. This frictional resistance becomes the tension in the slab on the section Y–Y.

Let f_{rc} = Residual compressive stress required

$$\therefore \text{ The pre–stressing force required longitudinally} = F_o = Bt \times f_{rc} + \mu\frac{W}{2} \qquad \text{...(14.6)}$$

The residual compressive stress recommend is $1 N/mm^2$. The coefficient of friction 'μ' depends on the nature of surface on which the road slab rests. In a natural condition, μ is taken from 1 to 1.5.

In another alternative method of design, the pre–stress in the slab gradually reduces towards the centre from the edges due to the friction and as shown in Fig. 14.3.

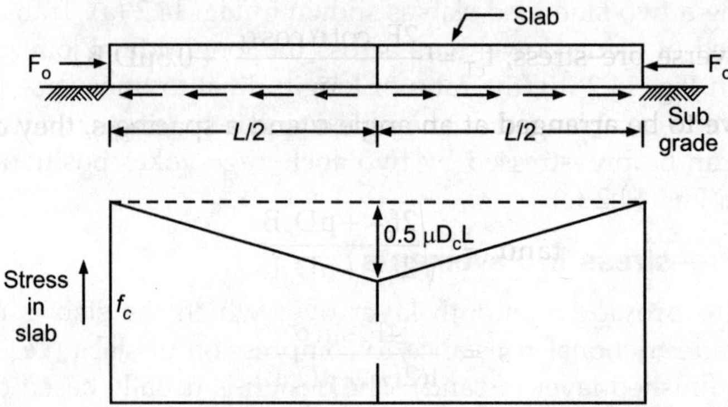

Fig. 14.3 Loss of Pre–stress in Slab due to Friction of Subgrade

If f_c = Compressive pre–stress in concrete at the ends of the slabs

μ = Coefficient of friction between subgrade and pavement

D_c = Density of pavement

L = Length of pavement

The minimum pre–stress available at the centre of the slab = $f_c + 0.5\mu D_c L$ ...(14.7)

In case of slabs with oblique cables, the arrangement is as shown in Fig. 14.4, the longitudinal and transverse pre–stress are expressed as components of the forces in the oblique cable system in the direction of the length and the width of the slab.

If, B = Width of the pavement slab

t = Thickness of the pavement slab

F_o = Pre–stressing force in each cable

s = Spacing of cables

L = Length of the pavement slab

α = Obliquity of the cables

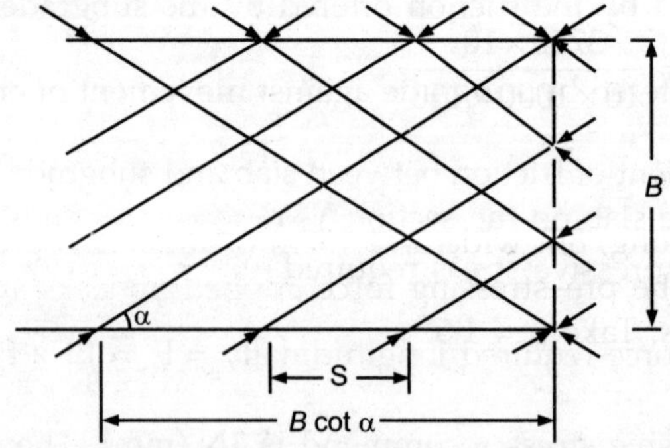

Fig. 14.4 Oblique Cable System for PSC Pavements

The minimum longitudinal and transverse pre–stress at the centre of the pavement slab, allowing for losses due to friction is given as,

Minimum longitudinal pre–stress, $f_{cL} = \dfrac{2F_o \cot\alpha \cos\alpha}{ts} - 0.5\mu D_c L$ ...(14.8)

Minimum transverse pre–stress, $f_{cT} = \dfrac{2F_o \cot\alpha \cos\alpha}{ts} - 0.5\mu D_c B$ ...(14.9)

If the cables have to be arranged at an angle α and a spacing s, they can be evaluated as follows.

$$\tan\alpha = \sqrt{\frac{2f_{cT} + \mu D_c B}{2f_{cL} + \mu D_c L}} \quad ...(14.10)$$

$$s = \frac{4F_o \sin\alpha}{t(2f_{cT} + \mu D_c B)} \quad ...(14.11)$$

14.2.3 Solved Examples

Problem 14.2

A road slab is 125m long, 10m wide and 150mm thick. Calculate the longitudinal pre–stressing force required. Take μ = 1, residual stress = 1N/mm². Find also the stress at edges.

Solution:

Weight of slab = 125 × 10 × 0.15 × 24 = 4500kN

Pre–stressing force required 'F_o' $= Bt \times f_{rc} + \mu \dfrac{W}{2}$

$$= 1 \times 10 \times 10^3 \times 0.15 \times 10^3 + 1 \times \frac{4500 \times 10^3}{2}$$

$$= 1.5 \times 10^6 + 2250000 = 3.75 \times 10^6 N$$

$$= 3750kN$$

$\therefore$ Stress at the edge $= \dfrac{3750 \times 10^3}{10 \times 1000 \times 150} = 2.5N/mm^2$

Problem 14.3

A road slab is 100m long, 8m wide and 150mm thick. It is pre–stressed longitudinally with a force of 5000kN. The pre–stressing force applied by jacks at two ends. Calculate the residue compressive stress. Take μ = 1.5.

Solution:

Weight of slab = 100 × 8 × 0.15 × 24000 = 2.88 × 10⁶N

Pre–stressing force, F_o = 5000kN, $F_o = f_{rc} \times Bt + \mu \dfrac{W}{2}$

$$5 \times 10^6 = f_{rc} \times 8 \times 10^3 \times 150 + 1.5 \times \frac{2.88 \times 10^6}{2}$$

$$\therefore \quad f_{rc} = 2.37\text{N/mm}^2$$

Problem 14.4

Design a suitable longitudinal and transverse pre–stressing system for a two way 7.5m wide and 100m long. Thickness of the slab = 150mm, $\mu = 0.15$, Freyssinate cables of 12–5mmϕ are used. A minimum longitudinal pre–stress of 2N/mm².

Solution:

The longitudinal pre–stress required at the ends of the slab after allowing frictional loss

$$= f_c + 0.5\mu D_c L = 2 + \frac{0.5 \times 1.5 \times 24 \times 10^3 \times 100}{10^6} = 3.8\text{N/mm}^2$$

$$\text{Total pre–stressing force, } F_o = \frac{3.8 \times 7.5 \times 1000 \times 150}{1000} = 4300\text{kN}$$

$$\text{Force in each cable} = \frac{20 \times 1200 \times 12}{1000} = 288\text{kN}$$

$$\text{Spacing of longitudinal cables} = \frac{7.5 \times 1000 \times 288}{4300} = 500\text{mm}$$

Transverse pre–stress = 0.7 × 3.80 = 2.66N/mm²

$$\therefore \text{ Force per m} = \frac{2.66 \times 1000 \times 150}{1000} = 399\text{kNm}$$

$$\text{Spacing of transverse cables} = \frac{288 \times 1000}{399} = 720\text{mm}$$

Problem 14.4

Design an oblique cable system for an airport runway of 20m wide and 150m long using Freyssinate cables of 12–5mm diameter, $\mu = 1.0$. Determine the obliquity and the spacing of the cables. The thickness of the pavement = 200mm and the minimum longitudinal and transverse pre–stress should not less than 2 and 1.5MPa respectively.

Solution:

t = 200mm	B = 20m
L = 150m	μ = 1.0
f_{cL} = 2N/mm²	f_{cT} = 1.5N/mm²

$$\text{Force in each cable, } F_o = \frac{20 \times 1200 \times 12}{1000} = 288\text{kN}$$

Obliquity of the cables,

$$\tan\alpha = \sqrt{\frac{2f_{cT} + \mu D_c B}{2f_{cL} + \mu D_c L}}$$

$$\tan\alpha = \sqrt{\frac{2\times 1.5 + 1\times 24\times 10^3 \times 20}{2\times 2 + 1\times 24\times 10^3 \times 150}} = 0.68,$$

$$\alpha = 34^{o}13'$$

$$\text{Spacing of the cables, } s = \frac{4F_o \sin\alpha}{t(2f_{cT} + \mu D_c B)}$$

$$= \frac{4\times 288\times \sin 34°13'}{200(2\times 1.5 + 1\times 24\times 10^3 \times 20)} = 930\text{mm}$$

14.3 PRE–STRESSED SLEEPERS

Due to the shortage of timber during Second World War, British railways developed pre–stressed concrete sleepers. The Dow–mac sleeper is pre–tensioned with 5mm diameter wire placed in pairs. For bond, a length of 700mm to 800mm is to be provided as transmission length. Bonding may be improved by looping the wires at the ends. This necessitates the sleepers to be pre–stressed individually rather than by long line method.

Factors influencing the design of sleepers are (1) Static and dynamic loads and (2) Ballast reaction.

The static loading in case of sleepers depends in the axle load. Lateral pressure on curves and accurate calculation of bending moments and shear forces will be difficult. In order to keep the maximum bending moment as small as possible, the projection of sleeper on either side is taken as $\frac{b}{2\sqrt{2}}$, where, b is the gauge length.

The static and dynamic loads imposed on rail seats depends upon the type of track (straight or curved), its combination and standard of maintenance, the axle loads and their spacing, the running characteristics, speed and standard of maintenance of vehicle.

The ballast reaction on the sleeper, which is influenced by the shape of the sleeper, its flexibility and spacing, the unit weight of the rail, the standard of maintenance of track and the characteristics of the ballast.

14.3.1 Types of Pre–stressed Sleepers

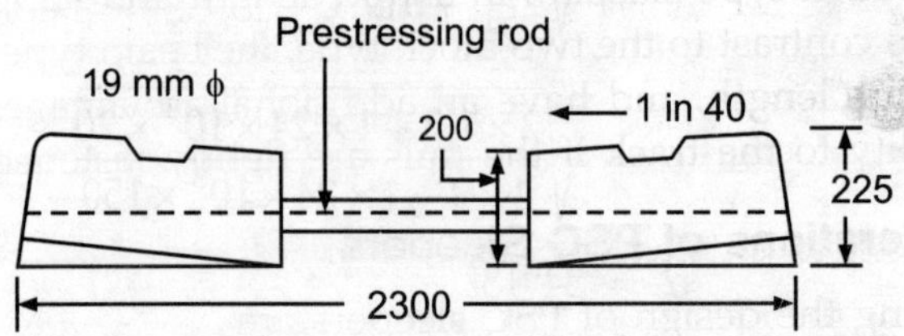

(a) SJ-101 two block R.C.C. with prestressed tie

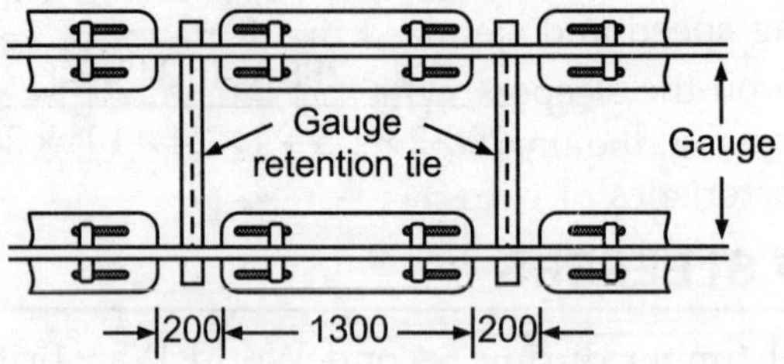

(b) Longrine-laval longitudinal blocks

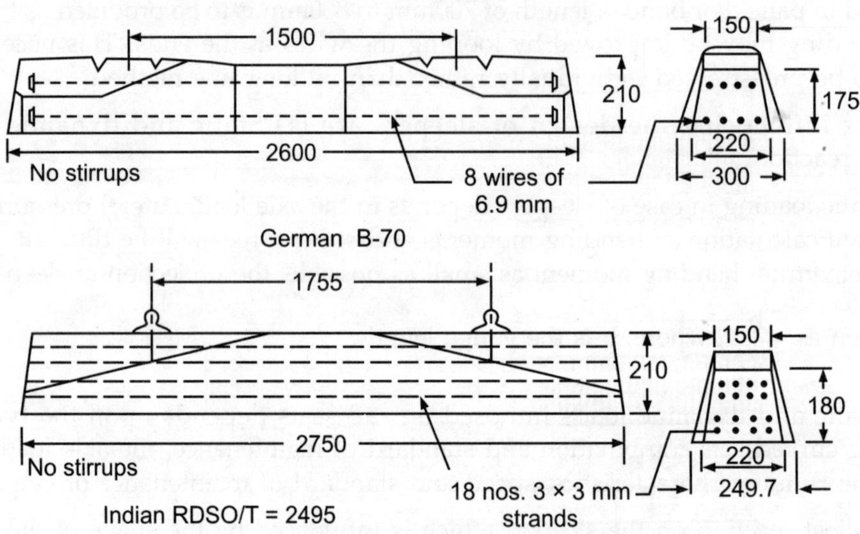

(c) Beam type sleepers

Fig. 14.5 Different Types of PSC Sleepers

The most commonly used pre–stressed concrete sleepers around the world are

i. Two block sleepers connected by a pipe filled with concrete and containing high tensile bars for compressing the concrete in the blocks. The Swedish SJ–101 tie is shown in Fig. 14.5 (a).

ii. Longitudinal sleepers located continuously under the rails and connected by flexible tie bars for gauge retention as shown on Fig. 14.5 (b).

iii. Beam type single piece pre-stressed concrete sleepers, which are quite similar to the conventional wooden type sleepers in shape, length and supporting area as shown on Fig. 14.5 (c). In contrast to the two block type, the beam type sleepers are flexurally stiff over the entire length, and have an additional advantages of providing greater measure of rigidity to the track if the rails are tightly fastened to the sleepers.

14.3.2 Design Considerations of PSC Sleepers

The factors influencing the design of PSC sleepers are

1. The static and dynamic loads imposed on rails depend upon the type of track (straight or curved), its construction and standard of maintenance, the axle load and their spacing, the running speed and standard maintenance of vehicles.
2. The ballast reaction on the sleepers, which is influenced by the shape of the sleeper, its flexibility and spacing, the unit weight of rail, the standard of maintenance of the track and the characteristics of ballast.

The loading standards and the hypothetical ballast pressure distribution have been evolved by various railway organizations based on laboratory and field experimentations. The loading criteria and the different loading conditions adopted by Indian railways is shown in Fig. 14.6.

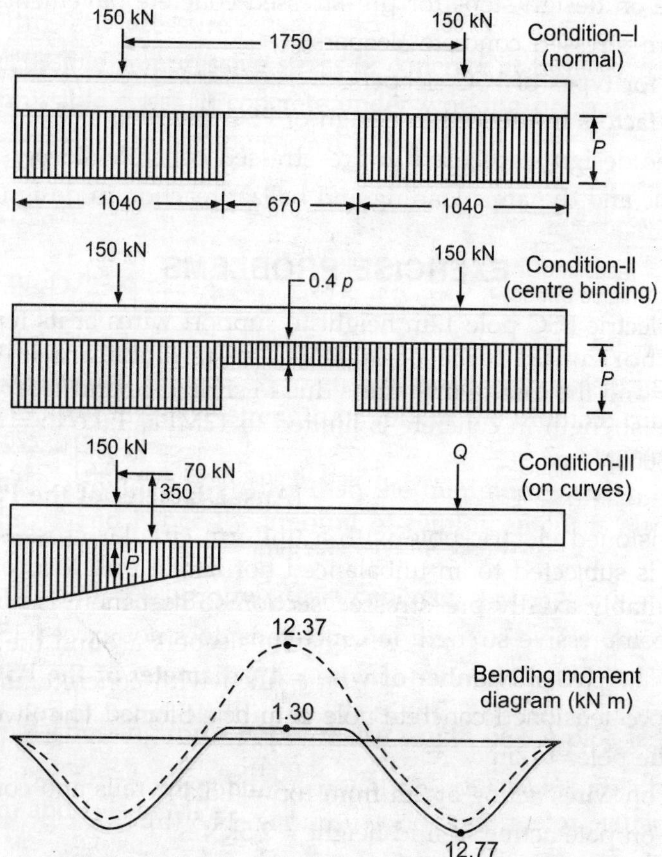

Fig. 14.6. Loading and BMD of Indian PCS–2A Sleeper

• ***Static and Dynamic Loading***

To allow the dynamic loading effects, the Indian railways department has adopted a design load of 150kN at each rail seat based on the static load of 60kN which ids amplified by a dynamic factor of 150% under different conditions of loading.

• ***Ballast Reaction***

The centre binding coefficient, defined as the ratio of intensity of pressure at the centre to that under the rail seat depends on the maintenance standards, climatic conditions and length of the sleeper. Value of the centre binding coefficient for the Indian railways is 0.40 based on the broad gauge and no frost heaving conditions.

REVIEW QUESTIONS

14.1 What are the advantages of pre–stressed concrete poles?

14.2 Explain the different shapes of the PSC poles with neat sketches.

14.3 What are the design considerations for PSC poles?

14.4 Elaborate the design procedure of pre–stressed concrete poles.

14.5 Explain the various methods of pre–stressing for PSC pavements.

14.6 Write a note on design steps for pre–stressed concrete pavements.

14.7 What are pre–stressed concrete sleepers?

14.8 Add points for types of PSC sleepers.

14.9 Explain the factors influenced in design of PSC sleepers.

14.10 Elaborate the design steps used in pre–stressed concrete sleepers.

14.11 Explain static and dynamic loading and ballast reaction in design of PSC sleepers.

EXERCISE PROBLEMS

14.1 Design an electric PSC pole 12m height to support wires at its top which can exerts reversible horizontal force of 3000N. The tendons are initially stressed to 1000N/mm^2 and the loss of pre–stress due to shrinkage and creep is 15%. Maximum compressive strength in concrete is limited to 12MPa. Take m = 6 and ϕ = 30°. Soil weight = 18000N/m^2.

[Ans: Diameter of the PSC pole = 370mm]

14.2 The pre–tensioned electric pole with a uniform circular cross–section stands 10m height and is subjected to an unbalanced horizontal pull load of 2500N at the top design a suitably axially pre–stressed section so that there is no tensile stress and maximum compressive strength in concrete is does not exceed 15N/mm^2

[Ans: Number of wire = 13, diameter of the PSC pole = 333.8mm]

14.3 A suitable pre–tensioned concrete pole is to be designed to suit the following data:

Height of the pole = 12m

Wind force on wires acting at 2m from top = 1.5kN

Wind force on pole acting at mid height = 2.5kN

Permissible strength of concrete in compression and tension are 16MPa and 4MPa respectively, Loss of pre–stress = 20%
High tensile wires of 8mm diameter initially pre–stressed to 1200MPa
Design a suitable rectangular section for the pole and the number of high tensile wires at the base section.

[Ans: Cross–section of rectangular PSC pole = 200mm by 300mm, number of wire = 13]

14.4 A road slab is 100m long, 8m wide and 150mm thick. It is pre–stressed longitudinally with a force of 5000kN. The pre–stressing force applied by jacks at two ends. Calculate the residue compressive stress. Take $\mu = 1.0$

[Ans: $f_{rc} = 2.37 N/mm^2$]

14.5 Design a suitable longitudinal and transverse pre–stressing system for a two way 7.5m wide and 100m long. Thickness of the slab = 200mm, $\mu = 0.15$, Freyssinate cables of 12–5mmϕ are used. A minimum longitudinal pre–stress of $2 N/mm^2$.

[Ans: Spacing of longitudinal cables = 450mm, spacing of transverse cables = 380mm]

14.6 Design an oblique cable system for an airport runway of 25m wide and 150m long using Freyssinate cables of 12–5mm diameter, $\mu = 1.0$. Determine the obliquity and the spacing of the cables. The thickness of the pavement = 200mm and the minimum longitudinal and transverse pre–stress should not less than 2 and 1.5MPa respectively.

[Ans: $\alpha = 34^o 13'$, s = 890mm]

OBJECTIVE QUESTIONS

14.1 The load factor required forstrength and serviceability criteria for PSC poles are prescribed in ....

a. IS: 1678–1998 b. IS: 1688–1998

c. IS: 1698–1998 d. IS: 1708–1998

14.2 Maximum compressive stress in pre–stressed concrete poles is given by ....

a. $f'_c = f_c (1 - \eta)$ b. $f'_c = f_c (1 + \eta)$

c. $f'_c = f_c \eta$ d. $f'_c = f_c (1/\eta)$

14.3 Depth of the Cross Section of PSC pole is given by ....

a. $D = \sqrt{\frac{M}{Bf_{cb}}}$ b. $D = \sqrt{\frac{2M}{Bf_{cb}}}$

c. $D = \sqrt{\frac{4M}{Bf_{cb}}}$ d. $D = \sqrt{\frac{6M}{Bf_{cb}}}$

14.4 Among the following which is not the method of pavements pre–stressing?

a. Arrangement of longitudinal and transverse cables

b. The oblique cables system

c. Four anchorage yokes position system

d. Two anchorage yokes position system

14.5 Weight of PSC pavement slab is ....

a. $W = LBD_c$ b. $W = LBtD_c$

c. $W = LtD_c$ d. $W = BtD_c$

14.6 The pre–stressing force required longitudinally in pre–stressed concrete pavements is ....

a. $Bt \times f_{rc} + \mu\dfrac{W}{2}$ b. $Bt \times f_{rc} + \mu\dfrac{W}{4}$

c. $Bt \times f_{rc} + \mu\dfrac{W}{6}$ d. $Bt \times f_{rc} + \mu\dfrac{W}{8}$

14.7 If f_c = Compressive pre–stress in concrete at the ends of the slabs, μ = Coefficient of friction between subgrade and pavement, D_c = Density of pavement, L = Length of pavement, the minimum pre–stress available at the centre of the slab = ...

a. $f_c + 0.3\mu D_c L$ b. $f_c + 0.4\mu D_c L$

c. $f_c + 0.5\mu D_c L$ d. $f_c + 0.6\mu D_c L$

14.8 The minimum longitudinal pre–stress in oblique method of pre–stressing of pavements is ...

a. $f_{cL} = \dfrac{2F_o \cot\alpha \cos\alpha}{ts} - 0.5\mu D_c L$ b. $f_{cL} = \dfrac{4F_o \cot\alpha \cos\alpha}{ts} - 0.5\mu D_c L$

c. $f_{cL} = \dfrac{2F_o \cot\alpha \cos\alpha}{ts} + 0.5\mu D_c L$ d. $f_{cL} = \dfrac{4F_o \cot\alpha \cos\alpha}{ts} + 0.5\mu D_c L$

14.9 The minimum transverse pre–stress in oblique method of pre–stressing of pavements is ...

a. $f_{cT} = \dfrac{4F_o \sin\alpha}{ts} + 0.5\mu D_c B$ b. $f_{cT} = \dfrac{2F_o \sin\alpha}{ts} + 0.5\mu D_c B$

c. $f_{cT} = \dfrac{4F_o \sin\alpha}{ts} - 0.5\mu D_c B$ d. $f_{cT} = \dfrac{2F_o \sin\alpha}{ts} - 0.5\mu D_c B$

14.10 The obliquity of the cables in PSC pavements is given by ....

a. $\tan\alpha = \sqrt{\dfrac{2f_{cT} - \mu D_c B}{2f_{cL} - \mu D_c L}}$ b. $\tan\alpha = \sqrt{\dfrac{2f_{cT} + \mu D_c B}{2f_{cL} + \mu D_c L}}$

c. $\tan\alpha = \sqrt{\dfrac{2f_{cT} - \mu D_c B}{2f_{cL} + \mu D_c L}}$ d. $\tan\alpha = \sqrt{\dfrac{2f_{cT} + \mu D_c B}{2f_{cL} - \mu D_c L}}$

14.11 The spacing of oblique cables in PSC pavements is.......

a. $s = \dfrac{8F_o \sin\alpha}{t(2f_{cT} + \mu D_c B)}$ b. $s = \dfrac{6F_o \sin\alpha}{t(2f_{cT} + \mu D_c B)}$

c. $s = \dfrac{4F_o \sin\alpha}{t(2f_{cT} + \mu D_c B)}$ d. $s = \dfrac{2F_o \sin\alpha}{t(2f_{cT} + \mu D_c B)}$

14.12 Which among the following is not the type of PSC sleeper?

a. Two block sleepers connected by a pipe filled with concrete and containing high tensile bars for compressing the concrete in the blocks.

b. Longitudinal sleepers located continuously under the rails and connected by flexible tie bars.

c. Beam type single piece pre–stressed concrete sleepers.

d. Beam type two piece pre–stressed concrete sleepers.

14.13 Value of the centre binding coefficient for PSC sleepers in the Indian railways is .... based on the broad gauge and no frost heaving conditions.

a. 0.40 b. 0.50 c. 0.60 d. 0.70

Answers

14.1 a	14.2 b	14.3 d	14.4 c	14.5 b
14.6 a	14.7 c	14.8 a	14.9 b	14.10 d
14.11 c	14.12 d	14.13 a		

❑

CHAPTER 15

Design of Pre–stressed Concrete Bridges

15.1 SCOPE

Bridges are structures that perform at least one single but major function that of providing a simple means to cross or reach between two points separated by a deep valley or river, a highway, or the like. Bridges offers structural engineers the opportunities to express simultaneously functionally and beauty while being unique; they have generated a continuous fascination in people, regardless of place and time. This is so much so that, symbolically the term "bridge" is used to describe connectivity between nations, races, religions, communities etc. and implies reaching out to the other as a sign of friendship and peace.

Bridges must be carefully planned and engineered before they can be constructed. Their final design must satisfy not only the technical constraints related to strength, safety and provision for future expansion, but also environmental, aesthetic, financial and political constraints. A bribed is a key element in a transportation system for three reasons.

- It controls the capacity of the system
- It is the highest cost per mile of the system
- If it fails, the system fails.

Moreover the bridge controls both the volume and the weight of the traffic carried by the transportation system, in the present and the foreseeable future.

There are many advantages in using pre–stressed concrete for bridges. Among them are minimum maintenance, increased durability, good aesthetics and when factory precast elements are used, assured plant quality, fast and easy construction, and low initial cost.

15.2 TYPES OF BRIDGES

Bridges can be classified in many ways, such as by type of structural material (steel v/s concrete), shape (arch v/s suspended bridge), continuity (simple v/s multiple span) and the like. However, for this section, we will classify pre–stressed concrete bridges according to their span length, namely short, medium and long spans; moreover, the range of spans is different for precast pre–stressed bridges versus cast–in–place post–tensioned bridges.

1. Precast pre–tensioned
   - Short spans: up to 50ft (16m)
   - Medium spans: 50 to 100ft (16 to 33m)
   - Long spans, include splicing: up to 180ft (55m)

2. Post–tensioned

   - Short spans such as slab type decks: up to 60ft (16m)
   - Medium spans: up to 180ft (55m)
   - Long spans such as segmental or extrados: up to 500ft (152m)
   - Very long spans such as cable stayed bridged: above 500ft (>152m)

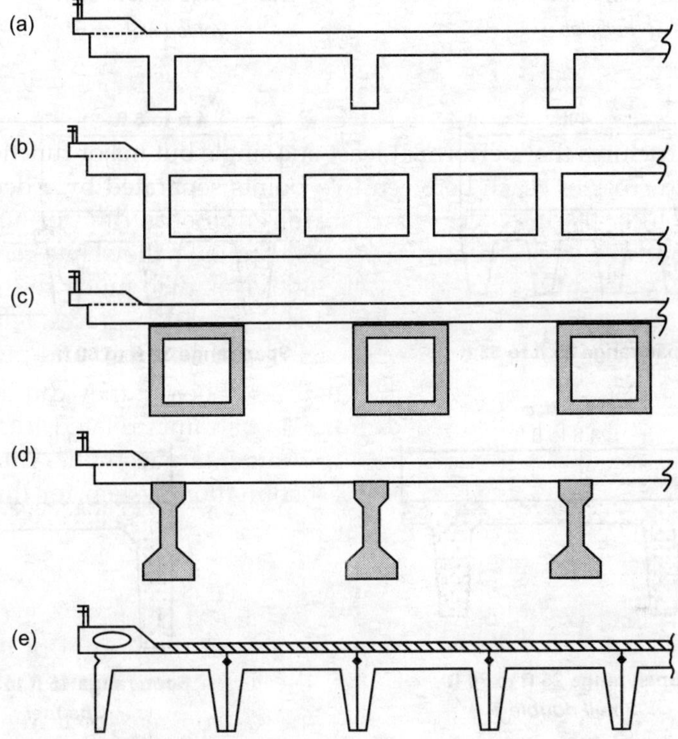

Fig. 15.1 Typical Cross–sections of Bridge Decks (a) T–Beams, (b) Box Beams, (c) Spread Box Beams, (d) I–Beams, (e) Adjacent Channel Beams

15.2.1 Short–span Bridges

These bridge decks are usually built with precast pre–tensioned beams that are transported to site and erected. The beams which have built in shear keys are generally placed adjacent to each other as shown in Fig. 15.1 (e). The shear keys are then filled with a mortar grout to provide the lateral resistance and load transfer. A topping of bituminous or equivalent, of about 2–3 inch (5–7cm) thickness, is generally added to provide a wearing surface and for the purpose of levelling. If a concrete topping is used, the deck will act as a composite deck.

Typical cross–sections of precast pre–tensioned elements used for common bridges and the corresponding ranges of dimensions and spans are shown in Fig. 15.2.

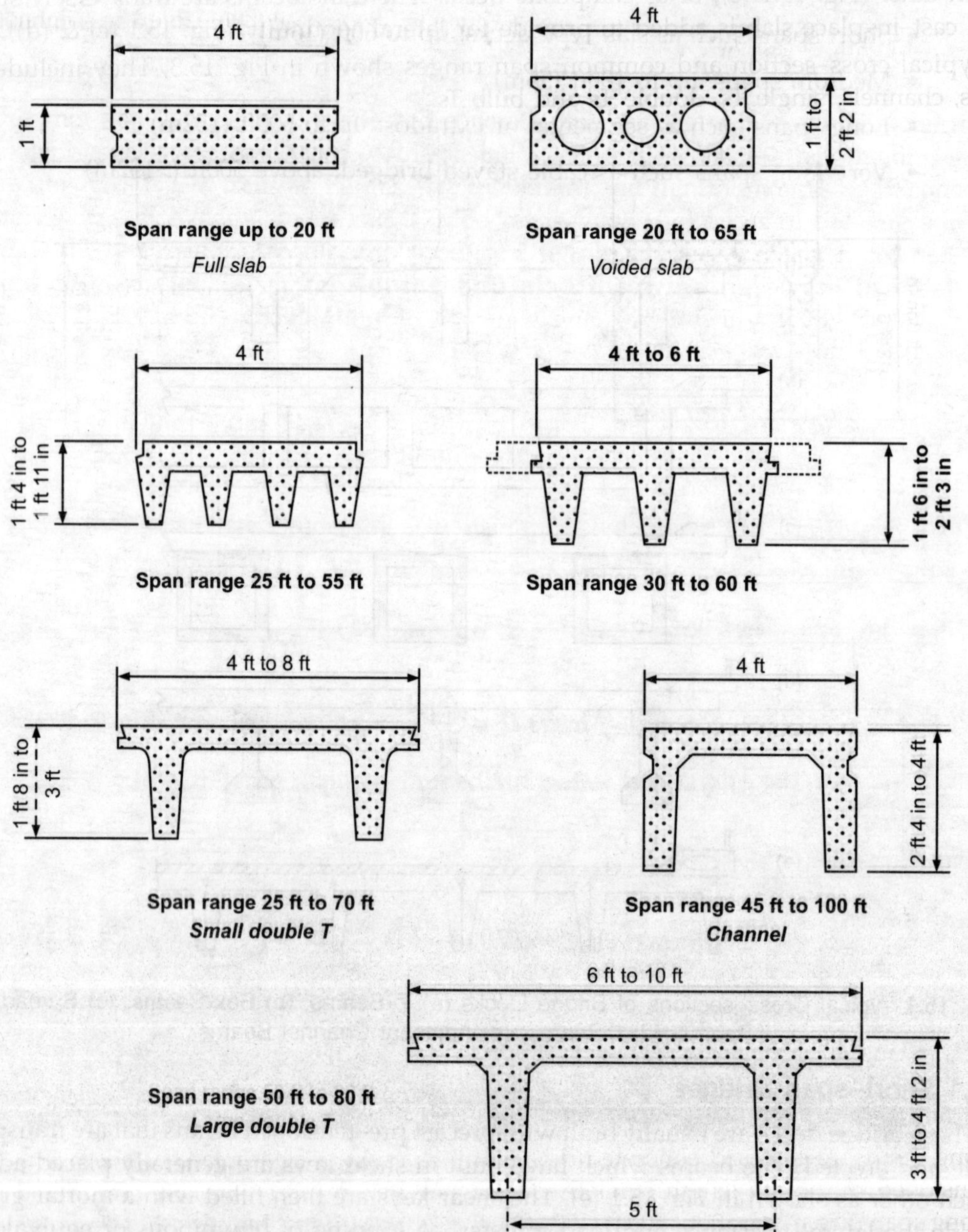

Fig. 15.2 Typical Precast Sections for Short to Medium–span Bridges

15.2.2 Medium and Long–span Bridges using Precast Beams

The decks of medium–span bridges using precast pre–tensioned beams are generally built either in a manner similar to short–span bridges where the beams are placed adjacent to each other (Fig. 15.1 (e)) or as composite decks where the beams are transversely spaced and a cast–in–place slab is added to provide for lateral continuity (Fig. 15.1 (c) & (d)).

Typical cross–section and common span ranges shown in Fig. 15.3. They include box beams, channels, single Ts, double Ts and bulb Ts.

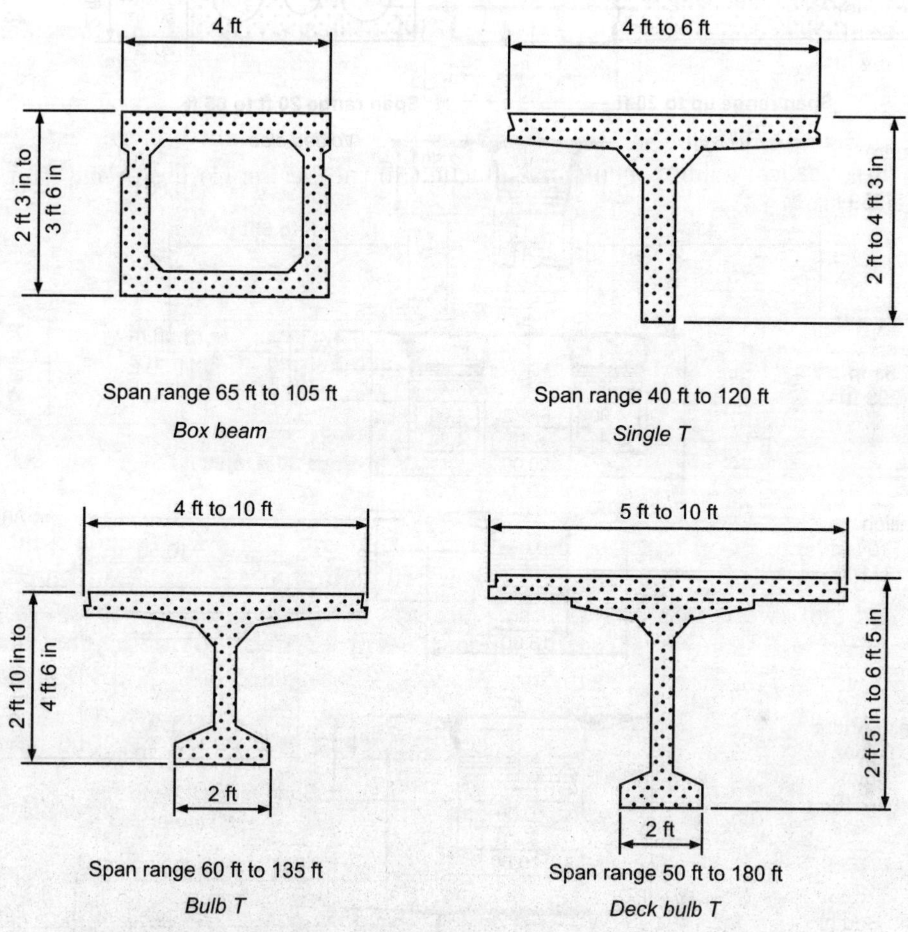

Fig. 15.3 Typical Precast Sections for Medium to Long–span Bridges

15.2.3 Long and Very Long–span Bridges

Long–span pre–stressed concrete bridges are generally cast–in–place or site precast and pre–tensioned. Although length is relative, spans are above 180ft (55m) are considered long. Very long spans are spans about 500 ft (152m). Long–span bridges are often built by a segmental construction technique. In segmental construction, the deck is built by segments, once at a time. Segments can be precast or cast–in–place. Box beams are considered best suited for this type of construction. They offer superior torsional rigidity and stability during construction and service. Their onside openings offers a platform for work and communication during construction and thereafter. Typical cross–sections and their evolutions are as shown in Fig. 15.4.

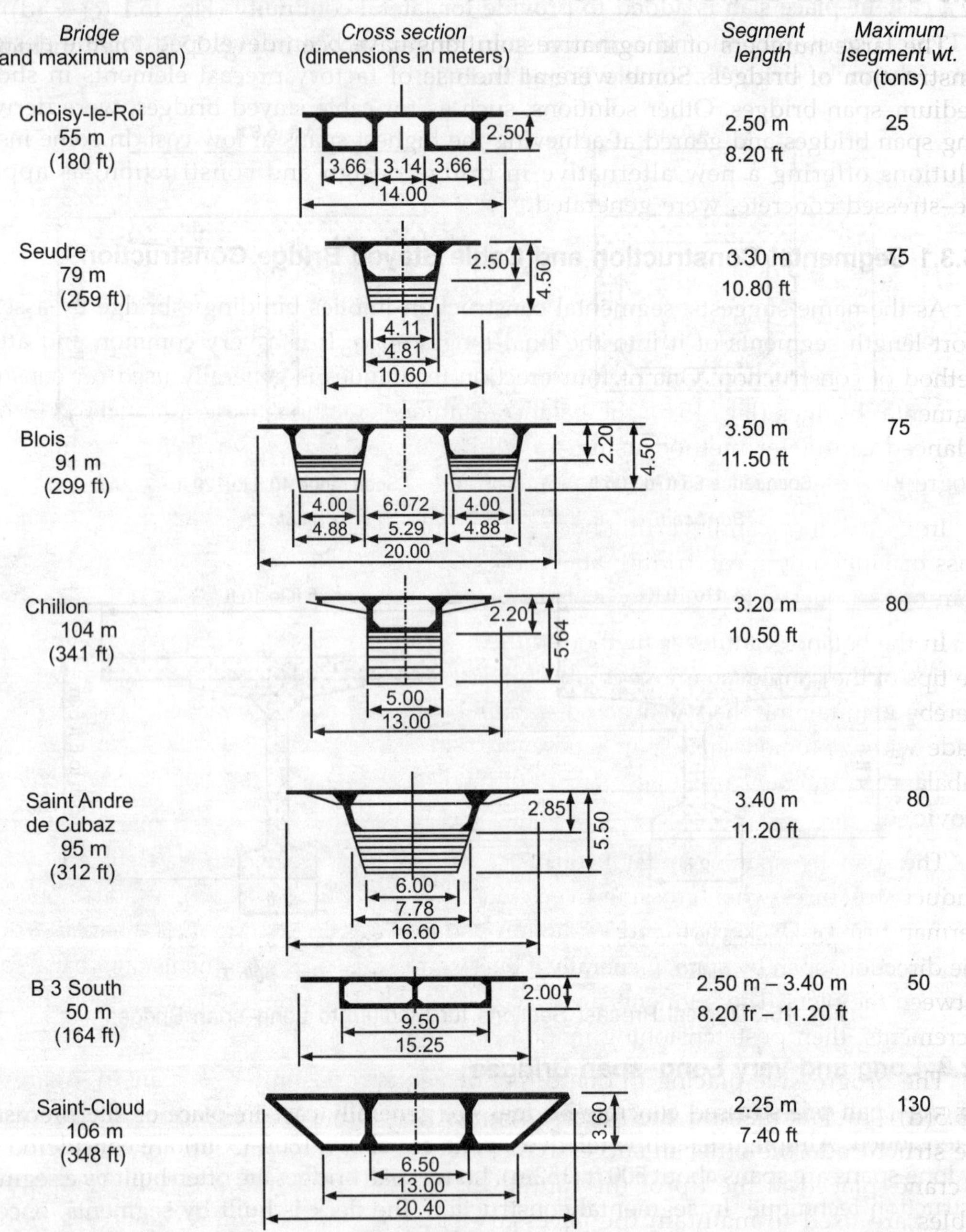

Fig. 15.4 Evolution of Typical Sections for Segmental Bridges in France

15.3 SPECIAL CONSTRUCTION TECHNIQUES FOR BRIDGES

The large numbers of imaginative solutions have been developed for the design and construction of bridges. Some were at the use of factory precast elements in short and medium–span bridges. Other solutions, such as for cable stayed bridges, were derived for long span bridges and geared at achieving the highest spans at low cost. In some instances, solutions offering a new alternative in bridge design and construction, as applied to pre–stressed concrete, were generated.

15.3.1 Segmental Construction and Cable Stayed Bridge Construction

As the name suggests, segmental construction implies building a bridge by assembling short length segments of it into the final longer span. It is a very common and attractive method of construction. One of four erection techniques is generally used for constructing segmental bridges (Fig. 15.5): the balance cantilever method using a launching girder, the balanced cantilever method using travelling forms, the span–by–span method, and the progressive placement in one direction method.

In the balance cantilever method using a launching girder (Fig. 15.5 (a)), an overhead truss or launching girder riding above the super structure places the segments alternating from one cantilever to the other to balance the load.

In the balance cantilever method with travelling forms (Fig. 15.5(b)), travelling forms at the tips of the cantilevers move in opposite directions at the same rate to cast new segments, thereby maintaining the balance and stability until mid–span is reached. Then the closure is made with a previous half span cantilever from the preceding pier. To account for possible imbalance, a moment resisting pier of temporary clamps between the deck and the pier are provided.

The span–by–span segmental construction method (Fig. 15.5(c)), is used mostly for long viaduct structures when crossing over water. The technique was initially developed by the German firm of Dyckerhoff and Widmann and features the execution of a super structure in one direction, span by span. Generally a barge crane places the segments on a truss spanning between the piers. The segments are placed progressively from one end to the other in span increments, then post–tensioning tendons are installed and stressed.

The progressive placing in one direction method is similar to span–by–span method (15.5(d)). In this method the precast segments are placed continuous from one end of the structure to the other in successive cantilevers on the same side of the various piers. A crane, placed at the tip of the cantilever, lowers them into position. Temporary stay cables are used to maintain the necessary balance. As the cantilever grows, additional temporary cables are installed for stability. A typical isometric view of the procedure is as shown in Fig. 15.6 and a photograph showing a segmentally built bridge during construction as shown in Fig. 15.7.

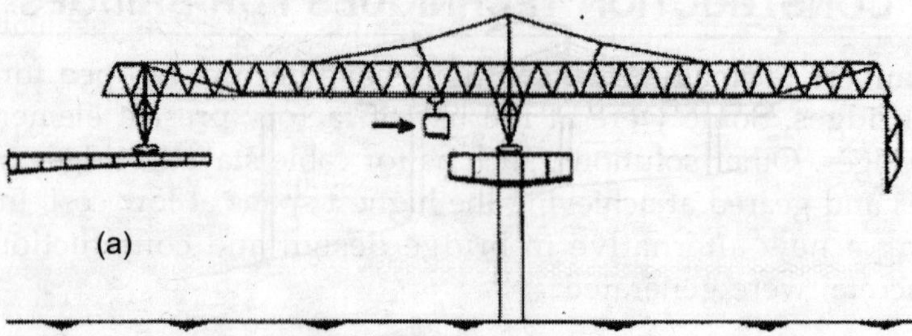

Balanced Cantilever Using Launching Girder. *An overhead truss or "launching girder," riding above the superstructure, places segments — alternating from one cantilever to the other — to balance loads.*

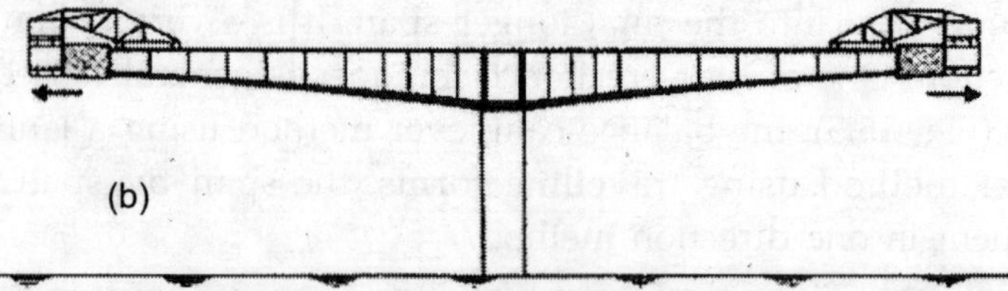

Balanced Cantilever with Travelling Forms. *Travelling forms at the tips of the cantilevers move in opposite directions at the same rate to cast new segments, there by maintaining balance and stability*

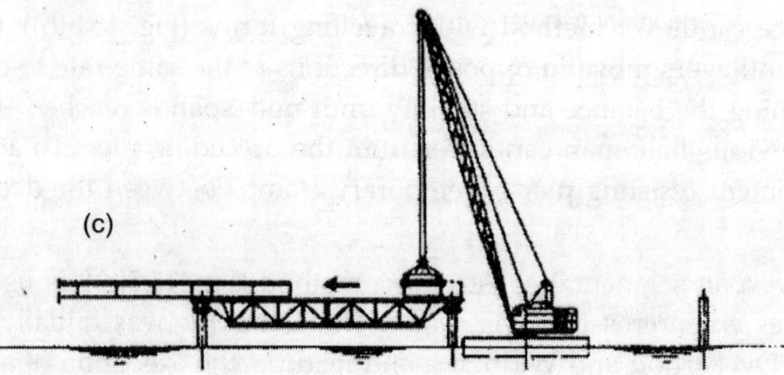

Span-by-Span Method. *A barge crane places segments on a truss spanning between piers, then post-tensioning is installed and stressed.*

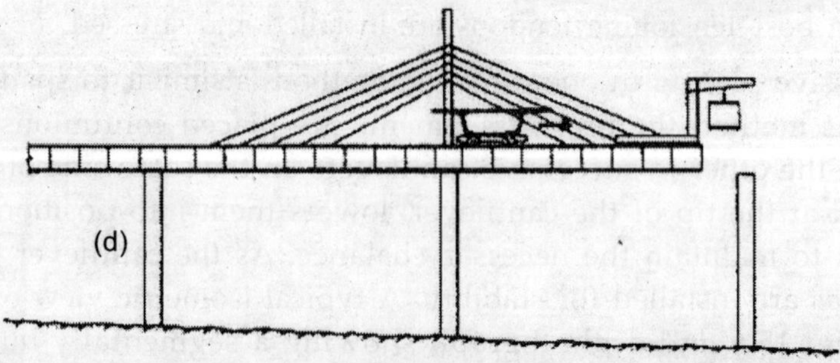

Progressive Placement in One Direction. *Segments are transported out to the tip of the cantilever where a crane places them in position. As the cantilever grows, additional temporary cables are installed for stability.*

Fig. 15.5 Representative Erection Techniques for Segmental Construction

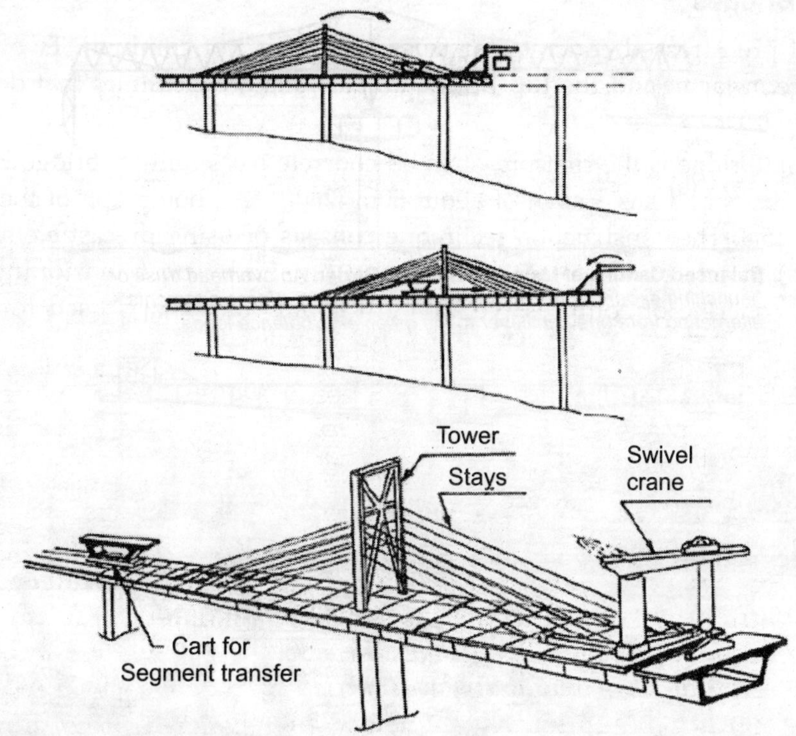

Fig. 15.6 Isometric View of Progressive Placing Technique

Fig. 15.7 Chesapeake and Delaware Canal Bridge, Delaware: All Precast Pre–stressed Concrete Segmental Cable Stayed Bridge with a Main Span of 229m (750ft)

15.3.2 Truss Bridges

The use of pre–stressed concrete for truss bridges is not common. Two bridges, the Iwahana Bridge in Japan and the Rip Bridge in Australia are examples that deserves to be mentioned.

The Iwahana Bridge is the first pre–stressed concrete truss railway bridge in Japan. It is a Warren–type truss and has a span of about 80m (260ft). A photograph of the bridge is as shown in Fig. 15.8. The construction technique consists of using precast elements (chords and diagonals) match–cast in plant, then joined on site by post–tensioning. Note that high–strength concrete of compressive strength about 93MPa was used for the precast elements.

Fig. 15.8 The Iwahana Bridge, Japan

The Rip Bridge (Fig. 15.9) is a three–span, 1082ft (330m) long structure with a central span of 600ft (183m). It spans the entrance to Brisbane waters, north of Sydney, Australia. It consists of three arch shaped cantilever truss system assembled from large precast concrete elements joined together on site by post–tensioning. The upper chord forming the deck essentially acts as a tensile member while the lower chord acts as a compression member. Each panel of the upper and lower chords of the truss was assembled from five precast elements across the width. The cross–sections of these elements were adequately designed so as to create the formwork for the in–situ cast–in–place concrete used for joining the elements and for embedding part of the post–tensioning tendons.

Fig. 15.9 The Rip Bridge, Australia

15.3.3 Stress Ribbon or Inverted Suspension Bridges

The concept of using pre–stressed concrete to carry the load mostly in tension instead of in flexure has always been considered as a very desirable alternative. However in practice, only a small proportion of bridges take advantage of such potential among which the trusses, as mentioned above, and stress–ribbon or inverted suspension bridges. Examples of stress–ribbon bridges include the Rio Colorado Bridge in Costa Rica, (Fig. 15.10) and a pedestrian bridge, the Shosai Bridge in Japan (Fig. 15.12).

Fig. 15.10 The Rio Colorado Stress–ribbon Bridge in Costa Rica

The photograph of the completed Rio Colorado Bridge is shown in Fig. 15.10. The bridge is 669ft (204m) long with a main span of 280ft (116m). The construction sequence is described in Fig. 15.11 and comprises the following stages: (1) The foundations, the vertical piers, and the abutments for the approach spans are cast; (2) the precast T–beams of the approach spans are erected and the main piers are cast vertically on hinges; (3) the main piers are leaned downward 30°, by adjusting anchorage cables, and their base is fixed; (4) the main span cables are strung and the topping of the approach spans completed; (5) precast sections forming the lower platform of the center span are fastened to the cables symmetrically from each end, then vertical column bents are attached to the cables; (6) the precast T–beams of the roadway are lowered into place and post–tensioned for continuity; (7) the structure is complete.[1]

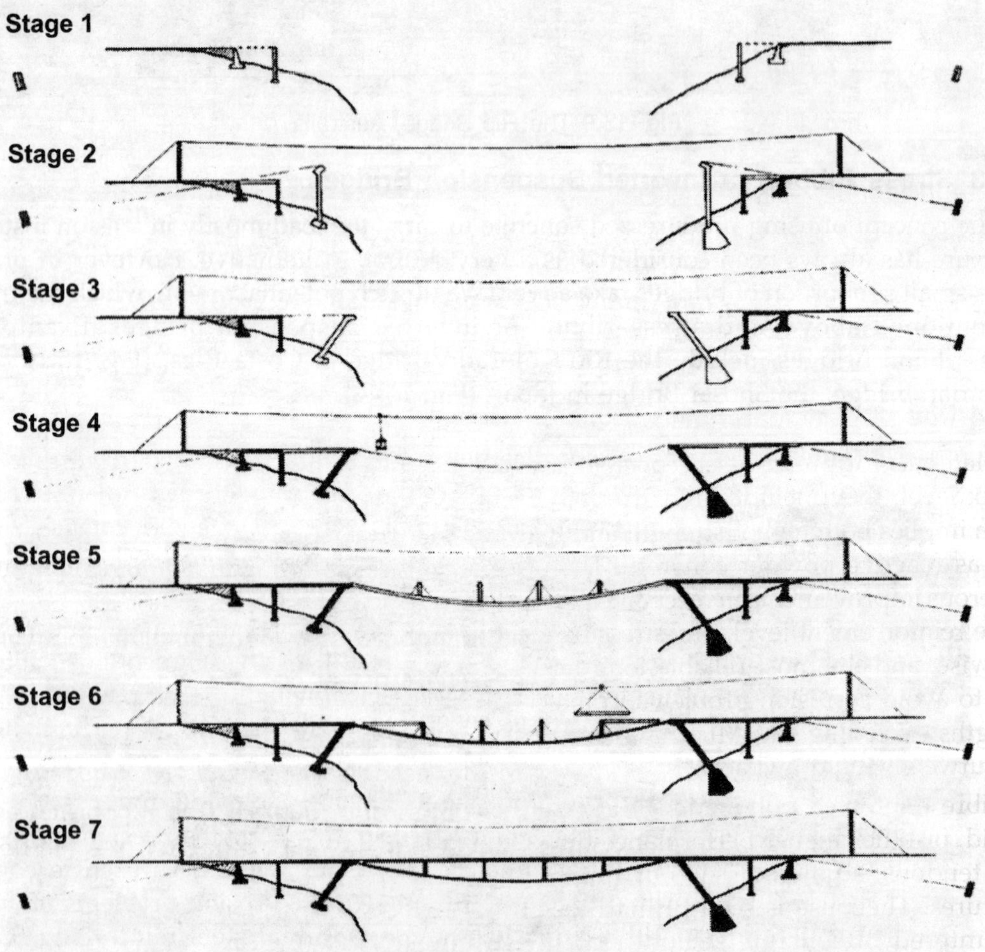

Fig. 15.11 Construction Sequence of the Rio Colorado Bridge, Costa Rica

Fig. 15.12 Shosai Bridge: Example of Stress–ribbon Pedestrian Bridge with a Main Span of 61m, Shizouka, Japan

15.3.4 Use of New Materials

Materials for pre–stressed concrete structures are continuously evolving. At the concrete matrix level, they include high strength concrete (50–100MPa), high performance concrete which implies both higher strength and high durability, and ultra–high performance concrete such as reactive powder concrete. The addition of fibres in the concrete matrix leads to numerous improvements in mechanical properties, particularly ductility and impact resistance. At the reinforcement level, pre–stressing steels remains the most competitive reinforcements, cost wise, and the most reliable long term. Stainless steel pre–stressing tendons have been used to avoid corrosion problems in marine applications although pre–stressing steels with strengths exceeding 2100MPa have been used, the benefits to move to higher strengths do not outweigh the drawbacks.

Fibre reinforced polymeric (FRP) reinforcements (Fig. 15.13), made from carbon, glass, aramid, or other higher performance fibres embedded in polymeric matrices, in the form of bars, tendons, strands, or two or three dimensional meshes, have been used in concrete structures. Their use is seen primarily as a means to avoid corrosion problems otherwise encountered in concrete structures when using conventional steel reinforcing bars or pre–stressing tendons, FRP reinforcement are suitable for use in new structures, as well as for repair or strengthening of the existing structures. Their applications in bridge decks, marine structures, railway ties and structures exposed to chemicals or harsh environment is bound to increase. Moreover their nonmagnetic properties make them ideal for special applications such as radar stations, special medical facilities (X–ray rooms), and structures for magnetic levitation trains.

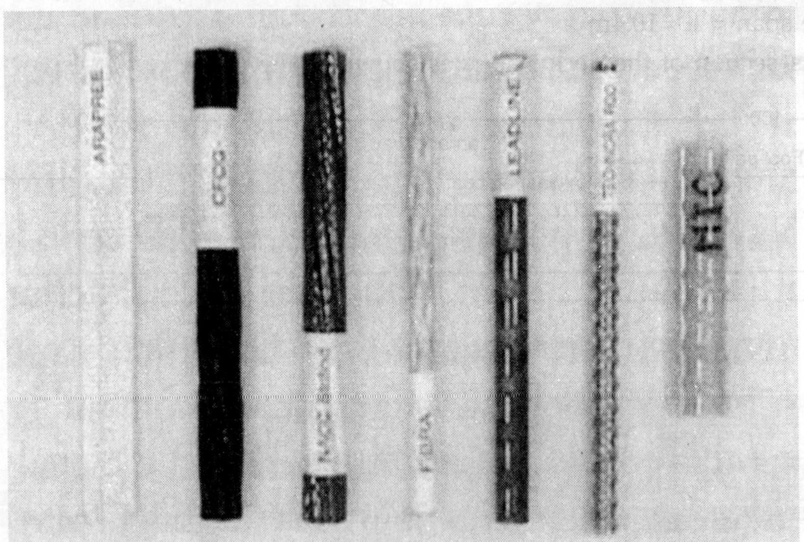

Fig. 15.13 Typical FRP Reinforcements for Pre–stressing

15.4 SOLVED EXAMPLES

Problem 15.1

Design a post–tensioned pre–stressed concrete slab bridge deck for a national highway crossing to suit the following data: clear span = 10m, width of bearing = 400mm, clear width of roadway = 7.5m, footpath = 1m from either side, kerbs = 600mm wide, thickness of wearing coat = 80mm, live load = IRC class AA tracked vehicle, class of structure = class 1 type. Materials: M40 grade concrete and 7mm diameter high tensile wires with an ultimate tensile strength of 1500N/mm^2 housed in cables with 12 wires and anchored by Freyssinet anchorages of 150mm diameter. For supplementary reinforcement, adopt Fe 415 grade HYSD bars. Compressive strength at transfer = f_{ci} = 35N/mm^2, loss ratio = 0.8. The design should confirm to the recommendations of the codes IRC: 6, IRC: 18, IRC: 21 and IS: 1343.

Solution:

- ***Permissible Stresses***

The permissible compressive stresses in concrete at transfer and working loads, as recommended in IRC–18 are as follows:

$$f_{ct} = 15\text{N/mm}^2 < 0.45 f_{ci} = (0.45 \times 35) = 15.75\text{N/mm}^2$$

Loss ratio η = 0.80

$$f_{cw} = 12\text{N/mm}^2 < 0.33 f_{ck} = (0.33 \times 40) = 13.2\text{N/mm}^2$$

$$f_{tt} = f_{tw} = 0$$

• Depth of Slab and Effective Span

Assuming the thickness of slab at 50mm per meter of span for highway bridge decks.

The overall thickness of the slab = (10 × 50) = 500m

Width of bearing = 400mm

Effective span = L =10.4m

The cross–section of the deck slab is as shown in Fig. 15.14

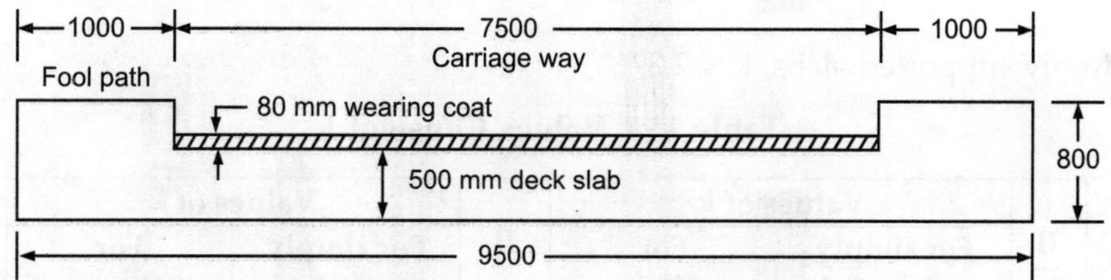

Fig. 15.14 Cross–section of the Deck Slab

• Dead Load Bending Moments

Dead weight of slab = (0.5 × 24) = 12kN/m^2

Dead weight of wearing coat = (0.08 × 22) = 1.76kN/m^2

Total dead load = g = 14.00N/mm^2

Dead load bending moment (M_G) = (g × L^2/8) (14 × 10.4^2)/8 = 190kNm

• Live Load Bending Moments

Generally, the bending moment due to live load will be maximum for IRC class AA tracked vehicle. Impact factor for the class a tracked vehicle is 25 percent for the 5m span, which decreases linearly to 10 percent for the 9m span.

∴ Impact factor = 10 percent for the span of 10.4m

The tracked vehicle is placed centrally on the span.

Effective length of the road = [3.6 + 2(0.5 + 0.08)] = 4.76m

Effective width of the slab perpendicular to the is expressed as,

$$b_e = kx\left(1 - \frac{x}{L}\right) + b_w$$

Referring to Fig. 15.15

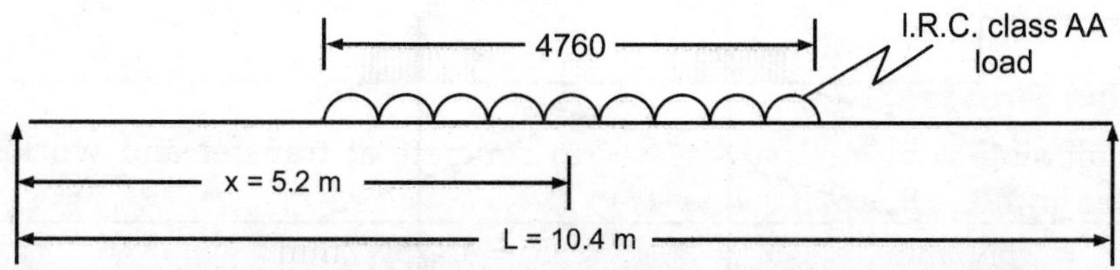

Fig. 15.15 Position of Load for Maximum Bending Moment

$x = 5.2\text{m},\ L = 10.4\text{m},\ B = 9.5\text{m}$

$$\therefore \quad \left(\frac{B}{L}\right) = \left(\frac{9.5}{10.4}\right) = 0.913$$

and $b_w = (0.85 + 2 \times 0.08) = 1.01\text{m}$

from Table 15.1 for $\left(\frac{B}{L}\right) = 0.913$

and simply supported slabs, k = 2.37

Table 15.1 Values Constant k

B/L	Values of k: For simply supported slabs	Values of k: For continuous slabs	B/L	Values of k: For simply supported slabs	Values of k: For continuous slabs
0.1	0.40	0.40	1.1	2.60	2.28
0.2	0.80	0.80	1.2	2.64	2.36
0.3	1.16	1.16	1.3	2.72	2.40
0.4	1.48	1.44	1.4	2.80	2.48
0.5	1.72	1.68	1.5	2.84	2.48
0.6	1.96	1.84	1.6	2.88	2.52
0.7	2.12	1.96	1.7	2.92	2.56
0.8	2.24	2.08	1.8	2.96	2.60
0.9	2.36	2.16	1.9	3.00	2.60
1.0	2.48	2.24	2.0 and above	3.00	2.60

$$b_e = 2.37 \times 5.2\left(1 - \frac{5.2}{10.4}\right) + 1.01 = 7.172\text{m}$$

The tracked wheel is placed closed to the kerb with the required minimum clearance as shown in Fig. 15.16.

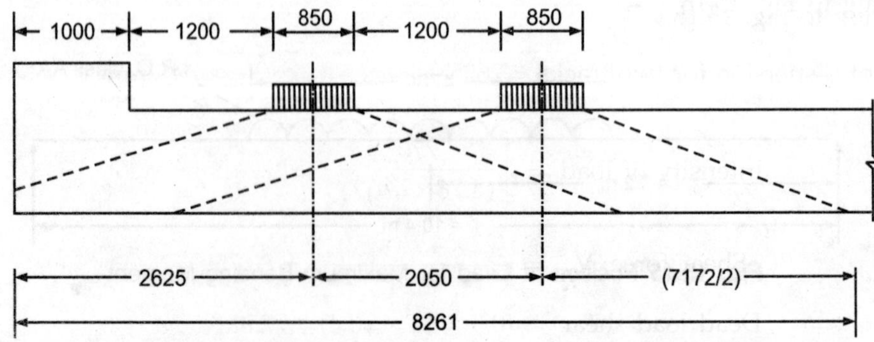

Fig. 15.16 Effective Width of Dispersion for IRC Class AA Tracked Vehicle

Net effective width of dispersion = 8.261m

Total load of two tracks with impact = (700 × 1.10) = 770kN

$$\text{Average intensity of load} = \frac{770}{4.76 \times 8.261} = 19.58\text{N/mm}^2$$

Maximum bending moment due to live load is given by

M_L = [(19.58 × 4.76)0.5 × 5.2] – [(19.58 × 4.76)0.5 × 0.25 × 4.76] = 187kNm

• *Shear due to Class AA Tracked Vehicle*

For maximum shear force at the support section the IRC class AA tracked vehicle is arranged as shown in Fig. 15.17.

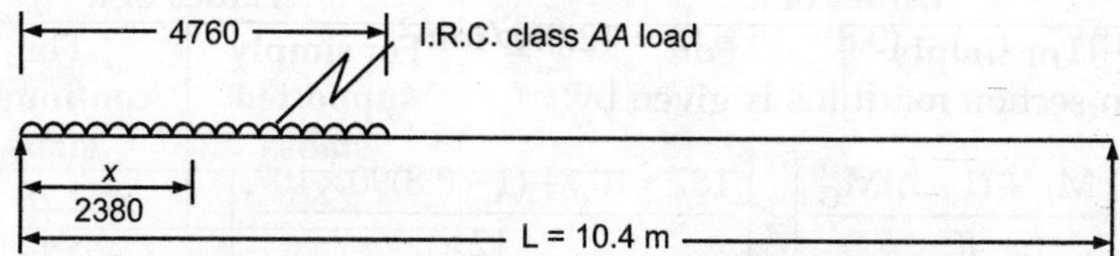

Fig. 15.17 Position of Load for Maximum Shear

Effective width of dispersion is given by,

$$b_e = kx\left(1 - \frac{x}{L}\right) + b_w$$

where, x = 2.38, L = 10.4m, B = 9.5m, b_w = 1.01m

$$\therefore \quad \left(\frac{B}{L}\right) = \left(\frac{9.5}{10.4}\right) = 0.913$$

and $\quad b_w = (0.85 + 2\ 0.08) = 1.01\text{m}$

∴ From Table 15.1 for $\left(\frac{B}{L}\right)$ = 0.913, the value of k = 2.37

$$\therefore \quad b_e = 2.37 \times 2.38\left(1 - \frac{2.38}{10.4}\right) + 1.01 = 5.364\text{m}$$

Referring to Fig. 15.16,

$$\text{Width of dispersion for two tracks} = 2625 + 2050 + \left(\frac{5364}{2}\right) = 7357\text{mm}$$

$$\text{Intensity of load} = \left[\frac{770}{(4.76 \times 7.357)}\right] = 22\text{kN/m}^2$$

$$\text{Shear force, } V_A = \frac{(22 \times 4.76 \times 8.02)}{10.4} = 80.75\text{kN}$$

$$\text{Dead load shear} = (0.5 \times 14 \times 10.4) = 72.8\text{kN}$$

∴ Total design shear = (80.75 + 72.80) = 153.55kN

- ***Check for Minimum Section Modulus***

Dead load moment, M_G = 190kNm

Live load moment, M_L = 187kNm

$$\text{Section modulus} = Z_t = Z_b = Z = \left(\frac{1000 \times 500^2}{6}\right) = 41.66 \times 10^6 \text{mm}^3$$

The permissible stress in concrete at transfer (f_{ct}), is obtained from IRC: 18.

f_{ct} = 15N/mm², f_{cw} = 12 N/mm², f_{tw} = 0

Loss ratio, η = 0.80

$$f_{br} = (\eta f_{ct} - f_{tw}) = (0.8 \times 15 - 0) = 12.0\text{N/mm}^2$$

The minimum section modulus is given by

$$Z_b \geq \left[\frac{M_L + (1-\eta)M_G}{f_{br}}\right] \geq \left[\frac{187 \times 10^6 + (1-0.8)90 \times 10^6}{12}\right]$$

$$\geq 18.75 \times 10^6 < 41.66 \times 10^6 \text{mm}^3 \text{ provided}$$

Hence, the section selected is adequate to resist the service loads without exceeding the permissible stresses.

- ***Minimum Pre–stressing Force***

The minimum pre–stressing force required is compute by using the relation

$$F_o = \left[\frac{A(f_{inf} Z_b + f_{sup} Z_t)}{Z_b + Z_t}\right]$$

Where,
$$f_{sup} = \left(f_{tt} - \frac{M_G}{Z_t}\right) = \left(0 - \frac{190 \times 10^6}{41.66 \times 10^6}\right) = -4.56\text{N/mm}^2$$

and
$$f_{inf} = \left(\frac{f_{tw}}{\eta} - \frac{M_G + M_L}{\eta Z_b}\right) = \left(0 - \frac{(187+190) \times 10^6}{0.8 \times 41.66 \times 10^6}\right)$$

$$= -11.31\text{N/mm}^2$$

$$F_o = \left[\frac{1000 \times 500 \times 41.66 \times 10^6 (11.31 - 4.56)}{2 \times 41.66 \times 10^6}\right]$$

$$= 1687.5 \times 10^3 \text{N} = 1687.5\text{kN}$$

Using Fressynet cables containing 12 wires of 7mm diameter which are stressed to 1200N/mm²,

$$\text{Force in each cable} = \frac{(12 \times 38.5 \times 1200)}{1000} = 554\text{kN}$$

$$\text{Spacing of the cables} = \left(\frac{1000 \times 554}{1687.5}\right) = 328\text{mm}$$

• ***Eccentricity of Cables***

The eccentricity of the cables at the centre of the span is is obtained from the relation

$$e = \left[\frac{Z_t Z_b (f_{inf} - f_{sup})}{A(f_{sup} Z_t + f_{inf} Z_b)}\right] = \left[\frac{(41.66)^2 \times 10^{12}\,(11.31 + 4.56)}{1000 \times 500 \times 41.66 \times 10^6\,(-4.56 + 11.31)}\right] = 195\text{mm}$$

The cables are arranged in a parabolic profile with a maximum eccentricity of 195mm at the centre of the span which reduces to zero (concentric) at supports.

• ***Check for Stresses at Service Loads***

$$P = 1687.5\text{kN},\ e = 195\text{mm}$$

$$A = (1000 \times 500) = 5 \times 10^5\text{mm}^2$$

$$Z_t = Z_b = Z = 41.66 \times 10^6\text{mm}^3$$

$$M_G = 190\text{kNm} \quad M_L = 187\text{kNm}$$

$$\left(\frac{P}{A}\right) = \left(\frac{1687.5 \times 10^3}{5 \times 10^5}\right) = 3.375\text{N/mm}^2$$

$$\left(\frac{Pe}{Z}\right) = \left(\frac{1687.5 \times 10^3 \times 195}{41.66 \times 10^6}\right) = 7.89\text{N/mm}^2$$

$$\left(\frac{M_G}{Z}\right) = \left(\frac{190 \times 10^6}{41.66 \times 10^6}\right) = 4.56\text{N/mm}^2$$

$$\left(\frac{M_L}{Z}\right) = \left(\frac{187 \times 10^6}{41.66 \times 10^6}\right) = 4.48\text{N/mm}^2$$

Stress at transfer

At top of the slab = (3.375 – 7.89 + 4.56) = 0.045N/mm^2

At bottom of the slab = (3.375 +7.89 – 4.56) = 6.705N/mm^2

Stress at working loads

At top of the slab = 0.8(3.375 – 7.89) + 4.56 + 4.48 = 5.428N/mm^2

At bottom of the slab = 0.8(3.375 + 7.89) – 4.56 – 4.48 = –0.028N/mm^2

The actual stresses developed are within the permissible limits.

• ***Check for Ultimate Strength (IRC: 18–2000)***

Considering 1m width of the slab, b = 1000mm

$$A_p = \frac{(12 \times 38.5 \times 1000)}{328} = 1408\text{mm}^2$$

$$d = 445\text{mm},\quad f_p = 1500\text{N/mm}^2$$

(a) Failure by Yielding of Steel

$M_u = 0.9dA_pf_p = (0.9 \times 445 \times 1408 \times 1500) = 846 \times 10^6 Nmm = 846kNm$

(b) Failure by Crushing of Concrete

$M_u = 0.176bd^2f_{ck} = (0.176 \times 1000 \times 445^2 \times 40) = 1394 \times 10^6 Nmm = 1394kNm$

According to IRC: 18–2000,

$$\begin{aligned}\text{Required ultimate moment} &= 1.5\,M_G + 2.5\,M_L\\ &= (1.5 \times 190) + (2.5 \times 187)\\ &= 285 + 467.5\\ &= 752.5kNm\end{aligned}$$

Hence ultimate moment capacity of the section (M_u = 846kNm) is greater than the required ultimate moment (752.5kNm)

- ***Check for Ultimate Shear Strength***

Ultimate shear force, $V_u = (1.5V_G + 2.5V_L) = (1.5 \times 72.8) + (2.5 \times 87.72) = 328.5kN$

According to IRC: 18–2000,

The ultimate shear resistance of the support sections uncracked in flexure is given by

$$V_{cw} = 0.67bh\sqrt{f_t^2 + 0.8f_{cp}f_t} + \eta P\sin\theta$$

Where, b = Width of slab = 1000mm

h = Overall depth = 500mm

f_t = Principal tensile stress = $0.24\sqrt{f_{ck}} = 0.24\sqrt{40} = 1.51N/mm^2$

f_{cp} = Compressive stress at the centroidal axis

$$= \left(\frac{\eta F_o}{A}\right) = \left(\frac{0.8 \times 1687.5 \times 10^3}{1000 \times 500}\right) = 2.7N/mm^2$$

Eccentricity of the cable at the centre of the span, e = 195mm

The cables are concentric at the support section.

$$\text{Slope of the cable} = \theta = \left(\frac{4e}{L}\right) = \left(\frac{4 \times 195}{(10.4 \times 1000)}\right) = 0.075$$

$$\therefore \quad V_{cw} = 0.67 \times 1000 \times 500\sqrt{1.51^2 + 0.8 \times 2.7 \times 1.51} + (0.8 \times 1687.5 \times 10^3 \times \sin 0.075)$$

$$= 889817N = 890kN$$

Since the ultimate shear V_u is less than 50 percent of the ultimate shear resistance V_{cw}, no shear reinforcement is required.

- ***Supplementary Reinforcement***

Supplementary reinforcement = 0.15% of the gross cross sectional area

$= (0.0015 \times 1000 \times 500) = 750mm^2$

Provide 10mm diameter, Fe 415 HYSD bars at 200mm centers, both at top and bottom faces in the longitudinal and transverse directions.

• *Design of End Block Reinforcement*

At the support sections, the concentric cables carrying a force of 554kN are spaced at intervals of 328mm. The bursting tension is computed using Table 15.2 as recommended in IRC: 18–2000. In the present design problem,

$P_k = 554kN$, $2y_{po} = 150mm$, $2y_o = 328mm$

Table 15.2 Design Bursting Tensile Force in End Blocks (IRC: 18–2000)

$\left(\frac{y_{po}}{y_o}\right)$	0.3	0.4	0.5	0.6	0.7
$\left(\frac{F_{bst}}{P_k}\right)$	0.23	0.20	0.17	0.14	0.11
F_{bst} = Bursting tensile force P_k = Tendon force, $2y_{po}$ = Side of loaded area, $2y_o$ = Side of end block					

$$\therefore \left(\frac{y_{po}}{y_o}\right) = \left(\frac{150}{328}\right) = 0.457$$

Interpolating the value of $\left(\frac{F_{bst}}{P_k}\right)$ and $\left(\frac{y_{po}}{y_o}\right) = 0.457$ from Table 15.2, we have

$$F_{bst} = (0.185 \times 554) = 103kN$$

Using 10mm diameter, Fe 250 grade mild steel bars as end block reinforcement.

$$\text{Area of steel} = \left(\frac{103 \times 10^3}{0.87 \times 250}\right) = 474mm^2$$

Provide 10mm diameter bars at 100mm centres in the vertical and horizontal directions in two planes in front of anchorages at 100 and 200mm respectively.

The longitudinal and cross–section of the deck slab is shown in Fig. 15.18 and 15.19.

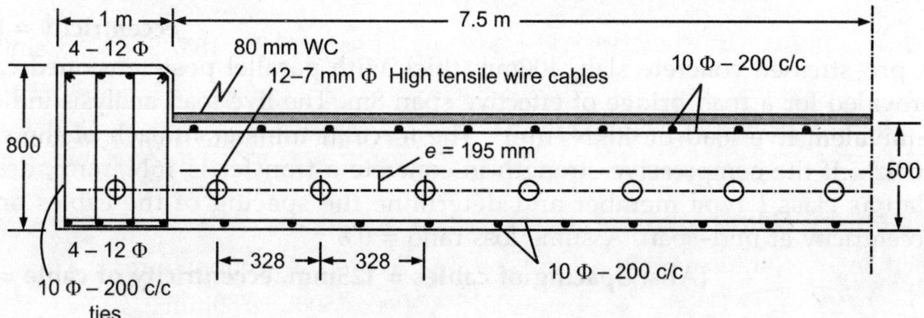

Fig. 15.18 Cross–section of Deck Slab at Centre of Span

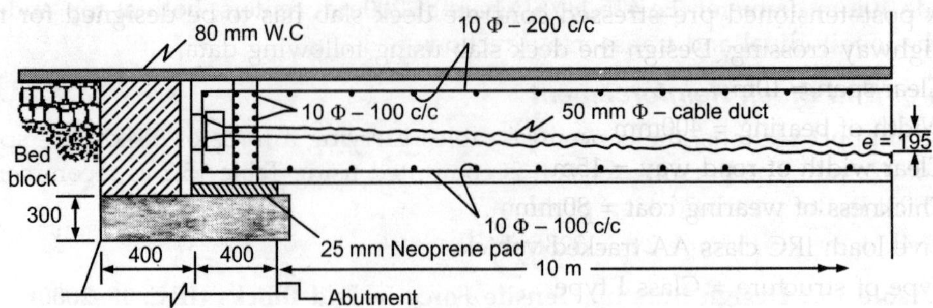

Fig. 15.19 Longitudinal Section of Deck Slab

REVIEW QUESTIONS

15.1 Define bridge and mention its scope.

15.2 Explain types of bridges with suitable sketches.

15.3 Describe short span bridges, medium span bridges, and long span and very long span bridges with beautiful diagrams.

15.4 Elaborate the special construction techniques for bridges

15.5 Explain truss bridge construction.

15.6 Explain construction of cable stayed bridge.

15.7 Describe the use of new materials in bridge construction.

15.8 Explain the steps used in the design of bridges.

EXERCISE PROBLEMS

15.1 A box girder of a pre–stressed concrete bridge of span 40m has overall dimensions of width 1200mm and depth 1800mm. The uniform thickness of walls is 200mm. The live load moment is 2000kNm. The strength of concrete at transfer is 16MPa. The loss ratio is 0.8. Cables consisting of 12 high tensile wires of 8mm diameter initially stressed to 1000MPa. Design the box girder as class I type structure and determine the number of cables required at the centre of the span and their positions in the centroidal axis.

[Ans: Pre–stressing force = 7090kN, number of cables = 12, eccentricity = 800mm]

15.2 A pre–stressed concrete slab, 400mm thick with parallel post–tensioned cables, is provided for a road bridge of effective span 8m. The live load analysis indicates an equivalent live load of 40kN/mm^2. The force at transfer in each of the cables is 400kN. If the compressive strength in concrete a transfer is 16N/mm^2, design the slab as class I type member and determine the spacing of the cables and their eccentricity at mid–span. Assume loss ratio = 0.8.

[Ans: Spacing of cables = 125mm, eccentricity of cable = 95mm]

15.3 A post–tensioned pre–stressed concrete deck slab has to be designed for national highway crossing. Design the deck slab using following data:
Clear span = 10m
Width of bearing = 400mm
Clear width of road way = 15m
Thickness of wearing coat = 80mm
Live load: IRC class AA tracked vehicle
Type of structure = Class I type
M40 grade concrete and cables containing 12 high tensile wires of 7mm diameter initially stressed to 1200N/mm^2 and Fe 415 bars used. Compressive strength in concrete a transfer = 35N/mm^2, percentage loss of pre–stress = 20%
Also find the spacing of cables.

[Ans: Thickness of Slab = 500mm, spacing of cables = 300mm]

15.4 Design a post–tensioned pre–stressed concrete T–beam and deck slab for a bridge which suits to the following data:
Effective span 24m, width of carriageway = 7.5m, kerbs 600mm wide on either side of road. Spacing of cross and main girders are 4 and 2m respectively. Loading is IRC class AA. Adopt M45 grade concrete and high tensile steel strands and supplementary reinforcement comprising Fe 415 bars. Permissible stress, as specified in IRC: 18–2000. Loss ratio is 0.85.

[Ans: Thickness of slab = 200mm, width of bottom flange = 500mm and depth of bottom flange = 400mm, overall depth of girder = 1400mm, thickness of web = 200mm, Freyssinet cables in 4 Nos. (7ply – 15.2mm strands) at an eccentricity of 705mm at the centre of span]

OBJECTIVE QUESTIONS

15.1 The short span of precast pre–tensioned bridge is up to......

a. 50ft b. 75ft
c. 100ft d. 125ft

15.2 The medium span of precast pre–tensioned bridge is in the range......

a. 25ft to 50ft b. 50 to 100ft
c. 75ft to 100ft d. 100ft to 125ft

15.3 The long span of precast pre–tensioned bridge is up to......

a. 150ft b. 160ft
c. 180ft d. 200ft

15.4 The short span of post-tensioned bridge is up to......

a. 50ft b. 60ft
c. 70ft d. 80ft

15.5 The medium span of post–tensioned bridge is up to……

a. 150ft b. 160ft

c. 180ft d. 200ft

15.6 The long span of post–tensioned bridge is up to……

a. 350ft b. 400ft

c. 450ft d. 500ft

15.7 The very long span of post–tensioned bridge is ……

a. >350ft b. >520ft

c. >450ft d. >600ft

Answers

15.1 a	15.2 b	15.3 c	15.4 b	15.5 c
15.6 d	15.7 b			

□□□

Index

A

B

C

D

E

F

G

H

I

L

M

N

O

P

R

S

T

U

V

W

Z

❑❑❑